Selected Titles in This Series

25 **Decio Levi and Orlando Ragnisco, Editors,** SIDE III—Symmetries and integrability of difference equations, 2000

24 **B. Brent Gordon, James D. Lewis, Stefan Müller-Stach, Shuji Saito, and Noriko Yui, Editors,** The arithmetic and geometry of algebraic cycles, 2000

23 **Pierre Hansen and Odile Marcotte, Editors,** Graph colouring and applications, 1999

22 **Jan Felipe van Diejen and Luc Vinet, Editors,** Algebraic methods and q-special functions, 1999

21 **Michel Fortin, Editor,** Plates and shells, 1999

20 **Katie Coughlin, Editor,** Semi-analytic methods for the Navier-Stokes equations, 1999

19 **Rajiv Gupta and Kenneth S. Williams, Editors,** Number theory, 1999

18 **Serge Dubuc and Gilles Deslauriers, Editors,** Spline functions and the theory of wavelets, 1999

17 **Olga Kharlampovich, Editor,** Summer school in group theory in Banff, 1996, 1998

16 **Alain Vincent, Editor,** Numerical methods in fluid mechanics, 1998

15 **François Lalonde, Editor,** Geometry, topology, and dynamics, 1998

14 **John Harnad and Alex Kasman, Editors,** The bispectral problem, 1998

13 **Michel C. Delfour, Editor,** Boundaries, interfaces, and transitions, 1998

12 **Peter C. Greiner, Victor Ivrii, Luis A. Seco, and Catherine Sulem, Editors,** Partial differential equations and their applications, 1997

11 **Luc Vinet, Editor,** Advances in mathematical sciences: CRM's 25 years, 1997

10 **Donald E. Knuth,** Stable marriage and its relation to other combinatorial problems: An introduction to the mathematical analysis of algorithms, 1997

9 **D. Levi, L. Vinet, and P. Winternitz, Editors,** Symmetries and integrability of difference equations, 1996

8 **J. Feldman, R. Froese, and L. M. Rosen, Editors,** Mathematical quantum theory II: Schrödinger operators, 1995

7 **J. Feldman, R. Froese, and L. M. Rosen, Editors,** Mathematical quantum theory I: Field theory and many-body theory, 1994

6 **Guido Mislin, Editor,** The Hilton Symposium 1993: Topics in topology and group theory, 1994

5 **D. A. Dawson, Editor,** Measure-valued processes, stochastic partial differential equations, and interacting systems, 1994

4 **Hershy Kisilevsky and M. Ram Murty, Editors,** Elliptic curves and related topics, 1994

3 **Rémi Vaillancourt and Andrei L. Smirnov, Editors,** Asymptotic methods in mechanics, 1993

2 **Philip D. Loewen,** Optimal control via nonsmooth analysis, 1993

1 **M. Ram Murty, Editor,** Theta functions: From the classical to the modern, 1993

SIDE III—Symmetries and Integrability of Difference Equations

Volume 25

C R M

CRM PROCEEDINGS & LECTURE NOTES

Centre de Recherches Mathématiques
Université de Montréal

SIDE III—Symmetries and Integrability of Difference Equations

Decio Levi
Orlando Ragnisco
Editors

American Mathematical Society
Providence, Rhode Island USA

The production of this volume was supported in part by the Fonds pour la Formation de Chercheurs et l'Aide à la Recherche (Fonds FCAR) and the Natural Sciences and Engineering Research Council of Canada (NSERC).

2000 *Mathematics Subject Classification.* Primary 22Exx, 33–XX, 42Cxx, 39Axx, 53–XX.

Library of Congress Cataloging-in-Publication Data

SIDE III : symmetries and integrability of difference equations / Decio Levi, Orlando Ragnisco.
p. cm. — (CRM proceedings & lecture notes ; ISSN 1065-8580 ; v. 25)
Includes bibliographical references.
ISBN 0-8218-2128-8 (softcover : alk. paper)
1. Difference equations—Congresses. 2. Hypergeometric functions—Congresses. 3. Symmetry (Physics)—Congresses. I. Title: SIDE 3. II. Title: SIDE three. III. Title: Symmetries and integrability of difference equations. IV. Levi, D. V. Ragnisco, O. (Orlando), 1946– VI. Series.
QA431 .S46 2000
515′.625—dc21 00-038087

∞ The paper used in this book is acid-free and falls within the guidelines
established to ensure permanence and durability.
This volume was typeset using AMS-TeX,
the American Mathematical Society's TeX macro system,
and submitted to the American Mathematical Society in camera ready
form by the Centre de Recherches Mathématiques.
Visit the AMS home page at URL: `http://www.ams.org/`

10 9 8 7 6 5 4 3 2 1 05 04 03 02 01 00

Contents

LIST OF PARTICIPANTS

Mark Alber
Department of Mathematics
University of Notre Dame
Notre Dame IN 46556-5683 (USA)

Natig Atakishiyev
Instituto de Matematicas UNAM
Apartado Postal 273-3
C.P. 62250 Cuernavaca
Morelos (Mexico)

Mariusz Bialecki
Department of Physics
Institute of Theoretical Physics
Warsaw University
ul. Hoza 69
00-681 Warsaw (Poland)

Maciej Blaszak
Physics Department
A. Mickiewicz University
Umultowska 85
61-614 Poznan (Poland)

Alexander Bobenko
Fachbereich mathematik
Technische Universitaet Berlin
Strasse des 17 Juni 136
D-10623 Berlin (Germany)

Leonid Bogdanov
L. D. Landau ITP
Kosygina str. 2
Moscow 117940 (Russia)

Claude Brezinski
Laboratoire d'Analyse Numérique et
d'Optimisation
Université des Sciences et
Technologies de Lille
59655–Villeneuve d'Ascq cedex (France)

Francesco Calogero
Dipartimento di Fisica
Universita' di Roma I "La Sapienza"
Piazzale Aldo Moro
200185 Roma (Italy)

Adrian Stefan Cârstea
Department of Theoretical Physics
Institute of Atomic Physics
P.O. Box MG-6
Bucharest (Romania)

Peter A. Clarkson
Head of Mathematics Group &
Chair of Mathematics
Institute of Mathematics & Statistics
University of Kent
Canterbury, CT2 7NF (U.K)

Clio Cresswell
The University of New South Wales
School of Mathematics
Sydney NSW 2052 (Australia)

Giuseppe Dattoli
ENEA—Dipartimento Innovazione
Divisione Fisica Applicata
Centro Ricerche di Frascati
C.P. 65
00044 Frascat, Roma (Italy)

Antonio Degasperis
Dipartimento di Fisica
Universita' di Roma I "La Sapienza"
Piazzale Aldo Moro 2
00185 Roma (Italy)

Adam Doliwa
Instytut Fizyki Teoretycznej
Uniwersytet Warszawski
Hoża 69
00-681 Warszawa (Poland)

Viktor Z. Enolskii
Theoretical Physics Division
NASU Institute of Magnetism
36–b Vernadsky str.
Kiev-680, 252142 (U.K)

Garcia Estevez
Area de Fisica Teorica
Edificio de Fisicas
Universidad de Salamanca
Salamanca 37008 (Spain)

Yuri Fedorov
Department of Mathematics
and Mechanics
Moscow Lomonosov University
Moscow 119899 (Russia)

Maria Luz Gandarias
Departamento de Matematicas
Universidad de Cadiz
P. O. Box 40 11510 Puerto Real
Cadiz (Spain)

David Gomez-Ullate Oteiza
Departamento de Fisica Teorica II
Facultad de CC. Fisicas
Universidad Complutense
28040 Madrid (Spain)

Vittorio Grassi
Dipartimento di Fisica
Universita' degli Studi di Lecce
ex Collegio Fiorini
Via per Arnesano
73100 Lecce (Italy)

Ismagil Habibullin
Institute of Mathematics of
Russian Academy of Sciences
Chernyshevskaya str. 112
Ufa 450000 (Russia)

John Harnad
Centre de Recherches Mathématiques
Université de Montréal
C.P. 6128 succ. Centre-ville
Montréal, Qc, H3C 3J7 (Canada)

Rafael Hernandez Heredero
Departamento de Fisica Teorica II
Facultad de Ciencias Fisicas
Ciudad Universitaria s/n
Universidad Complutense de Madrid
28040 Madrid (Spain)

Jarmo Hietarinta
Department of Physics
University of Turku
FIN-20014 Turku (Finland)

Ryogo Hirota
Department of Information and Computer Science
Waseda University
Tokyo 169 (Japan)

Andrew Hone
Dipartimento di Fisica "E. Amaldi"
Universita' di Roma Tre
Via della Vasca Navale 84
00146 Roma (Italy)

Mourad Ismail
University of South Florida
Department of Mathematics
4202 East Fowler Avenue, PHY 114
Tampa Florida 33620 (USA)

Nalini Joshi
Department of Pure Mathematics
University of Adelaide
Adelaide 5005 (Australia)

Kenji Kajiwara
Laboratory of Applied Mathematics
Department of Electrical Engineering
Doshisha University
Tanebe Kyoto 610-03 (Japan)

Pavel Klimczewski
Department of Physics
Institute of Theoretical Physics
Warsaw University
ul. Hoza 69
00-681 Warsaw (Poland)

Boris Konopelchenko
European Institute for
Nonlinear Studies via
Transnationally Extended Interchanges
Dipartimento di Fisica
dell'Università di Lecce
and Sezione INFN
73100 Lecce (Italy)

Igor V. Krasovski
Max-Planck-Institut fuer
Phisik Komplexer Systeme
Noethnitzer Str. 38
01187 Dresden (Germany)

Martin Kruskal
Department of Mathematics
Hill Center Busch Campus
Rutgers University
New Brunswick NJ 0893 (USA)

Nicolai A. Kudryashov
Department of Applied Mathematics
Moscow Engineering Physics Institute
31 Kashirskoe Shosse
Moscow 115409 (Russia)

Stephane Lafortune
Centre de Recherches Mathématiques
Université de Montréal
C.P. 6128, succ. Centre-ville
Montréal, Qc, H3C 3J7 (Canada)

Wolfgang Lay
Institut fuer Theoretische und
Angewandte Physik
Universitaet Stuttgart
Pfaffenwaldring 57
70550 Stuttgart (Germany)

Decio Levi
Dipartimento di Fisica "Edoardo Amaldi"
Universitá di Roma Tre and INFN
Sezione di Roma Tre
Via della Vasca Navale 84
00146 Roma (Italy)

D. V. Leykin
Theoretical Physics Division
NASU Institute of Magnetism
36–b Vernadsky str.
Kiev-680, 252142 (U.K)

Luigi Martina
Dipartimento di fisica
Universita' di Lecce
CP 193 73100 Lecce (Italy)

Ken-ichi Maruno
Department of Earth System Science
and Technology
Interdisciplinary Graduate School of
Engineering Sciences
Kyushu University
Kasuga Fukuoka 816 (Japan)

Tetsuji Miwa
Research Institute for
Mathematical Sciences
University of Kyoto
Kyoto 606-8502 (Japan)

Fabio Musso
Dipartimento di Fisica "E. Amaldi"
Universita' di Roma Tre
Via della Vasca Navale 84
00146 Roma (Italy)

Atsushi Nagai
Graduate School of
Mathematical Sciences
University of Tokyo
Komaba 3-8-1 Meguro-ku
Tokyo 153-0031 (Japan)

Yoshimasa Nakamura
Department of Mathematical Science
Graduate School of Engineerinfg Science
Osaka University
Toyonaka 560 (Japan)

Maciej Nieszporski
Department of Physics
Institute of Theoretical Physics
Warsaw University
ul. Hoza 69
00-681 Warsaw (Poland)

Luis M. Nieto
Departamento de Fisica Teorica
Facultad de Ciencias
Universidad de Valladolid
47011 Valladolid (Spain)

Frank Nijhoff
Department of Applied Mathematics
University of Leeds
Leeds LS2 9JT (U.K)

Yasuhiro Ohta
Department of Applied Mathematics
Faculty of Engineering
Hiroshima University
1-4-1 Kagamiyama
Higashi-Hiroshima 739 (Japan)

Vassilios Papageorgiou
Department of Mathematics
University of Patras
University Campus
GR-26500 Rio Patras (Greece)

F. Pempinelli
Dipartimento di Fisica dell'Università
and sezione INFN
73100 Lecce (Italy)

Andrew Pickering
Area de Fisica Teórica
Facultad de Ciencias
Edificio de Fisica
Universidad de Salamanca
37008 Salamanca, (Spain)

Barbara Prinari
Dipartimento di Fisica dell'Università
and sezione INFN
73100 Lecce (Italy)

Robert Prus
Department of Physics
Institute of Theoretical Physics
Warsaw University
ul. Hoza 69
00-681 Warsaw (Poland)

Reinout Quispel
Department of Mathematics
La Trobe University
Bundoora, Victoria 3083 (Australia)

Orlando Ragnisco
Dipartimento di Fisica "E. Amaldi"
Universita' di Roma Tre
Via della Vasca Navale 84
00146 Roma (Italy)

Alfred Ramani
CPT
École Polytechnique
CNRS, UPR 14
91128 Palaiseau (France)

Stefan Rauch-Wojciechowski
Department of Mathematics
Linkoping University
S-58183 Linkoping (Sweden)

Paolo M. Santini
Istituto Nazionale di Fisica Nucleare
Sezione di Roma and
Dipartimento di Fisica
Università di Roma "La Sapienza"
P.le Aldo Moro 2
I–00185 Roma (Italy)

Junkichi Satsuma
School of Mathematical Sciences
University of Tokyo
3-8-1 Komaba Meguro-ku
Tokyo 153-8914 (Japan)

Wolfgang K. Schief
School of Mathematics
The University of New South Wales
Sydney, NSW 2052 (Australia)

M. Senthilvelan
Dipartimento di Matematica
Citta' Universitaria
Viale Andrea Doria 6
I-95125 Catania (Italy)

Vyacheslav Spiridonov
Bogoliubov Laboratory of Theoretical Physics
Joint Inst. for Nuclear Research
Dubna, Moscow Region 141980
(Russia)

Yuri B. Suris
Fachbereich Mathematik
TU Berlin, Str. 17 Juni 136
10623 Berlin (Germany)

Marco Tarlini
I.N.F.N.
Largo E. Fermi 2
50125 Florence (Italy)

Piergiulio Tempesta
Dipartimento di Fisica
Universita' degli Studi di Lecce
c/o ex Collegio Fiorini
Via per Arnesano
73100 Lecce (Italy)

Tetsuji Tokihiro
Graduate School of
Mathematical Sciences
University of Tokyo
8-3-1 Komaba Meguro-ru
Tokyo 153-8914 (Japan)

Satoshi Tsujimoto
Waseda University
Shinjuku-ku Okubo 3-4-1
Tokyo 169 (Japan)

Alexander Turbiner
Instituto Ciencias Nucleares
UNAM A.P. 70-543
Mexico D. F.
04510 (Mexico)

Walter Van Assche
Department of Mathematics
Katholieke Universiteit Leuven
Celestijnenlaan 200 B
B-3001 Heverlee
Leuven (Belgium)

Valery Verbus
Institute for Physics of Microstructures
of the Russian Accademy of Sciences
46 Ul' yanov Street Nizhny Novgorod
603600 GSP-105 (Russia)

Vadim Vereschagin
Irktusk Computing Center
Siberian Branch
of the Russian Accademy of Sciences
P.O. Box 1233
Irkutsk 664033 (Russia)

Claude Viallet
CUniversité Paris VI
Boîte 126
4 place Jussieu
F–75252 Paris Cedex 05 (France)

Ralph Willox
Department of Theoretical Physics
Vrije Universiteit Brussel
Dienst Tena Pleinlaan 2
1050 Brussels (Belgium)

Pavel Winternitz
Centre de recherches mathématiques
Université de Montréal
C.P. 6128, succ. Centre-ville
Montréal, Qc, H3C 3J7 (Canada)

Preface

The third meeting on "Symmetries and Integrability of Difference Equations" (SIDE III) was held in Sabaudia, a beautiful natural resort by the sea about 100 km south of Rome, Italy, from May the 16th till May the 22nd, 1998. This series of international meetings started in 1994: SIDE I was held in Estérel, Québec (near Montréal, Canada) and was organized under the auspices of the CRM (Centre de Recherches Mathématiques) of the Université de Montréal. A second meeting took place in 1996 at the University of Kent in Canterbury (UK). Information on the first SIDE meeting has been reported in *Symmetries and Integrability of Difference Equations*, edited by D. Levi, L. Vinet and P. Winternitz, AMS, 1996, and on the second in *Symmetries and Integrability of Difference Equations*, edited by P. A. Clarkson and F. W. Nijhoff, C.U.P., 1999. These meetings are intended to bring together specialists from various disciplines, all working or using methods from discrete integrable systems, i.e., systems that can be described by ordinary or partial difference equations and that allow for exact methods for their solutions. This domain forms the core of a great variety of fields, including classical and quantum physics, numerical analysis, discrete geometry, computer science, mathematical biology, economics, and so on. The subject of Discrete Systems and, in particular, of Discrete Integrable Systems has undergone a remarkable development in recent years: much progress has been made in the comprehension of the related mathematical structures. Moreover, among physicists and applied scientists in general there has been an increasing interest in discrete models, ranging from partial to ordinary difference equations to cellular automata. Accordingly, the Sabaudia meeting was pretty interdisciplinary in nature, and so is the present volume, which collects the lectures and talks therein delivered. We hope that it will provide a clear picture of the state of the art and give some insight into the most advanced research topics in the area.

The meeting attracted 72 participants from 19 countries (ordered according to the number of participants: Italy (13), Japan (10), Poland, Russia and Spain (6), Australia and Germany (4), Canada, France, Mexico, U.K and U.S.A. (3), Belgium (2), Finland, Greece, India, Romania, Sweden and Ukraine (1). Forty-five talks were delivered (among them, 12 1-hour "review lectures") and 15 posters were presented. A large part of their content can be found in this volume, which includes contributions, covering the following interconnected and even overlapping subjects:

- *Lie-type symmetries of differential-difference and difference-difference equations*: M. L. Gandarias, D. Gomez-Ullate, R. H. Heredero, L. Martina, L. M. Nieto, M. Sentihilvelan;
- *Integrable differential-difference equations*: M. Alber, M. Blaszak, C. Brezinski, A. Cârstea, I. Habibullin, B. Prinari;

• *Fully discrete integrable equations*: *from multidimensional lattices to cellular automata*: Y. Fedorov, R. Hirota, A. Hone, K. Kashiwara, K. Maruno, Y. Ohta, V. Verbus;

• *Discrete geometry and integrability*: L. Bogdanov, A. Doliwa, B. Konopelchenko, P. M. Santini. W. Schieff;

• *Continuous and discrete Painlevé equations and related problems*: P. G. Estevez, N. Joshi, N. Kudriashov, S. Lafortune, V. Papageorgiou, R. Quispel, A. Ramani;

• *Isomonodromy, spectral theory and related problems*: C. Cresswell, V. Enolskii, J. Harnad, W. Lay, V. Vereschagin;

• *Discrete dynamics: integrability and chaos*: J. Hietarinta, C. Viallet;

• *Difference and q-difference equations, special functions and orthogonal polynomials*: N. Atakashiev, G. Dattoli, V. Spiridonov, A. Turbiner, W. Van Assche;

• *Difference equations and quantum groups*: T. Miwa.

The success of the meeting has been confirmed by the unanimous decision to organize a SIDE IV workshop, which is planned to take place in Japan in fall 2000.

D. Levi
O. Ragnisco

Montréal, February 2000

Centre de Recherches Mathématiques
CRM Proceedings and Lecture Notes
Volume **25**, 2000

Billiard Weak Solutions of Nonlinear PDE's and Toda Flows

Mark S. Alber, Roberto Camassa, and Michael Gekhtman

ABSTRACT. A certain class of partial differential equations possesses singular solutions having discontinuous first derivatives ("peakons"). The time evolution of peaks of such solutions is governed by a finite dimensional completely integrable system. Explicit solutions of this system are constructed by using algebraic-geometric method which casts it as a flow on an appropriate Riemann surface and reduces it to a classical Jacobi inversion problem. The algebraic structure of the finite dimensional flow is also examined in the context of the Toda flow hierarchy. Generalized peakon systems are obtained for any simple Lie algebra and their complete integrability is demonstrated.

1. Introduction

Camassa and Holm [**6**] introduced and studied classes of soliton-type weak solutions, for an integrable nonlinear equation derived in the context of a shallow water model. In particular, they described the soliton dynamics in terms of a system of Hamiltonian equations for the locations of the "peaks" of the solution, the points at which its spatial derivative changes sign. In other words, each peaked solution, or peakon, can be associated with a mechanical system of moving particles. New systems of this type were obtained in [**7, 8**]. The r-matrix approach was applied to the Lax pair formulation of an n-peakon system by Ragnisco and Bruschi [**17**], who also pointed out the connection of this system with the classical Toda lattice. A discrete version of the Adler–Kostant–Symes factorization method was used by Suris [**20**] to study a discretization of the peakons lattice, realized as a discrete integrable system on a certain Poisson submanifold of $\mathrm{gl}(n)$ equipped with r-matrix Poisson bracket.

In [**1, 3, 4**] the existence of peakons was linked to the presence of poles in the energy dependent Schrödinger operators associated with integrable evolution equations. Namely, it was shown that the presence of a pole in the potential is essential for a special limiting procedure which allows for the formation of "billiard" weak solutions. By using the algebraic-geometric method, these billiard solutions

---

2000 *Mathematics Subject Classification.* 58F07, 70H99, 76B15.

The research of the first author is partially supported by NSF grant DMS 9626672 and NATO grant CRG 950897.

The research of the second author is partially supported by DOE grant BES Applied Mathematical Sciences KC-07-01-01 and NATO grant CRG 950897.

This is the final form of the paper.

are related to finite dimensional integrable dynamical systems with reflections. In this way, profiles of billiard weak solutions are associated with billiard motion inside quadrics with and without the presence of a Hooke's potential. The points of impact on the quadrics correspond to the peaks of the profiles of weak solutions of the nonlinear PDE. Billiard solutions include new quasi-periodic and soliton-like solutions, as well as peaked solitons with compact support. This method can be used for a number of equations including the shallow water equation, the Dym type equation, as well as N-component systems with poles, together with all the equations in their respective hierarchies.

In Section 2 of this paper we derive equations for the motion of the peaks of billiard solutions in the context of the algebraic-geometric approach. We construct solutions of these equations by connecting them to a flow on an appropriate Riemann surface which leads to a classical Jacobi inversion problem.

In Section 3 we show that orbits of the Toda flow in $\mathrm{sl}(n)$ into which the peakons lattice is mapped by means of the Lax matrix introduced in [**17**] admits a natural generalization to the case of an arbitrary Lie algebra $\mathfrak{g}$. Restrictions of the Toda flow in $\mathfrak{g}$ to these orbits can be viewed as generalized peakons lattices. Using the methods developed in [**11**] we give an intrinsic description of these orbits and prove complete integrability of the corresponding flows. Unlike in the $\mathrm{sl}(n)$ case, Chevalley invariants of $\mathfrak{g}$ are not sufficient to ensure integrability in the general situation, and so we produce the requisite number of additional first integrals. Lastly, we explain how to construct Darboux coordinates for generalized peakon orbits and consider $\mathrm{sl}(n)$ and G_2 as examples.

2. The Dynamics of the Peak Points

As shown in [**1, 3**], quasi-periodic solutions of the shallow water equation

$$U_t + 3UU_x = U_{xxt} + 2U_xU_{xx} + UU_{xxx}, \tag{2.1}$$

on the infinite line can be represented in the form

$$U(x,t) = \mu_1 + \cdots + \mu_n - \mathfrak{m}, \tag{2.2}$$

where μ's are solutions of the following systems of equations in x and t,

$$\mu_i' = \frac{\partial \mu_i}{\partial x} = \mathrm{Sign}(\mu_i, m_{2i-1}, m_{2i}, s_i)\frac{\sqrt{R(\mu_i)}}{\mu_i \prod_{j\neq i}^n (\mu_i - \mu_j)}, \tag{2.3}$$

and

$$\dot{\mu}_i = \frac{\partial \mu_i}{\partial t} = \mathrm{Sign}(\mu_i, m_{2i-1}, m_{2i}, d_i)\frac{(\mu_i - \Sigma)\sqrt{R(\mu_i)}}{\mu_i \prod_{j\neq i}^n (\mu_i - \mu_j)}, \quad i = 1, \ldots, n, \quad \Sigma = \mu_1 + \cdots + \mu_n, \tag{2.4}$$

with

$$R(\mu) = \mu \prod_{r=1}^{2n+1} (\mu - m_r). \tag{2.5}$$

Here the constants m_r depend on the initial condition for the equation (2.1), $\mathfrak{m} = \sum_r^{2n+1} m_r$, and $\mathrm{Sign}(y, l, r, s)$ is a function on the Riemann surface, $\Gamma = \{w^2 = R(\mu)\}$, which switches from -1 to 1 and back each time y reaches l or r, endpoints of a cut on the Riemann surface. These ordinary differential equations (ODE's) for the

μ-variables provide half of the equations of a finite dimensional Hamiltonian system. Notice that periodic solutions form a subclass of the quasi-periodic solutions. Also, soliton solutions can be obtained from quasi-periodic solutions by shrinking pairwise the roots m_r of the spectral polynomial $R(E)$.

Peaked quasi-periodic solutions of equation (2.1) can be constructed from solutions of (2.3) and (2.4) by using the limiting process $m_1 \to 0$ (see [**1, 3**]) and the trace formula(2.2). In what follows ODE's governing the time evolution of the peak locations are obtained and a connection with the canonical variables of Camassa and Holm [**6**] is described.

The x and t evolution of the μ variables can also be connected to a constrained motion of a particle on an n-dimensional hypersurface imbedded in $\mathbb{R}^{n+1}$ which is parameterized by the constants m_r, under the action of a harmonic force field (see[**3**]). The limit $m_1 \to 0$ corresponds to "flattening" this surface along one direction which results in harmonically forced motion in a region of $\mathbb{R}^n$ confined by an $(n-1)$-dimensional boundary. The particle collides with the boundary and moves along segments of an n-dimensional billiard. The reflection on the boundary causes, after using the trace formula (2.2), the appearance of peaks in the corresponding PDE solutions of equation (2.1).

On the level of (2.3) and (2.4), the boundary of the billiard can be shown to be described by the following condition: $\mu_1(x,t) = 0$. Reflection on the boundary is described by the sign switch from 1 to -1 (or from -1 to 1) of first derivative of μ_1 in (2.3) and (2.4).

The periodic case. The main steps in deriving equations for the peak locations can be demonstrated by considering the following basic example. Let us assume $n = 2$, and $m_1 = 0$. Equation $\mu_1(x,t) = 0$ defines a function of time $x = q(t)$. An equation for the time dependence of $q(t)$ can be found by differentiating $\big(\mu_1\big(q(t),t\big) = 0\big)$ with respect to time. We have

$$\frac{d}{dt}\big(\mu_1(q(t),t)\big) = 0 = \frac{\partial \mu_1}{\partial x}\big(q(t),t\big)\dot{q} + \frac{\partial \mu_1}{\partial t}\big(q(t),t\big). \tag{2.6}$$

The commuting x- and t-flows of μ_1 are described by

$$\begin{gathered}\frac{\partial \mu_1}{\partial x} = \operatorname{Sign}(\mu_1)\frac{\sqrt{C_4(\mu_1)}}{\mu_1 - \mu_2}, \quad \frac{\partial \mu_1}{\partial t} = \operatorname{Sign}(\mu_1)\frac{(-\mu_2)\sqrt{C_4(\mu_1)}}{\mu_1 - \mu_2}, \\ C_4(\mu) = (\mu - m_2)(\mu - m_3)(\mu - m_4)(\mu - m_5).\end{gathered} \tag{2.7}$$

Notice that the spatial derivative of μ_1 is not defined at $x = q(t)$, so that in the above formula $\partial\mu_1/\partial x$ evaluated at $\big(q(t),t\big)$ needs to be found in the limit as $x \to q$ either from the left or from the right. The sign uncertainty connected with this choice factors out from the final formula for $\dot{q}$. This combined with (2.6) and the billiard condition at the boundary $\mu_1\big(q(t),t\big) = 0$ gives

$$\dot{q} = \mu_2\big(q(t),t\big). \tag{2.8}$$

We also need an evolution equation for the quantity $\mu_2\big(q(t),t\big)$. Notice that from the trace formula (2.2) and the expression for the boundary $\mu_1\big(q(t),t\big) = 0$, this is the quantity that determines the amplitude of the peak $U\big(q(t),t\big)$. The time derivative of $\mu_2\big(q(t),t\big)$ is

$$\frac{d}{dt}\big(\mu_2(q(t),t)\big) = \frac{\partial \mu_2}{\partial x}\big(q(t),t\big)\dot{q} + \frac{\partial \mu_2}{\partial t}\big(q(t),t\big), \tag{2.9}$$

which used with the commuting x- and t-flows equations for μ_2 yields, in analogy with system (2.8),

$$\begin{aligned} \frac{dy}{dt}(t) &= -\operatorname{Sign}(y)\frac{\sqrt{C_4(y)}}{y}(\dot{q}+B_1(y)), \\ &= -\operatorname{Sign}(y)\frac{\sqrt{C_4(y)}}{y}\dot{q}. \end{aligned} \tag{2.10}$$

Here we have introduced the notation

$$y(t) = \mu_2\big(q(t), t\big),$$

and used again the definition of $q(t)$ so that

$$B_1\big(\mu_2(q(t)\big) \equiv B_1(y) = -\mu_1\big(q(t), t\big) = 0.$$

Combining this equation for $\dot{y}$ with the evolution equation (2.8) for q gives the following system for the variables y and q,

$$\dot{y} = -\operatorname{Sign}(y)\sqrt{(y-m_2)(y-m_3)(y-m_4)(y-m_5)}, \quad \dot{q} = y. \tag{2.11}$$

Notice that the evolution for y is decoupled from that of q. The inversion problem corresponding to the equation for y determines a periodic function of t with nonzero average. The evolution of q is then given by the combination of a linear growth in time with slope given by the average of y, plus periodic oscillations. We also remark that for weak solutions of equation (2.1) having isolated discontinuities in the first derivative, the jump condition yields

$$\frac{dq}{dt} = U\big(q(t), t\big),$$

and using the trace formula (2.2) this can be seen to coincide with the second equation in system (2.11).

The genus 2 quasi-periodic case. We describe a billiard genus 2 quasi-periodic solution. The spectral polynomial in this case is of the 5th order and will be denoted as $C_5(\mu)$. The boundary is again described by $\mu_1\big(q(t), t\big) = 0$. We shall write

$$y_1 = \mu_2\big(q(t), t\big) \quad \text{and} \quad y_2 = \mu_3\big(q(t), t\big).$$

From the weak solution approach it follows that $\dot{q} = U(q_1) = y_1 + y_2$. We differentiate y_1 and y_2 to obtain

$$\begin{aligned} \dot{y}_1 &= \frac{d}{dt}\big[\mu_2(q,t)\big] = \frac{\partial \mu_2}{\partial x}\dot{q} + \frac{\partial \mu_2}{\partial t}, \\ &= \frac{\operatorname{Sign}(\mu_2)\sqrt{C_6(\mu_2)}}{\mu_2(\mu_2-\mu_3)}(\dot{q}-\mu_3) = \frac{\operatorname{Sign}(y_1)\sqrt{C_6(y_1)}}{y_1-y_2}, \\ \dot{y}_2 &= \frac{d}{dt}\big[\mu_3(q,t)\big] = \frac{\partial \mu_3}{\partial x}\dot{q} + \frac{\partial \mu_3}{\partial t}, \\ &= \frac{\operatorname{Sign}(\mu_3)\sqrt{C_6(\mu_3)}}{\mu_3(\mu_3-\mu_2)}(\dot{q}-\mu_2) = \frac{\operatorname{Sign}(y_2)\sqrt{C_6(y_2)}}{y_2-y_1}. \end{aligned} \tag{2.12}$$

This is a well-defined system associated with a genus 2 Riemann surface. The corresponding problem of inversion involves only holomorphic differentials and therefore the quantities y_1, y_2 and $q(t)$ can be expressed in terms of standard θ-functions on the Riemann surface.

Peakon solutions. We now specialize the above formalism to the limiting case of soliton solutions of equation (2.1) on the real line, in which each pair m_{2i}, m_{2i+1} is taken to limit to the a_i, $i = 1, \ldots, n$. For just one μ-variable, using the trace formula $U = \mu - a$ at the billiard boundary results in $U\big(q(t), t\big) = -a$. The single peakon solution determined by this procedure is therefore

$$U(x,t) = -ae^{-|x+at|}, \tag{2.13}$$

which is a traveling wave soliton-type solution. Camassa and Holm [**6**] found

$$U(x,t) = p(t)e^{-|x-q(t)|}, \tag{2.14}$$

with an additional link between $p(t)$ and $q(t)$ (Hamiltonian structure) which in the 1-peakon case results in $p(t)$ being a constant and $q(t)$ being a linear function.

2-peakon solution. In this case, one obtains the following system that describes 2-peakon profiles,

$$\frac{\partial \mu_1}{\partial x} = \operatorname{Sign}(\mu_1)\frac{(\mu_1 - a_1)(\mu_1 - a_2)}{(\mu_1 - \mu_2)}, \quad \frac{\partial \mu_2}{\partial x} = \operatorname{Sign}(\mu_2)\frac{(\mu_2 - a_1)(\mu_2 - a_2)}{(\mu_2 - \mu_1)}, \tag{2.15}$$

where μ_1 and μ_2 are evaluated between a_1 and 0 and a_2 and 0, respectively. The corresponding time-flow is

$$\begin{aligned} \frac{\partial \mu_1}{\partial t} &= \operatorname{Sign}(\mu_1)B_1(\mu_1)\frac{(\mu_1 - a_1)(\mu_1 - a_2)}{(\mu_1 - \mu_2)}, \\ \frac{\partial \mu_2}{\partial t} &= \operatorname{Sign}(\mu_2)B_1(\mu_2)\frac{(\mu_2 - a_1)(\mu_2 - a_2)}{(\mu_2 - \mu_1)}, \end{aligned} \tag{2.16}$$

and the first order polynomial B_1 for equation (2.1) gives

$$B_1(\mu_1) = -\mu_2 + a_1 + a_2, \qquad B_1(\mu_2) = -\mu_1 + a_1 + a_2.$$

Define the boundary by introducing functions $q_1(t)$ and $q_2(t)$ such that

$$\mu_1\big(q_1(t), t\big) = 0, \mu_2\big(q_2(t), t\big) = 0. \tag{2.17}$$

Define functions

$$y_1 = \mu_2\big(q_1(t), t\big), y_2 = \mu_1\big(q_2(t), t\big). \tag{2.18}$$

Differentiating $\mu_1\big(q_1(t), t\big)$ and $\mu_2\big(q_2(t), t\big)$ in (2.17) results in

$$\begin{aligned} \frac{d\mu_1}{dt} &= 0 = \frac{\partial \mu_1}{\partial x}\left(\frac{dq_1}{dt} + B_1(\mu_1)\right) = \frac{\partial \mu_1}{\partial x}\left(\frac{dq_1}{dt} - y_1 + a_1 + a_2\right), \\ \frac{d\mu_2}{dt} &= 0 = \frac{\partial \mu_2}{\partial x}\left(\frac{dq_2}{dt} + B_1(\mu_2)\right) = \frac{\partial \mu_2}{\partial x}\left(\frac{dq_2}{dt} - y_2 + a_1 + a_2\right), \end{aligned} \tag{2.19}$$

which coincides with the jump conditions for weak solutions

$$\begin{aligned} \frac{dq_1}{dt} &= U(q_1) = (\mu_1 + \mu_2 - a_1 - a_2)\big|_{q_1} = \mu_2(q_1) - a_1 - a_2 = y_1 - a_1 - a_2, \\ \frac{dq_2}{dt} &= U(q_2) = (\mu_1 + \mu_2 - a_1 - a_2)\big|_{q_2} = \mu_1(q_2) - a_1 - a_2 = y_2 - a_1 - a_2. \end{aligned} \tag{2.20}$$

Finally, differentiate y_1 and y_2 to find

$$\frac{dy_1}{dt} = \frac{\partial \mu_2}{\partial x}\frac{dq_1}{dt} + \frac{\partial \mu_2}{\partial t} = \frac{\partial \mu_2}{\partial x}\left(\frac{dq_1}{dt} + B_1(\mu_2)\right),$$
$$\frac{dy_2}{dt} = \frac{\partial \mu_1}{\partial x}\frac{dq_2}{dt} + \frac{\partial \mu_1}{\partial t} = \frac{\partial \mu_1}{\partial x}\left(\frac{dq_2}{dt} + B_1(\mu_1)\right),$$

and so

$$\frac{dy_1}{dt} = \mathrm{Sign}(y_1)(y_1 - a_1)(y_1 - a_2), \quad \frac{dy_2}{dt} = \mathrm{Sign}(y_2)(y_2 - a_1)(y_2 - a_2). \tag{2.21}$$

Thus, the equations of evolution for y_1 and y_2 decouple from those of q_1 and q_2. The decoupled equations (2.21) can be solved first and q_1, q_2 subsequently determined from (2.20) by quadratures.

It is interesting to examine the connection of the set of variables q_i, y_i, $i = 1, 2$, with the q_i, p_i, $i = 1, 2$, introduced by [**6**]. By definition, the q's are the same in both sets (the (q, y) system does have a canonical Hamiltonian form). As to the y's, notice that

$$\begin{aligned} U(q_1) &= y_1 - a_1 - a_2 = p_1 + p_2 e^{-|q_1 - q_2|}, \\ U(q_2) &= y_2 - a_1 - a_2 = p_2 + p_1 e^{-|q_1 - q_2|}, \end{aligned} \tag{2.22}$$

which provides a definition of the transformation of the y variables in terms of p's and q's. This together with (2.20) yields the first set of equations for the Hamiltonian system derived by [**6**],

$$\frac{dq_1}{dt} = p_1 + p_2 e^{-|q_1 - q_2|}, \quad \frac{dq_2}{dt} = p_2 + p_1 e^{-|q_1 - q_2|}. \tag{2.23}$$

The constants of motion a_1 and a_2 can be expressed in terms of p's and q's via the first integrals

$$P_{12} = p_1 + p_2 = -(a_1 + a_2),$$

and

$$H_{12} = \frac{1}{2}(p_1^2 + p_2^2) + p_1 p_2 e^{-|q_1 - q_2|} = \frac{1}{2}(a_1^2 + a_2^2),$$

which are the total momentum and Hamiltonian for the (q, p)-flow, respectively. Using these expressions, the constants of motion a_1 and a_2 can be eliminated from (2.22) and the explicit variable transformation of variables for y_1, y_2 in terms of the (q, p) system is

$$y_1 = p_2(e^{-|q_1 - q_2|} - 1), \quad y_2 = p_1(e^{-|q_1 - q_2|} - 1). \tag{2.24}$$

Differentiating these equations with respect to time, using (2.21), (2.20), and the transformation (2.24) itself, results in a system of equations for $\dot{p}_1$ and $\dot{p}_2$, which can be solved to yield

$$\frac{dp_1}{dt} = \mathrm{sgn}(q_2 - q_1) p_1 p_2 e^{-|q_1 - q_2|}, \quad \frac{dp_2}{dt} = \mathrm{sgn}(q_1 - q_2) p_2 p_1 e^{-|q_1 - q_2|}. \tag{2.25}$$

Equations (2.23) and (2.24) are precisely those that follow from H_{12} with a canonical Hamiltonian structure in the (p, q) variables.

The case of three or more derivative-shock singularities x_i, y_i, $i = 1, 2, \ldots, N \geq 3$ proceeds in complete analogy with the case $N = 2$ above. Once again, the q-flows decouple from those of the y's, however the equations in the system governing the y-flow are now coupled and it is not immediately obvious that this system is integrable. A closer inspection however reveals that the y-flow shares the same structure as that of the μ-variables flow analogous to (2.12), and is therefore integrable by a similar argument leading to a solution in terms of standard τ-functions [**3**].

3. Generalized Peakons Lattices

In this section we discuss the connection between the peakons lattice and the Toda flows. The fact that the peakons lattice can be realized as a special case of the Toda flows in gl(n) was discovered in [**17**] and later used by Suris [**20**] to introduce a discrete time peakons lattice. Let us review these results, concentrating, for simplicity, on a particular case of the peakons lattice with a Hamiltonian

$$H(p,q) = \frac{1}{2}\sum_{i=1}^{n} p_i^2 + \sum_{1\le i<j\le n} p_i p_j \exp(q_j - q_i), \tag{3.1}$$

that corresponds to the case when particles q_i are ordered as follows: $q_1 > q_2 > \cdots > q_n$. It turns out that (3.1) possesses a Lax representation with the Lax matrix

$$L = L(p,q) = \sum_{i=1}^{n} p_i E_{ii} + \sum_{1\le i<j\le n} \sqrt{p_i p_j}\exp\Big(\frac{1}{2}(q_j - q_i)\Big)(E_{ij} + E_{ji}), \tag{3.2}$$

and an auxiliary matrix equal to the half of the skew-symmetric part of L, the peakons lattice thus being a restriction of the Toda flow to the set of matrices of the form (3.2). It was shown that this set forms a Poisson submanifold w.r.t. the associated r-matrix bracket on gl(n). Complete integrability of (3.1) is provided by the Chevalley invariants of gl(n), i.e. spectral invariants of L.

In what follows we give a uniform Lie-algebraic procedure which allows us to construct peakon-type orbits in any simple Lie algebra. On each of these orbits the Toda flow is shown to be a completely integrable system. Notice that for proving this we provide additional first integrals which supplement the Chevalley invariants of the algebra. We also indicate a convenient way of constructing Darboux coordinates for orbits under consideration.

Let us first recall the Hamiltonian formalism for the generalized (symmetric) Toda flows (cf. [**12, 15**]). Let $\mathfrak{g}$ be the normal real form of a simple Lie algebra of rank r, $\mathbf{G}$ a corresponding Lie group and $\mathfrak{h}$ a Cartan subalgebra of $\mathfrak{g}$. We denote by Φ the root system of $\mathfrak{g}$, and by Φ^+ (resp. Φ^-) the set of all positive (resp. negative) roots. We also fix a Chevalley basis $\{e_\alpha, \alpha\in\Phi; h_i, i = 1,\dots,r\}$ in $\mathfrak{g}$. All properties of the root systems and Chevalley bases that we need can be found in [**13, 16**].

Consider a direct sum decomposition $\mathfrak{g} = \mathfrak{k} + \mathfrak{b}_+$, where $\mathfrak{b}_+$ is the upper Borel subalgebra and $\mathfrak{k}$ is the maximal compact subalgebra of $\mathfrak{g}$. The dual space $\mathfrak{b}_+^*$ of $\mathfrak{b}_+$ can be identified with the space S of "symmetric" elements of $\mathfrak{g}$ that have a form

$$L = \sum_{j=1}^{r} b_j h_j + \sum_{\alpha\in\Phi} a_\alpha(e_\alpha + e_{-\alpha}). \tag{3.3}$$

S is an annihilator of $\mathfrak{k}$ w.r.t. the Killing form $\langle\, ,\, \rangle$.

The pull-back of the Lie–Poisson bracket on $\mathfrak{b}_+^*$ equips S with the Poisson bracket

$$\{f_1, f_2\}_S(L) = \frac{1}{2}\big\langle L, \big[\pi_+\nabla f_1(L), \pi_+\nabla f_2(L)\big]\big\rangle, \tag{3.4}$$

where gradients are defined w.r.t. the Killing form and π_+ is a projection on $\mathfrak{b}_+$ parallel to $\mathfrak{k}$. Equations of motion of the generalized Toda flow on S are generated by a Hamiltonian $H(L) = 1/2\langle L, L\rangle$ and have the Lax form

$$\dot{L} = \left[L, \frac{1}{2}\pi_+(L)\right] = \left[\frac{1}{2}\pi_-(L), L\right], \tag{3.5}$$

where $\pi_- = Id - \pi_+$. The bracket (3.4) is a restriction to S of the so called r-matrix bracket on $\mathfrak{g}$:

$$\{f_1, f_2\}_r(X) = \frac{1}{2}\big\langle X, \big[\pi_+\nabla f_1(X), \pi_+\nabla f_2(X)\big] - \big[\pi_-\nabla f_1(X), \pi_-\nabla f_2(X)\big]\big\rangle. \tag{3.6}$$

Symplectic leaves of the Poisson manifold $\big(S, \{\ ,\ \}_S\big)$ coincide with the orbits of the coadjoint action of the upper Borel subgroup $\mathbf{B}_+$ of $\mathbf{G}$:

$$\mathrm{Ad}^*_b(L) = \pi_S \mathrm{Ad}_{b^{-1}}(L), \tag{3.7}$$

where π_S is a projection on S parallel to $\mathfrak{b}_+$. On every indecomposable symplectic leaf, restrictions of the Chevalley invariants $I_1 = H, I_2, \dots, I_r$ of $\mathfrak{g}$ form a family of independent Poisson commuting integrals for the Toda flow. This family is maximal, however, only for a few distinguished orbits, that were classified in [**12, 14**].

To describe a particular type of coadjoint orbits, which in the sl(n) case will be shown to be associated with the peakons lattice, we need the following definitions introduced in [**11, 14**]. Let m be the maximal positive root and

$$h_m = [e_m, e_{-m}]. \tag{3.8}$$

Define

$$\begin{aligned} \mathfrak{g}' &= \mathrm{Ker}_{\mathrm{ad}_{e_m}} \cap \mathrm{Ker}_{\mathrm{ad}_{e_{-m}}}, \quad F = \mathrm{Span}\big\{e_\alpha \colon \alpha \in \Phi^- \text{ and } (m, \alpha) \neq 0\big\}, \\ \widetilde{F} &= \mathrm{Span}\{F, h_m\}, \quad V = F - \mathbf{R}\langle e_{-m}\rangle. \end{aligned} \tag{3.9}$$

Then $\mathfrak{g}'$ is a semisimple subalgebra of $\mathfrak{g}$ and F is a Heisenberg subalgebra of $\mathfrak{g}$, i. e. V is spanned by root vectors e_{α_i}, $e_{-m-\alpha_i}$, $i = 1, \dots N$, such that

$$\begin{aligned} [e_{\alpha_i}, e_{\alpha_j}] = [e_{-m-\alpha_i}, e_{-m-\alpha_j}] &= [e_{-m}, e_{\alpha_i}] = [e_{-m}, e_{-m-\alpha_i}] = 0, \\ [e_{\alpha_i}, e_{-m-\alpha_j}] &= c_i \delta_i^j e_{-m}, \end{aligned} \tag{3.10}$$

where c_i are some positive constants. Note also that

$$[h_m, e_{\alpha_i}] = -e_{\alpha_i}, [h_m, e_{-m-\alpha_i}] = -e_{-m-\alpha_i}, [h_m, e_{\pm m}] = \pm 2 e_{\pm m}. \tag{3.11}$$

Therefore (3.4) and (3.11) yield the following Poisson brackets

$$\{y_0, x_0\}_S = x_0, \quad \{y_0, x_i\}_S = \frac{1}{2}x_i, \quad \{y_0, y_i\}_S = \frac{1}{2}y_i, \quad \{y_i, x_i\}_S = \frac{1}{2}x_0, \quad i = 1, \dots N, \tag{3.12}$$

for the linear functions

$$x_i = \langle L, e_{-\alpha_i}\rangle, \quad y_i = \frac{1}{c_i}\langle L, \quad e_{m+\alpha_i}\rangle, \quad x_0 = \langle L, e_m\rangle, \quad y_0 = \langle L, h_m\rangle. \tag{3.13}$$

All other brackets between functions (3.13) are zero.

Let $h_0 \in \mathfrak{h}$ be orthogonal to h_m w.r.t. the Killing form and let $\mathfrak{O}_{m,h_0}$ be the coadjoint orbit of $\mathbf{B}_+$ through $(e_{-m} + h_0 + e_m)$.

THEOREM 3.1. *$\mathfrak{O}_{m,h_0}$ can be parameterized by elements of $\widetilde{F}$:*

$$\mathfrak{O}_{m,h_0} = \left\{ L = \pi_S \left(\zeta - \frac{1}{2x(\zeta)} \left[ad_{e_m} v(\zeta), v(\zeta) \right] \right) : \zeta \in \widetilde{F}_-, x(\zeta) > 0 \right\}, \tag{3.14}$$

where element $\zeta \in \widetilde{F}_-$ is decomposed as

$$\zeta = x(\zeta) e_{-m} + y(\zeta) h_m + v(\zeta),$$

and $v(\zeta) \in V$.

The Toda flow (3.5) is completely integrable on $\mathfrak{O}_{m,h_0}$ with a maximal family of functions in involution given by the coefficients of polynomials

$$I_j \left(L + \frac{\lambda}{x(\zeta)} e_m \right) = \sum_k I_{jk}(L) \lambda^k, \quad j = 1, \dots r, \tag{3.15}$$

where I_j are the Chevalley invariants of $\mathfrak{g}$.

The proof makes use of the so-called 1-chop map, a Poisson map from $\mathfrak{g}$ to $\mathfrak{g}'$, which was introduced for sl(n) in [**18, 19**] and generalized for an arbitrary Lie algebra in [**11**]. This map and the functions (3.15) play a key role in the proof of the complete integrability of the Toda flows on generic orbits (see [**9, 10**] for the sl(n) case and [**11**] for a general case).

It also follows from Theorem 3.1 that for constructing Darboux coordinates on $\mathfrak{O}_{m,h_0}$, it is sufficient to construct Darboux coordinates for a rather simple Poisson algebra generated by (3.12).

The following example demonstrates the connection with the peakons lattice.

EXAMPLE 3.1. Let $\mathfrak{g} = \mathrm{sl}(n)$. Then $e_m = E_{1n}, e_{-m} = E_{n1}, h_m = E_{11} - E_{nn}$,

$$\widetilde{F} = \left\{ \zeta = \begin{bmatrix} y & 0 & 0 \\ v_1 & 0 & 0 \\ x & v_2^T & -y \end{bmatrix} : x, y \in \mathbf{R}, v_1, v_2 \in \mathbf{R}^{n-2} \right\}.$$

Since Tr(L) is invariant under the action (3.7), we can define S to be the space of all symmetric matrices, not necessarily with zero trace. One can choose h_0 to be any diagonal matrix of the form $h_0 = \mathrm{diag}(\kappa, d, \kappa)$, where d is a $(n-2) \times (n-2)$ diagonal matrix. It follows from (3.14) that

$$\mathfrak{O}_{m,h_0} = \left\{ L = \begin{bmatrix} y + \kappa & v_1^T & x \\ v_1 & d + \pi_S(x^{-1} v_1 v_2^T) & v_2 \\ x & v_2^T & \kappa - y \end{bmatrix} \right. \\ \left. : x > 0, \quad y \in \mathbf{R}, \ v_1, v_2 \in \mathbf{R}^{n-2} \right\}. \tag{3.16}$$

If $d = 0$, an open set $\{-\kappa < y < \kappa; v_{1i} > 0, v_{2i} > 0, i = 1, \dots, n-2\} \subset \mathfrak{O}_{m,h_0}$ admits Darboux coordinates p_i and q_i defined as follows:

$$p_1 = y + \kappa, \quad p_n = \kappa - y, \quad p_{i+1} = \frac{v_{1i} v_{2i}}{x}, \qquad i = 1, \dots, n-2,$$

and

$$\exp(q_n - q_1) = \frac{x^2}{\kappa^2 - y^2}, \quad \exp(q_i - q_1) = \frac{x v_{1i}}{(y+\kappa) v_{2i}}, \qquad i = 2, \dots, n-1.$$

With this parameterization matrix L in (3.16) coincides with the Lax matrix (3.2). Note that $p_1 + \cdots + p_n = \mathrm{const}$ on $\mathfrak{O}_{m,h_0}$ and that coordinates q_i are defined only up to a translation. Note that a different choice of Darboux coordinates for $\mathfrak{O}_{m,0}$ was proposed in [**14**] as an example of the general construction of canonical coordinates on coadjoint orbits.

EXAMPLE 3.2. Let $\mathfrak{g}$ be an exceptional algebra of type G_2. Denote the short and long simple roots of $\mathfrak{g}$ by ν_1, ν_2 resp., and root vector corresponding to a negative root $\alpha = -(i\nu_1 + j\nu_2)$ by e_{ij}. Then F is spanned by vectors $e_{-m} = e_{32}$ and $e_{i1}, i = 1, \ldots, 4$. Thus, for any h_0 the orbit $\mathfrak{O}_{m,h_0}$ is 6-dimensional, while there are only two independent Chevalley invariants. This illustrates the necessity of using functions (3.15) to establish complete integrability.

Put $h_0 = 0$ and choose Darboux coordinates q_i, p_i for Poisson algebra (3.12) with $N = 3$ in the form:

$$q_1 = x_0, \quad p_1 = -\frac{y_0}{x_0}, \quad q_2 = -\frac{x_1}{3\sqrt{x_0}}, \quad p_2 = \frac{6y_1}{\sqrt{x_0}}, \quad q_3 = -\frac{y_2}{\sqrt{x_0}}, \quad p_3 = -\frac{2x_2}{\sqrt{x_0}}.$$

Then Theorem 3.1 gives the following integrable polynomial Hamiltonian quadratic in momenta:

$$H(p,q) = \frac{1}{12}(p_3q_2 + 6q_3^2)^2 + \frac{1}{8}(p_3q_3 + p_2q_2)^2 + \frac{q_1}{12}(p_3^2 + 12p_2^2 + 3p_1^2) + q_1^2 + 3q_3^2q_1 + 3q_3^4. \tag{3.17}$$

A detailed analysis of the system generated by (3.17), as well as integrable systems associated with peakon-type orbits in other simple Lie algebras will be given in a forthcoming publication.

References

1. M. S. Alber, R. Camassa, D. D. Holm, and J. E. Marsden, *The geometry of peaked solitons and billiard solutions of a class of integrable* PDE's, Lett. Math. Phys. **32** (1994), no. 2, 137–151.
2. ———, *On the link between umbilici geodesics and soliton solutions of nonlinear* PDE's, Proc. Roy. Soc. London Ser. A **450** (1995), no. 1940, 677–692.
3. M. S. Alber, R. Camassa, Yi. Fedorov, D. D. Holm, and J. E. Marsden, *The geometry of new classes of weak billiard solutions of nonlinear* PDE's, preprint LAUR, 1999; Los Alamos Nat. Lab. (submitted).
4. M. S. Alber, G. G. Luther, and J. E. Marsden, *Energy dependent Schrődinger operators and complex Hamiltonian systems on Riemann surfaces*, Nonlinearity **10** (1997), no. 1, 223–241.
5. E. D. Belòkolos, A. I. Bobenko, V. Z. Enol'sii, A. R. Its, and V. B. Matveev, *Algebro-geometric approach to nonlinear integrable equations*, Springer Ser. Nonlinear Dynam., vol. XII, 1994.
6. R. Camassa and D. D. Holm, *An integrable shallow water equation with peaked solitons*, Phys. Rev. Lett. **71** (1993), no. 11, 1661–1664.
7. F. Calogero, *An integrable Hamiltonian system*, Phys. Lett. A. **201** (1995), no. 4, 306–310.
8. F. Calogero and J. P. Françoise, *Solvable quantum version of an integrable Hamiltonian system*, J. Math. Phys. **37** (1996), no. 9, 4243–4251.
9. P. Deift, L. Li, T. Nanda, and C. Tomei, *The Toda flow on a generic orbit is integrable*, Comm. Pure Appl. Math. **39** (1986), no. 2, 183–232.
10. N. Ercolani, H. Flaschka, and S. F. Singer, *The geometry of the full Kostant–Toda lattice*, Progr. Math. **115** (1993), 181–225.
11. M. I. Gekhtman and M. Z. Shapiro, *Noncommutative and commutative integrability of generic Toda flows in simple Lie algebra*, Comm. Pure Appl. Math. **52** (1999), no. 1, 53–84.
12. R. Goodman and N. R. Wallach, *Classical and quantum mechanical systems of Toda-lattice type.* II. *Solutions of the classical flows*, Comm. Math. Phys. **94** (1984), no. 2, 177–217.
13. J. E. Humphreys, *Introduction to Lie algebras and representation theory*, Springer-Verlag, Berlin, 1980.

14. S. A. Kamalin and A. M. Perelomov, *Construction of canonical coordinates on polarized coadjoint orbits of Lie groups*, Comm. Math. Phys. **97** (1985), no. 4, 553–568.
15. B. Kostant, *The solution to a generalized Toda lattice and representation theory*, Adv. in Math. **34** (1979), no. 3, 195–338.
16. A. L. Onishchik and E. B. Vinberg, *Lie groups and algebraic groups*, Springer-Verlag, Berlin, 1990.
17. O. Ragnisco and M. Bruschi, *Peakons, r-matrix and Toda lattice*, Physica A **228** (1996), no. 1–4, 150–159.
18. S. Singer, Ph.D., Courant Institute of Mathematical Sciences, New York Univ., 1990.
19. ———, *Some maps from the full Toda lattice are Poisson*, Phys. Lett. A **174** (1993), no. 1–2, 66–70.
20. Y. B. Suris, *A discrete time peakons lattice*, Phys. Lett. A **217** (1996), 321–329.

Department of Mathematics, University of Notre Dame, Notre Dame, IN 46556, USA.

E-mail address: Mark.S.Alber.1@nd.edu

Department of Mathematics, University of North Carolina, Chapel Hill, NC 27599, USA.

E-mail address: camassa@math.unc.edu

Department of Mathematics, University of Notre Dame, Notre Dame, IN 46556, USA.

E-mail address: Michael.Gekhtman.1@nd.edu

Centre de Recherches Mathématiques
CRM Proceedings and Lecture Notes
Volume **25**, 2000

Fourier-Gauss Transforms of Some q-Spectral Functions

N. M. Atakishiyev

ABSTRACT. We discuss the transformation properties with respect to the classical Fourier-Gauss integral transform of some q-polynomial families, q-exponential and q-Bessel functions.

Although the classical Fourier-Gauss integral transform is based on the q-independent exponential kernel, various q-special functions exhibit simple behaviour with respect to it. The goal of the present contribution is to discuss instances of already known integral transforms of this type. They include the relations between some polynomial families (such as q-Hermite and q^{-1}-Hermite, Rogers-Szegö and Stieltjes-Wigert, and q-Laguerre and Wall polynomials), as well as various q-exponential and q-Bessel functions.

1. q-Hermite Polynomials

The continuous q-Hermite polynomials $H_n(x \mid q)$ of Rogers [**5**] are generated by the three-term recurrence relation

$$H_{n+1}(x \mid q) = 2xH_n(x \mid q) - (1-q^n)H_{n-1}(x \mid q), \quad n = 0,1,2,\dots, \tag{1.1}$$

with the initial condition $H_0(x \mid q) = 1$. The explicit form of $H_n(x \mid q)$ is given by their Fourier expansion

$$H_n(\cos\theta \mid q) = \sum_{k=0}^{n} \begin{bmatrix} n \\ k \end{bmatrix}_q e^{i(n-2k)\theta} = e^{in\theta}\, {}_2\phi_0(q^{-n}, 0; q, q^n e^{-2i\theta}), \tag{1.2}$$

where ${}_2\phi_0$ is the basic hypergeometric series and $\begin{bmatrix} n \\ k \end{bmatrix}_q$ is the q-binomial coefficient (see, for example, [**19**] or [**27**]),

$$\begin{bmatrix} n \\ k \end{bmatrix}_q := \frac{(q;q)_n}{(q;q)_k (q;q)_{n-k}} = \begin{bmatrix} n \\ n-k \end{bmatrix}_q. \tag{1.3}$$

The key point in deriving a Fourier-Gauss transform of $H_n(x \mid q)$ (and in all other cases when one considers the q-quadratic lattice) is to use in (1.2) the parametrization $x_q(s) = \sin\kappa s, q = \exp(-2\kappa^2)$, which is equivalent to the change of variables

---

2000 *Mathematics Subject Classification.* Primary: 26C15, 33D10; Secondary: 44A20.

Discussions with F. Leyvraz, P. Winternitz and K. B. Wolf are gratefully acknowledged. This work was partially supported by the UNAM-DGAPA project IN106595.

This is the final form of the paper.

$\theta = \pi/2 - \kappa s$. This means that we represent (1.2) as

$$(1.4)\quad H_n\big(\sin\kappa s \mid q\big) = i^n \sum_{k=0}^{n} (-1)^k \begin{bmatrix} n \\ k \end{bmatrix}_q e^{i(2k-n)\kappa s} = i^n e^{-in\kappa s}\,{}_2\phi_0(q^{-n}, 0; q, -q^n e^{2i\kappa s}).$$

In the limit when the parameter q tends to 1 (and, consequently, $\kappa \to 0$), $H_n\big(\sin\kappa s \mid q\big)$ converge to the ordinary Hermite polynomials $H_n(s)$, namely

$$(1.5)\quad \lim_{q\to 1} \kappa^{-n} H_n\big(\sin\kappa s \mid q\big) = H_n(s).$$

One can also consider the continuous q^{-1}-Hermite polynomials [**3**]

$$(1.6)\quad h_n\big(x \mid q\big) := i^{-n} H_n\big(ix \mid q^{-1}\big),$$

by transforming $q \to q^{-1}$ in (1.2). Because of the inversion formula

$$(1.7)\quad \begin{bmatrix} n \\ k \end{bmatrix}_{q^{-1}} = q^{k(k-n)} \begin{bmatrix} n \\ k \end{bmatrix}_q,$$

for the q-binomial coefficient (1.3), from (1.4) it follows that their explicit form is

$$(1.8)\quad h_n\big(\sinh\kappa s \mid q\big) = \sum_{k=0}^{n} (-1)^k q^{k(k-n)} \begin{bmatrix} n \\ k \end{bmatrix}_q e^{(n-2k)\kappa s} = e^{n\kappa s}\,{}_1\phi_1(q^{-n}; 0; q, -e^{-2\kappa s}).$$

The continuous q-Hermite (1.4) and q^{-1}-Hermite (1.8) polynomials are related to each other by the Fourier-Gauss integral transform [**14**]

$$(1.9)\quad \frac{1}{\sqrt{2\pi}} \int_{-\infty}^{\infty} e^{isr - s^2/2} H_n\big(\sin\kappa s \mid q\big)\, ds = i^n q^{n^2/4} h_n\big(\sinh\kappa r \mid q\big) e^{-r^2/2}.$$

To verify (1.9), substitute the explicit form of the q-Hermite polynomials (1.4) into the left-hand side of (1.9) and integrate it by the aid of the well-known Fourier transform

$$(1.10)\quad \frac{1}{\sqrt{2\pi}} \int_{-\infty}^{\infty} e^{isr - s^2/2}\, ds = e^{-r^2/2},$$

for the Gauss exponential function $\exp(-s^2/2)$.

As is evident from (1.5), in the limit when $q \to 1$ the integral transform (1.9) reduces to that for the Hermite polynomials $H_n(s)$:

$$(1.11)\quad \frac{1}{\sqrt{2\pi}} \int_{-\infty}^{\infty} e^{isr - s^2/2} H_n(s)\, ds = i^n H_n(r) e^{-r^2/2}.$$

It is essential to note that one may consider another parametrization $x = y_q(s) := \cos\kappa s$ for the argument of the q-Hermite polynomials in (1.2), which is linearly independent from $x_q(s) = \sin\kappa s$. The corresponding Fourier-Gauss transform has the form [**9**]

$$(1.12)\quad \frac{1}{\sqrt{2\pi}} \int_{-\infty}^{\infty} e^{isr - s^2/2} H_n\big(\cos\kappa s \mid q\big)\, ds = q^{n^2/4} H_n\big(\cosh\kappa r \mid q^{-1}\big) e^{-r^2/2}.$$

Since $\lim_{q\to 1} H_n\big(\cos\kappa s \mid q\big) = 2^n$, when the limit as $q \to 1$ is taken the integral transform (1.12) only reproduces the particular case of (1.11) with $n = 0$, that is, the Fourier transform (1.10) for the Gauss exponential function $\exp(-s^2/2)$. This circumstance explains why it is more interesting to consider the parametrization $x_q(s) = \sin \kappa s$.

2. Rogers-Szegö and Stieltjes-Wigert Polynomials

The q-analogue of Hermite polynomials on the unit circle are the Rogers-Szegö polynomials [**33, 1, 17**]

$$H_n(z;q) := \sum_{k=0}^{n} \begin{bmatrix} n \\ k \end{bmatrix}_q z^k = {}_2\phi_0(q^{-n}, 0; q, zq^n). \tag{2.1}$$

The Stieltjes-Wigert polynomials are defined [**19**] as

$$S_n(z;q) := H_n(q^n z; q^{-1}) = \sum_{k=0}^{n} \begin{bmatrix} n \\ k \end{bmatrix}_q q^{k^2} z^k = {}_1\phi_1(q^{-n}; 0; q, zq^{n+1}). \tag{2.2}$$

These two families represent the lower level in the ladder of ${}_4\phi_3$ biorthogonal rational functions of Al-Salam and Ismail [**2**]. They are interrelated by the Fourier-Gauss transform [**15**]

$$\frac{1}{\sqrt{2\pi}} \int_{-\infty}^{\infty} H_n(te^{2i\kappa y}; q) e^{ixy - y^2/2}\, dy = S_n(te^{-2\kappa x}; q) e^{-x^2/2}, \tag{2.3}$$

where t is an arbitrary complex number.

3. q-Laguerre and Wall Polynomials

The q-Laguerre polynomials $L_n^{(\alpha)}(x;q)$ are defined [**20, 18, 28**] as

$$\begin{aligned} L_n^{(\alpha)}(x;q) &:= \frac{(q^{\alpha+1};q)_n}{(q;q)_n}\, {}_1\phi_1(q^{-n}; q^{\alpha+1}; q, -xq^{n+\alpha+1}), \\ &= \frac{(q^{\alpha+1};q)_n}{(q;q)_n} \sum_{k=0}^{n} \frac{(q^{-n};q)_k x^k}{(q^{\alpha+1}, q; q)_k} q^{k(n+\alpha)+k(k+1)/2}. \end{aligned} \tag{3.1}$$

In the limit when the parameter q tends to 1, they converge to the ordinary Laguerre polynomials $L_n^{(\alpha)}(x)$, namely

$$\lim_{q\to 1} L_n^{(\alpha)}\big((1-q)x; q\big) = L_n^{(\alpha)}(x). \tag{3.2}$$

Applying the inversion formula

$$(z;q^{-1})_n = q^{-n(n-1)/2}(-z)^n (z^{-1};q)_n, \tag{3.3}$$

to the definition (3.1), gives the relation

$$\begin{aligned} L_n^{(\alpha)}(x;q^{-1}) &= q^{-n\alpha} \frac{(q^{\alpha+1};q)_n}{(q;q)_n} \sum_{k=0}^{n} \frac{(q^{-n};q)_k}{(q^{\alpha+1}, q; q)_k} (-qx)^k, \\ &= q^{-n\alpha} \frac{(q^{\alpha+1};q)_n}{(q;q)_n} p_n\big(-x; q^\alpha \mid q\big), \end{aligned} \tag{3.4}$$

between the q-Laguerre and Wall (or little q-Laguerre) polynomials $p_n\big(x; a \mid q\big)$ [**27**]. The latter polynomials are defined [**34**] as

$$p_n\big(x; a \mid q\big) := {}_2\phi_1(q^{-n}, 0; aq; q, qx) = \sum_{k=0}^{n} \frac{(q^{-n};q)_k (qx)^k}{(aq, q; q)_k}. \tag{3.5}$$

In exactly the same way as in previous two sections, one can show that the polynomials (3.1) and (3.5) are related by the classical Fourier–Gauss transform [**10**]

$$(3.6)\quad \frac{1}{\sqrt{\pi}}\int_{-\infty}^{\infty} L_n^{(\alpha)}(te^{2\kappa s};q)e^{2irs-s^2}\,ds = \frac{(q^{\alpha+1};q)_n}{(q;q)_n}p_n\big(tq^{n+\alpha-1/2}e^{2i\kappa r};q^\alpha \mid q\big)e^{-r^2}.$$

Note that since the q-Charlier polynomials $C_n(x;-q^{-\alpha};q) = (q;q)_n L_n^{(\alpha)}(-x;q)$, one may regard (3.6) as a Fourier–Gauss transform between the q-Charlier and Wall polynomials (3.5).

4. q-Bessel Functions

Moak has proved that the q-Laguerre polynomials converge to the entire function

$$(4.1)\quad \lim_{n\to\infty} L_n^{(\alpha)}(x;q) = x^{-\alpha/2}J_\alpha^{(2)}(2\sqrt{x};q),$$

the roots of which are the mass points of an extreme measure of orthogonality [**28**]. Jackson's q-Bessel function $J_\alpha^{(2)}(z;q)$ in (4.1) is defined [**24, 21**] as

$$(4.2)\quad J_\alpha^{(2)}(z;q) := \frac{1}{(q;q)_\alpha}\sum_{n=0}^{\infty}\frac{(-1)^n q^{n(n+\alpha)}}{(q^{\alpha+1},q;q)_n}\left(\frac{z}{2}\right)^{2n+\alpha}.$$

One can similarly prove that the polynomials (3.5) also converge to the entire function [**10**]

$$(4.3)\quad \lim_{n\to\infty} p_n\big(q^n x;q^\alpha \mid q\big) = \frac{(q;q)_\alpha}{x^{\alpha/2}}J_\alpha^{(3)}(\sqrt{x};q),$$

where Jackson's third q-Bessel function $J_\alpha^{(3)}(z;q)$ is given [**25, 22**] by the relation

$$(4.4)\quad J_\alpha^{(3)}(z;q) := \frac{1}{(q;q)_\alpha}\sum_{n=0}^{\infty}\frac{(-1)^n q^{n(n+1)/2}}{(q^{\alpha+1},q;q)_n}z^{2n+\alpha}.$$

In view of the limit relations (4.1) and (4.3) we may let $n\to\infty$ on both sides of (3.6). After some simple manipulations, this results in the Fourier–Gauss transform

$$(4.5)\quad \frac{1}{\sqrt{\pi}}\int_{-\infty}^{\infty} J_\alpha^{(2)}(2te^{\kappa s};q)e^{2irs-s^2}\,ds = q^{\alpha(2-3\alpha)/8}J_\alpha^{(3)}(q^{(\alpha-1)/4}te^{i\kappa r};q)e^{-r^2},$$

relating Jackson's second and third q-Bessel functions (4.2) and (4.4), respectively. Integral transforms of this type for Jackson's q-Bessel functions $J_\alpha^{(i)}(z;q)$, $i=1$, 2, 3, have recently been discussed in [**6**]. It has been proved in particular that

$$(4.6)\quad J_\alpha^{(3)}(q^{(\alpha-1)/4}te^{-\kappa x};q)e^{-x^2} = q^{\alpha(\alpha-2)/8}\int_{-\infty}^{\infty}\frac{dy}{\sqrt{\pi}}J_\alpha^{(1)}(2te^{i\kappa y};q)e^{2ixy-y^2},$$

where [**24, 21**]

$$(4.7)\quad J_\alpha^{(1)}(z;q) := \frac{1}{(q;q)_\alpha}\sum_{n=0}^{\infty}\frac{(-1)^n(z/2)^{2n+\alpha}}{(q^{\alpha+1},q;q)_n}.$$

By the aid of the inversion formula (3.3) it is not hard to verify that the Fourier–Gauss transforms (4.5) and (4.6) are interrelated by a replacement of the base $q\to q^{-1}$.

5. Ramanujan's Integrals

Two more examples of the classical Fourier–Gauss transform for q-special functions are as follows. Ramanujan's legacy contains Gaussian integrals [**31**]

$$\int_{-\infty}^{\infty} E_q(aqe^{2\kappa x})E_q(bqe^{-2\kappa x})e^{2mx-x^2}\,dx, \qquad (5.1)$$
$$= \sqrt{\pi}e^{m^2}E_q(-abq)e_q(aq^{1/2}e^{2m\kappa})e_q(bq^{1/2}e^{-2m\kappa}),$$

$$\int_{-\infty}^{\infty} e_q(aq^{1/2}e^{2i\kappa x})e_q(bq^{1/2}e^{-2i\kappa x})e^{2mx-x^2}\,dx \qquad (5.2)$$
$$= \sqrt{\pi}e^{m^2}e_q(abq)E_q(aqe^{2im\kappa})E_q(bqe^{-2im\kappa}),$$

where $e_q(z)$ and $E_q(z)$ are Jackson's q-exponential functions [**26**]

$$e_q(z) := \sum_{n=0}^{\infty}\frac{z^n}{(q;q)_n} = (z;q)_\infty^{-1}, \quad E_q(z) := \sum_{n=0}^{\infty}\frac{q^{n(n-1)/2}}{(q;q)_n}z^n = (-z;q)_\infty. \qquad (5.3)$$

P. I. Pastro [**29**] has shown that (5.1) and (5.2) are q-extensions of Euler's beta integral

$$B(\alpha,\beta) = \int_0^{\infty}\frac{x^{\alpha-1}\,dx}{(1+x)^{\alpha+\beta}} = \frac{\Gamma(\alpha)\Gamma(\beta)}{\Gamma(\alpha+\beta)}, \qquad (5.4)$$

and of Cauchy's form

$$B^{-1}(\alpha,\beta) = \frac{\alpha+\beta-1}{\pi}2^{\alpha+\beta-2}\int_{-\infty}^{\infty}\frac{dt}{(1+it)^{\alpha}(1-it)^{\beta}}, \qquad (5.5)$$

of the beta integral $B(\alpha,\beta)$, respectively.

Derivations of Ramanujan's integrals (5.1) and (5.2) for real values of the parameter m have been given by R. Askey [**4**]. Later on it has become clear that these identities are in fact valid for arbitrary complex values of the parameter m and they are thus instances of the standard Fourier transform with the exponential kernel $\exp(ixy)$ [**12**]. This fact is a consequence of the Fourier transform (1.10) and the q-binomial theorem

$$\sum_{n=0}^{\infty}\frac{(a;q)_n}{(q;q)_n}z^n = e_q(z)E_q(-az). \qquad (5.6)$$

Details can be found in [**12**].

6. q-Exponential Functions

Jackson's q-extension (5.3) of the classical exponential function e^z is not the only one. One can presumably extend e^z in many other ways and at least some of these possibilities are already known (for a comprehensive discussion of this point see [**30**]). In particular, a two-parameter q-exponential function on the q-quadratic lattice is defined [**23**] as

$$\mathcal{E}_q^{IZ}(\cos\theta;a,b) := \sum_{n=0}^{\infty}\frac{q^{n^2/4}}{(q;q)_n}(aq^{1-n/2}e^{i\theta},\quad aq^{1-n/2}e^{-i\theta};q)_n b^n. \qquad (6.1)$$

Suslov proposed to consider it as a function of two independent variables $x=\cos\theta$ and $y=\cos\varphi$ and a parameter α by replacing $a=-e^{i\varphi}$ and $b=\alpha e^{-i\varphi}$ in (6.1)

[**32**]. Then the function

$$\begin{aligned}(6.2)\quad \mathcal{E}_q(x,y;\alpha) &:= e_{q^2}(q\alpha^2)E_{q^2}(-\alpha^2)\mathcal{E}_q^{IZ}(x;-e^{i\varphi},\alpha e^{-i\varphi}) \\ &= e_{q^2}(q\alpha^2)E_{q^2}(-\alpha^2)\sum_{n=0}^{\infty}\frac{q^{n^2/4}}{(q;q)_n}(-q^{1-n/2}e^{i(\theta+\varphi)},-q^{\frac{1-n}{2}}e^{i(\varphi-\theta)};q)_n(\alpha e^{-i\varphi})^n,\end{aligned}$$

is symmetric in the variables x and y. Moreover, these variables are separated, i.e.,

$$(6.3)\qquad \mathcal{E}_q(x,y;\alpha) = \mathcal{E}_q(x,0;\alpha)\mathcal{E}_q(0,y;\alpha).$$

The normalization constant $e_{q^2}(q\alpha^2)E_{q^2}(-\alpha^2)$ in (6.2) is chosen in such a way that $\mathcal{E}_q(0,0;\alpha) = 1$ [**32**].

As a q-analogue of the addition law for the exponential function e^z, the relation (6.3) reveals the importance of the particular case of (6.2), given explicitly by

$$(6.4)\qquad \mathcal{E}_q(x,0;\alpha) = \mathcal{E}_q(x;\alpha) = e_{q^2}(q\alpha^2)E_{q^2}(-\alpha^2)\mathcal{E}_q^{IZ}(x;-i,-i\alpha).$$

Once the relation between $\mathcal{E}_q(x;\alpha)$ and $\mathcal{E}_q^{IZ}(x;-i,-i\alpha)$ is clear, one can use the known properties of the latter [**23, 13**]. Firstly, $\mathcal{E}_q^{IZ}(x;-i,-i\alpha)$ can also be considered as a generating function for the continuous q-Hermite polynomials [**23**]. Therefore, as follows from (6.4),

$$(6.5)\qquad \mathcal{E}_q(x;\alpha) = e_{q^2}(q\alpha^2)\sum_{n=0}^{\infty}\frac{q^{n^2/4}\alpha^n}{(q;q)_n}H_n(x\mid q).$$

Secondly, the relation (6.5) defines the Fourier–Gauss transformation properties of $\mathcal{E}_q(x;\alpha)$. Indeed, substituting the inverse transform with respect to (1.9) into the right-hand side of (6.5) and summing over n the aid of the generating function

$$(6.6)\qquad \sum_{n=0}^{\infty}\frac{t^n q^{n(n-1)/2}}{(q;q)_n}h_n(\sinh\kappa r\mid q) = E_q(te^{\kappa r})E_q(-te^{-\kappa r}),$$

for the q^{-1}-Hermite polynomials (1.8), yields

$$\begin{aligned}(6.7)\quad &\mathcal{E}_q(\sin\kappa s;\alpha)e^{-s^2/2} \\ &\qquad = \frac{e_{q^2}(q\alpha^2)}{\sqrt{2\pi}}\int_{-\infty}^{\infty}e^{isr-r^2/2}E_q(i\alpha q^{1/2}e^{-\kappa r})E_q(-i\alpha q^{1/2}e^{\kappa r})\,dr.\end{aligned}$$

Observe that this integral representation may be considered as an analytic continuation of $\mathcal{E}_q(x;\alpha)$, initially defined by (6.4) for the values of $|\alpha| < 1$ only. Due to the factor $e_{q^2}(q\alpha^2)$, the q-exponential function $\mathcal{E}_q(x;\alpha)$ is meromorphic in the complex α plane with simple poles at $\alpha_n^{(\pm)} = \pm q^{-(n+1/2)}$, $n = 0, 1, 2, \cdots$ (cf., [**23**]).

The q-analogues of the trigonometric functions $\cos\omega x$ and $\sin\omega x$ are defined [**32**] as

$$(6.8\text{a})\qquad C_q(x;\omega) = \frac{1}{2}\big[\mathcal{E}_q(x;i\omega) + \mathcal{E}_q(x;-i\omega)\big],$$

$$(6.8\text{b})\qquad S_q(x;\omega) = \frac{1}{2i}\big[\mathcal{E}_q(x;i\omega) - \mathcal{E}_q(x;-i\omega)\big].$$

Combining equation (6.8) with the Fourier–Gauss transform (6.7), gives

$$\text{(6.9a)}\quad C_q(\sin\kappa s;\omega)e^{-s^2/2} = \frac{e_{q^2}(-q\omega^2)}{\sqrt{2\pi}}\int_{-\infty}^{\infty} E_q(\omega q^{1/2}e^{-\kappa r})E_q(-\omega q^{1/2}e^{\kappa r})e^{-r^2/2}\cos rs\,dr,$$

$$\text{(6.9b)}\quad S_q(\sin\kappa s;\omega)e^{-s^2/2} = \frac{e_{q^2}(-q\omega^2)}{\sqrt{2\pi}}\int_{-\infty}^{\infty} E_q(\omega q^{1/2}e^{\kappa r})E_q(-\omega q^{1/2}e^{-\kappa r})e^{-r^2/2}\sin rs\,dr$$

In the case of the q-linear lattice $x_q(s) = e^{i\kappa s}$ the q-analogue of the exponential function e^z has the form

$$\text{(6.10)}\quad \varepsilon_q(z) := \sum_{n=0}^{\infty}\frac{q^{n(n-1)/4}}{(q;q)_n}z^n.$$

The relations between (6.10) and the q-exponential functions $E_q(z)$ and $e_q(z)$ are [**8**]

$$\text{(6.11)}\quad \varepsilon_q(\alpha e^{i\kappa s})e^{-s^2/2} = \frac{1}{\sqrt{2\pi}}\int_{-\infty}^{\infty} e^{isr-r^2/2}E_q(q^{1/4}\alpha e^{\kappa r})\,dr,$$

$$\text{(6.12)}\quad \varepsilon_q(\alpha e^{-\kappa s})e^{-s^2/2} = \frac{1}{\sqrt{2\pi}}\int_{-\infty}^{\infty} e^{isr-r^2/2}e_q(q^{-1/4}\alpha e^{i\kappa r})\,dr.$$

Since $e_{1/q}(z) = E_q(-qz)$ and $\varepsilon_{1/q}(z) = \varepsilon_q(-qz)$, the Fourier-Gauss transforms (6.11) and (6.12) are interrelated by the substitution $q \to 1/q$ (i.e., $\kappa \to i\kappa$).

Analogues of the trigonometric functions $\cos\omega x$ and $\sin\omega x$ on the q-linear lattice $c_q(\omega x)$ and $s_q(\omega x)$ [**16, 32**] are defined in terms of $\varepsilon_q(z)$ by the relations

$$\text{(6.13a)}\quad c_q(\omega x) = \frac{1}{2}\left[\varepsilon_q(i\omega x) + \varepsilon_q(-i\omega x)\right],$$

$$\text{(6.13b)}\quad s_q(\omega x) = \frac{1}{2i}\left[\varepsilon_q(i\omega x) - \varepsilon_q(-i\omega x)\right],$$

respectively. Therefore, from (6.13) it follows that

$$\text{(6.14a)}\quad c_q(\omega e^{i\kappa s})e^{-s^2/2} = \frac{1}{\sqrt{2\pi}}\int_{-\infty}^{\infty} e^{isr-r^2/2}\,\mathrm{Cos}_q(q^{1/4}\omega e^{\kappa r})\,dr,$$

$$\text{(6.14b)}\quad s_q(\omega e^{i\kappa s})e^{-s^2/2} = \frac{1}{\sqrt{2\pi}}\int_{-\infty}^{\infty} e^{isr-r^2/2}\,\mathrm{Sin}_q(q^{1/4}\omega e^{\kappa r})\,dr.$$

Here the q-trigonometric functions $\mathrm{Cos}_q(z)$ and $\mathrm{Sin}_q(z)$ of Jackson are defined [**26**] as

$$\text{(6.15)}\quad \mathrm{Cos}_q(z) = \frac{1}{2}\left[E_q(iz) + E_q(-iz)\right],\quad \mathrm{Sin}_q(z) = \frac{1}{2i}\left[E_q(iz) - E_q(-iz)\right].$$

In a like manner, for the linear lattice $x_{1/q}(s) = e^{-\kappa s}$ from (6.12) we have

$$\text{(6.16a)}\quad c_q(\omega e^{-\kappa s})e^{-s^2/2} = \frac{1}{\sqrt{2\pi}}\int_{-\infty}^{\infty} e^{isr-r^2/2}\cos_q(q^{-1/4}\omega e^{i\kappa r})\,dr,$$

$$\text{(6.16b)}\quad s_q(\omega e^{-\kappa s})e^{-s^2/2} = \frac{1}{\sqrt{2\pi}}\int_{-\infty}^{\infty} e^{isr-r^2/2}\sin_q(q^{-1/4}\omega e^{i\kappa r})\,dr,$$

where [26]

$$\cos_q(z) = \frac{1}{2}\big[e_q(iz) + e_q(-iz)\big], \quad \sin_q(z) = \frac{1}{2i}\big[e_q(iz) - e_q(-iz)\big]. \tag{6.17}$$

We note in closing that the above cases, which comprise some polynomial families and various q-extensions of the exponential e^z and Bessel $J_\alpha(z)$ functions, exhibit the remarkable ability of the classical Fourier integral transform to reveal close relations between q-special functions. There are many ways to increase the number of such examples. For instance, one can combine the integral transform (1.9) for the continuous q-Hermite polynomials $H_n(x \mid q) := p_n(x; 0, 0, 0, 0 \mid q)$ with connection coefficients formulae for the Askey–Wilson polynomials $p_n(x; a, b, c, d \mid q)$ to obtain Fourier–Gauss transforms for the whole Askey–Wilson hierarchy [**9, 11, 7**]. Recall that the particular cases of this family include the big q-Hermite polynomials $H_n(x; a \mid q) := p_n(x; a, 0, 0, 0 \mid q)$ and the Al–Salam–Chihara polynomials $p_n(x; a, b \mid q) := p_n(x; a, b, 0, 0 \mid q)$.

Also, since the Rogers–Szegö (2.1) and Stieltjes–Wigert (2.2) polynomials represent the lower level in another ladder of ${}_4\phi_3$ biorthogonal rational functions of Al-Salam and Ismail [**2**], an interesting research direction is to extend the Fourier–Gauss transform (2.3) to the next levels of the Al-Salam–Ismail family.

References

1. W. A. Al-Salam and L. Carlitz, *A q-analog of a formula of Toscano*, Boll. Un. Mat. Ital. (3) **12** (1957), 414–417.
2. W. A. Al-Salam and M. E. H. Ismail, *A q-beta integral on the unit circle and some biorthogonal rational functions*, Proc. Amer. Math. Soc. **121** (1994), no. 2, 553–561.
3. R. Askey, *Continuous q-Hermite polynomials when q > 1*, q-Series and Partitions (Minneapolis, 1988) (D. Stanton, ed.), IMA Vol. Math. Appl., vol. 18, Springer, New York-Berlin, 1989, pp. 151–158.
4. ———, *Two integrals of Ramanujan*, Proc. Amer. Math. Soc. **85** (1982), no. 2, 192–194.
5. R. Askey and M. E. H. Ismail, *A generalization of ultraspherical polynomials*, Studies in Pure Mathematics (P. Erdös, ed.), Birkhäuser, Basel-Boston, Mass., 1983, pp. 55–78.
6. N. M. Atakishiyev, *Fourier–Gauss transforms of q-exponential and q-Bessel functions*, Symmetries and Integrability of Difference Equations (P.A. Clarkson and F.W. Nijhoff, eds.), London Mathematical Society Lecture Note Series, vol. 255, University Press, Cambridge, 1999, pp. 345–352.
7. ———, *Fourier–Gauss transforms of the Askey–Wilson polynomials*, J. Phys. A **30** (1997), no. 24, L815–L820.
8. ———, *On a one-parameter family of q-exponential functions*, J. Phys. A **29** (1996), no. 10, L223–L227.
9. M. K. Atakishiyeva and N. M. Atakishiyev, *Fourier–Gauss transforms of the continuous big q-Hermite polynomials*, J. Phys. A **30** (1997), L559–L565.
10. ———, *q-Laguerre and Wall polynomials are related by the Fourier–Gauss transform*, J. Phys. A **30**, (1997), L429–L432.
11. ———, *Fourier–Gauss transforms of the Al-Salam–Chihara polynomials*, J. Phys. A **30** (1997), L655–L661.
12. N. M. Atakishiyev and P. Feinsilver, *Two Ramanujan's integrals with a complex parameter*, Proceedings of the IV Wigner Symposium (Guadalajara, 1995) (N. M. Atakishiyev, T. H. Seligman, and K. B. Wolf, eds.) World Scientific, Singapore, 1996, pp. 406–416.
13. ———, *On the coherent states for the q-Hermite polynomials and related Fourier transformation*, J. Phys. A **29** (1996), 1659–1664.
14. N. M. Atakishiyev and Sh. M. Nagiyev, *On the wave functions of a covariant linear oscillator*, Theoret. Math. Phys. **98** (1992), 162–166.
15. ———, *On the Rogers–Szegö polynomials*, J. Phys. A **27** (1994), L611–L615.

16. N. M. Atakishiyev and S. K. Suslov, *Difference hypergeometric functions*, Progress in Approximation Theory (Tampa, 1990) (A. A. Gonchar and E. B. Staff, eds.), Springer Ser. Comput. Math., vol. 19, Springer, New York, 1992, pp. 1–35,
17. L. Carlitz, *Note on orthogonal polynomials related to theta functions*, Publ. Math. Debrecen **5** (1958), 222–228.
18. H. Exton, *Basic Laguerre polynomials*, Pure Appl. Math. Sci. **5** (1977), 19–24.
19. G. Gasper and M. Rahman, *Basic hypergeometric series*, Encyclopedia Math. Appl., vol. 35, Cambridge Univ. Press, Cambridge, 1990.
20. W. Hahn, *Über Orthogonalpolynome, die q-Differenzengleichunger genügen*, Math. Nachr. **2** (1949), 4–34.
21. M. E. H. Ismail, *The zeros of basic Bessel functions, the functions* $J_{\nu+ax}(z)$ *and associated orthogonal polynomials*, J. Math. Anal. Appl. **86** (1982), 1–19.
22. ———, *Some properties of Jackson's third q-Bessel function*, J. Approx. Theory (to appear).
23. M. E. H. Ismail and R. Zhang, *Diagonalization of certain integral operators*, Adv. Math. **109** (1994), 1–33.
24. F. H. Jackson, *On generalized functions of Legendre and Bessel*, Trans. Roy. Soc. Edinburgh **41** (1904), 1–28.
25. ———, *The application of basic numbers to Bessel's and Legendre's functions*, Proc. Lond. Math. Soc. (2) **2** (1905), 192–220.
26. ———, *A basic-sine and cosine with symbolical solutions of certain differential equations*, Proc. Edinburgh Math. Soc. (2) **22** (1904), 28–38.
27. R. Koekoek and R. F. Swarttouw, *The Askey-scheme of hypergeometric orthogonal polynomials and its q-analogue*, Report 94–05, Delft Univ. of Technology, 1994.
28. D. S. Moak, *The q-analogue of the Laguerre polynomials*, J. Math. Anal. Appl. **81** (1981), 20–471.
29. P. I. Pastro, *Orthogonal polynomials and some q-beta integrals of Ramanujan*, J.Math. Anal. Appl. **112** (1985), 517–540.
30. M. Rahman, *The q-exponential functions, old and new*, Proc. Internat. Workshop Finite Dimensional Integrable Systems (Dubna, 1994) (A. N. Sissakian and G. S. Pogosyan, eds.), JINR, Dubna, 1995, pp. 161–170
31. S. Ramanujan, *Collected papers* (G. H. Hardy, P. V. Seshu Aiyar, and B. M. Wilson, eds.), Cambridge Univ. Press, Cambridge, 1927; reprinted by Chelsea, New York, 1959.
32. S. K.Suslov, *"Addition" theorems for some q-exponential and q-trigonometric functions*, Methods Appl. Anal. **4** (1997), 11–32.
33. G. Szegö, *Ein Beitrag zur Theorie der Thetafunktionen*, Sitz. Preuss. Akad. Wiss., Phys.-Math. Klasse **19** (1926), 242–252; reprinted in *Collected Papers* (R. Askey, ed.), **1**, 795–805, Birkhäuser, Boston, MA, 1982.
34. H. S. Wall, *A continued fraction related to some partition formulas of Euler*, Amer. Math. Monthly **48** (1941), 102–108.

Institute of Physics, Azerbaijan Academy of Sciences, H. Javid Prospekt 33, Baku 370143, Azerbaijan.

Current address: Instituto de Matematicas, UNAM, Apartado Postal 273–3, CP 62250 Cuernavaca, Morelos, Mexico.

E-mail address: natig@matcuer.unam.mx

Centre de Recherches Mathématiques
CRM Proceedings and Lecture Notes
Volume **25**, 2000

Integrable Differential-Difference Systems in (2 + 1)-Dimensions and Their Hamiltonian Structure

Maciej Błaszak and Andrzej Szum

ABSTRACT. We present two different methods of extension of integrable $(1+1)$-dimensional bi-Hamiltonian lattice systems to integrable $(2+1)$-dimensional lattice-field systems. The first method is related to the concept of central extension, the second one to the so-called operand formalism, i.e. $\mathcal{R}$-matrix approach for operator variables.

1. Introduction

In this work we present two different methods for constructing integrable lattice-field systems in $(2+1)$-dimensions. The first one is based on the so-called central extension approach, the second one is connected with the operator valued variables approach. Both methods present an extension of the standard $\mathcal{R}$-matrix formalism in $(1+1)$-dimension for lattice systems, which was developed in the series of works [**2–4, 7**]. The construction of Hamiltonian equations and the related hierarchy of symmetries, becomes quite transparent when using the terminology of the $\mathcal{R}$-structure. This terminology was applied to constructing lattice systems in $(1+1)$-dimensions in papers [**4, 7**]. Our task is to extend the results of [**2–4, 7**] onto $(2+1)$-dimensional space with one discrete and one continuous variables. At first we apply the central extension approach, considered for the first time in early works of Reiman and Semenov-Tian-Shansky [**10, 11**] and by Prikarpatsky [**8, 9**], and derive a hierarchy of commuting lattice-field vector fields and their Hamiltonian representation. Further we implement the ideas of Magri–Morosi–Tondo [**6**], Athrone–Dorfman [**1**] and Dorfman–Fokas [**5**] of operand approach into the frame of the $\mathcal{R}$-matrix formalism for construction of bi-Hamiltonian representation of integrable differential-difference systems in $(2+1)$-dimensions. Both methods give the same results. We illustrate the theory with many examples, including these which are not reduced to $(1+1)$-dimensions.

2000 *Mathematics Subject Classification.* Primary: 35, 37; Secondary: 37K10, 35Q.
Supported by KBN Research Grant No. 2 P03B 113 13.
We would like to thank prof. A. K. Prykarpatsky for stimulating discussions.
This is the final form of the paper.

2. Basic Considerations

Let $\mathcal{G}$-be an arbitrary continuous associative algebra over $\mathbf{C}$ and $T\colon \mathcal{G} \to \mathcal{G}$ some its automorphism:

$$T(k_1a_1 + k_2a_2) = k_1T(a_1) + k_2T(a_2),\\ T(ab) = T(a)T(b), \tag{2.1}$$

for all $a_1,\ a_2 \in \mathcal{G}$ and $k_1,\ k_2 \in \mathbf{C}$. Assume also that on the algebra $\mathcal{G}$ a linear mapping $\Gamma\colon \mathcal{G} \to \mathbf{C}$ is defined such that:

$$\Gamma(ab) = \Gamma(ba), \quad \Gamma(Ta) = \Gamma(a), \tag{2.2}$$

for $a,\ b \in \mathcal{G}$. Define now an associative operator algebra g having such a structure: $A \in g$ if

$$A(\mathcal{T}) = \sum_{i \ll \infty} u_i \mathcal{T}^i, \tag{2.3}$$

where $u_i \in \mathcal{G}$, $i \in \mathbf{Z}$ and $\mathcal{T}$ is an operator satisfying the commutation rule $\mathcal{T}a = (Ta)\mathcal{T}$. On the associative operator algebra g one can define the following linear *trace*-operation:

$$\operatorname{tr} A(\mathcal{T}) := \Gamma(u_0), \tag{2.4}$$

such that

$$\operatorname{tr}\big(A(\mathcal{T})B(\mathcal{T})\big) = \operatorname{tr}\big(B(\mathcal{T})A(\mathcal{T})\big), \tag{2.5}$$

for all $A(\mathcal{T}),\ B(\mathcal{T}) \in g$. It is a standard procedure for conversion of the associative algebra g into a Lie algebra g with the Lie product chosen as the commutator in g:

$$\big[A(\mathcal{T}), B(\mathcal{T})\big] := A(\mathcal{T})B(\mathcal{T}) - B(\mathcal{T})A(\mathcal{T}), \tag{2.6}$$

for $A(\mathcal{T})$ and $B(\mathcal{T}) \in g$. In addition we assume that the scalar product

$$\big\langle A(\mathcal{T}), B(\mathcal{T})\big\rangle := \operatorname{tr}\big(A(\mathcal{T})B(\mathcal{T})\big), \tag{2.7}$$

is nondegenerate on g and from (2.5) it is symmetric and ad-invariant:

$$\big\langle A(\mathcal{T}), \big[B(\mathcal{T}), C(\mathcal{T})\big]\big\rangle = \big\langle \big[A(\mathcal{T}), B(\mathcal{T})\big], C(\mathcal{T})\big\rangle, \tag{2.8}$$

for any $A(\mathcal{T})$, $B(\mathcal{T})$ and $C(\mathcal{T}) \in g$. Assume that the basic associative algebra $\mathcal{G}$ depends effectively on an independent parameter $x \in S^1$, which allows a definition of the new associative current algebra $C(\mathcal{G}) = C^{(\infty)}(S^1; \mathcal{G})$ which naturally generates, via the construction above, the corresponding current operator Lie algebra $C(g) = C^{(\infty)}(S^1; g)$ with the following modified Tr-operation:

$$\operatorname{Tr} A(\mathcal{T}) := \int_{S^1} dx\, \Gamma(u_0)(x), \tag{2.9}$$

and scalar product:

$$\big(A(\mathcal{T}), B(\mathcal{T})\big) := \operatorname{Tr}\big(A(\mathcal{T})B(\mathcal{T})\big), \tag{2.10}$$

for $A(\mathcal{T})$ and $B(\mathcal{T}) \in C(g)$.The current Lie algebra $C(g)$ can be naturally extended via its central extension procedure: $C(g) \to \overline{C}(g) = C(g) \oplus \mathbf{C}$ with the following Lie product:

$$\big[(A(\mathcal{T}), \alpha), (B(\mathcal{T}), \beta)\big] := \big(\big[A(\mathcal{T}), B(\mathcal{T})\big], \omega_2\big(A(\mathcal{T}), B(\mathcal{T})\big)\big), \tag{2.11}$$

where by definition, $\omega_2 : C(g)\times C(g) \to \mathbf{C}$ is the standard Maurer–Cartan 2-cocycle on $C(g)$, α, $\beta \in \mathbf{C}$ and for any $A(\mathcal{T})$ and $B(\mathcal{T}) \in C(g)$

$$\omega_2\big(A(\mathcal{T}), B(\mathcal{T})\big) := \int_{S^1} dx\, \langle A(\mathcal{T}), dB(\mathcal{T})/dx\rangle = \mathrm{Tr}\Big(A(\mathcal{T})\frac{dB(\mathcal{T})}{dx}\Big). \tag{2.12}$$

3. $\mathcal{R}$-Matrix Formalism for (2 + 1)-Lattice-Field Systems

3.1. The central extension approach. Consider a nonlinear lattice-field dynamical system

$$u_t = K[u], \tag{3.1}$$

on a Poisson manifold $M \subset C^{(\infty)}(\mathbf{Z}\times S^1; \mathbf{C})$, $S^1 \equiv \mathbf{R}/2\pi\mathbf{Z}$, where K means the corresponding vector field on M. Let the automorphism T be a shift operation E:

$$E^\alpha u(n) = u(n+\alpha), \quad \alpha \in \mathbf{Z}, \tag{3.2}$$

g be the shift algebra

$$g \ni L = \sum_{i\ll\infty} u_i(n)\mathcal{E}^i, \tag{3.3}$$

where $\mathcal{E}$ is the shift operator:

$$\mathcal{E}^\alpha u = (E^\alpha u)\mathcal{E}^\alpha, \tag{3.4}$$

and $\mathcal{G}$ be a commutative algebra of difference functions of $u_i(n,x;t)$; for any $a \in \mathcal{G}$ we put:

$$\Gamma(a) = \sum_{n\in\mathbf{Z}} a(n). \tag{3.5}$$

Then, let $C(\mathcal{G}) := C^\infty(S^1, \mathcal{G})$ be the related associative current algebra which generates the corresponding current shift Lie algebra $C(g) := C^\infty(S^1, g)$ with the following Tr-operation:

$$\mathrm{Tr}\, A(L) := \int_0^{2\pi} dx \sum_{n\in\mathbf{Z}} u_0(n,x;t), \quad A \in C(g). \tag{3.6}$$

Finally we pass to the centrally extended current shift Lie algebra $\overline{C}(g) = C(g) \oplus \mathbf{C}$ with the commutator supplemented with 2-cocycle ω_2. Define *a priori* some momentum mapping: $\bar{L} : M \to \overline{C}(g^*)$ associated with the corresponding current Lie algebra action of $\overline{C}(g)$ on M, considered as Poissonian, that is the Poisson structure $\{.\,,.\}$ on M is invariant with respect to this action, and the diagram

$$\begin{array}{ccc} M & \xrightarrow{\bar{L}} & \overline{C}(g^*) \\ \bar{A}(u)\ \downarrow & & \downarrow\ \mathrm{ad}^*_{\bar{A}(\bar{L})} \\ M & \xrightarrow{\bar{L}} & \overline{C}(g^*) \end{array} \tag{3.7}$$

is a commutative one for any $\bar{A}(\bar{L}) = (A,\alpha) \in \overline{C}(g)$. As a result of the above construction one can represent the Poisson structure on the manifold M as the standard reduction on M of the canonical Lie–Poisson structure on $\overline{C}(g^*) \cong \overline{C}(g)$:

$$\{h_i, h_j\}(u) := \{H_i, H_j\}(L) = \big\langle L, \big[\nabla H_i(L), \nabla H_j(L)\big]\big\rangle, \tag{3.8}$$

for any pair of smooth functionals H_i, $H_j \in F(\overline{C}(g))$. Concerning the integrable Poissonian flows (3.1) on M we intend to construct a hierarchy of Poisson commuting functionals $H_{i|L:=L(u)} := h_i \in F(M)$, $i \in \mathbf{Z}_+$, which can be produced here via the standard $\mathcal{R}$-matrix approach. Consider a Casimir functional $H \in F(\overline{C}(g))$ of the Lie–Poisson bracket (3.8), satisfying the functional equation:

$$\frac{d\nabla H(L)}{dx} = [L, \nabla H(L)], \tag{3.9}$$

for all $\bar{L} := (L, 1) \in \overline{C}(g^*) \cong \overline{C}(g)$. Assume also that the Lie algebra $\overline{C}(g)$ admits a standard $\mathcal{R}$-structure, that is the new Lie bracket

$$[(A,\alpha),(B,\beta)]_{\overline{\mathcal{R}}} := ([A,B]_{\mathcal{R}}, \omega_2^{\mathcal{R}}(A,B)), \tag{3.10}$$

where

$$\omega_2^{\mathcal{R}}(A,B) := \omega_2(\mathcal{R}A, B) + \omega_2(A, \mathcal{R}B), \tag{3.11}$$

and

$$[A,B]_{\mathcal{R}} := [A, \mathcal{R}B] + [\mathcal{R}A, B], \tag{3.12}$$

are defined by some linear homomorphism $\mathcal{R}\colon C(g) \to C(g)$.

THEOREM 1. *Given a hierarchy of Casimir functionals $H_i \in F(\overline{C}(g^*))$ satisfying equation* (3.9), *all the reduced functionals $h_i := H_{i|\bar{L}:=\bar{L}(u)}$, $u \in M$ are in involution with respect to the Poisson bracket*

$$\{H_i, H_j\}_{\mathcal{R}}(L) = (L, [\nabla H_i(L), \nabla H_j(L)]_{\mathcal{R}}) + \omega_2^{\mathcal{R}}(\nabla H_i, \nabla H_j) = 0, \tag{3.13}$$

reduced from $\overline{C}(g^*)$ on M, with respect to the diagram (3.7). *Moreover, the momentum map $\bar{L}\colon M \to \overline{C}(g^*)$ satisfies with respect to the bracket* (3.13) *and to the Casimir functional H_i the following evolution equation*:

$$\frac{d}{dt_i}L = [L, \mathcal{R}\nabla H_i(L)] - \frac{d(\mathcal{R}\nabla H_i(L))}{dx} = [L - \partial_x, \mathcal{R}\nabla H_i(L)], \tag{3.14}$$

where $[\partial_x, a] = da/dx = a_x$, which is just a Lax representation of the dynamical system (3.1).

Let us pass to examples. The admissible n-field Lax operators read

$$L = \mathcal{E}^{n+\alpha} + u_{n+\alpha-1}\mathcal{E}^{n+\alpha-1} + \cdots + u_\alpha \mathcal{E}^\alpha, \quad -n < \alpha \le 0, \quad \alpha \in \mathbf{Z}, \tag{3.15}$$

where $u_i = u_i(n, x; t)$. The difference, when compared to the $(1+1)$-dimensional case [**4**] is that now $\alpha = 0$ is included. Let $P_{\ge 0}$ and $P_{<0}$ be projections onto Lie subalgebras $C(g_{\ge 0})$ and $C(g_{<0})$, where

$$g_{\ge 0} = \Big\{\sum_{i\ge 0} u_i \mathcal{E}^i\Big\}, \quad g_{<0} = \Big\{\sum_{i<0} u_i \mathcal{E}^i\Big\}. \tag{3.16}$$

The L-linear Poisson structure related with $\mathcal{R}_1 = \frac{1}{2}(P_{\ge 0} - P_{<0})$ reads:

$$\begin{aligned}\Theta_1(L, \partial_x)\nabla H &= -\mathrm{ad}_{(L-\partial_x)}(\nabla H)_{\ge 0} + (\mathrm{ad}_{(L-\partial_x)}\nabla H)_{>0},\\ &= -\mathrm{ad}_L(\nabla H)_{\ge 0} + (\mathrm{ad}_L \nabla H)_{>0} - (D_x \nabla H)_{>0} + D_x(\nabla H)_{\ge 0},\end{aligned} \tag{3.17}$$

where

$$D_x a = a_x, \quad \nabla H = \mathcal{E}^{-\alpha}\gamma_\alpha + \cdots + \mathcal{E}^{-n-\alpha+1}\gamma_{n+\alpha-1}, \quad \gamma_i = \frac{\delta h_i}{\delta u}, \tag{3.18}$$

and is equal to the centrally extended $(L \to L - \partial_x)$ linear Poisson structure $\Theta_1(L)$ for $(1+1)$-lattice systems [**2–4**]. The quadratic Poisson structure $\Theta_2(L)$ related to

$\mathcal{R}_2 = \frac{1}{2}\mathcal{R}_1 \mathrm{ad}_L^+$ [**2–4, 7**], where ad^+ denotes the anticommutator, does not survive the central extension as $\Theta_2(L - \partial_x)$ does not fulfill the Jacobi identity. Eq. (3.9) for Casimir functionals can be solved putting

$$\nabla H_m(L) := \sum_{m-j\in\mathbf{Z}_+} a_j \mathcal{E}^j, \tag{3.19}$$

where the functional parameters a_j are subsequently obtained from (3.9) via a recursive procedure.

EXAMPLE 1. The lattice-field Benjamin–Ono equation: $n = 1$, $\alpha = 0$, $L = \mathcal{E} + u$ (it does not exist in the one dimensional lattice case).

The first equations of the hierarchy read:

$$\begin{aligned} u(n)_{t_1} &= 2uu_x + \mathcal{H}u_{xx}, \\ u(n)_{t_2} &= \frac{1}{4}u_{xxx} + 3u^2u_x + \frac{3}{2}\mathcal{H}(uu_x)_x + \frac{3}{2}(u\mathcal{H}u_x)_x + \frac{3}{2}\mathcal{H}^2u_{xxx}, \\ &\ \vdots \end{aligned} \tag{3.20}$$

where $\mathcal{H} = (E+1)/(E-1)$ and $\theta_1 = D_x$.

EXAMPLE 2. The lattice-field Toda equation: $n = 2$, $\alpha = -1$, $L = \mathcal{E} + p + v\mathcal{E}^{-1}$.

The first equations of the hierarchy read:

$$\begin{aligned} v(n)_{t_1} &= v(n)[p(n) - p(n-1)], \\ p(n)_{t_1} &= v(n+1) - v(n) + p(n)_x, \\ v(n)_{t_2} &= v(n)[p^2(n) - p^2(n-1) + v(n+1) \\ &\qquad - v(n-1) + p(n)_x + p(n-1)_x], \\ p(n)_{t_2} &= v(n+1)[p(n+1) + p(n)] - v(n)[p(n-1) + p(n)] + v(n)_x \\ &\qquad + v(n+1)_x + \mathcal{H}p(n)_{xx} + 2p(n)p(n)_x, \\ &\ \vdots \end{aligned} \tag{3.21}$$

and

$$\theta_1 = \begin{pmatrix} 0 & v(1-E^{-1}) \\ (E-1)v & D_x \end{pmatrix}. \tag{3.22}$$

EXAMPLE 3. Consider the case $n = 2$, $\alpha = 0$: $L = \mathcal{E}^2 + u\mathcal{E} + w$ (it does not exist in the one dimensional lattice case).

The first equation of the hierarchy reads:

$$\begin{aligned} u(n)_t &= u(n)\mathcal{H}^{-1}u(n) - w(n+1) + w(n), \\ w(n)_t &= -(E+1)^{-1}u(n)_x, \end{aligned} \tag{3.23}$$

and

$$\theta_1 = \begin{pmatrix} D_x & 0 \\ 0 & E - E^{-1} \end{pmatrix}. \tag{3.24}$$

Notice that eliminating the w variables we get

$$u(n)_{tt} = \mathcal{H}^{-1}u_x + (u\mathcal{H}^{-1}u)_t, \tag{3.25}$$

and then interchanging the variables $x \leftrightarrow t$ and putting $\mathcal{H}^{-1}u := v$, we get

$$v(n)_t = \mathcal{H}v_{xx} - (v\mathcal{H}v)_x. \tag{3.26}$$

EXAMPLE 4. Our last example is the case of $n = 3$, $\alpha = -1$: $L = \mathcal{E}^2 + p\mathcal{E} + v + u\mathcal{E}^{-1}$.

The first equations and the Poisson structure read:

$$\begin{aligned} u(n)_{t_1} &= u(n)\mathcal{H}^{-1}p(n-1), \\ v(n)_{t_1} &= u(n+1) - u(n) + (E+1)^{-1}p(n)_x, \\ p(n)_{t_1} &= v(n+1) - v(n) - p(n)\mathcal{H}^{-1}p(n), \end{aligned} \tag{3.27}$$

$$\begin{aligned} u(n)_{t_2} &= u(n)\big[v(n) - v(n-1)\big], \\ v(n)_{t_2} &= p(n)u(n+1) - p(n-1)u(n) + v(n)_x, \\ p(n)_{t_2} &= u(n+2) - u(n) + p(n)_x, \ldots, \end{aligned} \tag{3.28}$$

$$\theta_1 = \begin{pmatrix} 0 & u(1-E)^{-1} & 0 \\ (E-1)u & D_x & 0 \\ 0 & 0 & E - E^{-1} \end{pmatrix}. \tag{3.29}$$

3.2. The operator valued variables approach. In this Section we implement the ideas of Magri–Morosi–Tondo [**6**] and Athorne–Dorfman–Fokas [**1, 5**] of operand approach in the frame of the $\mathcal{R}$-matrix formalism. This gives a powerful tool for construction of multi-Hamiltonian bilocal formalism of integrable systems in $(2+1)$-dimensions.

Let us consider a situation when the functional element $u \equiv \widehat{u} \in \widehat{M} \subset C^{(\infty)}\big(\mathbf{Z} \times (R/2\pi\mathbf{Z}); \mathcal{B}\big)$ with $\mathcal{B}$ being some associative algebra, for instance, the algebra of pseudodifferential operators. Thus, we pass from field variables to operator variables. All considerations for the $(1+1)$ case [**2–4, 7**] can be extended to the operator level with the only modification related to noncommutativity of $\mathcal{G}$. As $\widehat{M}$ is some operator manifold, we produce bi-Hamiltonian operator vector fields with two Poisson operand structures. To compare the results with the central extension approach let us consider the case :

$$\hat{L} = \sum_{i \ll \infty} \widehat{u}_i(n,x)\mathcal{E}^i, \quad \widehat{u}_i = u_i, \quad i \neq 0, \quad \widehat{u}_0 = u_0 - \partial_x, \tag{3.30}$$

where the operator ∂_x is such that $\mathrm{ad}_{\partial_x} = D_x$. So we have $\hat{L} = L - \partial_x$. On the operator-field Lie algebra $\widehat{g}$ we define an invariant nondegenerate scalar product

$$(\widehat{a}, \hat{b}) = (\hat{b}, \widehat{a}) = \mathrm{Tr}(\widehat{a} \cdot \hat{b}) = \int_0^{2\pi} dx \sum_{n \in \mathbf{Z}} \widehat{\mathrm{Sp}}(\widehat{a} \cdot \hat{b})_0, \quad \widehat{a},\ \hat{b} \in \widehat{g}, \tag{3.31}$$

where $\widehat{\mathrm{Sp}}(\widehat{a}\cdot\hat{b}\cdot\widehat{c}) := \widehat{a}\cdot\hat{b}\cdot\widehat{c}$, and $\widehat{\mathrm{Sp}}(\widehat{a}\cdot\hat{b}\cdot\widehat{c}) = \widehat{\mathrm{Sp}}(\widehat{c}\cdot\widehat{a}\cdot\hat{b}) = \widehat{\mathrm{Sp}}(\hat{b}\cdot\widehat{c}\cdot\widehat{a})$ for any $\widehat{a}$, $\hat{b}$, $\widehat{c} \in \widehat{g}$. The bi-Hamiltonian formalism on $\widehat{M}$ produces, of course, hierarchies of commuting operator vector fields, i.e. vector fields which are polynomials in ∂_x operator. What we have to do, is to find through an appropriate linear combinations, ∂_x-free vector fields, i.e. "physical" objects.

THEOREM 2. *Let $\hat{\theta}_1$ and $\hat{\theta}_2$ be two compatible Poisson operands (operators acting in the space of operators) such that $\hat{\theta}_1$ is of f-type (i.e. ∂_x free) and $\hat{\theta}_2 = \bar{\theta} - \varepsilon R_{\partial_x}, \hat{\theta}_1$, where R_{∂_x} means right multiplication by ∂_x, $\bar{\theta}$ is f-type operator and*

$\varepsilon = \text{const}$. *Let* $A = \text{const.} \in \ker \hat{\theta}_1$, *then*

$$K_n := \hat{\theta}_1 \circ \sum_{k=0}^{n} \varepsilon^k \binom{n}{k} \widehat{\Psi}^{n-k-1} \circ A\partial_x^k, \tag{3.32}$$

where $\widehat{\Psi} = \hat{\theta}_1^{-1}\hat{\theta}_2 = \widetilde{\Psi} - \varepsilon R_{\partial_x}$ *is a recursion operand for covectors fields on* $\widehat{M}$, *are* f*-type vector fields.*

LEMMA 1. *If the basic operator vector fields* $\hat{\theta}_2 A\partial_x^k$ *commute*

$$[\hat{\theta}_2 A\partial_x^k, \hat{\theta}_2 A\partial_x^l] = 0, \tag{3.33}$$

and leave the recursion operand invariant with respect to the Lie derivative $\mathcal{L}$:

$$\mathcal{L}_{\hat{\theta}_2 A\partial_x^k} \widehat{\Phi} = 0, \tag{3.34}$$

where $\widehat{\Phi} = \hat{\theta}_2\hat{\theta}_1^{-1} = \overline{\Phi} - \varepsilon R_{\partial_x}$, *then* K_n *constitute a commuting hierarchy of Hamiltonian vector fields.*

There are two special cases of the general formula (3.32):

LEMMA 2. *Take the assumptions from Theorem 2.*

(i) *If* $A\partial_x^k \in \ker \hat{\theta}_1$, $k = 0, 1, 2, \ldots$, *then*

$$K_n := \sum_{k=0}^{n} \varepsilon^k \binom{n}{k} \widehat{\Phi}^{n-k} \circ \hat{\theta}_2 A\partial_x^k. \quad (\textit{Dorfman–Athorne } 1993), \tag{3.35}$$

(ii) *If* $A \in \ker \hat{\theta}_1$, $\hat{\theta}_2$ *then* $\hat{\theta}_2 A\partial_x^k \equiv \widehat{\Phi} \circ \hat{\theta}_1 A\partial_x^k$ *and*

$$K_n := \sum_{k=0}^{n} \varepsilon^k \binom{n}{k} \widehat{\Phi}^{n-k} \circ \hat{\theta}_1 A\partial_x^k. \quad (\textit{Magri–Morosi–Tondo } 1988). \tag{3.36}$$

Let us concentrate on our previous examples.

LEMMA 3. *Let*

$$\Theta_1(\hat{L})\nabla\widehat{H} = -\mathrm{ad}_{\hat{L}}(\nabla\widehat{H})_{\geq 0} + (\mathrm{ad}_{\hat{L}}\nabla\widehat{H})_{>0}, \tag{3.37}$$

$$\Theta_2(\hat{L})\nabla\widehat{H} = -\frac{1}{2}\mathrm{ad}_{\hat{L}}(\mathrm{ad}_{\hat{L}}^{+}\nabla\widehat{H})_{\geq 0} + \frac{1}{2}\mathrm{ad}_{\hat{L}}^{+}(\mathrm{ad}_{\hat{L}}\nabla\widehat{H})_{>0} + \frac{1}{2}\mathrm{ad}_{\hat{L}}\Pi_{n+\alpha}(\mathrm{ad}_{\hat{L}}\nabla\widehat{H})_0, \tag{3.38}$$

where

$$\Pi_{n+\alpha} = (E^{n+\alpha} + 1)(E^{n+\alpha} - 1)^{-1},$$

be two compatible Poisson operands for

$$\hat{L} = \mathcal{E}^{n+\alpha} + \widehat{u}_{n+\alpha-1}\mathcal{E}^{n+\alpha-1} + \cdots + \widehat{u}_\alpha\mathcal{E}^\alpha, \quad -n < \alpha \leq 0, \quad \alpha \in \mathbf{Z}. \tag{3.39}$$

If $\hat{L} = L - \partial_x$ *then*

(i) $\hat{\theta}_1$ *is of* f*-type.*

(ii) $\hat{\theta}_2 = \bar{\theta} - R_{\partial_x}\hat{\theta}_1$, $\bar{\theta}$ – f*-type.*

Proof follows immediately from the fact that $\mathrm{ad}_{\hat{L}} \Leftrightarrow f$-type operator and $\mathrm{ad}_{\hat{L}}^{+} \Leftrightarrow$ f-type $-2R_{\partial_x}$. Notice that $\hat{\theta}_1$ operand when applied to the f-type covectors (i.e. when reduce to $M \subset \widehat{M}$), is equal to $\hat{\theta}_1$ constructed via the central extension.

EXAMPLE 5. The case $n = 1$, $\alpha = 0$: $\hat{L} = \mathcal{E} + u - \partial_x$:

$$\hat{\theta}_1 = \mathrm{ad}_u - D_x, \tag{3.40}$$

$$\begin{aligned}\hat{\theta}_2 = &\frac{1}{2}\mathrm{ad}_u^+\mathrm{ad}_u - \frac{1}{2}\mathrm{ad}_u^+ D_x - \frac{1}{2}D_x\mathrm{ad}_u + \frac{1}{2}D_x^2 - \mathrm{ad}_u\mathcal{H}\mathrm{ad}_u \\ &+ \frac{1}{2}D_x\mathcal{H}\mathrm{ad}_u + \frac{1}{2}\mathrm{ad}_u\mathcal{H}D_x - \frac{1}{2}\mathcal{H}D_x^2 - R_{u_x} - (\mathrm{ad}_u - D_x)R_{\partial_x},\end{aligned} \tag{3.41}$$

$$\widehat{\Phi} = \frac{1}{2}\mathrm{ad}_u^+ - \frac{1}{2}D_x - \frac{1}{2}\mathrm{ad}_u\mathcal{H} + \frac{1}{2}\mathcal{H}D_x - R_{\partial_x}, \tag{3.42}$$

deduced for the first time by Morosi and Tondo in 1989 [**11**] through a completely different approach. Taking $A = 1 \in \ker\hat{\theta}_1, \hat{\theta}_2$ and applying the reduced formula (3.36) we get the hierarchy of Example 1.

EXAMPLE 6. The lattice-field Toda: $n = 2$, $\alpha = -1$, $\hat{L} = \mathcal{E} + p - \partial_x + v\mathcal{E}^{-1}$:

$$\hat{\theta}_1 = \begin{pmatrix} 0 & R_v - vE^{-1} \\ ER_v - v & -(\mathrm{ad}_p - D_x) \end{pmatrix}, \tag{3.43}$$

$$\begin{aligned}\hat{\theta}_2 &= \bar{\theta} - R_{\partial_x}\hat{\theta}_1, \\ \bar{\theta}_{11} &= (R_vE - vE^{-1})(E-1)^{-1}(ER_v - v), \\ \bar{\theta}_{12} &= (R_v - vE^{-1})(E-1)^{-1}(ER_p - p + D_x) + R_{v_x}, \\ \bar{\theta}_{21} &= (R_pE - p + D_x)(E-1)^{-1}(ER_v - v), \\ \bar{\theta}_{22} &= ER_v - vE^{-1} - (R_pE - p + D_x)(E-1)^{-1}(\mathrm{ad}_p - D_x).\end{aligned} \tag{3.44}$$

The recursion operand for covector fields:

$$\widehat{\Psi} = \hat{\theta}_1^{-1} \circ \hat{\theta}_2 = \widetilde{\Psi} - R_{\partial_x}, \tag{3.45}$$

where

$$\begin{aligned}\widetilde{\Psi}_{11} = &\{[E^1R_v - v]^{-1}[\mathrm{ad}_p - D_x][vE^{-1} - R_v]^{-1}[R_vE^1 - vE^{-1}] \\ &+ [v - E^1R_v]^{-1}[R_pE^1 - p + D_x]\}[E^1 - 1]^{-1}[v - E^1R_v] \\ &- [v - E^1R_v]^{-1}E^1R_{v_x}, \\ \widetilde{\Psi}_{12} = &[E^1R_v - v]^{-1}[\mathrm{ad}_p - D_x][E^1 - 1]^{-1}[E^1R_p - p + D_x] + [v - E^1R_v]^{-1} \\ &\times \{[vE^{-1} - E^1R_v] - R_{p_x} + [R_pE^1 - p + D_x][E^1 - 1]^{-1}[\mathrm{ad}_p - D_x]\}, \\ \widetilde{\Psi}_{21} = &[vE^{-1} - R_v]^{-1}[R_vE^1 - vE^{-1}][E^1 - 1]^{-1}[v - E^1R_v], \\ \widetilde{\Psi}_{22} = &[E^1 - 1]^{-1}[E^1R_p - p + D_x].\end{aligned}$$

Taking $A = \binom{1}{0} \in \ker\hat{\theta}_1$, the general formula (3.32) gives the hierarchy of Example 2.

Further examples and details of the present theory will be given in a forthcoming paper.

References

1. C. Athorne and I. Dorfman, *The Hamiltonian structure of the* $(2+1)$*-dimensional Ablowitz–Kaup–Newell–Segur hierarchy*, J. Math. Phys. **34** (1993), 3507–3517.
2. M. Błaszak, *R-matrix approach to multi-Hamiltonian Lax dynamics*, Rep. Math. Phys. **40** (1997), 395–430.
3. ———, *Multi-Hamiltonian theory of dynamical systems*, Springer-Verlag, 1998.

4. M. Błaszak and K. Marciniak, *R-matrix approach to lattice integrable systems*, J. Math. Phys. **35** (1994), 4661–4682.
5. I. Dorfman and A. S. Fokas, *Hamiltonian theory over noncommutative rings and integrability in multidimensions*, J. Math. Phys **33** (1992), 2504–2514.
6. F. Magri, C. Morosi, and G. Tondo, *Nijenhuis G-manifolds and Lenard bicomplexes: A new approach to* KP *systems*, Comm. Math. Phys. **116** (1988), 457–475.
7. W. Oevel, *Poisson bracket if integrable lattice systems*, Algebraic Aspects of Integrable systems (A. S. Fokas and I. M. Gelfand, eds.) Progr. Nonlinear Differential Equations Appl., vol. 26, Birkhäuser, Boston, 1996, pp. 261–283.
8. A. K. Prykarpatsky, V. Hr. Samoilenko, and R. I. Andrushkiw, *Algebraic structure of the gradient-holonomic algorithm for Lax integrable nonlinear dynamical systems.* I, J. Math. Phys. **35** (1994), 1763–1777.
9. ______, *Algebraic structure of the gradient-holonomic algorithm for Lax integrable nonlinear dynamical systems.* II. *The reduction via Dirac and canonical quantization procedure*, J. Math. Phys. **35** (1994), 4088–4116.
10. A. G. Reiman and M. A. Semenov-Tian-Shansky, *Current algebras and nonlinear partial differential equations*, Dokl. Akad. Nauk **251** (1980), 1310–1313. (Russian)
11. ______, *Hamiltonian structure of Kadomtsev–Petviashvilii type equations*, Zap. Nauchn. Sem. Leningrad. Otdel. Mat. Inst. Steklov. (LOMI) **133** (1984), 212–227. (Russian)

Physics Department, A. Mickiewicz Univ., Umultowska 85, 61–614 Poznan, Poland.
E-mail address, Maciej Błaszak: blaszakm@amu.edu.pl
E-mail address, Andrzej Szum: szum@main.amu.edu.pl

Centre de Recherches Mathématiques
CRM Proceedings and Lecture Notes
Volume **25**, 2000

Möbius Invariant Integrable Lattice Equations Associated with the Generalized KP Hierarchy

L. V. Bogdanov and B. G. Konopelchenko

ABSTRACT. Möbius invariant integrable lattice equations arising in the context of singular manifold equations for scalar and multicomponent generalized KP hierarchy and 2D Toda lattice hierarchy are considered. These equations generate the corresponding continuous hierarchy of singular manifold equations, its Bäcklund transformations and different forms of superposition principles for them. They possess rather special form of compatibility representation. Geometric interpretation of these discrete equations is given.

1. Introduction

The equations we are going to discuss here were derived by the authors in the frame of the analytic-bilinear approach to integrable hierarchies [**1**], [**2**]. They arise on the third level of the hierarchy, i.e. as a discrete singular manifold equations (or, in other words, in the context of the hierarchy in Schwarzian form). These equations are highly symmetric and possess very peculiar properties.

First, they have a very special compatibility condition representation, symmetric with respect to lattice variables and indicating duality between wave functions and potential (providing in fact equations for both objects in a similar way).

Second, they possess (even in the matrix case, corresponding to multicomponent KP hierarchy) a Möbius symmetry, which is typical for equations in Schwarzian form and which is deeply connected to projective geometry. Similar equations for the KdV case were interpreted by Bobenko and Pinkall [**4**] as equations of discrete isothermic surfaces. We hope that equations we discuss may also be interpreted in terms of some special classes of discrete surfaces, though the interpretation is yet unknown for the matrix case. In the scalar case, however, there are at least two ways to connect the equations to discrete geometry: as a reduction of Darboux system and in terms of the three-dimensional lattice on the complex plane with some special geometrical properties.

2000 *Mathematics Subject Classification.* Primary: 35Q58; Secondary: 52C99.

The first author (LB) is grateful to the Dipartimento di Fisica dell'Università and Sezione INFN, Lecce, for hospitality and support; (LB) also acknowledges partial support from the Russian Foundation for Basic Research under grants No 98-01-00525 and 96-15-96093 (scientific schools).

This is the final form of the paper.

Third, for the basic hierarchy we have *one* lattice equation of the considered type, and it generates by expansion in the parameters the continuous hierarchy itself, its Bäcklund transformations and different types of superposition formulae for Bäcklund transformations. Thus it encodes all the information about the hierarchy (more or less in the way of Hirota bilinear discrete equation, with which it is closely connected, but not in terms of the τ-function, rather in terms of wave functions and potentials).

2. The Basic Equations

In this section we introduce basic equations and their representations in the form of compatibility conditions without the reference to the original scheme of derivation of these equations (a sketch of it will be given later).

In the context of the scalar KP hierarchy the basic equation reads

$$(T_2\Delta_1\Phi)(T_3\Delta_2\Phi)(T_1\Delta_3\Phi) = (T_2\Delta_3\Phi)(T_3\Delta_1\Phi)(T_1\Delta_2\Phi). \tag{2.1}$$

Here

$$\Phi = \Phi(n_1, n_2, n_3), \quad T_1\Phi = \Phi(n_1 + 1, n_2, n_3), \quad \Delta_1 = T_1 - 1.$$

In the case of multicomponent KP hierarchy Φ is $N \times N$ matrix-valued function and the basic equation looks like

$$(T_1\Delta_3\Phi)\cdot(T_3\Delta_1\Phi)^{-1}\cdot(T_3\Delta_2\Phi)\cdot(T_2\Delta_3\Phi)^{-1}\cdot(T_2\Delta_1\Phi)\cdot(T_1\Delta_2\Phi)^{-1} = 1. \tag{2.2}$$

Equation (2.1) may be treated as compatibility condition for the following set of equations for the function f

$$\frac{a_iT_if}{a_jT_jf} = \frac{\Delta_i\Phi}{\Delta_j\Phi} \quad i,j,k \in \{1,2,3\}, \tag{2.3}$$

where a_i are some constants. To demonstrate this, we break the symmetry of the relation (2.3) and distinguish one of the shifts (say, T_1).

REMARK 1. Unfortunately, we don't know a symmetric derivation starting from relation (2.3), though the initial equation and the final equation are symmetric. Symmetric way of deriving (2.1) in another context will be given later in this paper.

Then we rewrite (2.3) in the form

$$T_2f = \frac{a_1\Delta_2\Phi}{a_2\Delta_1\Phi}T_1f = UT_1f, \quad T_3f = \frac{a_1\Delta_3\Phi}{a_3\Delta_1\Phi}T_1f = VT_1f.$$

In this form the system looks more familiar. Then, as usual, taking cross-shifts of the first and the second equations (which should give the same results) and using the equations themselves to get rid of the shifts T_2, T_3 acting on f, one obtains the following equation (compatibility condition)

$$(T_3U)(T_1V) = (T_2V)(T_1U).$$

Substituting the expressions of U, V through Φ, one gets exactly equation (2.1).

It is possible to start also with the following ('dual') system to get equation (2.1)

$$\frac{a_iT_i^{-1}\widetilde{f}}{a_jT_j^{-1}\widetilde{f}} = \frac{\widetilde{\Delta}_i\Phi}{\widetilde{\Delta}_j\Phi} \quad i,j,k \in \{1,2,3\}, \tag{2.4}$$

where

$$\widetilde{\Delta}_i = T_i^{-1} - 1.$$

On the other hand, it is possible to treat the system (2.3) as a set of linear equations for the function Φ. To do it, we rewrite the system in the form

$$\Delta_2\Phi = \frac{a_2T_2f}{a_1T_1f}\Delta_1\Phi = U\Delta_1\Phi, \quad \Delta_3\Phi = \frac{a_3T_3f}{a_1T_1f}\Delta_1\Phi = V\Delta_1\Phi.$$

The compatibility condition for this system gives two equations

$$(T_3U)(T_1V) = (T_2V)(T_1U),$$
$$\Delta_3U + (T_3U)\Delta_1V = \Delta_2V + (T_2V)\Delta_1U.$$

The first of these equations is resolved by the substitution of expressions for the functions U, V in terms of f, the second gives the following equation for f

$$\sum_{(ijk)} \varepsilon_{ijk}a_ja_kT_j\left(\frac{f}{T_if}\right) = 0, \tag{2.5}$$

summation here is over different permutations of indices.

Starting from the dual linear system (2.4), one obtains an equation for the function $\widetilde{f}$

$$\sum_{(ijk)} \varepsilon_{ijk}a_ja_kT_k\left(\frac{T_i\widetilde{f}}{\widetilde{f}}\right) = 0. \tag{2.6}$$

Both these equation are connected with the modified KP hierarchy (the equation for $\widetilde{f}$ with the mKP hierarchy itself and the equation for f with the dual (adjoint) hierarchy). Similar equations were derived by Nijhoff *et al.* [**7**].

The matrix case (2.2) can be treated the same way. Equations (2.4) and the dual system for this case read

$$(\Delta_j\Phi)^{-1}\Delta_i\Phi = (A_jT_jf)^{-1}(A_iT_if),$$
$$\widetilde{\Delta}_i\Phi(\widetilde{\Delta}_j\Phi)^{-1} = (T_i^{-1}\widetilde{f}A_i)(T_j^{-1}\widetilde{f}A_j)^{-1},$$

where A_i are some diagonal matrices; in what follows we suggest that the determinant of A_1 is not equal to zero.

The linear system for the function f takes the form

$$A_1^{-1}A_2T_2f = (T_1f)(\Delta_1\Phi)^{-1}\Delta_2\Phi = (T_1f)U,$$
$$A_1^{-1}A_3T_3f = (T_1f)(\Delta_1\Phi)^{-1}\Delta_3\Phi = (T_1f)V.$$

The compatibility condition for this system gives equation (2.2) in the form

$$(T_1\Delta_3\Phi)\cdot(T_3\Delta_1\Phi)^{-1}\cdot(T_3\Delta_2\Phi) = (T_1\Delta_2\Phi)\cdot(T_2\Delta_1\Phi)^{-1}\cdot(T_2\Delta_3\Phi).$$

Then, the linear system for the function Φ reads

$$\Delta_2\Phi = \Delta_1\Phi(A_1T_1f)^{-1}A_2T_2f = \Delta_1\Phi U,$$
$$\Delta_3\Phi = \Delta_1\Phi(A_1T_1f)^{-1}A_3T_3f = \Delta_1\Phi V,$$

and a compatibility condition gives a matrix version of equation (2.5)

$$\sum_{(ijk)} \varepsilon_{ijk}A_jT_j\big(f\cdot(T_if)^{-1}\big)A_k = 0.$$

A matrix version of (2.6) is

$$\sum_{(ijk)} \varepsilon_{ijk}A_kT_k\big(\widetilde{f}^{-1}\cdot(T_i\widetilde{f})\big)A_j = 0.$$

3. The Möbius Symmetry

It is easy to check that equation (2.1) possesses a Möbius symmetry transformation, i.e. if function Φ is a solution to this equation, then the function Φ',

$$\Phi' = \frac{a\Phi + b}{c\Phi + d},$$

where a, b, c, d are arbitrary constants, is also a solution. For the matrix equation (2.2) the symmetries include inversion, left and right matrix multiplication and shift

$$\Phi' = \Phi^{-1}, \quad \Phi' = A\Phi, \quad \Phi' = \Phi B, \quad \Phi' = \Phi + C,$$

where A, B, C are some matrices (determinant of A and B is not equal to zero).

The structure of equation (2.2) resembles the structure of quartic equations given by Bobenko and Pinkall [**4**] for the quaternion description of discrete isothermic surfaces in $\mathbf{R}^3$. Probably equation (2.2) also has a geometric interpretation in terms of quaternion construction.

It is interesting to note that due to the inversion symmetry, the equation (2.2) admits a reduction

$$\Phi\Phi = -1,$$

that for Φ belonging to the field of quaternions means that the vector in $\mathbf{R}^3$ corresponding to the (imaginary) quaternion Φ has a unit length, and so the equation (2.2) in this case defines a three-dimensional lattice on the unit sphere.

Equations (2.5), (2.6) are connected by the transformation

$$f' = \widetilde{f}^{-1}, \quad \widetilde{f}' = f^{-1},$$

that keeps in the matrix case.

4. The Hierarchy of Equations

Let us suggest that the function Φ in equation (2.1) is a function of a standard infinite set of KP variables $\mathbf{x}$, $\mathbf{x} = \{x_k\}$, $1 \le k < \infty$ (we will also use notations $x = x_1$, $y = x_2$, $t = x_3$) and the shift operators T_i are realized as

$$T_i \colon \mathbf{x} \to \mathbf{x} + [a_i], \tag{4.1}$$

$$[a_i] = \left\{ \frac{1}{k} a_i^k \right\}.$$

This suggestion is justified by the explicit construction given in [**1,2**]. Then we can introduce the lattice with infinitely many lattice variables, each of the variables is characterized by the lattice parameter a_i. For each three variables with the different parameters a_i, a_j, a_k we will get the equation (2.1) (the indices 1, 2, 3 should be changed for i, j, k). Thus the formula (2.1) encodes not a single equation, but the *hierarchy* of discrete equations labelled by three continuous parameters (similar ideology was developed in the other context by Nijhoff [**7**]). Each representative of the hierarchy possesses infinite number of commuting discrete symmetries given by the set of other representatives of the hierarchy, and it is very easy to extract from the formula (2.1) the equations of the Bäcklund transformations for the representative of the hierarchy and different forms of their superposition principles. However, we will not do it in detail here, but will rather concentrate on the continuous symmetries of the hierarchy (2.1).

Expanding the equation (2.1) written in terms of the shifts (4.1) into the powers of a_1 and taking the first term, we get

$$(T_2\Phi_x)(T_3\Delta_2\Phi)\Delta_3\Phi = (T_3\Phi_x)(T_2\Delta_3\Phi)\Delta_2\Phi. \tag{4.2}$$

The next step is to expand this equation in a_2. The first nonvanishing term arises at the second order in a_2. After simple transformations we get

$$\frac{\partial}{\partial x}\ln\left(\frac{1}{\Phi_x}\frac{\Delta\Phi}{a}\right) = \frac{1}{2}\Delta\left(\frac{\Phi_y + \Phi_{xx}}{\Phi_x}\right), \tag{4.3}$$

where we are left with one discrete variable and parameter a, so we omit the index 3. And finally, expanding in parameter and taking the first nonvanishing term, which corresponds to the third order in a_3 in the initial equation (2.1), we obtain

$$\Phi_t = \frac{1}{4}\Phi_{xxx} + \frac{3}{8}\frac{\Phi_y^2 - \Phi_{xx}^2}{\Phi_x} + \frac{3}{4}\Phi_x W_y, \quad W_x = \frac{\Phi_y}{\Phi_x}. \tag{4.4}$$

This equation first arose in Painlevé analysis of the KP equation as a singular manifold equation [**8**]. The higher terms of expansion of equation (2.1) will lead to the hierarchy of singular manifold equations.

So we have the following objects:

(1) Lattice equation (2.1).
(2) Equation with two discrete and one continuous variables (4.2).
(3) Equation with one discrete and two continuous variables (4.3).
(4) PDE (4.4) with three continuous variables (we would like to emphasize that for the cases 2, 3, 4 we have an infinite hierarchy of equations). All these equations are the symmetries of each other by construction, and can be interpreted in different ways (continuous symmetries of discrete equations, discrete symmetries of continuous equations and mixed cases). The interpretation depends on the choice of the basic equation (i.e. in some sense on the point of reference).

A standard way is to take continuous equation (4.4) as a basic system. Then the interpretation of the other objects is the following:

3. Equation (4.3) defines a Bäcklund transformation for the equation (4.4).
2. Equation (4.2) is a superposition principle for two Bäcklund transformations.
1. Lattice equation (2.1) provides an *algebraic* superposition principle for three Bäcklund transformations.

On the other hand, the discrete equation (2.1) may be treated as a basic system as well. Then the formula (4.2) is a lowest order continuous symmetry for this system, the equation (4.3) is a superposition principle for two continuous symmetries, and the equation (4.4) is a superposition principle for three continuous symmetries of different orders (we should have written down the explicit formulae for the continuous symmetries of the higher order to complete the picture).

5. The Basic Equations in the Framework of the Generalized KP Hierarchy

The derivation of the basic equations. In [**1, 2**] the following equations were derived for the scalar KP Cauchy-Baker-Akhiezer (CBA) function from the

Hirota bilinear identity for this function

$$\Delta_i \psi(\lambda, \mu, \mathbf{x}) = a_i \widetilde{\psi}(\mu, \mathbf{x}) T_i \psi(\lambda, \mathbf{x}), \tag{5.1}$$

where $\psi(\lambda, \mu, \mathbf{x})$ is a CBA function,

$$\psi(\lambda, \mu, \mathbf{x}) = g(\lambda, \mathbf{x}) \chi(\lambda, \mu, \mathbf{x}) g^{-1}(\mu, \mathbf{x}), \quad g(\lambda, \mathbf{x}) = \exp\left(\sum_{n=1}^{\infty} x_n \lambda^{-n}\right),$$

$\chi(\lambda, \mu, \mathbf{x})$ is defined in the unit disc in λ, μ, is analytic in these variables for $\lambda \neq \mu$ and for $\lambda \to \mu$ behaves like $(\lambda - \mu)^{-1}$; $\psi(\lambda, \mathbf{x})$ and $\widetilde{\psi}(\mu, \mathbf{x})$ are respectively Baker-Akhiezer (BA) and dual (adjoint) BA functions

$$\psi(\lambda, \mathbf{x}) = g(\lambda, \mathbf{x}) \chi(\lambda, 0, \mathbf{x}), \quad \widetilde{\psi}(\mu, \mathbf{x}) = \chi(0, \mu, \mathbf{x}) g^{-1}(\mu, \mathbf{x}),$$

the shifts T_i are realized in the form (4.1).

For the multicomponent KP case CBA and BA functions are $N \times N$ matrix-valued functions, we have N infinite sets of KP variables

$$\mathbf{x} = (\mathbf{x}_1, \dots, \mathbf{x}_N),$$

and the action of three shift operators T_i is defined as

$$T_i \colon \mathbf{x} \to \mathbf{x} + [\mathbf{a}_i], \quad [\mathbf{a}_i] = \big([a_{i(1)}], \dots, [a_{i(N)}]\big),$$

$\mathbf{a}_i$ are some constant N-dimensional vectors. The equation (5.1) for the multicomponent case reads

$$\Delta_i \psi(\lambda, \mu, \mathbf{x}) = \widetilde{\psi}(\mu, \mathbf{x}) A_i T_i \psi(\lambda, \mathbf{x}), \tag{5.2}$$

$$A_i = \operatorname{diag}\big(a_{i(1)}, \dots, a_{i(N)}\big). \tag{5.3}$$

Now we will perform the derivation of the equations of the generalized hierarchy for the scalar KP case; the multicomponent case is completely analogous.

In terms of the Cauchy kernel $\chi(\lambda, \mu, \mathbf{x})$ having simple analytic properties the equation (5.1) reads

$$\frac{\Delta_i}{a_i} \chi(\lambda, \mu, \mathbf{x}) - \frac{1}{\mu} T_i \chi(\lambda, \mu, \mathbf{x}) + \frac{1}{\lambda} \chi(\lambda, \mu, \mathbf{x}) = T_i\big(\chi(\lambda, 0, \mathbf{x})\big) \chi(0, \mu, \mathbf{x}), \tag{5.4}$$

or, equivalently,

$$\frac{\widetilde{\Delta}_i}{a_i} \chi(\lambda, \mu, \mathbf{x}) - \frac{1}{\mu} \chi(\lambda, \mu, \mathbf{x}) + \frac{1}{\lambda} \widetilde{T}_i \chi(\lambda, \mu, \mathbf{x}) = \chi(\lambda, 0, \mathbf{x}) \widetilde{T}_i \chi(0, \mu, \mathbf{x}), \tag{5.5}$$

where

$$\widetilde{T}_i = T_i^{-1}, \quad \widetilde{\Delta}_i = 1 - \widetilde{T}_i.$$

Let us take equation (5.4) at $\lambda = 0$

$$\left(\frac{\Delta_i}{a_i} - \frac{\Delta_j}{a_j}\right) \chi(0, \mu, \mathbf{x}) + \frac{1}{\mu}\big((T_i - T_j)\chi(0, \mu, \mathbf{x})\big) = \chi(0, \mu, \mathbf{x})(T_i - T_j) u(g), \tag{5.6}$$

and respectively equation (5.5) at $\mu = 0$

$$\left(\frac{\widetilde{\Delta}_i}{a_i} - \frac{\widetilde{\Delta}_j}{a_j}\right) \chi(\lambda, 0, \mathbf{x}) - \frac{1}{\lambda}\big((\widetilde{T}_i - \widetilde{T}_j)\chi(\lambda, 0, \mathbf{x})\big) = \big((\widetilde{T}_i - \widetilde{T}_j) u(g)\big) \chi(\lambda, 0, \mathbf{x}), \tag{5.7}$$

where $u = \chi_{\mathrm{r}}(0, 0, \mathbf{x})$, which $\chi_{\mathrm{r}}(\lambda, \mu, \mathbf{x}) = \chi(\lambda, \mu, \mathbf{x}) - (\lambda - \mu)^{-1}$.

Then, in terms of the Baker–Akhiezer and dual Baker-Akhiezer functions we obtain

$$\left(\frac{\Delta_i}{a_i}-\frac{\Delta_j}{a_j}\right)\widetilde{\psi}(\mu,\mathbf{x})=\widetilde{\psi}(\mu,\mathbf{x})(T_i-T_j)u(g), \tag{5.8}$$

and

$$\left(\frac{\widetilde{\Delta}_i}{a_i}-\frac{\widetilde{\Delta}_j}{a_j}\right)\psi(\lambda,\mathbf{x})=\big((\widetilde{T}_i-\widetilde{T}_j)u(g)\big)\psi(\lambda,\mathbf{x}). \tag{5.9}$$

Finally, we integrate the equation (5.8) over the unit circle in μ with arbitrary weight function $\widetilde{\rho}(\mu)$ and the equation (5.9) over the unit circle in λ with a weight function $\rho(\lambda)$ to get

$$\left(\frac{\Delta_i}{a_i}-\frac{\Delta_j}{a_j}\right)\widetilde{f}(g)=\widetilde{f}(g)(T_i-T_j)u(g), \tag{5.10}$$

and

$$\left(\frac{\widetilde{\Delta}_i}{a_i}-\frac{\widetilde{\Delta}_j}{a_j}\right)f(g)=\big((\widetilde{T}_i-\widetilde{T}_j)u(g)\big)f(g), \tag{5.11}$$

where we have introduced the wave functions

$$f(\mathbf{x})=\int\psi(\lambda,\mathbf{x})\rho(\lambda)\,d\lambda,\quad \widetilde{f}(\mathbf{x})=\int\widetilde{\rho}(\mu)\widetilde{\psi}(\mu,\mathbf{x})\,d\mu.$$

The equations (5.10) with different substitutions of indices gives three linear equations for the wave function $\widetilde{f}$ with the potential u (only two of them are linearly independent), and the compatibility condition for these equations generates the equation for the potential

$$\sum_{(ijk)}\varepsilon_{ijk}T_k\left(\frac{\Delta_i}{a_i}u-uT_iu\right)=0, \tag{5.12}$$

where the summation goes over different permutations of indices. This equation represents a discrete version of the KP hierarchy in the form of algebraic superposition principle for three Bäcklund transformations. Similar equation can be found in [**7**].

The dual equations (5.11) generate the same equation (5.12) as a compatibility condition.

Integrating the equation (5.1) with two arbitrary functions $\rho(\lambda)$, $\widetilde{\rho}(\mu)$ over the unit circle in λ, μ, we get

$$\Delta_i\Phi(\mathbf{x})=a_i\widetilde{f}(\mathbf{x})T_if(\mathbf{x}), \tag{5.13}$$

or, equivalently,

$$\widetilde{\Delta}_i\Phi(\mathbf{x})=a_i\big(\widetilde{T}_i\widetilde{f}(\mathbf{x})\big)f(\mathbf{x}), \tag{5.14}$$

where

$$\Phi(\mathbf{x})=\oint\oint\big(\psi(\lambda,\mu;\mathbf{x})-(\lambda-\mu)^{-1}\big)\rho(\lambda)\widetilde{\rho}(\mu)\,d\lambda\,d\mu.$$

Using this equation, it is easy to check that Φ satisfies equation (2.1).

Also taking cross-differences of equations (5.13) (or (5.14)) and combining them, one gets equations (2.5), (2.6) in a symmetric algebraic manner, without reference to compatibility conditions.

Indeed, let us take a pair of equations (5.13), (5.14)

$$\frac{\Delta_i}{a_i}\Phi = (T_i f)\widetilde{f}, \quad \frac{\Delta_j}{a_j}\Phi = (T_j f)\widetilde{f},$$

and

$$\frac{\widetilde{\Delta}_i}{a_i}\Phi = f\widetilde{T}_i\widetilde{f}, \quad \frac{\widetilde{\Delta}_j}{a_j}\Phi = f\widetilde{T}_j\widetilde{f}.$$

Taking cross-differences and equating them, one gets the equations

$$\widetilde{f}\left(\frac{\Delta_j}{a_j}T_i - \frac{\Delta_i}{a_i}T_j\right)f = (T_iT_jf)\left(\frac{\Delta_i}{a_i} - \frac{\Delta_j}{a_j}\right)\widetilde{f},$$

and

$$f\left(\frac{\widetilde{\Delta}_j}{a_j}\widetilde{T}_i - \frac{\widetilde{\Delta}_i}{a_i}\widetilde{T}_j\right)\widetilde{f} = (\widetilde{T}_i\widetilde{T}_j\widetilde{f})\left(\frac{\widetilde{\Delta}_i}{a_i} - \frac{\widetilde{\Delta}_j}{a_j}\right)f.$$

Acting on these equations by the shift operator T_k ($\widetilde{T}_k$), after simple transformations one obtains

$$\left(\frac{\Delta_j}{a_j}T_iT_k - \frac{\Delta_i}{a_i}T_jT_k\right)f = (T_iT_jT_kf)T_k\widetilde{f}^{-1}\left(\frac{\Delta_i}{a_i} - \frac{\Delta_j}{a_j}\right)\widetilde{f},$$

and

$$\left(\frac{\widetilde{\Delta}_j}{a_j}\widetilde{T}_i\widetilde{T}_k - \frac{\widetilde{\Delta}_i}{a_i}\widetilde{T}_j\widetilde{T}_k\right)\widetilde{f} = (\widetilde{T}_i\widetilde{T}_j\widetilde{T}_k\widetilde{f})T_kf^{-1}\left(\frac{\widetilde{\Delta}_i}{a_i} - \frac{\widetilde{\Delta}_j}{a_j}\right)f.$$

Multiplying these equations over ε_{ijk} and taking the sum over different permutations of indices, we cancel the left-hand sides of both equations, and the right-hand sides give just

$$(T_iT_jT_kf)\sum_{(ijk)}\varepsilon_{ijk}T_k\left(\widetilde{f}^{-1}\left(\frac{\Delta_i}{a_i}\widetilde{f}\right)\right) = 0,$$

and

$$(\widetilde{T}_i\widetilde{T}_j\widetilde{T}_k\widetilde{f})\sum_{(ijk)}\varepsilon_{ijk}\widetilde{T}_k\left(f^{-1}\left(\frac{\widetilde{\Delta}_i}{a_i}f\right)\right) = 0,$$

so we have reproduced the equations (2.5), (2.6) in a symmetric algebraic manner from the essentially different starting point. In this context it is also possible to introduce the representation of the equations for the wave functions in the form of compatibility condition. Indeed, taking ratios of the equations (5.13), (5.14) with different indices, one obtains a set of relations of the form

$$\frac{a_iT_if}{a_jT_jf} = \frac{\Delta_i\Phi}{\Delta_j\Phi}, \tag{5.15}$$

and

$$\frac{a_i\widetilde{T}_i\widetilde{f}}{a_j\widetilde{T}_j\widetilde{f}} = \frac{\widetilde{\Delta}_i\Phi}{\widetilde{\Delta}_j\Phi}. \tag{5.16}$$

Thus we have derived the equations (2.3), (2.4) starting from the equation for the CBA function (5.13), (5.14).

The multicomponent case (2.2) is analogous.

Taking the weight functions on the unit circle in the form $\rho(\lambda) = \delta(\lambda_0 - \lambda)$, $\widetilde{\rho}(\mu) = \delta(\mu_0 - \mu)$, one arrives to the conclusion that the CBA function itself satisfies the equation (2.1) (or (2.2) in the matrix case). Then, using the analytic properties of the CBA function and going to the limit $\lambda \to \mu$, one obtains for the function $u(\mathbf{x}) = \chi_{\mathrm{r}}(0, 0, \mathbf{x})$ the equation (5.12). It is also possible to reproduce in this way equations (2.5), (2.6) taking only one weight function as δ-function and going to the limit $\lambda \to 0$ or $\mu \to 0$.

Thus the equations (2.1), (2.2) encode all the three levels of generalized KP hierarchy: hierarchy of singular manifold equations, modified hierarchy and the basic KP hierarchy.

The general picture of the generalized KP hierarchy. We have obtained three levels of the hierarchy of discrete equations corresponding to the same hierarchy of transformations defined by the elementary rational loops. The first level is formed by the equations for the potential u (5.12)

$$\sum_{(ijk)} \varepsilon_{ijk} T_k \left(\frac{\Delta_i}{a_i} u - u T_i u \right) = 0,$$

the second level by the equations for the wave functions of linear operators producing the equations of the first level as compatibility conditions (the modified equations), and it splits into two parts: equations for the dual wave function $\widetilde{f}$ (2.5)

$$\sum_{(ijk)} \varepsilon_{ijk} a_j a_k T_k \big(\widetilde{f}^{-1} (T_i \widetilde{f}) \big) = 0,$$

and equations for the wave functions f (2.6)

$$\sum_{(ijk)} \varepsilon_{ijk} a_j a_k \widetilde{T}_k \big(f^{-1} (\widetilde{T}_i f) \big) = 0.$$

Finally, the third level represents the equations for the wave functions Φ of the linear operators of the second level (the singular manifold type equations)

$$(T_j \Delta_i \Phi)(T_k \Delta_j \Phi)(T_i \Delta_k \Phi) = (T_j \Delta_k \Phi)(T_k \Delta_i \Phi)(T_i \Delta_j \Phi).$$

One could expect this chain to continue, but the wave functions of linear operators of the third level coincide with the wave functions of linear operators of the first level, and so the chain closes.

Connection between three different levels of the hierarchy of discrete equations may be described in terms of Miura maps and Combescure symmetry transformations, which are in some sense complementary.

Miura maps transform the solutions of the higher level of the hierarchy to the solution of the lower level.

Let us start from the Miura maps connecting the second and the first levels of the hierarchy. In fact we have already written down these maps in the process of derivation of the modified equations:

$$(T_i - T_j) u = \frac{1}{\widetilde{f}} \left(\frac{\Delta_i}{a_i} - \frac{\Delta_j}{a_j} \right) \widetilde{f} \quad \text{and} \quad (\widetilde{T}_i - \widetilde{T}_j) u = \frac{1}{f} \left(\frac{\widetilde{\Delta}_i}{a_i} - \frac{\widetilde{\Delta}_j}{a_j} \right) f,$$

(see the formulae (5.8), (5.9)).

As for the Miura maps from the third level to the second, we also have the formulae for them at hand

$$\frac{a_i T_i f}{a_j T_j f} = \frac{\Delta_i \Phi}{\Delta_j \Phi} \quad \text{and} \quad \frac{a_i \widetilde{T}_i \widetilde{f}}{a_j \widetilde{T}_j \widetilde{f}} = \frac{\widetilde{\Delta}_i \Phi}{\widetilde{\Delta}_j \Phi},$$

(see (5.15) and (5.16)).

Another way to describe the connections between the different levels of the hierarchy is in terms of the Combescure transformations.

Integrable equations for the wave functions (i.e. modified equations and singular manifold type equations) possess the symmetry transformations corresponding to the freedom of choice of the wave function corresponding to the same Cauchy kernel. In spectral terms this symmetry is described by the change of weight functions $\rho(\lambda)$, $\widetilde{\rho}(\mu)$, which enter the definitions of the wave functions f, $\widetilde{f}$, Φ. In terms of equations themselves this symmetry is represented via the Combescure invariants. If the solutions are Combescure equivalent (i.e. correspond to the same Cauchy kernel), the Combescure invariants for them are the same. In terms of the Miura transformation the Combescure equivalence class is a class of solutions having the same image under the Miura map, and it is not amazing that we have already written down different types of the Combescure invariants, they are given by the right-hand sides of our formulae for the Miura transformations.

6. Equation (2.1) and Quadrilateral Lattices

A more general version of equation (5.1) reads (we need only scalar case here) [**1**]

$$\psi(\lambda, \mu, \mathbf{x} + [b]) - \psi(\lambda, \mu, \mathbf{x} + [a]) = (b - a)\widetilde{\psi}_a(\mu, \mathbf{x} + [a])\psi_a(\lambda, \mathbf{x} + [b]), \tag{6.1}$$

where $\psi_a(\lambda, \mathbf{x})$, $\widetilde{\psi}_a(\mu, \mathbf{x})$ are respectively BA and dual BA function associated with the point a rather then the point zero for equation (5.1),

$$\psi_a(\lambda, \mathbf{x}) = \chi(\lambda, a, \mathbf{x}) g(\lambda, \mathbf{x}) \quad \text{and} \quad \widetilde{\psi}_a(\mu, \mathbf{x}) = g^{-1}(\mu, \mathbf{x})\chi(a, \mu, \mathbf{x}).$$

For $b = 0$ the equation (6.1) gives

$$\Delta_a \psi(\lambda, \mu, \mathbf{x}) = a\psi_a(\lambda, \mathbf{x}) T_a \widetilde{\psi}_a(\mu, \mathbf{x}). \tag{6.2}$$

Choosing a set of three point a_i, a_j, a_k and introducing corresponding lattice variables, we get

$$\Delta_i \psi(\lambda, \mu, \mathbf{n}) = a_i \psi_{a_i}(\lambda, \mathbf{n}) T_i \widetilde{\psi}_{a_i}(\mu, \mathbf{n}). \tag{6.3}$$

The idea now is to take the equation (6.3) at the set of points λ, $\mu \in a_i, a_j, a_k$ to get a system of equations for the functions depending only on the lattice variables. To make the following derivation more clear, it is convenient to rewrite the equation (6.3) in terms of the Cauchy kernel $\chi(\lambda, \mu, \mathbf{x})$ possessing simple analytic properties

$$\left(\frac{\lambda}{\lambda - a_i}\right) T_i \chi(\lambda, \mu, \mathbf{n}) - \left(\frac{\mu}{\mu - a_i}\right) \chi(\lambda, \mu, \mathbf{n}) \;=\; a_i \chi(\lambda, a_i, \mathbf{n}) T_i \chi(a_i, \mu, \mathbf{n}). \tag{6.4}$$

First, we take the equation (6.4) at $\lambda = a_j$, $\mu = a_k$ and get

$$\left(\frac{a_j}{a_j - a_i}\right) T_i \beta_{jk} - \left(\frac{a_k}{a_k - a_i}\right) \beta_{jk} = a_i \beta_{ji} T_i \beta_{ik}, \tag{6.5}$$

where the functions β_{ij} depend only on the lattice variables,

$$\beta_{ij}(\mathbf{n}) = \chi(a_i, a_j, \mathbf{n}).$$

Substituting $\lambda = 0$, $\mu = a_k$, we obtain the equation

$$T_i\beta_{ik} = \frac{1}{a_i - a_k}\frac{a_k}{a_i}\frac{\beta_{0k}}{\beta_{0i}}, \tag{6.6}$$

where

$$\beta_{0i}(\mathbf{n}) = \chi(0, a_i, \mathbf{x}).$$

For $\lambda = a_j$, $\mu = a_p$ we get

$$T_i^{-1}\beta_{ji} = \frac{1}{a_j - a_i}\frac{a_j}{a_i}\frac{\beta_{j0}}{\beta_{i0}}, \tag{6.7}$$

where

$$\beta_{i0}(\mathbf{n}) = \chi(a_i, 0, \mathbf{x}).$$

The last type of equations arises when we take $\lambda = 0$, $\mu = 0$:

$$\beta_{0i}T_i\beta_{i0} = -a_i^{-2}. \tag{6.8}$$

Formula (6.5) gives a closed set of six equations for the six functions β_{ij}, β_{ji}, β_{ik}, β_{kj}, β_{jk}, β_{ki}, which, after some modification of the definition of the functions β_{ij},

$$\beta_{ij} = \sqrt{a_i a_j}\frac{g_j(a_i)g_k(a_i)}{g_i(a_j)g_k(a_j)}\chi(a_i, a_j),$$

with

$$g_i(\lambda, n_i) = g(\lambda, n_i[a_i]) = \left(\frac{\lambda}{\lambda - a_i}\right)^{n_i},$$

are equivalent to the discrete Darboux equations in terms of rotation coefficients [**3**]

$$\Delta_i\beta_{jk} = \beta_{ji}T_i\beta_{ik}. \tag{6.9}$$

As it was recently discovered [**5**], equations (6.9) describe a system of planar quadrilateral lattices.

We have derived equations (6.9) in some very special setting, i.e. in the context of the scalar KP hierarchy, while the standard context for them is the *matrix* KP hierarchy, so we could expect to have some additional constraints. Indeed, equations (6.6), (6.7) produce the following constraints for the system (6.9)

$$\begin{aligned}
&\phi_{ij}\phi_{ji} = 1,\\
&\phi_{ij}\phi_{jk}\phi_{ki} = \phi_{ik}\phi_{kj}\phi_{ji} = 1,\\
&\phi_{ij} = T_i\beta_{ij},
\end{aligned} \tag{6.10}$$

and

$$\begin{aligned}
&\widetilde{\phi}_{ij}\widetilde{\phi}_{ji} = 1,\\
&\widetilde{\phi}_{ij}\widetilde{\phi}_{jk}\widetilde{\phi}_{ki} = \widetilde{\phi}_{ik}\widetilde{\phi}_{kj}\widetilde{\phi}_{ji} = 1,\\
&\widetilde{\phi}_{ij} = T_j^{-1}\beta_{ij}.
\end{aligned} \tag{6.11}$$

The first equations in these pairs of constraints are equivalent, while the second are not.

These constraints look a bit mysterious at first sight, and one wonders how is it possible to satisfy them. But recalling the expression for $\chi(a_i, a_j, \mathbf{n})$ in terms of the τ-function [**1**]

$$\chi(a_i, a_j, \mathbf{n}) = \frac{1}{a_i - a_j} \frac{T_j T_i^{-1} \tau(\mathbf{n})}{\tau(\mathbf{n})},$$

for ϕ_{ij}, $\widetilde{\phi}_{ij}$ one gets

$$\phi_{ij} = \sqrt{\frac{a_i}{a_j} \frac{g_j(a_i) g_k(a_i)}{g_i(a_j) g_k(a_j)}} \frac{T_j \tau(\mathbf{n})}{T_i \tau(\mathbf{n})},$$

and

$$\widetilde{\phi}_{ij} = \sqrt{\frac{a_i}{a_j} \frac{g_j(a_i) g_k(a_i)}{g_i(a_j) g_k(a_j)}} \frac{T_i^{-1} \tau(\mathbf{n})}{T_j^{-1} \tau(\mathbf{n})}.$$

It is easy to check that all the constraints are satisfied by the τ-functional substitution. Moreover, if we use this substitution for all Darboux equations (6.9), we get just one standard Hirota bilinear difference equation

$$a_j(a_k - a_i)(T_i T_k \tau) T_j \tau + a_k(a_i - a_j)(T_i T_j \tau) T_k \tau + a_i(a_j - a_k)(T_j T_k \tau) T_i \tau = 0.$$

REMARK 2. The existence of constraints (6.10), (6.11) is not limited by the case of one-component KP hierarchy and discrete times labeled by the continuous parameter (a_i); in fact what we really need to get these constraints is common zero of the functions $g_i(\lambda)$ corresponding to discrete shifts T_i (see [**2**] for more details).

REMARK 3. It is interesting to note that the set of relations of the type (6.10), (6.11) arises also in projective geometry in the description of coordinate systems in the incidence space ([**6**, pp. 260-262]).

There is an important question: which system of planar quadrilateral lattices is described by this special reduction of the discrete Darboux system (corresponding to the one-component KP case). To understand it, we should recall that the function Φ satisfying the equation (2.1) is connected with the *radius-vector* of the related system of lattices. More explicitly,

$$\mathbf{r}(\mathbf{n}) = (\Phi_1, \Phi_2, \Phi_3),$$

where Φ_1, Φ_2, Φ_3 are solutions to (2.1) obtained from the same CBA function integrated with different weight functions.

So we get the following geometrical characterization of this special system of planar quadrilateral discrete surfaces connected with the one-component KP hierarchy: it is the three-dimensional lattice built of planar quadrilaterals such that the projection of the radius-vector to each coordinate axis satisfies equation (2.1).

The case of equations (6.9) + constrains (6.10), (6.11) corresponding to the multicomponent KP hierarchy will lead us to the matrix equation (2.2), where the matrix contains several radius-vectors as columns, and we do not have clear geometric interpretation for this case yet.

There is another geometric interpretation for equation (2.1) if the function Φ is complex-valued. Then this equation defines two conditions for the system of hexagons in the complex plane with the vertices

$$(T_1\Phi, T_1T_3\Phi, T_3\Phi, T_3T_2\Phi, T_2\Phi, T_1T_2\Phi).$$

These hexagons are not necessarily convex and may even have self-intersections. The first condition is that the sum of the angles at the vertices $T_1\Phi$, $T_3\Phi$, $T_2\Phi$ (and also at the vertices $T_1T_3\Phi$, $T_3T_2\Phi$, $T_1T_2\Phi$) is $2\pi n$. The second is that the products of lengths of three nonadjacent sides are equal

$$|T_1\Phi, T_1T_3\Phi||T_3\Phi, T_3T_2\Phi||T_2\Phi, T_1T_2\Phi| = |T_1T_3\Phi, T_3\Phi||T_3T_2\Phi, T_2\Phi||T_1T_2\Phi T_1\Phi|.$$

References

1. L. V. Bogdanov and B. G. Konopelchenko, *Analytic-bilinear approach to integrable hierarchies.* I. *Generalized* KP *hierarchy*, J. Math. Phys. **39** (1998), no. 9, 4683–4700.
2. ———, *Analytic-bilinear approach to integrable hierarchies.* II. *Multicomponent* KP *and* 2D *Toda lattice hierarchies*, J. Math. Phys. **39** (1998), no. 9, 4701–4728.
3. ———, *Lattice and q-difference Darboux-Zakharov-Manakov systems via $\overline{\partial}$-dressing method*, J. Phys. A **28** (1995), no. 5, L173–L178.
4. A. Bobenko and U. Pinkall, *Discrete isothermic surfaces*, J. Reine Angew. Math. **475** (1996), 187–208.
5. A. Doliwa and P. M. Santini, *Multidimensional quadrilateral lattices are integrable*, Phys. Lett. A **233** (1997), no. 4-6, 365–372.
6. W. V. D. Hodge and D. Pedoe, *Methods of algebraic geometry*, Cambridge Univ. Press, vo. I, 1994.
7. F. W. Nijhoff and H. W. Capel, *The direct linearisation approach to hierarchies of integrable* PDEs *in* $2+1$ *dimensions.* I. *Lattice equations and the differential-difference hierarchies*, Inverse Problems **6** (1990), no. 4, 567–590.
8. J. Weiss, *The Painlevé property for partial differential equations.* II. *Bäcklund transformation, Lax pairs, and the Schwarzian derivative*, J. Math. Phys. **24** (1983), no. 6, 1405–1413.

L. D. Landau Institute for Theoretical Physics, Kosygin str. 2, Moscow 117940, GSP1, Russia.

Current address: European Institute for Nonlinear Studies via Transnationally Extended Interchanges, Dipartimento di Fisica dell'Università di Lecce and Sezione INFN, 73100 Lecce, Italy.

E-mail address: `leonid@eandau.ac.ru`

European Institute for Nonlinear Studies via Transnationally Extended Interchanges, Dipartimento di Fisica dell'Università di Lecce and Sezione INFN, 73100 Lecce, Italy.

E-mail address: `konopel@le.infn.it`

Centre de Recherches Mathématiques
CRM Proceedings and Lecture Notes
Volume **25**, 2000

Wave Soliton Solutions on a Generic Background for KPI Equation

M. Boiti, F. Pempinelli, A. K. Pogrebkov, and B. Prinari

1. Formulation of the Problem

Our aim is to extend the spectral transform for the KPI equation

$$(u_t - 6uu_x + u_{xxx})_x = 3u_{yy}, \tag{1.1}$$

to the case of potentials nondecaying along a finite number of directions in the plane, that is we are interested in a potential u having nontrivial limits

$$u_{n,\pm}(x,y) = \lim_{y'\to\pm\infty} u\bigl(x + 2\mu_n(y-y'), y'\bigr), \quad n = 1, 2, \dots, N, \tag{1.2}$$

for N real constants μ_n and otherwise vanishing at large distances.

Such extension of the admissible potentials is of high mathematical relevance and, above all, is crucial in physical applications since solitons in two dimensions are in general not localized in space and also in the case of Davey Stewartson equation, for which localized solitons have been discovered, the dynamics is governed by the inverse scattering theory of the KPI equation.

Some general results were obtained in [**2, 3**] by using a new approach in the Inverse Scattering Method, called resolvent approach. The validity of our approach is general and in [**7, 8**] it was also tested in the discrete case.

Here we present the case of a potential obtained by superimposing a soliton to a generic rapidly decaying background via Bäcklund transformations. We are able to construct the resolvent, the Jost solutions and the spectral data and we present a complete study of their properties, showing that they meet completely our scheme. From one side this demonstrates that the theory is not void of interest since we found that it works well in a sufficiently generic case, and from the other side it allows us to get some experience with a case less involved from a mathematical point of view than the generic case.

2. The General Scheme of the Method

A large class of nonlinear equations can be linearized by using the well known Inverse Spectral (or Scattering) Transform technique (IST). KPI equation is integrable in the framework of IST method, since it can be associated to the linear

2000 *Mathematics Subject Classification.* 35Q53, 35Q51.
This is the final form of the paper.

spectral problems

$$(i\partial_y + \partial_x^2 - u)\Phi = 0, \tag{2.1}$$
$$(\partial_t + 4\partial_x^3 + 6u\partial_x + 3u_x - 3i\partial_x^{-1}u_y)\Phi = 0,$$

the inverse problem being formulated as a nonlocal Riemann-Hilbert problem.

We work in the space of Fourier trasform of solutions of the nonstationary Schrödinger equation (2.1). In particular to each differential operator

$$\mathcal{L}(x; \partial_x, \partial_y) = \sum_{n_1,n_2} a_{n_1,n_2}(x,y)\partial_x^{n_1}\partial_y^{n_2}, \tag{2.2}$$

we associate an "image"

$$L(p; \mathbf{q}) = \sum_{n_1,n_2} a_{n_1,n_2}(p)(-i\mathbf{q}_1)^{n_1}(-i\mathbf{q}_2)^{n_2}, \tag{2.3}$$

where

$$a_{n_1,n_2}(p) = \frac{1}{(2\pi)^2}\int dx \int dy e^{i(p_1x+p_2y)} a_{n_1,n_2}(x,y). \tag{2.4}$$

We consider this space of "images" as a subspace of the Schwartz space of distributions depending on two real and two complex variables, $p = (p_1, p_2)$ and $\mathbf{q} = (\mathbf{q}_1, \mathbf{q}_2)$ respectively, which we denote $\mathcal{S}'_{p,\mathbf{q}}$.

For any $A \in \mathcal{S}'_{p,\mathbf{q}}$ we introduce its Hermitian conjugate

$$A^{\dagger}(p; \mathbf{q}) = \overline{A(-p; p + \overline{\mathbf{q}})}, \tag{2.5}$$

and we also define in $\mathcal{S}'_{p,\mathbf{q}}$ a composition law by the shifted convolution

$$(AB)(p; \mathbf{q}) = \int dp' A(p - p'; \mathbf{q} + p')B(p'; \mathbf{q}), \tag{2.6}$$

if the integral exists in the sense of distributions.

Then the image of the nonstationary Schrödinger operator is

$$L = L_0 - v, \tag{2.7}$$

where

$$L_0(p, \mathbf{q}) = (\mathbf{q}_2 - \mathbf{q}_1^2)\delta(p), \tag{2.8}$$

and v is the Fourier transform of potential u

$$v(p) = \frac{1}{(2\pi)^2}\int dx \int dy e^{i(p_1x+p_2y)} u(x,y). \tag{2.9}$$

The resolvent M is defined as the left and right inverse, with respect to the composition (2.6) of the spectral operator L, that is

$$LM = ML = I, \tag{2.10}$$

where $I(p, \mathbf{q}) = \delta(p)$ is the image of 1.

Equations (2.10) are equivalent to the following dual integral equations for M

$$M = M_0 + M_0vM, \quad M = M_0 + MvM_0.$$

The main advantage with respect to the standard integral equations defining the Jost solution is that the integrals in the right hand sides are well defined also for potentials u of the form (1.2) and therefore we have a well defined quantity to be used for defining the Jost solutions also in the non regular case.

The main technical tool in the study of resolvent is the Hilbert identity. A pair of resolvents M and M', relative to spectral operators L and L' satisfies

$$M' - M = -M'(L' - L)M. \tag{2.11}$$

All relevant quantities of IST, in particular Jost solutions and spectral data, are obtained from the resolvent via a reduction procedure. In particular, the analytical properties of the Jost solutions and characterization equations for the spectral data are derived as consequences of the Hilbert identity (2.11).

In the general case of nondecaying potentials of the type (1.2), the Jost solution $|\nu\rangle(p;\mathbf{q})$ is obtained as the residuum of the resolvent on the singular curve $\mathbf{q}_2 = \mathbf{q}_1^2$ and results to be defined by the integral equation

$$|\nu\rangle(p;\mathbf{q}) = \delta(p) + \frac{1}{p_1 + i0\mathbf{q}_{1\Im}}\left(\frac{p_1(v|\nu\rangle)(p;\mathbf{q})}{p_2 - p_1(p_1 + 2\mathbf{q}_1)}\right),$$

where the products of distributions have to be performed in the order indicated. Function $|\nu\rangle(p;\mathbf{q})$ is, up to a gauge shift, the Fourier transform of standard Jost solution of (2.1).

The study of the leading singularities of $|\nu\rangle(p;\mathbf{q})$ allows us to define the transmission coefficient $t(\mathbf{q}_1)$

$$T^{-1}(p;\mathbf{q}) \equiv \frac{\delta(p)}{t(\mathbf{q}_1)} = \delta(p) - 2i\pi \operatorname{sgn} \mathbf{q}_{1\Im}\delta(p_1)\left(\frac{p_1(v|\nu\rangle)(p;\mathbf{q}_1)}{p_2 - p_1(p_1 + 2\mathbf{q}_1)}\right), \tag{2.12}$$

and to introduce the proper modifications of spectral data

$$R_{\pm}^{\sigma}(p;\mathbf{q}) = \big(|\nu^{\sigma}\rangle T^{\sigma}\big)(p;\mathbf{q}) - \frac{((p_2 - p_1(p_1 + 2\mathbf{q}_{1\Re}))(|\nu^{\sigma}\rangle T^{\sigma})(p;\mathbf{q}))}{p_2 - p_1(p_1 + 2\mathbf{q}_{1\Re}) \pm i0}, \tag{2.13}$$

where $|\nu^{\sigma}\rangle$ and T^{σ} are the values of the Jost solutions and of the transmission coefficient in the limits $\mathbf{q}_{1\Im} \to \sigma 0$, $\sigma = +, -$.

We proved that the spectral data satisfy the modified characterization equations (see [**2, 3**])

$$(R_{\mp}^{-\sigma})^{\dagger} R_{\pm}^{\sigma} = T^{\sigma}, \tag{2.14}$$

$$R_{\pm}^{\sigma}(T^{\sigma})^{-1}(R_{\mp}^{-\sigma})^{\dagger} = I, \tag{2.15}$$

$$(R_{+}^{\sigma})^{\dagger} R_{+}^{\sigma} = (R_{-}^{\sigma})^{\dagger} R_{-}^{\sigma}, \tag{2.16}$$

and finally we formulated the inverse problem by means of the data of the discrete spectrum and the discontinuity of the Jost solution on the real axis

$$|\nu^{\sigma}\rangle T^{\sigma} = |\nu^{-\sigma}\rangle (R_{\pm}^{\sigma})^{\dagger} R_{\pm}^{\sigma},$$

which allows, by the standard $\overline{\partial}$-method, to reconstruct, once given the spectral data, the Jost solutions and the potential $v(p)$.

3. A Relevant Example: One Soliton on a Generic Background

As it was briefly shown above, the analysis of the most general potential of the kind (1.2) is particularly involved; so we consider here a potential, describing a soliton on a generic background, obtained by performing a Bäcklund transformation on a rapidly decaying potential.

Let the potential $v(p)$ in the nonstationary Schrödinger operator L introduced in (2.7), (2.8) be regular, that is the Fourier transform of a potential u rapidly decaying at large distances in the x, y-plane. A new potential $v'(p)$, corresponding to the nonstationary Schrödinger operator L', can be generated by using a gauge transformation B as follows

$$L'B = BL. \tag{3.1}$$

The particularly simple gauges B of the form

$$B = Q_1 + \alpha, \quad \alpha(p; \mathbf{q}) \equiv \alpha(p), \quad Q_1 = \mathbf{q}_1 \delta(p), \tag{3.2}$$

are usually called elementary Bäcklund gauges. Without loss of generality (for details see [**7, 8**]) one can rewrite α as

$$\alpha = -[Q_1, \mathrm{LOG}\,\beta] - \lambda I, \tag{3.3}$$

where $\mathrm{LOG}\,\beta$ denotes the "image" of some function $\log b(x, y)$, that is

$$(\mathrm{LOG}\,\beta)(p) = \frac{1}{2\pi} \int dx \int dy\, e^{i(p_1 x + p_2 y)} \log b(x, y),$$

and λ is the spectral parameter

$$\lambda = \mu + i\kappa, \quad \kappa > 0.$$

We can choose $\beta(p) = |\nu\rangle(p; \lambda)$, the value of the Jost solution of the original potential v at $\mathbf{q}_1 = \lambda$.

We will obtain a soliton superimposed to a generic background by performing a binary Bäcklund transformation (see [**6**]), that is the elementary Bäcklund gauge (3.2) and an inverse elementary Bäcklund gauge

$$\widetilde{B} = Q_1 - [Q_1, \mathrm{LOG}\,\widetilde{\beta}] - \widetilde{\lambda} I,$$

with parameter $\widetilde{\lambda} = \overline{\lambda}$.

In constrast with β, $\widetilde{\beta}$ is neither the value of the Jost solution at $\mathbf{q}_1 = \widetilde{\lambda}$, nor admits inverse in the space of Schwartz distributions. Anyway we are able to express both $\widetilde{\beta}$ and the potential $\widetilde{v}$ in terms of β and of the spectral parameter λ; in particular

$$\tilde{v} = v + 2i[Q_1, \beta^\dagger \tilde{\beta}].$$

Let now U denotes the Darboux transformation that generates this binary Bäcklund transformation from v to $\widetilde{v}$, i.e.

$$U = B^{-1}\widetilde{B}. \tag{3.4}$$

$U^\dagger$ is right inverse of U, that is

$$UU^\dagger = I, \tag{3.5}$$

but

$$\Pi = U^\dagger U, \tag{3.6}$$

is a nontrivial orthogonal projector,

$$\Pi^\dagger = \Pi, \quad \Pi^2 = \Pi. \tag{3.7}$$

Π has a simple expression in terms of β and of the parameter of the soliton (for details see [**4**]).

We use for the resolvent the following orthogonal decomposition:

$$\widetilde{M} = U^\dagger M U + \widetilde{M}(I - \Pi), \tag{3.8}$$

where M and U in the first term are considered given data together with the generic background v. By these means we are able to analyze the analytic properties of the resolvent and we find that the second term is discontinous at $\mathbf{q}_{2\Im} = 2\mu\mathbf{q}_{1\Im}$ (which is a manifestation of the solitonic content of the potential); moreover at the end points of this cut, that is

$$\mathbf{q}_{1\Im} \to \pm\kappa, \quad \mathbf{q}_{2\Im} \pm 2\mu\kappa,$$

the resolvent has logarithmic singularities.

We also get the expression of the Jost solution of potential $\widetilde{v}$ in terms of U and of the Jost solution of the original potential v, that is

$$|\widetilde{\nu}\rangle = \begin{cases} U^{\dagger}|\nu\rangle & \mathbf{q}_{1\Im} > 0, \\ U^{\dagger}|\nu\rangle \frac{Q_1 - \overline{\lambda} I}{Q_1 - \lambda I} & \mathbf{q}_{1\Im} < 0. \end{cases}$$

Taking into account the explicit formulae for $U^{\dagger}$ and the analytic properties of the Jost solution of the decaying potential, it is easy to check that $|\widetilde{\nu}\rangle$ is sectionally analytic in the plane of the complex spectral parameter $\mathbf{q}_1$ with a cut on the real axis, in spite of the involved properties of the resolvent.

4. Discussion of the Main Results

We believe that all this construction proves that the general theory developed in [**2, 3**] is not void since at least in the case $N = 1$ (one soliton present) we found an explicit rather general example satisfying all the required properties.

Obviously the case we discussed above is not the most general case of potentials (1.2), even in the case $N = 1$. A more general example can be obtained by a perturbation of $\widetilde{u}$. In this case the additional cut of the extended resolvent which we spoke about leads also to a cut of the Jost solutions accross the segment connecting points λ and $\overline{\lambda}$ on the plane of complex spectral parameter. This suggests that for a general potential of the kind (1.2) additional spectral data have to be introduced and so the generalization can be significantly involved.

Another interesting result comes from the study of the behaviour of potential $\widetilde{u}$ at large distances. Let us choose some direction on the x, y-plane

$$\xi' = x + 2\mu' y = \text{const},$$

with μ' a real parameter. We find that for $\mu' = \mu$ we have, as expected, a pure one soliton behaviour corrected, however, by terms $O(1/y)$.

Along a direction $\xi' = \text{const}$ with $\mu' \neq \mu$ and $(\mu - \mu')y \to -\infty$ the behavior of $\widetilde{u}$ with respect to that one of the background potential u is corrected by an additional term which is exponentially decreasing, while for $(\mu - \mu')y \to +\infty$ the correction decreases algebraically.

So the direction $\xi = \text{const}$ of the wave soliton emerging from the background at large distances divide the x-plane into two regions with different asymptotic behavior. In the first region (to the left of the soliton) the original potential u is modified at large distances by an exponentially decreasing term, while in the second region (to the right) by a term decreasing as $1/y^3$. If one considers the case $\kappa < 0$ the role of the two regions is exchanged.

References

1. M. Boiti, B. Konopelchenko, and F. Pempinelli, *Bäcklund transformations via gauge transformations in* $2+1$ *dimensions*, Inverse Problems **1** (1985), 33–56.

2. M. Boiti, F. Pempinelli, A. K. Pogrebkov, *Spectral theory of solitons on a generic background for the KPI equation*, Nonlinear Physics, Theory and Experiment (Lecce, 1995) (E. Alfinito, M. Boiti, L. Martina, and F. Pempinelli, eds.), World Sci. Publishing, River Edge, NJ, 1996, pp. 37–52
3. ———, *Solving the Kadomtsev-Petviashvili equation with initial data not vanishing at large distances*, Inverse Problems **13** (1997), L7–L10.
4. M. Boiti, F. Pempinelli, A. K. Pogrebkov, and M. C. Polivanov, *New features of Bäcklund and Darboux transformations in* $2 + 1$ *dimensions*, Inverse problems **7** (1991), 43–56.
5. M. Boiti, F. Pempinelli, A. Pogrebkov and B. Prinari, *Towards an inverse scattering theory for two-dimensional nondecaying potentials*, Theoret. Math. Phys. **116** (1998), 741–781.
6. V. B. Matveev and M. A. Salle, *Darboux transformations and solitons*, Springer, Berlin, 1991.
7. A. K. Pogrebkov and M. C. Prati, *Resolvent approach to the Ablowitz-Ladik linear system*, Nuovo Cimento B **111** (1996), 1495–1505.
8. ———, *Ablowitz-Ladik system with discrete potential.* I. *Extended resolvent*, Theoret. Math. Phys. (to appear).

Dipartimento di Fisica dell'Università and sezione INFN, 73100 Lecce, Italy.
E-mail address: `boiti@le.infn.it`

Dipartimento di Fisica dell'Università and sezione INFN, 73100 Lecce, Italy.
E-mail address: `pempi@le.infn.it`

Steklov Mathematical Institute, Gubkin str., 8, Moscow, 117966, GSP-1, Russia.
Current address: Dipartimento di Fisica dell'Università and sezione INFN, 73100 Lecce, Italy.
E-mail address: `pogreb@mi.ras.ru`

Dipartimento di Fisica dell'Università and sezione INFN, 73100 Lecce, Italy.
E-mail address: `prinari@le.infn.it`

Centre de Recherches Mathématiques
CRM Proceedings and Lecture Notes
Volume **25**, 2000

Difference and Differential Equations and Convergence Acceleration Algorithms

C. Brezinski

Convergence acceleration methods are closely connected to integrable systems as explained, for example, in [**18, 19**].

Thus, the aim of this paper is to give a review on convergence acceleration algorithms for those who want a quick introduction to the subject.

When a sequence of numbers (S_n) converges too slowly to its limit S, it can be transformed into a new sequence (T_n) by a sequence transformation T. Under some assumptions (T_n) converges to the same limit S and, under supplementary assumptions, it converges faster, that is

$$\lim_{n\to\infty} \frac{T_n - S}{S_n - S} = 0.$$

In that case, the transformation T is said to accelerate the convergence of (S_n).

In this paper, we will review the main methodology for constructing sequence transformations and the mechanism of the corresponding algorithms for the implementation of these transformations. We will also discuss partial differential and differential equations connected to some convergence acceleration algorithms. More details on these algorithms could be found in [**11**]. Since space is limited here, quite a large bibliography has been included. See [**9**] for more references.

1. Construction of Sequence Transformations

Let us first explain how to construct a sequence transformation. Let N_T^k be the set of sequences (u_n) such that, for all n,

$$u_n = R(n, u, a_1, \dots, a_k), \tag{1.1}$$

where u and $a_1, \dots, a_k$ are parameters. The function R can also depend on some auxiliary sequences whose terms are known. These auxiliary sequences can be related, in some way, to the sequence (u_n) itself in which case (1.1) is a difference equation. The idea is now to construct a sequence transformation $T\colon (S_n) \mapsto (T_n)$ such that $\forall n$, $T_n =$ constant if (and sometimes only if) $(S_n) \in N_T^k$.

This is possible by imposing the interpolation conditions

$$S_{n+i} = R(n+i, u, a_1, \dots, a_k), \quad \text{for } i = 0, \dots, k.$$

2000 *Mathematics Subject Classification.* Primary: 65B05; Secondary: 65Q05, 39A10.
This is the final form of the paper.

Then, this system of $k+1$ equations in $k+1$ unknowns is solved and, for an arbitrary sequence (S_n), its solution depends on n. This is why we set $T_n = u$. Obviously, by construction, $\forall n$, $T_n =$ constant if $(S_n) \in N_T^k$. The set N_T^k is called the *kernel* of the transformation T.

Let us give a first example. We consider the kernel $N_T^k = \{(u_n) \mid \forall n, u_n = u + a_1 x_n + \cdots + a_k x_n^k\}$ where (x_n) is a known auxiliary sequence. Imposing the preceding interpolation conditions, we see that u is the value at the point zero of the polynomial $P_k(x) = a_0 + a_1 x + \cdots + a_k x^k$ satisfying $S_{n+i} = P_k(x_{n+i})$ for $i = 0, \ldots, k$. From the well known theory of polynomial interpolation, we have

$$T_n = P_k(0) = \begin{vmatrix} S_n & \cdots & S_{n+k} \\ x_n & \cdots & x_{n+k} \\ \vdots & & \vdots \\ x_n^k & \cdots & x_{n+k}^k \end{vmatrix} \Bigg/ \begin{vmatrix} 1 & \cdots & 1 \\ x_n & \cdots & x_{n+k} \\ \vdots & & \vdots \\ x_n^k & \cdots & x_{n+k}^k \end{vmatrix}.$$

Since T_n also depends on k, we will denote it by $T_k^{(n)}$. The next problem is now a purely computational one: how to compute recursively these $T_k^{(n)}$ for increasing values of k and for all n? Again, the answer is given by a known result about polynomial interpolation. Since the interpolation polynomials P_k also depend on n, the Neville-Aitken scheme allows to compute them recursively. Setting $x = 0$ in this scheme leads to the following recursive algorithm, called the *Richardson extrapolation process*,

$$T_{k+1}^{(n)} = \frac{x_{n+k+1} T_k^{(n)} - x_n T_k^{(n+1)}}{x_{n+k+1} - x_n}, \qquad k, n = 0, 1, \ldots,$$

with $T_0^{(n)} = S_n$.

Let us now give a second example. We consider the kernel $N_T^k = \{(u_n) \mid \forall n, a_0(u_n - u) + \cdots + a_k(u_{n+k} - u) = 0\}$ with the condition $a_0 + \cdots + a_k \neq 0$. Thus, the generality is not restricted by assuming that $a_0 + \cdots + a_k = 1$ and it follows that N_T^k only depends on u and $a_1, \ldots, a_k$. The interpolation conditions write

$$S_{n+i} = u + a_1(S_{n+i+1} - u) + \cdots + a_k(S_{n+i+k} - u), \quad i = 0, \ldots, k.$$

So, it holds

$$\Delta S_{n+i} = a_1 \Delta S_{n+i+1} + \cdots + a_k \Delta S_{n+i+k}, \quad i = 0, \ldots, k-1.$$

Solving this system gives $a_1, \ldots, a_k$ and, then, we obtain $u = a_0 S_n + \cdots + a_k S_{n+k}$ with $a_0 + \cdots + a_k = 1$. From the well-known determinantal formula for the solution of a system of linear equations, it is easy to see that the numbers $e_k(S_n) = u$ are given by

$$\begin{aligned} e_k(S_n) &= \begin{vmatrix} S_n & \cdots & S_{n+k} \\ \Delta S_n & \cdots & \Delta S_{n+k} \\ \vdots & & \vdots \\ \Delta S_{n+k-1} & \cdots & \Delta S_{n+2k-1} \end{vmatrix} \Bigg/ \begin{vmatrix} 1 & \cdots & 1 \\ \Delta S_n & \cdots & \Delta S_{n+k} \\ \vdots & & \vdots \\ \Delta S_{n+k-1} & \cdots & \Delta S_{n+2k-1} \end{vmatrix} \\ &= H_{k+1}(S_n)/H_k(\Delta^2 S_n), \end{aligned}$$

where $H_k(u_n)$ denotes the Hankel determinant corresponding to an arbitrary sequence (u_n) and is defined by

$$H_k(u_n) = \begin{vmatrix} u_n & u_{n+1} & \cdots & u_{n+k-1} \\ u_{n+1} & u_{n+2} & \cdots & u_{n+k} \\ \vdots & \vdots & & \vdots \\ u_{n+k-1} & u_{n+k} & \cdots & u_{n+2k-2} \end{vmatrix}.$$

The sequence transformation $T\colon (S_n) \mapsto \big(e_k(S_n)\big)$ is due to Shanks [**23**] and, according to the theory of linear difference equations of order k, it corresponds to extrapolation at infinity by a combination of exponential functions. Let us mention that the kernel of Shanks transformation can also be written as

$$N_T^k = \big\{(u_n) \mid \forall n, u_n = u + a_1 \Delta u_n + \cdots + a_k \Delta^k u_n\big\}. \tag{1.2}$$

The Hankel determinants $H_k^{(n)} = H_k(u_n)$ satisfy the recurrence relationship

$$H_{k+2}^{(n)} H_k^{(n+2)} = H_{k+1}^{(n)} H_{k+1}^{(n+2)} - [H_{k+1}^{(n+1)}]^2,$$

with $H_0^{(n)} = 1$ and $H_1^{(n)} = u_n$. So, the numerators and the denominators of the quantities $e_k(S_n)$ can be recursively computed by this relation.

Shanks transformation can be also implemented by Wynn's ε-algorithm [**25**]

$$\varepsilon_{k+1}^{(n)} = \varepsilon_{k-1}^{n+1)} + \frac{1}{\varepsilon_k^{(n+1)} - \varepsilon_k^{(n)}}, \qquad k, n = 0, 1, \ldots,$$

with $\forall n, \varepsilon_{-1}^{(n)} = 0$ and $\varepsilon_0^{(n)} = S_n$. Then we get $\varepsilon_{2k}^{(n)} = e_k(S_n)$. The quantities with an odd lower index are also related to Shanks transformation by

$$\varepsilon_{2k+1}^{(n)} = 1/e_k(\Delta S_n) = H_k(\Delta^3 S_n)/H_{k+1}(\Delta^2 S_n).$$

Let us mention that, for $k = 1$, the well known Aitken Δ^2 process is recovered.

We also have

$$\varepsilon_{2k+2}^{(n)} = \varepsilon_{2k}^{(n)} - \frac{[H_{k+1}(\Delta S_n)]^2}{H_k(\Delta^2 S_n) H_{k+1}(\Delta^2 S_n)}.$$

The numbers $\varepsilon_k^{(n)}$ are displayed in a two-dimensional table as follows

$$\begin{array}{ccc} & \varepsilon_k^{(n-1)} = N, & \\ \varepsilon_{k-1}^{(n+1)} = W, & \varepsilon_k^{(n)} = C, & \varepsilon_{k+1}^{(n-1)} = E, \\ & \varepsilon_k^{(n+1)} = S. & \end{array}$$

Eliminating either all the quantities with an odd lower index in the rule of the ε-algorithm or all those with an even lower index, leads to the so-called *cross rule* due to Wynn [**32**]

$$(N - C)^{-1} + (S - C)^{-1} = (W - C)^{-1} + (E - C)^{-1},$$

and to the *inverse cross rule*

$$(N^{-1} - C^{-1})^{-1} + (S^{-1} - C^{-1})^{-1} = (W^{-1} - C^{-1})^{-1} + (E^{-1} - C^{-1})^{-1},$$

with the convention that $\varepsilon_{-2}^{(n)} = \infty$ for all n.

Obviously, a division by zero can occur in the rule of the ε-algorithm when two neighboring quantities are equal. Let us examine this case in details since it is connected to the problem of singularity confinement. Let us assume that

$$\varepsilon_k^{(n-1)} \neq \varepsilon_k^{(n)} = \varepsilon_k^{(n+1)} \neq \varepsilon_k^{(n+2)}.$$

Let us denote by C this common value. Then $\varepsilon_{k+1}^{(n)} = \infty$ and it follows that $\varepsilon_{k+1}^{(n-1)} = \varepsilon_{k+1}^{(n)} = C$ and no other adjacent quantity is equal to C. So, considering only the ε's with a lower index of the same parity (usually even), we see that we have 4 identical adjacent ε's equal to C and that they are located inside a square of side 2. More generally, it can be proved that identical adjacent quantities can only be located inside square blocks. Denoting by C the common value of these quantities, by N_i, $i = 1, \ldots, N$ the quantities on the north side of the block beginning with $i = 1$ at the north-west corner, by W_i those on the west side from the north-west corner, by S_i those on the south side starting with $i = 1$ from the south-east corner, and by E_i those on the east side numbered from the south-east corner, then it was proved by Cordellier [**16**] that we have the following extended cross rule

$$(N_i - C)^{-1} + (S_i - C)^{-1} = (W_i - C)^{-1} + (E_i - C)^{-1}, \quad i = 1, \ldots, N.$$

Shanks transformation is related to Padé approximation. Indeed, if it is applied to the sequence $(S_n = c_0 + \cdots + c_n t^n)$ of the partial sums of a series f, then $\varepsilon_{2k}^{(n)} = e_k(S_n) = [n + k/k]_f(t)$. Thus, the ε-algorithm can be used for computing the upper half of the Padé table. Let us also remind that there is a close connection to formal orthogonal polynomials as explained in [**7**]. Reciprocally, algorithms for the recursive computation of Padé approximants, continued fractions or formal orthogonal polynomials lead to algorithms for implementing Shanks transformation or are related to them. This is the case, in particular, of the *qd*-algorithm of Rutishauser [**21**] and the η algorithm and the g transformation of Bauer [**1–3**]. Such connections are described in details in [**6**].

Shanks transformation can also be implemented by the ω-algorithm [**35**]

$$\omega_{2k+1}^{(n)} = \omega_{2k-1}^{(n+1)} + \frac{\omega_{2k}^{(n)}}{\omega_{2k}^{(n+1)}}, \quad \omega_{2k+2}^{(n)} = \omega_{2k}^{(n+1)} \left(\omega_{2k+1}^{(n)} - \omega_{2k+1}^{(n+1)} \right),$$

with $\omega_{-1}^{(n)} = 0$ and $\omega_0^{(n)} = \Delta^n S_i$ where i is a fixed index. We obtain $\omega_{2k}^{(0)} = e_k(S_i)$ and $\omega_{2k}^{(1)} = 1/\varepsilon_{2k+1}^{(i)}$. The quantities ω with an odd lower index can be eliminated, thus leading to a kind of cross rule.

Let us now consider the kernel $N_T^k = \{(u_n) \mid \forall n, u_n = (ux_n^k + a_1 x_n^{k-1} + \cdots + a_k)/(x_n^k + a_{k+1} x_n^{k-1} + \cdots + a_{2k})\}$ where (x_n) is an auxiliary sequence. Writing down the interpolation conditions, solving the system obtained for the unknowns and setting $\rho_{2k}^{(n)} = u$, we obtain

$$\rho_{2k}^{(n)} = \frac{\begin{vmatrix} 1 & S_n & x_n & x_n S_n & \cdots & x_n^{k-1} & x_n^{k-1} S_n & x_n^k S_n \\ \vdots & \vdots & \vdots & \vdots & & \vdots & \vdots & \vdots \\ 1 & S_{n+2k} & x_{n+2k} & x_{n+2k} S_{n+2k} & \cdots & x_{n+2k}^{k-1} & x_{n+2k}^{k-1} S_{n+2k} & x_{n+2k}^k S_{n+2k} \end{vmatrix}}{\begin{vmatrix} 1 & S_n & x_n & x_n S_n & \cdots & x_n^{k-1} & x_n^{k-1} S_n & x_n^k \\ \vdots & \vdots & \vdots & \vdots & & \vdots & \vdots & \vdots \\ 1 & S_{n+2k} & x_{n+2k} & x_{n+2k} S_{n+2k} & \cdots & x_{n+2k}^{k-1} & x_{n+2k}^{k-1} S_{n+2k} & x_{n+2k}^k \end{vmatrix}}.$$

This sequence transformation corresponds to rational extrapolation at infinity. The numbers $\rho_{2k}^{(n)}$ can the recursively computed by the ρ-algorithm whose rules are

$$\rho_{k+1}^{(n)} = \rho_{k-1}^{(n+1)} + \frac{x_{n+k+1} - x_n}{\rho_k^{(n+1)} - \rho_k^{(n)}}, \qquad k, n = 0, 1, \ldots,$$

with $\forall n$, $\rho_{-1}^{(n)} = 0$ and $\rho_0^{(n)} = S_n$. This rule is nothing else than the relation for the reciprocal differences which are commonly used for interpolation by a continued fraction using Thiele formula.

The most general kernel considered so far is $N_T^k = \{(u_n) \mid \forall n, u_n = u + a_1 g_1(n) + \cdots + a_k g_k(n)\}$ where the $(g_i(n))$ are auxiliary known sequences (whose terms can depend on some terms of the sequence (u_n)) [**8**]. Writing down the interpolation conditions and solving the corresponding system, we obtain, setting $E_k^{(n)} = u$,

$$(1.3) \qquad E_k^{(n)} = \begin{vmatrix} S_n & \cdots & S_{n+k} \\ g_1(n) & \cdots & g_1(n+k) \\ \vdots & & \vdots \\ g_k(n) & \cdots & g_k(n+k) \end{vmatrix} \Bigg/ \begin{vmatrix} 1 & \cdots & 1 \\ g_1(n) & \cdots & g_1(n+k) \\ \vdots & & \vdots \\ g_k(n) & \cdots & g_k(n+k) \end{vmatrix}.$$

These quantities can be recursively computed by the E-algorithm

$$E_{k+1}^{(n)} = \frac{g_{k,k+1}^{(n+1)} E_k^{(n)} - g_{k,k+1}^{(n)} E_k^{(n+1)}}{g_{k,k+1}^{(n+1)} - g_{k,k+1}^{(n)}}, \qquad k, n = 0, 1, \ldots,$$

with $E_0^{(n)} = S_n$. The quantities $g_{k,i}^{(n)}$ for $i > k$ are given by a ratio of determinants similar to the previous one after replacing, in the first row of the numerator, S_{n+j} by $g_i(n+j)$. They can be recursively computed by

$$g_{k+1,i}^{(n)} = \frac{g_{k,k+1}^{(n+1)} g_{k,i}^{(n)} - g_{k,k+1}^{(n)} g_{k,i}^{(n+1)}}{g_{k,k+1}^{(n+1)} - g_{k,k+1}^{(n)}}, \qquad k, n = 0, 1, \ldots, \quad i > k,$$

with $g_{0,i}^{(n)} = g_i(n)$. Let us mention that, from the numerical point of view, writing these rules under this form leads to an unstable algorithm.

The E-algorithm is the most general algorithm presently known. For the choice $g_i(n) = x_n^i$, Richardson process is recovered, while $g_i(n) = \Delta S_{n+i-1}$ leads to Shanks transformation. Other transformations, such as the ρ-algorithm, also enter into this framework. Thus, instead of a specific recursive algorithm for implementing each transformation separately, the E-algorithm can be used.

As we can see, all the sequence transformations given above have a common feature: they are all written as a ratio of two determinants and there exists a recursive algorithm for their implementation. Moreover, in all these transformations, the denominators are related to the numerators in a quite similar way. As proved by the general theory presented in [**14**], ratios of determinants of this type can always be computed by a triangular recursive scheme where, as in the case of the E-algorithm, $E_{k+1}^{(n)}$ is computed from $E_k^{(n+1)}$ and $E_k^{(n)}$. The reciprocal of this result is also true, namely that elements of a set of numbers which can be computed by a triangular recursive scheme can be written as a ratio of two determinants. Other transformations of this kind can be found in [**11**].

There exists some sequence transformations which do not enter into the previous framework, namely those which were not built from a given kernel. This is, for example, the case of the θ-algorithm [4] which was obtained by a modification of the rule of the ε-algorithm. It consists in using the same rule as in the ε-algorithm for computing the quantities with an odd lower index and by introducing a factor into the rule for the even quantities. It is

$$\theta_{2k+1}^{(n)} = \theta_{2k-1}^{(n+1)} + D_{2k}^{(n)}, \quad \theta_{2k+2}^{(n)} = \theta_{2k}^{(n+1)} - \frac{\theta_{2k}^{(n+2)} - \theta_{2k}^{(n+1)}}{D_{2k+1}^{(n+1)} - D_{2k+1}^{(n)}} D_{2k+1}^{(n)},$$

$$D_k^{(n)} = (\theta_k^{(n+1)} - \theta_k^{(n)})^{-1},$$

with $\theta_{-1}^{(n)} = 0$ and $\theta_0^{(n)} = S_n$, $n = 0, 1, \ldots$

This algorithm is among the most powerful convergence acceleration algorithms but results on it are lacking. This is due to the fact that, although it is known to exist [**14**], the expression for the quantities $\theta_k^{(n)}$ as a ratio of two determinants is not known. The problem of division by zero in this algorithm has been treated by Redivo Zaglia [**20**].

Other algorithms entering into this category are the so-called first and second generalizations of the ε-algorithm [**5**]. As seen above, the rules of the ε and the ρ algorithms are quite similar. The ρ-algorithm is obtained by replacing the 1 in the numerator of the fractional part of the rule of the ε-algorithm by $x_{n+k+1} - x_n$. Replacing this 1 by $x_{n+1} - x_n$ gives the first generalization of the ε-algorithm, while the second one is obtained by replacing the 1 by $x_{n+k+1} - x_{n+k}$.

The new quantities $\varepsilon_k^{(n)}$ computed by these two generalizations can be written as ratios of two determinants. Writing that $\forall n$, $\varepsilon_{2k}^{(n)} = S$ leads to the difference equation defining the kernels of these transformations. This problem was solved in [**5**] for the first generalization, and in [**22**] for the second one.

2. The Mechanism of Extrapolation

In this section, we will give a different derivation of the E-algorithm which helps in understanding the mechanism of extrapolation algorithms. It is based on the concept of annihilation difference operator introduced by Weniger [**24**] and developed in [**10**]. Herein, we will only give the conclusions and refer the reader to [**12**] for more details.

Let $u = (u_n)$ be a sequence of numbers and let L be a linear difference operator. We will denote by $L(u_n)$ the n-th term of the sequence obtained by applying L to the sequence u. We say that L is an annihilation operator for u if, $\forall n$, $L(u_n) = 0$. For example, the difference operator Δ^k is an annihilation operator for $\big(u_n = p_{k-1}(n)\big)$

where p_{k-1} is a polynomial of degree $k-1$. The divided difference operator of order k is an annihilation operator for $\big(u_n = p_{k-1}(x_n)\big)$ where (x_n) is an auxiliary known sequence.

The E-algorithm is constructed by eliminating the successive terms in the kernel $S_n = S + a_1 g_1(n) + a_2 g_2(n) + \dots$. Dividing both sides of the previous relation by $g_1(n)$, we have

$$\frac{S_n}{g_1(n)} - S\frac{1}{g_1(n)} = a_1 + a_2\frac{g_2(n)}{g_1(n)} + \dots.$$

Applying the annihilation operator Δ for the constant sequence (a_1), we obtain

$$\Delta\big(S_n/g_1(n)\big) = S\Delta\big(1/g_1(n)\big) + a_2\Delta\big(g_2(n)/g_1(n)\big) + \dots.$$

So, setting

$$E_1^{(n)} = \frac{\Delta(S_n/g_1(n))}{\Delta(1/g_1(n))},$$

we recover the first step of the E-algorithm. The other steps are recovered similarly by repeating the same process.

Let us now give an interpretation of this approach in terms of difference operators. We define the sequence (N_k) of difference operators by

$$N_{k+1}^{(n)}(u) = \Delta\left(\frac{N_k^{(n)}(u)}{N_k^{(n)}(g_{k+1})}\right), \tag{2.1}$$

with $N_0^{(n)}(u) = u_n$. $N_k^{(n)}(u)$ denotes the n-th term of the sequence $\big(N_k(u)\big)$ and $g_i = \big(g_i(n)\big)$. Let 1 denote the sequence whose terms are all equal to 1. We have

$$E_k^{(n)}(u) = N_k^{(n)}(u)/N_k^{(n)}(1). \tag{2.2}$$

Conversely, we also have

$$N_k^{(n)}(u) = \Delta\left(\frac{E_{k-1}^{(n)}(u)}{E_{k-1}^{(n)}(g_k)}\right).$$

From (2.1), we see that the quantities $N_k^{(n)}(u)$ are computed by a triangular recursive scheme and, so, by the theory developed in [**14**], they can be expressed as a ratio of two determinants. It follows that the quantities $E_k^{(n)}(u)$ can also be written under that form and (1.3)) is recovered. Moreover, using (2.1) in the numerator and the denominator of (2.2), leads to the rule of the E-algorithm.

Instead of computing recursively the quantities $E_k^{(n)}$ by the E-algorithm, an algorithm due to Ford and Sidi [**17**] can also be used. This algorithm can be recovered using the preceding difference operators. The same is true for a generalization of the ε-algorithm due to Carstensen [**15**].

This theory has been generalized to the vector and matrix cases in [**13**].

3. Associated Partial Differential Equations

As seen above, the quantities involved in convergence acceleration algorithms depend on two indexes. So, they can be considered as first order finite difference approximations of some partial differential equations, an idea introduced by Wynn [**31**] and exploited by him in [**37,38**] where some properties of these partial differential equations and their solutions are studied.

In order to explain how these partial differential equations are derived, let us consider the case of the ε-algorithm. We saw that the quantities $\varepsilon_k^{(n)}$ are displayed in a two dimensional array where the lower index k refers to a column and the upper index n to a descending diagonal. Let us introduce the continuous coordinates x and y and the stepsize h and display them as follows

$$\begin{array}{ccccc} & & x, y-2h & & \\ & x-h, y-h & & x+h, y-h & \\ x-2h, y & & x, y & & x+2h, y \\ & x-h, y+h & & x+h, y+h & \\ & & x, y+2h & & \end{array} .$$

The array of the ε-algorithm is transformed into this new array by the change of coordinates

$$x = a + 2kh, \quad y = b + 2(n+k)h,$$

where a and b are two constants. We set

$$\varepsilon_{2k-1}^{(n)} = h^{-1}\tilde{\varepsilon}(x,y), \quad \varepsilon_{2k}^{(n)} = h^{-1}\hat{\varepsilon}(x,y).$$

Letting h tend to zero leads to

$$\tilde{\varepsilon}_x \hat{\varepsilon}_y = 1, \quad \hat{\varepsilon}_x \tilde{\varepsilon}_y = 1,$$

where the lower index x or y indicates partial differentiation with respect to this variable. Thus, the functions $\tilde{\varepsilon}$ and $\hat{\varepsilon}$ both satisfy the same partial differential equation

$$\left(\frac{1}{\varepsilon_x}\right)_x = \left(\frac{1}{\varepsilon_y}\right)_y .$$

For the qd-algorithm [**21**], the partial differential equation is

$$q_y = e_x, \quad e q_x = q e_y.$$

The functions q and e satisfy

$$q_y = \left(\frac{q_{yy} - (\ln q)_x q_y}{(\ln q)_{xx}}\right)_x, \quad e_x = \left(\frac{e_{xx} - (\ln e)_y e_x}{(\ln e)_{yy}}\right)_y .$$

For the g-algorithm [**1–3**], setting $g_{2k-1}^{(n)} = \widetilde{g}(x,y)$ and $g_{2k}^{(n)} = \widehat{g}(x,y)$, we have

$$\widetilde{g}\widehat{g}_x = -(1-\widehat{g})\widetilde{g}_x, \quad \left(S(y) - \widetilde{g}\right)_y \widehat{g} = \left(S(y) - \widetilde{g}\right)\widehat{g}_x,$$

where S is some function.

For the η-algorithm [**1–3**], we set $\eta_{2k-1}^{(n)} = \widetilde{\eta}(x,y)$ and $\eta_{2k}^{(n)} = \widehat{\eta}(x,y)$, and we have

$$\widehat{\eta}_x = \widetilde{\eta}_y, \quad (1/\widetilde{\eta})_x = (1/\widehat{\eta})_y.$$

The functions $\widetilde{\eta}$ and $\widehat{\eta}$ satisfy

$$\widehat{\eta}_x = \left(\frac{-\widehat{\eta}_x(1/\widehat{\eta})_y \pm \sqrt{(\widehat{\eta}_x)^2\widehat{\eta}^{-2} + (1/\widehat{\eta})_{yy}\widehat{\eta}_{xx}}}{(1/\widehat{\eta})_{yy}}\right)_x ,$$

$$\widetilde{\eta}_y = \left(\frac{-\widetilde{\eta}_y(1/\widetilde{\eta})_x \pm \sqrt{(\widetilde{\eta}_y)^2\widetilde{\eta}^{-2} + (1/\widetilde{\eta})_{xx}\widetilde{\eta}_{yy}}}{(1/\widetilde{\eta})_{xx}}\right)_y .$$

For the ρ-algorithm, we set $\rho_{2k-1}^{(n)} = h^{-2}\tilde{\rho}(x,y)$ and $\rho_{2k}^{(n)} = h^{-1}\hat{\rho}(x,y)$, and we obtain

$$\tilde{\rho}_x \hat{\rho}_y = x - a, \quad \tilde{\rho}_y \hat{\rho}_x = x - a,$$

and both functions satisfy

$$\left(\frac{x-a}{\rho_y}\right)_y = \left(\frac{x-a}{\rho_x}\right)_x.$$

4. Confluent Algorithms

The algorithms studied in the previous sections are aimed at accelerating the convergence of a sequence (S_n) when n tends to infinity. This is the discrete case. We will now consider the continuous case of a function f of the variable t tending to a limit when t tends to infinity. Similarly to the discrete case, this function will be transformed into a new function tending faster to the same limit when t tends to infinity. The construction of such function transformations is again based on the notion of kernel. We want to construct a function transformation $f \mapsto g$ such that $\forall t$, $g(t) = S$ if (and only if) $f \in N_T$, where N_T is the kernel of the transformation. The procedure for constructing such a transformation is completely similar to what we did in the discrete case.

Let us, for example, consider the kernel $N_T^k = \{u \mid \forall t, u(t) = S + a_1 u'(t) + \cdots + a_k u^{(k)}(t)\}$. This kernel is the continuous analog of the kernel (1.2) of Shanks transformation. We set

$$H_k^{(n)}(t) = \begin{vmatrix} f^{(n)}(t) & f^{(n+1)}(t) & \cdots & f^{(n+k-1)}(t) \\ \vdots & \vdots & & \vdots \\ f^{(n+k-1)}(t) & f^{(n+k)}(t) & \cdots & f^{(n+2k-2)}(t) \end{vmatrix},$$

and $H_0^{(n)}(t) = 1$. The transformation $f \mapsto \varepsilon_{2k}$ where $\varepsilon_{2k}(t) = H_{k+1}^{(0)}(t)/H_k^{(2)}(t)$ has N_T^k as its kernel. These functions can be recursively computed by the so-called *first confluent form* of the ε-algorithm [**26**]

$$\varepsilon_{k+1}(t) = \varepsilon_{k-1}(t) + \frac{1}{\varepsilon_k'(t)}, \quad \forall t,$$

with $\varepsilon_{-1}(t) = 0$ and $\varepsilon_0(t) = f(t)$. The functions with an odd lower index are intermediate computations and $\varepsilon_{2k+1}(t) = H_k^{(3)}(t)/H_{k+1}^{(1)}(t)$.

Confluent forms of the other algorithms described in Section 1 can also be defined [**26**]. Their rules are, as for the confluent form of the ε-algorithm, difference-differential equations. They can be obtained by a limiting process similar to the approach described in Section 3, but acting on only one of the two variables. Such equations have been extensively studied by Wynn [**27–30, 33–36**].

References

1. F. L. Bauer, *The g-algorithm*, J. SIAM **8** (1960), 1–17.
2. ———, *The quotient-difference and epsilon algorithms*, On Numerical Approximation (R. E. Langer, ed.), The University of Madison Press, Madison, 1959, pp. 361–370.
3. ———, *Nonlinear sequence transformations*, Approximation of Functions (P. Garabedian, ed.), Elsevier, Amsterdam, 1965, pp. 134–151.
4. C. Brezinski, *Accélération des suites à convergence logarithmique*, C. R. Acad. Sci. Paris A **273** (1971), 727–730.
5. ———, *Conditions d'application et de convergence de procédés d'extrapolation*, Numer. Math. **20** (1972), 64–79.
6. ———, *Sur le calcul de certains rapports de déterminants*, Padé Approximation and its Applications (Antwerp, 1979) (L. Wuytack ed.), Lecture Notes in Math., vol. 765, Springer, Berlin-New York, 1979, pp. 184–210.

7. ———, *Padé-type approximation and general orthogonal Polynomials*, Birkhäuser, Basel, 1980.
8. ———, *A general extrapolation algorithm*, Numer. Math. **35** (1980), 175–187.
9. ———, *A bibliography on continued fractions, Padé approximation, sequence transformation, and related subjects*, Ciencias, vol. 3, Prensas, Univ. de Zaragoza, Zaragoza, 1991.
10. C. Brezinski and A. C. Matos, *A derivation of extrapolation algorithms based on error estimates*, J. Comput. Appl. Math. **66** (1996), 5–26.
11. C. Brezinski and M. Redivo Zaglia, *Extrapolation methods. Theory and practice*, North-Holland, Amsterdam, 1991.
12. C. Brezinski and M. Redivo Zaglia, *A general extrapolation procedure revisited*, Adv. Comput. Math. **2** (1994), 461–477.
13. C. Brezinski and A. Salam, *Matrix and vector sequence transformations revisited*, Proc. Edinburgh Math. Soc. **38** (1995), 495–510.
14. C. Brezinski and G. Walz, *Sequences of transformations and triangular recursion schemes with applications in numerical analysis*, J. Comput. Appl. Math. **34** (1991), 361–383.
15. C. Carstensen, *On a general epsilon algorithm*, Numerical and Applied Mathematics (C. Brezinski, ed.), Baltzer, Basel, 1989, pp. 437–441.
16. F. Cordellier, *Démonstration algébrique de l'extension de l'identité de Wynn aux tables de Padé non normales*, Padé Approximation and its Applications (Antwerp, 1979) (L. Wuytack, ed.), Lecture Notes in Math., vol. 765, Springer, Berlin-New York, 1979, pp. 36–60.
17. W. F. Ford and A. Sidi, *An algorithm for a generalization of the Richardson extrapolation process*, SIAM J. Numer. Anal. **24** (1987), 1212–1232.
18. A. Nagai and J. Satsuma, *Discrete soliton equations and convergence acceleration algorithms*, Phys. Lett. A **209** (1995), 305–312.
19. V. Papageorgiou, B. Grammaticos, and A. Ramani, *Integrable lattices and convergence acceleration algorithms*, Phys. Lett. A **179** (1993), 111–115.
20. M. Redivo Zaglia, *Particular rules for the θ-algorithm*, Numer. Algorithms **3** (1992), 353–370.
21. H. Rutishauser, *Der Quotienten-Differenzen Algorithmus*, Birkhäuser, Basel, 1957.
22. A. Salam, *On a generalization of the ε-algorithm*, J. Comput. Appl. Math. **46** (1993), 455–464.
23. D. Shanks, *Nonlinear transformations of divergent and slowly convergent sequences*, J. Math. Phys. **34** (1955), 1–42.
24. E. J. Weniger, *Nonlinear sequence transformations for the acceleration of convergence and the summation of divergent series*, Comput. Phys. Rep. **10** (1989), 189–371.
25. P. Wynn, *On a device for computing the $e_m(S_n)$ transformation*, Math. Tables Aids Comput. **10** (1956), 91–96.
26. ———, *Confluent forms of certain nonlinear algorithms*, Arch. Math. **11** (1960), 223–234.
27. ———, *A note on a confluent form of the ε-algorithm*, Arch. Math. **11** (1960), 237–240.
28. ———, *Upon a second confluent form of the ε-algorithm*, Proc. Glasgow Math. Ass. **5** (1962), 160–165.
29. ———, *Una nota su un analogo infinitesimale des $q-d$ algoritmo*, Rend. Mat. Roma (5) **21** (1962), 77–85.
30. ———, *On a connection between the first and the second confluent forms of the ε-algorithm*, Nieuw Arch. Wiskd. (3) **11** (1963), 19–21.
31. ———, *Partial differential equations associated with certain nonlinear algorithms*, Z. Angew. Math. Phys. **15** (1964), 273–289.
32. ———, *Upon systems of recursions which obtain among the quotients of the Padé table*, Numer. Math. **8** (1966), 264–269.
33. ———, *Difference-differential recursions for Padé quotients*, Proc. London Math. Soc. (3) **23** (1971), 283–300.
34. ———, *Hierarchies of arrays and function sequences associated with the epsilon algorithm and its first confluent form*, Rend. Mat. Roma (4) **5** ser. VI (1972), 819–852.
35. ———, *Upon some continuous prediction algorithms.* I, Calcolo **9** (1972), 197–234; II, Calcolo **9** (1972), 235–278.
36. ———, *Invariants associated with the epsilon algorithm and its first confluent form*, Rend. Circ. Mat. Palermo (2) **21** (1972), 31–41.
37. ———, *A note on a partial differential equation*, CRM-22, Univ. de Montréal, Québec, 1970.
38. ———, *Sur l'équation aux dérivées partielles de la surface de Padé*, C. R. Acad. Sci. Paris A **278** (1974), 847–850.

Laboratoire d'Analyse Numérique et d'Optimisation, Université des Sciences et Technologies de Lille, 59655–Villeneuve d'Ascq cedex, France.

E-mail address: Claude.Brezinski@univ-lille1.fr

Centre de Recherches Mathématiques
CRM Proceedings and Lecture Notes
Volume **25**, 2000

On the Dynamics of Rational Solutions for 1-D Volterra System

A. S. Cârstea

ABSTRACT. The Hirota formalism and soliton solutions for 1-D Volterra system is presented. Also, starting from the soliton solutions, we obtain a class of nonsingular rational solutions using the "long wave limit" procedure of Ablowitz and Satsuma and appropriate "gauge" transformations. Their properties are also discussed and it is shown that these solutions interact elastically with no phase shift.

1. Introduction

It is known that with the development of soliton theory a significant progress has been made in finding special solutions of integrable nonlinear evolution equations which include soliton solutions, rational solutions, similarity solutions and so on, by means of Inverse Scattering Transform[**3**], Hirota bilinear formalism [**8**], Bäcklund transformations [**13**] and Wronskian formalism [**7,11**].

The class of rational solutions was first investigated for KdV equation [**4**]. A very simple method to find them was developed by Ablowitz and Satsuma, [**2**], and it consists in taking the "long wave" limit in the multisoliton solution. This method was used successfully to find also the rational solutions for other completely integrable systems. Although, for continuous case, many results have been obtained with respect to finding them, a relatively less work for differential-difference systems has been done [**5,9**].

Unlike the $(2+1)$-dimensional continuous systems where the dynamics and phenomenology of rational nonsingular solutions ("lump" solutions) are well known [**6,10**], the dynamics of $(1+1)$-dimensional nonsingular rational solutions was very little investigated. In this paper the bilinear formalism for a $(1+1)$-dimensional Volterra system, which represents a coupled system of generalized Volterra equations proposed by M. Wadati [**14,15**] is derived. From its soliton solutions, a class

2000 *Mathematics Subject Classification.* 37K10, 37K40.

The author would like to thank M. J. Ablowitz, R. Conte and J. Hietarinta for valuable discussions. Moreover, he thanks D. Grecu and A. Visinescu for reading the manuscript and for their helpful suggestions and remarks.

This is the final form of the paper.

of real and complex rational nonsingular solutions is obtained using the Ablowitz–Satsuma limiting procedure. Here the crucial step is the existence of many nontrivial arbitrary constants and "gauge" transformations. Studying a "head-on" collision of two 1-rational nonsingular solutions it is shown that these solutions collide elastically with no phase shift. Also their real and complex structure is preserved. The properties of these special solutions are also discussed.

The paper is organized as follows. In Section 2 the bilinear form of the system by a reduction to a variant of Ablowitz–Ladik system, [**1**], with nonzero boundary condition is constructed. In Section 3 the construction of rational solutions is presented. In Section 4 properties and dynamics are discussed and, finally, conclusions are given in Section 5.

2. Bilinear Formalism

The 1-D Volterra system under consideration is:

$$\begin{aligned} \dot{Q}_n &= (a + bQ_n + cQ_n^2)(R_{n+1} - R_{n-1}), \\ \dot{R}_n &= (a + bR_n + cR_n^2)(Q_{n+1} - Q_{n-1}), \end{aligned} \tag{2.1}$$

with $Q_n, R_n \to 0$ as n goes to infinity and a, b, c as real constants.

For waves propagating in one direction we may put $Q_n = R_n$ and the Volterra equation proposed by M. Wadati [**14, 15**] is found.

By the following transformations:

$$\begin{aligned} u_n &= \frac{2c}{\sqrt{4ac - b^2}}(Q_n + b/2c), \\ v_n &= \frac{2c}{\sqrt{4ac - b^2}}(R_n + b/2c), \\ t &\to t\left(\frac{4ac - b^2}{4c}\right), \end{aligned} \tag{2.2}$$

and putting

$$\alpha_0 = \frac{b}{\sqrt{4ac - b^2}},$$

the system (2.1) becomes:

$$\begin{aligned} \dot{u}_n &= (1 + u_n^2)(v_{n+1} - v_{n-1}), \\ \dot{v}_n &= (1 + v_n^2)(u_{n+1} - u_{n-1}), \end{aligned} \tag{2.3}$$

with $u_n, v_n \to \alpha_0$ as n goes to infinity. In order to have α_0 real we consider that c, b are nonzero and $4ac - b^2$ is positive. The system (2.3) represents a variant of Ablowitz–Ladik system [**1**] with nonvanishing boundary conditions.

The bilinear form of this system is constructed by taking:

$$\begin{aligned} u_n &= \alpha_0 - \frac{i}{2}\frac{d}{dt}\log\left(\frac{f_n}{g_n}\right), \\ v_n &= \alpha_0 - \frac{i}{2}\frac{d}{dt}\log\left(\frac{f'_n}{g'_n}\right), \end{aligned} \tag{2.4}$$

where is assumed that f_n/g_n and $f'_n/g'_n \to$ const as n goes to infinity.

Using (2.4) in the system (2.3) integrating with respect to t and taking into account the nonzero boundary conditions we get:

$$\tan^{-1} u_n = \tan^{-1}\alpha_0 - \frac{i}{2}\log\left(\frac{f'_{n+1}g'_{n-1}}{g'_{n+1}f'_{n-1}}\right),$$
$$\tan^{-1} v_n = \tan^{-1}\alpha_0 - \frac{i}{2}\log\left(\frac{f_{n+1}g_{n-1}}{g_{n+1}f_{n-1}}\right).$$

Using (2.4) again, after some algebra, it is found the following system

$$\begin{aligned} D_t f_n g_n &= 2(1+\alpha_0^2)\sinh D_n f'_n g'_n, \\ D_t f'_n g'_n &= 2(1+\alpha_0^2)\sinh D_n f_n g_n, \\ (\cosh D_n + i\alpha_0 \sinh D_n) f'_n g'_n &= f_n g_n, \\ (\cosh D_n + i\alpha_0 \sinh D_n) f_n g_n &= f'_n g'_n, \end{aligned} \tag{2.5}$$

which represents the Hirota form of the system (2.3), D_t and D_n being the classical bilinear Hirota operators [**8**]. One can see for $\alpha_0 = 0$ the self-dual network form proposed by Hirota is recovered [**8**].

The 1-soliton solution is given by:

$$\begin{aligned} f_n &= 1 + \exp(\eta_1 + \phi_1), \\ f'_n &= 1 + \exp(\eta_1 + \phi'_1), \\ g_n &= 1 + \exp(\eta_1 + \psi_1), \\ g'_n &= 1 + \exp(\eta_1 + \psi'_1), \end{aligned} \tag{2.6}$$

where

$$\begin{aligned} \eta_j &= 2\varepsilon_j(1+\alpha_0^2)\sinh P_j t + P_j n + \eta_j^0, \\ \exp\phi_j &= \frac{\alpha_0}{2}K_j\frac{\varepsilon_j + \cosh P_j}{\sinh P_j} + \frac{i}{2}K_j, \\ \exp\psi_j &= \frac{\alpha_0}{2}K_j\frac{\varepsilon_j + \cosh P_j}{\sinh P_j} - \frac{i}{2}K_j, \\ \exp\phi'_j &= \frac{\alpha_0}{2}K_j\frac{1 + \varepsilon_j\cosh P_j}{\sinh P_j} + \frac{i\varepsilon_j}{2}K_j, \\ \exp\psi'_j &= \frac{\alpha_0}{2}K_j\frac{1 + \varepsilon_j\cosh P_j}{\sinh P_j} - \frac{i\varepsilon_j}{2}K_j. \end{aligned} \tag{2.7}$$

The index j labels the number of the soliton i.e. for 1-soliton solution $j = 1$, for 2-soliton $j = 1, 2$ and so on, $\varepsilon_j = \pm 1$ indicates the direction of propagation. On the other hand we have chosen the phases such that any multisoliton solution to be real. K_j represent arbitrary real constants which do not depend on n or t. The procedure of finding rational solutions relies mainly on this arbitrariness.

Now, proceeding to the 2-soliton solution, the following form is found:

$$\begin{aligned} f_n &= 1 + \exp(\eta_1+\phi_1) + \exp(\eta_2+\phi_2) + \exp(\eta_1+\eta_2+\phi_1+\phi_2+A_{12}), \\ f'_n &= 1 + \exp(\eta_1+\phi'_1) + \exp(\eta_2+\phi'_2) + \exp(\eta_1+\eta_2+\phi'_1+\phi'_2+A_{12}), \\ g_n &= 1 + \exp(\eta_1+\psi_1) + \exp(\eta_2+\psi_2) + \exp(\eta_1+\eta_2+\psi_1+\psi_2+A_{12}), \\ g'_n &= 1 + \exp(\eta_1+\psi'_1) + \exp(\eta_2+\psi'_2) + \exp(\eta_1+\eta_2+\psi'_1+\psi'_2+A_{12}), \end{aligned} \tag{2.8}$$

where

$$\exp A_{12} = \left[\frac{\sinh(P_1 - P_2/2)}{\sinh(P_1 + P_2/2)}\right]^2,$$

for solitons propagating in the same directions ($\varepsilon_1\varepsilon_2 = 1$) and

$$\exp A_{12} = \left[\frac{\cosh(P_1 - P_2/2)}{\cosh(P_1 + P_2/2)}\right]^2,$$

for solitons propagating in opposite directions ($\varepsilon_1\varepsilon_2 = -1$).

Using the classical procedure [**8**], this construction can be extended to any N-soliton solution.

3. Rational Solutions

The "long wave" limit method of Ablowitz and Satsuma consists in three main steps. In order to obtain a N-rational solution from a N-soliton solution one has to:

- expand the Hirota form of the N-soliton solution in power series of P_j with $j = 1, \ldots, N$;
- choose the phase constants such that all $O(P_j^k)$ with k less than N to be zero;
- make $P_j \to 0$ and recover the N-rational solution from the $O(P_j^l)$ terms with $l \geq N$.

We are focusing on f_n and g_n and the procedure for f_n' and g_n' works in the same manner. For the beginning 1-rational solutions, f_n and g_n given by (2.6) are expanded. The following notations are used:

$$\theta_j = 2\varepsilon_j(1+\alpha_0^2)t + n, \quad \zeta_j = \exp \eta_j^0, \quad P_j = \delta \bar{p}_j,$$

and the expansions will be power series of δ, $\bar{p}_j$ being $O(1)$ terms.

So, up to $O(\delta)$:

$$f_n = 1 + \zeta_1\big(1 + \delta\bar{p}_1\theta_1 + O(\delta^2)\big)\left[\frac{\alpha_0(\varepsilon_1+1)}{2\bar{p}_1}K_1\delta^{-1} + \frac{i}{2}K_1 + \frac{K_1\bar{p}_1\alpha_0}{12}\delta + O(\delta^2)\right],$$

$$g_n = (f_n)^+,$$

where $(+)$ means changing the sign of the terms multiplicated by i (complex conjugate as if ζ_j would be real for $j = 1, 2$).

For $\varepsilon_1 = 1$ we *must* take K_1 proportional to δ. To cancel $O(1)$ terms. we choose $\zeta_1 = -1/\alpha_0$. Thus when $\delta \to 0$ we obtain:

$$u_n = \alpha_0 - \frac{i}{2}\frac{d}{dt}\log\left(\frac{\theta_1 + i/2\alpha_0}{\theta_1 - i/2\alpha_0}\right) = \alpha_0 - \frac{4\alpha_0(1+\alpha_0^2)}{4\alpha_0^2\theta_1^2 + 1}. \tag{3.1}$$

In the same manner $v_n = u_n$. Thus, up to a constant and time scaling factor,

$$Q_n = R_n \sim -\frac{4\alpha_0(1+\alpha_0^2)}{4\alpha_0^2\theta_1^2 + 1},$$

which is a nonsingular real rational solution.

For $\varepsilon_1 = -1$ the situation is drastically changed. In this case the expansions become:

$$f_n = 1 + \frac{i}{2}K_1\zeta_1 + \frac{i}{2}K_1\zeta_1\bar{p}_1(\theta_1 - i\alpha_0/2)\delta + O(\delta^2), \quad g_n = (f_n)^+.$$

There are no divergent terms in δ so, $K_1 \sim O(1)$. But in this case the $O(1)$ terms in f_n and g_n are complex conjugate each other. Accordingly, there is no ζ_1 to cancel both of them. To cope with this situation, one can see that u_n and v_n are invariant under the following "gauge" transformation:

$$\frac{f_n}{g_n} \to \frac{f_n}{g_n}\Lambda, \tag{3.2}$$

where Λ is an arbitrary complex function which does not depends on t.

Now, taking $\zeta_1 = 2i/K_1$ the $O(1)$ terms are cancelled in f_n only. So,

$$f_n = -\bar{p}_1(\theta_1 - i\alpha_0/2)\delta + O(\delta^2), \quad g_n = 2 + \bar{p}_1(\theta_1 + i\alpha_0/2)\delta + O(\delta^2).$$

Choosing $\Lambda = 1/\delta$ and then $\delta \to 0$ we shall find:

$$\frac{f_n}{g_n} = -\frac{1}{2}\bar{p}_1\left(\theta_1 - \frac{i\alpha_0}{2}\right),$$

and:

$$\begin{aligned} u_n &= \alpha_0 - \frac{i}{2}\frac{d}{dt}\log(\theta_1 - i\alpha_0/2), \\ v_n &= (u_n)^*, \end{aligned} \tag{3.3}$$

which represent complex solitary waves. The physical significance of this type of solutions is not clear. The form of Q_n and R_n will be:

$$Q_n = (R_n)^+ \sim -\frac{\alpha_0}{2}\frac{1+\alpha_0^2}{\theta_1^2 + \alpha_0^2/4} + i\frac{\theta_1(1+\alpha_0^2)}{\theta_1^2 + \alpha_0^2/4}.$$

In order to study the "head-on" collision of the solutions (3.1) and (3.3) one needs the 2-rational solution from 2-soliton solution with solitons having opposite directions i.e. f_n and g_n given by (2.8) and phase factors given by (2.7) with $\varepsilon_1\varepsilon_2 = -1$. Expanding f_n and g_n up to $O(\delta^2)$ we shall take $K_1 \sim O(\delta)$ in order to eliminate divergent behaviour of δ^{-1} type in $\exp\phi_1$

$$f_n = M_0 + M_1\delta + M_2\delta^2 + O(\delta^3), \quad g_n = (f_n)^+,$$

where:

$$\begin{aligned} M_0 &= 1 + \zeta_1\alpha_0 + \frac{i}{2}K_2\zeta_2 + \frac{i}{2}\alpha_0 K_2\zeta_1\zeta_2, \\ M_1 &= \frac{i}{2}\bar{p}_1\zeta_1 + \alpha_0\bar{p}_1\zeta_1\theta_1 + \frac{1}{4}\alpha_0\zeta_2 K_2\bar{p}_2 + \frac{i}{2}\bar{p}_2\zeta_2\theta_2 K_2 + \\ &\quad + \frac{K_2}{4}\left(\alpha_0^2\frac{\bar{p}_2}{\bar{p}_1} - 1\right)\bar{p}_1\zeta_1\zeta_2 + \frac{i}{2}\alpha_0 K_2\zeta_1\zeta_2(\bar{p}_1\theta_1 + \bar{p}_2\theta_2). \end{aligned} \tag{3.4}$$

The expression of M_2 is quite involved and it will not be given here. But, taking $\zeta_1 = -1/\alpha_0$, the $O(1)$ terms will be cancelled . Then taking $K_2 \sim O(1)$ and $\zeta_2 = 2i/K_2$ the $O(\delta)$ terms will be cancelled only in f_n. Introducing the "gauge" transform in the form $\Lambda = 1/\delta$, f_n and g_n are recovered from M_2 and M_1^+ respectively. So, the rational solution will be:

$$\begin{aligned} f_n &= \theta_1\theta_2 - \frac{3}{4} + \frac{i}{2}\left(\frac{\theta_2}{\alpha_0} - \alpha_0\theta_1\right), \\ g_n &= \theta_1 - \frac{i}{2\alpha_0}, \end{aligned} \tag{3.5}$$

and ε

$$f'_n = \theta_1 + \frac{i}{2\alpha_0},$$

$$g'_n = \theta_1\theta_2 - \frac{3}{4} - \frac{i}{2}\left(\frac{\theta_2}{\alpha_0} - \alpha_0\theta_1\right). \tag{3.6}$$

In order to see the the rational solutions in the solitonic backround we shall use the "long wave limit " method starting from the 2-rational solution and making the limit only in the first soliton, namely on the relation (2.8). Of course we have here 2 cases, $\varepsilon_1\varepsilon_2 = 1$ and $\varepsilon_1\varepsilon_2 = -1$. In the first case the we choose as before $K_1 \sim \delta$ and keep only $O(\delta)$ terms. Putting $\delta \longrightarrow 0$ we shall obtain the following nonsingular mixed rational -soliton solution:

$$f_n = (1 + e^{\eta_2+\phi_2})\left(\theta_1 + \frac{i}{2\alpha_0}\right) - 2\coth\left(\frac{P_2}{2}\right)e^{\eta_2+\phi_2}, \tag{3.7}$$

$$g_n = (1 + e^{\eta_2+\psi_2})\left(\theta_1 - \frac{i}{2\alpha_0}\right) - 2\coth\left(\frac{P_2}{2}\right)e^{\eta_2+\psi_2},$$

$$f'_n = f_n, \tag{3.8}$$

$$g'_n = g_n, \tag{3.9}$$

which represents the rational solution in interaction with a soliton.

In the case $\varepsilon_1\varepsilon_2 = -1$ we have the same situation, namely we cannot cancel the $O(1)$in f_n and g_n. Accordingly we use exactly the same gauge function and constant as in the case of pure rational solution and we shall find:

$$f_n = (\theta'_1 - i\alpha_0/2)(1 + e^{\eta_2+\phi_2}) - 2\tanh\left(\frac{P_2}{2}\right)e^{\eta_2+\phi_2} \tag{3.10}$$

$$g_n = 1 + e^{\eta_2+\psi_2},$$

$$f'_n = g^*_n,$$

$$g'_n = f^*_n,$$

a complex rational solution in interaction with the soliton.

4. Dynamics and Properties

The solutions (3.5) and (3.6) represent a "head-on" collision of two rational solitary waves one being real and the other one complex, having same velocities but opposite directions. To see this we are keeping θ_1 fixed in (3.5) and, for $t \to \pm\infty$, (3.5) becomes:

$$u_n \to \alpha_0 - \frac{i}{2}\frac{d}{dt}\log\left(\frac{\theta_1 + i/2\alpha_0}{\theta_1 - i/2\alpha_0}\right),$$

which represents exactly the real solitary wave (3.1).

Also, keeping θ_2 fixed, for $t \to \pm\infty$ we obtain:

$$u_n \to \alpha_0 - \frac{i}{2}\frac{d}{dt}\log(\theta_2 - i\alpha_0/2),$$

which means that there is no phase shift at this "head-on" collision so, we have a $(1+1)$-dimensional "lump-type" dynamics. Also, the real or the complex character is preserved. We do not know exactly if do exist real nonsingular rational solutions for both directions. Also if we want to see the interaction between the rational solutions and soliton we do the same but on the mixed rational soliton solutions. Thus keeping again θ_1 fixed we have 2 limits. The first one is related to the limit $t \to \infty$

$$\lim_{t\to\infty} \frac{f_n}{g_n} = \frac{\theta_1 + i/2\alpha_0 - 2\coth(P_2/2)}{\theta_1 - i/2\alpha_0 - 2\coth(P_2/2)}, \tag{4.1}$$

and the other limit is related to $t \to -\infty$ namely

$$\lim_{t\to-\infty} \frac{f_n}{g_n} = \frac{\theta_1 + i/2\alpha_0}{\theta_1 - i/2\alpha_0}, \tag{4.2}$$

and we can see that the phase shift is given by the soliton parameter P_2. In the reference frame of the soliton (η_2 fixed) one can easily observe that the soliton remains unperturbed

$$\lim_{t\to\pm\infty} \frac{f_n}{g_n} = \frac{1 + e^{\eta_2+\phi_2}}{1 + e^{\eta_2+\psi_2}}. \tag{4.3}$$

For the other case of a mixed complex rational and soliton solution we have the same situation. In the reference frame of the rational solution ,

$$\lim_{t\to\infty} \frac{f_n}{g_n} = \theta_1' - i\alpha_0/2 - 2\tanh\left(\frac{P_2}{2}\right), \tag{4.4}$$

$$\lim_{t\to-\infty} \frac{f_n}{g_n} = \theta_1' - i\alpha_0/2, \tag{4.5}$$

and in the reference frame of the soliton we have also no effect.

5. Conclusions

To summarise, we have shown that rational nonsingular solutions both real and complex exist even for differential-difference systems. Thus, the system (2.1) possesses both solitons and nonsingular $(1+1)$-dimensional discrete rational solutions decaying to zero as $O(n^{-2})$ and their character being real or complex, depending on the direction of propagation. So, although the system is reducible to a classical Ablowitz–Ladik system, the nonvanishing boundary conditions are essential for the existence of these nonsingular rational solutions.

The dynamics of these rational solutions was discussed and we have shown that there is no phase shift after their collisions, which is specific for the $(2+1)$-dimensional continuous "lump" solutions. Unfortunately we do not know any condition of existence for this type of solutions and why they occur in $(1+1)$-dimensional systems. Its continuous limit could be viewed as a coupled pair of mKdV equations with nonzero boundary conditions. One can see that the form of 1-rational solution (the real one) has the same structure as the rational solution of mKdV given by Ono [**12**] and Ablowitz–Satsuma [**2**] and thus the continuous limit does not affects it drastically. Probably a setting up of this type of solutions in an Inverse Scattering framework will clarify much more about their form, dynamics, stability.

References

1. M. J. Ablowitz and J. Ladik, *Nonlinear differential-difference equations*, J. Math. Phys. **16** (1975), 598–603.

2. M. J. Ablowitz and J. Satsuma, *Solitons and rational solutions of nonlinear evolution equations*, J. Math. Phys. **19** (1978), no. 10, 2180–2186; J. Satsuma and M. J. Ablowitz, *Two-dimensional lumps in nonlinear dispersive systems*, J. Math. Phys. **20** (1979), no. 7, 1496–1503.
3. M. J. Ablowitz and H. Segur, *Solitons and the inverse scattering transform*, SIAM, Philadelphia, 1981.
4. H. Airault, H. P. McKean, and J. Moser, *Rational and elliptic solution of* KdV *equation and related many-body problems*, Comm. Pure Appl. Math. **30** (1977), 95–198.
5. A. S. Cârstea and D. Grecu, *On a class of rational and mixed rational-soliton solutions for the Toda lattice*, Progr. Theoret. Phys. **96** (1996), 29–36.
6. H. C. Freeman, *Soliton interactions in two dimensions*, Adv. Appl. Mech. **20** (1980), 1–37.
7. D. Grecu and V. Cionga, *On the mixed rational-soliton solutions of the KdV equation*, Rev. Roumaine Phys. **33** (1988), 3–10.
8. R. Hirota and J. Satsuma, *A variety of nonlinear network equations generated from the Bäcklund transformation for the Toda lattice*, Progr. Theoret. Phys. Suppl. **59** (1976), 64–100.
9. X. B. Hu and P. Clarkson, *Rational solutions of a differential-difference* KdV *equation, the Toda equation and the discrete* KdV *equation*, J. Phys. A **28** (1995), 5009–5016.
10. B. D. Konopelchenko, *Introduction to multidimensional integrable systems*, Plenum Press, New York, 1990.
11. J. J. C. Nimmo, *Soliton solution of three differential-difference equations in Wronskian form*, Phys. Lett. A **99** (1983), 281–286.
12. H. Ono, *Algebraic soliton of the modified Korteweg-de Vries equation*, J. Phys. Soc. Japan **41** (1976), 1817–1818.
13. C. Rogers and W. Shadwick, *Bäcklund transformations and their applications*, Math. Sci. Eng., vol. 161, Academic Press, Inc., New York, 1982.
14. M. Wadati, *Transformation theories for nonlinear discrete systems*, Progr. Theoret. Phys. Suppl. **59** (1976), 36–63.
15. ———, *Generalized matrix form of the inverse scattering method*, Solitons (R. Bullough and P. J. Caudrey, eds.), Springer-Verlag, Berlin, 1980, pp. 287–299.

DEPARTMENT OF THEORETICAL PHYSICS, INSTITUTE OF ATOMIC, PHYSICS, P.O. BOX MG-6, BUCHAREST, ROMANIA.

E-mail address: `acarst@theor1.theory.nipne.ro`

Centre de Recherches Mathématiques
CRM Proceedings and Lecture Notes
Volume **25**, 2000

Monodromy Data for the Discrete First Painlevé Hierarchy

Clio Cresswell

Abstract. The discrete first Painlevé equation (dP_I) is a difference equation which has the classical first Painlevé equation (P_I) as a continuum limit. dP_I is believed to be integrable because it is the discrete isomonodromy condition for an associated (single-valued) linear problem. An infinite hierarchy of integrable difference equations has been derived from the linear deformation problem associated with dP_I. This dP_I hierarchy was generated by difference recursion operators. Linear deformation problems may be used to solve the isomonodromy conditions (for a large class of initial data). The first step is to isolate the monodromy data. The essential monodromy data required to find a solution plays the role of conserved quantities. In this paper, we derive the essential monodromy data for each member of the dP_I hierarchy by using the recursion operators. The number of essential monodromy data for each equation in the hierarchy and the order of each equation are equivalent. This provides further evidence that each equation in the hierarchy is a distinct integrable difference equation.

1. Introduction

The discrete first Painlevé equation (dP_I)

$$x_{n+1} + x_n + x_{n-1} = \frac{C_1 + C_2 n}{x_n} + C_3, \tag{1.1}$$

is believed to be an integrable difference equation because it is the isomonodromy condition for an associated (single-valued) linear system of equations [**15**]. An infinite hierarchy of nonlinear difference equations based on this linear system has also been derived [**2**]. In this paper, we find the essential monodromy data required to solve each equation in the hierarchy. The aim is to provide evidence that each equation is a distinct integrable difference equation.

Equation (1.1) is known as dP_I, because the scaled continuum limit $x_n = 1 + h^2 u(t)$, $t = nh$, with $C_1 = -3 + r_1 h^2$, $C_2 = r_2 h^5$, $C_3 = 3 - C_1 - r_3 h^4$, where the r_i ($i = 1, 2, 3$) are constants, yields a scaled and translated version of the classical P_I

$$u'' = 6u^2 + t,$$

2000 *Mathematics Subject Classification.* Primary: 39A10; Secondary: 34A34.
The research reported here was supported by the Australian Research Council. The author thanks Dr. N. Joshi for helpful conversations.
This is the final form of the paper.

as $h \to 0$.

P_{I} is the simplest of six well known nonlinear second-order ordinary differential equations (ODEs) in the complex plane called the Painlevé equations. Their characteristic property that all movable singularities of all solutions are poles is called the Painlevé property. Painlevé [**18**], Gambier [**9**], and Fuchs [**8**] identified them (under some mild conditions) as the only such equations with the Painlevé property whose general solutions are new transcendental functions. A discrete version of the Painlevé property called the singularity confinement property has also been proposed by Grammaticos, Ramani *et al.* [**10, 19**]. dP_{I} possesses this property.

P_{I} is believed to be integrable because it is the isomonodromy condition for an associated (single-valued) linear system of differential equations [**17**]. A hierarchy of integrable ODEs may be deduced from this linear system [**3**]. The linear system may be used to solve P_{I} for a large class of initial data [**7, 16**]. A complete description of the method by which to find solutions for nonlinear ODEs that are such isomonodromy conditions may be found in [**4, 11**]. The first step involves finding the monodromy data. Some details of this process are presented below.

The linear system associated with an integrable ODE generating $u(t)$ has the form

$$\frac{\partial \phi}{\partial \lambda} = A(\lambda, t)\phi, \quad \frac{\partial \phi}{\partial t} = B(\lambda, t)\phi,$$

where $\phi(\lambda, t)$ is an m component vector and A, B are $m \times m$ matrices with rational entries in λ dependent on $u(t)$ (and its derivatives). The isomonodromy method of solution considers the first equation in this linear system. Multivaluedness in the fundamental solution matrix Φ is determined by the singularities of A. When Φ is analytically continued along some contour that surrounds a singularity of A, the resulting fundamental solution matrix may be expressed as $M\Phi$, where M is referred to as the monodromy matrix and its trace and determinant as monodromy data.

Consider the first equation in the above linear system. If it possesses an irregular singularity (determined by a singularity of A), the asymptotic representation of Φ (in some neighbourhood of the singularity) may exhibit Stokes phenomenon [**20**]. This means that as Φ is analytically continued around the singularity, its asymptotic representation jumps discontinuously through sectors. The formal solution matrix in a particular sector will then be connected to the formal solution matrix from the previous sector by a *Stokes matrix.* The entries of the Stokes matrices are called the Stokes multipliers.

Compatibility of the linear problem implies that the monodromy data remains invariant (with respect to t) [**4, 12–14**]. For the isomonodromy method of solution, we loosely refer to the set consisting of the entries of the monodromy and Stokes matrices as the monodromy data since this set may be over-determined. That is, through the use of properties innate in the specific deformation problem (e.g. symmetries), it may be possible to reduce the set to essential monodromy data. The reduced set of monodromy data forms a complete set of first integrals of motion for the associated integrable compatibility condition.

Joshi *et al.* [**15**] showed that compatibility of linear differential-difference systems of equations is equivalent to the invariance of the monodromy data under discrete iteration. Monodromy data associated with dP_{I} has been studied in connection with 'similarity' solutions of the Kac–Moerbeke equation [**5, 6**]. In this paper, we study the monodromy data needed to solve the equations in the dP_{I} hierarchy.

The plan of the paper is as follows. In Section 2, we recall the derivation of the dP_{I} hierarchy in operator form [**2**]. We then write the associated linear problem in a form suitable for finding the monodromy data. In Section 3, we find the full set of monodromy data. We therefore examine solutions of the linear problem in the neighbourhood of the regular and irregular singular points and how these may be analytically continued throughout the complex plane. We conclude in Section 4, by showing that the full set of monodromy data is over determined. The number of data which is essential, is equivalent to the order of each equation in the hierarchy.

In the following, all occurrences of c_i denote constants.

2. The dP_{I} Hierarchy and Associated Linear Problem

In this section, we first recall the derivation of the dP_{I} hierarchy based on the linear problem associated with dP_{I} [**2**]. We then write the associated linear problem in a form suitable for finding the monodromy data.

The linear problem associated with dP_{I} [**15**] is

$$x_n\phi_{n+1} = \lambda\phi_n - \phi_{n-1}, \tag{2.1}$$

$$\frac{\partial\phi_n}{\partial\lambda} = a_n\phi_{n+1} + b_n\phi_n, \tag{2.2}$$

where ϕ, x, a, b depend on a discrete variable n; ϕ, a, b depend on the continuous variable λ, and a, b are rational in λ.

The compatibility conditions of Eqs. (2.1) and (2.2) are

$$b_{n+1} - b_{n-1} + \lambda\left(\frac{a_{n+1}}{x_{n+1}} - \frac{a_n}{x_n}\right) = 0, \tag{2.3}$$

$$\frac{\lambda^2}{x_n}\left(\frac{a_{n+1}}{x_{n+1}} - \frac{a_n}{x_n}\right) + \frac{\lambda}{x_n}(b_{n+1} - b_n) - \frac{1}{x_n}\left(x_n\frac{a_{n+1}}{x_{n+1}} - a_{n-1} + 1\right) = 0. \tag{2.4}$$

Define $p_n := a_n/x_n$ and $q_n := b_n + b_{n-1}$. Then b can be eliminated from Eq. (2.4) by using Eq. (2.3). The result is

$$-\lambda^2(p_{n+1} - p_n) + x_{n+1}p_{n+2} + x_np_{n+1} - x_np_n - x_{n-1}p_{n-1} + 2 = 0,$$

which may be written as

$$\begin{aligned}&-\lambda^2(\Delta^2p_n + \Delta p_n) + 3\Delta(x_n\Delta p_n) + 4\Delta(x_np_n)\\ &-2(\Delta x_n)\cdot(\Delta p_n + p_n) + \Delta^2x_n(\Delta p_n) + \Delta^2(x_np_n) + 2 = 0,\end{aligned}$$

where $\Delta p_n := p_{n+1} - p_n$, or

$$\left[\lambda^2(\Delta^2 + \Delta) - \mathcal{J}_n\right]p_n = 2, \tag{2.5}$$

where

$$\mathcal{J}_n := 3\Delta(x_n\Delta) + 4\Delta(x_n) - 2(\Delta x_n)\cdot(\Delta + 1) + \Delta^2x_n(\Delta) + \Delta^2(x_n),$$

(the subscript n refers to the subscript of the x iterate).

A nontrivial solution of Eq. (2.5) may be found by taking p_n to be the even polynomial

$$p_n = \sum_{k=0}^{m} P_{2k,n}\lambda^{2k}.$$

Substituting this expression into Eq. (2.5) and solving for the coefficients $P_{2k,n}$, we get

$$P_{2m,n} = c_{m+2}, \quad P_{2k-2,n} = \mathcal{S}_n P_{2k,n}, \quad \text{for } 1 \le k \le m,$$

with the compatibility condition $\mathcal{J}_n P_{0,n} = -2$, where $\mathcal{S}_n = (\triangle^2 + \triangle)^{-1}\mathcal{J}_n$. Recursive substitution of the $P_{k,n}$ leads to the compatibility condition

$$\mathcal{J}_n \mathcal{S}_n^m c_{m+2} = -2, \quad m \in \mathbb{N}. \tag{2.6}$$

Solving Eq. (2.6) for increasing m, yields each equation in the hierarchy. We refer to these as $\mathrm{d}_{2m}P_{\mathrm{I}}$, where $2m$ is the order of each equation.

Below we list the first three equations in the hierarchy with their continuum limits.

- $m = 0$ yields the trivial linear nonautonomous equation, $\mathrm{d}_0 P_{\mathrm{I}}$:

$$x_n c_2 = c_1 - c(-1)^n - n.$$

- $m = 1$ yields the second-order equation, $\mathrm{d}_2 P_{\mathrm{I}}$:

$$x_n c_3(x_{n-1} + x_n + x_{n+1}) + x_n c_2 = c_1 - c(-1)^n - n. \tag{2.7}$$

  For $c = 0$, we recover $\mathrm{d}P_{\mathrm{I}}$ with P_{I} as a continuum limit. For $c \neq 0$, Eq. (2.7) has P_{II} as a continuum limit [**2**].
- $m = 2$ yields the fourth-order equation, $\mathrm{d}_4 P_{\mathrm{I}}$:

$$\begin{aligned} x_n c_4(x_{n+1}x_{n+2} + x_{n+1}^2 + 2x_n x_{n+1} + x_{n-1}x_{n-2} + x_{n-1}^2 + 2x_n x_{n-1} \\ + x_n^2 + x_{n-1}x_{n+1}) + x_n c_3(x_{n-1} + x_n + x_{n+1}) + x_n c_2 \\ = c_1 - c(-1)^n - n. \end{aligned} \tag{2.8}$$

  For $c = 0$, Eq. (2.8) has the fourth-order member of the P_{I} hierarchy as a continuum limit. For $c \neq 0$, Eq. (2.8) has the fourth-order member of the P_{II} hierarchy as a continuum limit [**2**].

The linear system (2.1)–(2.2) is second order. In order to study the solutions of Eq. (2.2) we rewrite the problem as a two dimensional first order system

$$\frac{\partial}{\partial \lambda}\begin{pmatrix}\phi_{1,n}\\ \phi_{2,n}\end{pmatrix} = \begin{pmatrix} b_n & p_n x_n \\ -p_{n+1} & p_{n+1}\lambda + b_{n+1}\end{pmatrix}\begin{pmatrix}\phi_{1,n}\\ \phi_{2,n}\end{pmatrix}, \tag{2.9}$$

$$\begin{pmatrix}\phi_{1,n+1}\\ \phi_{2,n+1}\end{pmatrix} = \begin{pmatrix} 0 & 1 \\ -1/x_{n+1} & \lambda/x_{n+1}\end{pmatrix}\begin{pmatrix}\phi_{1,n}\\ \phi_{2,n}\end{pmatrix}, \tag{2.10}$$

where $\phi_{1,n} = \phi_n$ and $\phi_{2,n} = \phi_{n+1}$. We know that this system is compatible with $\mathrm{d}_{2m}P_{\mathrm{I}}$ as compatibility condition when

$$p_n = \sum_{k=0}^{m-1} P_{2k,n}\lambda^{2k} + c_{m+2}\lambda^{2m} = \sum_{k=0}^{m} \lambda^{2k}\mathcal{S}_n^{m-k} c_{m+2}. \tag{2.11}$$

The explicit dependence of b_n on λ needs to be found. Using (2.11) to solve the second order difference Eq. (2.3) for b_n, we find

$$b_n = \frac{c}{\lambda}(-1)^n - \sum_{k=1}^{m} \lambda^{2k-1}\mathcal{T}_n \mathcal{S}_n^{m-k} c_{m+2}, \tag{2.12}$$

where $\mathcal{T}_n = x_n(\triangle) + x_n + \triangle^{-1}(x_n\triangle)$ and c is a constant. The dependence of the constants of integration on λ is found by ensuring the compatibility of (2.9) and (2.10). Without loss of generality, one of the constants of integration has been taken to be zero.

From now on referring to Eq. (2.9) assumes p_n and b_n to be defined as in (2.11) and (2.12). By Eqs. (2.11) and (2.12), we see that the linear deformation Eq. (2.9) has two singularities; a regular singularity at $\lambda = 0$ and an irregular singularity at $\lambda = \infty$. We proceed by finding solutions in neighbourhoods of the singularities in order to isolate the monodromy data.

3. Solutions Near Singular Points

Solutions of Eq. (2.9) in neighbourhoods of its singular points $\lambda = 0$, $\lambda = \infty$, and how these may be analytically continued throughout the complex plane are presented.

In a neighbourhood of the regular singularity $\lambda = 0$, the standard Frobenius method [**1**]

$$\phi_{1,n}(\lambda) = \left[\kappa_1 \lambda^{c(-1)^n} + \kappa_2 \lambda^{1-c(-1)^n}\right]\left(1 + O(\lambda^2)\right), \tag{3.1}$$

$$\phi_{2,n}(\lambda) = \left[\kappa_3 \lambda^{1+c(-1)^n} + \kappa_4 \lambda^{-c(-1)^n}\right]\left(1 + O(\lambda^2)\right), \tag{3.2}$$

with

$$\kappa_2 = \frac{\kappa_4 x_n S_n^m c_{m+2}}{1 - 2c(-1)^n} \quad \text{and} \quad \kappa_3 = \frac{-\kappa_1 S_{n+1}^m c_{m+2}}{1 + 2c(-1)^n},$$

where the κ_i $(i = 1, \ldots, 4)$ are constant in λ, however may still depend on n as Eq. (2.10) must also be satisfied. In fact, we have $\kappa_{4,n} = \kappa_{1,n+1}$ and $x_{n+1}\kappa_{1,n+2} = -\kappa_{1,n}$.

We have restricted ourselves to the case $n \neq 1/(i\pi)\log\big(\mp j/(2c)\big)$ $\big(j \in \mathbb{N}/\{0\}\big)$ to avoid the break down of the Frobenius method. This occurs when the relations specifying the coefficients of the series solution become ill-defined. In such a case, a standard modification to the possible solutions is to introduce logarithmic terms [**1**]. Subsequent analysis is similar with the modified solution.

In a neighbourhood of the irregular singularity $\lambda = \infty$, the notation $(\phi_{1,n}, \phi_{2,n}) = (\psi_{1,n}, \psi_{2,n})$ is introduced. Formal asymptotic analysis [**1**] gives

$$\psi_{1,n}(\lambda) \sim \left[\kappa_5 \lambda^{n-c_1-1} e^{\vartheta} + \kappa_6 \lambda^{c_1-n}\right]\left(1 + O(\lambda^{-2})\right), \tag{3.3}$$

$$\psi_{2,n}(\lambda) \sim \left[\kappa_7 \lambda^{n-c_1} e^{\vartheta} + \kappa_8 \lambda^{c_1-n-1}\right]\left(1 + O(\lambda^{-2})\right), \tag{3.4}$$

where

$$\vartheta = \sum_{j=1}^{m+1} \frac{c_{j+1}}{2j} \lambda^{2j},$$

the κ_i $(i = 5, \ldots, 8)$ are constant in λ and the c_1 term in the exponent of λ appears after simplification using the equations in the hierarchy.

This solution exhibits Stokes phenomenon. If $\Re e(\vartheta) > 0$ the exponential term is dominant, however if $\Re e(\vartheta) < 0$ it becomes recessive. In order that (2.9) and (2.10) be satisfied we then also have that: if $\Re e(\vartheta) > 0$, $\kappa_{5,n} = x_n \kappa_{7,n}$, $\kappa_{7,n} = x_{n+1}\kappa_{7,n+1}$, and if $\Re e(\vartheta) < 0$, $\kappa_{6,n} = \kappa_{8,n} = \text{constant}$.

Determining where $\Re e(\vartheta) = 0$ determines where the dominance of a particular term alters. In some neighbourhood near $\lambda = \infty$ this is equivalent to determining where $\Re e(c_{m+2}\lambda^{2(m+1)}) = 0$. Therefore there exists $4(m+1)$ sectors

$$S_k = \left\{\lambda \in C\colon \frac{\pi}{4(m+1)}(2k-3) < \arg(\lambda) < \frac{\pi}{4(m+1)}(2k-1)\right\},$$

where $k = 1, \ldots, 4(m+1)$. Assume $c_{m+2} \in \mathbb{R}^+$. The exponential term will therefore be dominant in S_{2l-1} and recessive in S_{2l} where $l = 1, \ldots, 2(m+1)$. $c_{m+2} \in \mathbb{C}$ and $c_{m+2} \in \mathbb{R}^-$ are also valid assumptions but only change the nature of the dominant term in each sector. Subsequent analysis is similar.

Due to Stokes phenomenon, it is not possible to find the solution in a particular sector by analytic continuation of the solution in a neighbouring one. Stokes multiplier matrices are defined to take account of the dominance altering.

Let $\Psi_n^{(k)}(\lambda)$ be a fundamental solution in S_k, then

$$\Psi_n^{(2l)}(\lambda) = \Psi_n^{(2l-1)}(\lambda) \begin{pmatrix} 1 & 0 \\ s_l & 1 \end{pmatrix}, \tag{3.5}$$

in $\pi(4l-3)/(4m+4) \leq \arg(\lambda) < \pi(4l-1)/(4m+4)$, and

$$\Psi_n^{(2l+1)}(\lambda) = \Psi_n^{(2l)}(\lambda) \begin{pmatrix} 1 & r_l \\ 0 & 1 \end{pmatrix}, \tag{3.6}$$

in $\pi(4l-1)/(4m+4) \leq \arg(\lambda) < \pi(4l+1)/(4m+4)$, for $l = 1, \ldots, 2(m+1)$, where the s_l and r_l are the Stokes multipliers.

One further solution behaviour remains outstanding. We wish to know how the solution of Eq. (2.9) changes as $\lambda = \infty$ is encircled. On the Riemann sphere, a closed loop around $\lambda = \infty$ is deformable to one around $\lambda = 0$. $\Psi_n^{(1)}(\lambda)$ and $\Phi_n(\lambda)$ are therefore related by a constant matrix Γ (the connection matrix),

$$\Psi_n^{(1)}(\lambda) = \Phi_n(\lambda)\Gamma, \quad \text{where } \Gamma = \begin{pmatrix} \gamma_1 & \gamma_2 \\ \gamma_3 & \gamma_4 \end{pmatrix}. \tag{3.7}$$

4. The Essential Monodromy Data

From the previous section, we have the full set of $4(m+2)$ monodromy data $\{s_l, r_l, \gamma_1, \gamma_2, \gamma_3, \gamma_4\}$. The aim is to show that each compatibility condition $\mathrm{d}_{2m}P_{\mathrm{I}}$ of (2.9) and (2.10) is an integrable equation with $2m$ conserved quantities. Only $2m$ of the monodromy data should therefore be essential in order to construct a solution via the isomonodromy method. That is, specifying $2m$ of the monodromy data should be sufficient in order to deduce the full set and follow through with the problem.

To show this, we make use of a symmetry present in the system (2.9)–(2.10). If $\Phi_n(\lambda)$ is a solution then so is $\sigma\Phi_n(-\lambda)$, where

$$\sigma = \begin{pmatrix} 1 & 0 \\ 0 & -1 \end{pmatrix}. \tag{4.1}$$

The solutions (3.1), (3.2) and (3.3), (3.4) give

$$\sigma\Phi_n(\lambda e^{-i\pi}) = \Phi_n(\lambda)\sigma \begin{pmatrix} e^{-i\pi c(-1)^n} & 0 \\ 0 & e^{i\pi c(-1)^n} \end{pmatrix}, \tag{4.2}$$

and

$$\sigma\Psi_n^{(k)}(\lambda e^{-i\pi}) = \Psi_n^{(k+2(m+1))}(\lambda)\sigma \begin{pmatrix} e^{-i\pi(n-c_1-1)} & 0 \\ 0 & e^{-i\pi(c_1-n-1)} \end{pmatrix}. \tag{4.3}$$

For $k = 2l$, Eq. (4.3) becomes

$$\sigma\Psi_n^{(2l)}(\lambda e^{-i\pi}) = \Psi_n^{(2(l+m+1))}(\lambda)\sigma \begin{pmatrix} e^{-i\pi(n-c_1-1)} & 0 \\ 0 & e^{-i\pi(c_1-n-1)} \end{pmatrix}.$$

Using (3.5),

$$\sigma\Psi_n^{(2l-1)}(\lambda e^{-i\pi})\begin{pmatrix} 1 & 0 \\ s_l & 1 \end{pmatrix}$$
$$= \Psi_n^{(2(l+m)+1)}(\lambda)\begin{pmatrix} 1 & 0 \\ s_{l+m+1} & 1 \end{pmatrix}\sigma\begin{pmatrix} e^{-i\pi(n-c_1-1)} & 0 \\ 0 & e^{-i\pi(c_1-n-1)} \end{pmatrix}.$$

From Eq. (4.3) evaluated for $k = 2l-1$, we then have

$$\Psi_n^{(2(l+m)+1)}(\lambda)\sigma\begin{pmatrix} e^{-i\pi(n-c_1-1)} & 0 \\ 0 & e^{-i\pi(c_1-n-1)} \end{pmatrix}\begin{pmatrix} 1 & 0 \\ s_l & 1 \end{pmatrix}$$
$$= \Psi_n^{(2(l+m)+1)}(\lambda)\begin{pmatrix} 1 & 0 \\ s_{l+m+1} & 1 \end{pmatrix}\sigma\begin{pmatrix} e^{-i\pi(n-c_1-1)} & 0 \\ 0 & e^{-i\pi(c_1-n-1)} \end{pmatrix},$$

which implies $s_{l+m+1} = -s_l e^{-2i\pi(c_1-n)}$. A similar calculation beginning with Eq. (4.3) evaluated at $k = 2l+1$ gives $r_{l+m+1} = -r_l e^{-2i\pi(n-c_1)}$. Therefore half of the Stokes multipliers may be deduced from the other. We are left so far with an essential set of $2(m+1)+4$ monodromy data.

From Eq. (3.7) $\Psi_n^{(1)}(\lambda e^{-i\pi}) = \Phi_n(\lambda e^{-i\pi})\Gamma$. $\Psi_n^{(1)}(\lambda e^{-i\pi})$ and $\Phi_n(\lambda e^{-i\pi})$ may be written in terms of $\Psi_n^{(1)}(\lambda)$ and $\Phi_n(\lambda)$. Using Eq. (4.3) at $k=1$ and then Eqs. (3.5), (3.6) repeatedly,

$$\sigma\Psi_n^{(1)}(\lambda e^{-i\pi}) = \Psi_n^{(1+2(m+1))}(\lambda)\sigma\begin{pmatrix} e^{-i\pi(n-c_1-1)} & 0 \\ 0 & e^{-i\pi(c_1-n-1)} \end{pmatrix},$$
$$= \Psi_n^{(1)}(\lambda)P\sigma\begin{pmatrix} e^{-i\pi(n-c_1-1)} & 0 \\ 0 & e^{-i\pi(c_1-n-1)} \end{pmatrix},$$

where

$$P = \begin{pmatrix} p_1 & p_2 \\ p_3 & p_4 \end{pmatrix} = \begin{pmatrix} 1 & 0 \\ s_1 & 1 \end{pmatrix}\begin{pmatrix} 1 & r_1 \\ 0 & 1 \end{pmatrix}\cdots\begin{pmatrix} 1 & 0 \\ s_{m+1} & 1 \end{pmatrix}\begin{pmatrix} 1 & r_{m+1} \\ 0 & 1 \end{pmatrix}.$$

The entries in P are linearly independent polynomials in s_l and r_l. By multiplying both sides of Eq. (4.2) by Γ we have an expression for $\sigma\Phi_n(\lambda e^{-i\pi})\Gamma$. Therefore,

$$\text{(4.4)}\quad \Phi_n(\lambda)\Gamma P\sigma\begin{pmatrix} e^{-i\pi(n-c_1-1)} & 0 \\ 0 & e^{-i\pi(c_1-n-1)} \end{pmatrix}$$
$$= \Phi_n(\lambda)\sigma\begin{pmatrix} e^{-i\pi c(-1)^n} & 0 \\ 0 & e^{i\pi c(-1)^n} \end{pmatrix}\Gamma,$$

where $\Psi_n^{(1)}(\lambda)$ has been replaced by $\Phi_n(\lambda)\Gamma$.

Four relations come from Eq. (4.4):

$$\text{(4.5)}\quad \gamma_1 e^{-i\pi c(-1)^n} = (\gamma_1 p_1 + \gamma_2 p_3)e^{-i\pi(n-c_1-1)},$$

$$\text{(4.6)}\quad \gamma_2 e^{-i\pi c(-1)^n} = -(\gamma_1 p_2 + \gamma_2 p_4)e^{-i\pi(c_1-n-1)},$$

$$\text{(4.7)}\quad \gamma_3 e^{i\pi c(-1)^n} = -(\gamma_3 p_1 + \gamma_4 p_3)e^{-i\pi(n-c_1-1)},$$

$$\text{(4.8)}\quad \gamma_4 e^{i\pi c(-1)^n} = (\gamma_3 p_2 + \gamma_4 p_4)e^{-i\pi(c_1-n-1)}.$$

Eqs. (4.5) and (4.6) form a linear homogeneous system of equations in γ_1, γ_2. For $\gamma_1 \neq 0$, $\gamma_2 \neq 0$, the determinant of the coefficient matrix must vanish,

$$\text{(4.9)}\quad \det(P) + p_1 e^{-i\pi(n-c_1-1+c(-1)^n)} - p_4 e^{-i\pi(c_1-n-1+c(-1)^n)} - e^{-2i\pi c(-1)^n} = 0.$$

Similarly, Eqs. (4.7) and (4.8) form a linear homogeneous system of equations in γ_3, γ_4, and $\gamma_3 \neq 0$, $\gamma_4 \neq 0$, imply

$$\det(P) - p_1 e^{-i\pi(n-c_1-1-c(-1)^n)} + p_4 e^{-i\pi(c_1-n-1-c(-1)^n)} - e^{2i\pi c(-1)^n} = 0. \tag{4.10}$$

Eqs. (4.9) and (4.10) are independent. Once $2(m+1)-2$ Stokes multipliers are specified, the remaining two are given by these equations. The γ_i $(i = 1, \ldots, 4)$ may then be found from Eqs. (4.5) through to (4.8). Hence, the number of essential monodromy data in order to fully describe the problem is $2m$. This coincides with the order of the m'th equation in the $\mathrm{d}_{2m}P_{\mathrm{I}}$ hierarchy.

References

1. C. M. Bender and S. A. Orszag, *Advanced mathematical methods for scientists and engineers*, McGraw-Hill Internat., Auckland, Singapore, international student edition, 1984.
2. C. Cresswell and N. Joshi, *The discrete first, second and thirty-fourth Painlevé hierarchies*, J. Phys. A **32** (1999), no. 4, 655–669.
3. ———, *The discrete Painlevé* I. *Hierarchy*, Symmetries and integrability of difference equations (P. A. Clarkson and F. W. Nijhoff, eds.), London Mathematical Society, Lecture Note Series, Cambridge Univ. Press, vol. 25, 1999, pp. 197–205.
4. H. Flaschka and A. C. Newell, *Monodromy- and spectrum-preserving deformations* I, Commun. Math. Phys. **76** (1980), 65–116.
5. A. S. Fokas, A. R. Its, and A. V. Kitaev, *Discrete Painlevé equations and their appearance in quantum gravity*, Commun. Math. Phys. **142** (1991), 313–344.
6. ———, *The isomonodromy approach to matrix models in* 2D *quantum gravity*, Commun. Math. Phys. **147** (1992), 395–430.
7. A. S. Fokas, U. Mugan, and X. Zhou, *On the solvability of Painlevé.* I; III; *and* V, Inverse Problems **8** (1992), 757–785.
8. R. Fuchs, *Über lineare homogene differentialgleichungen zweiter or ordnung mit drei im endlich gelegene wesentlich singulären stellen*, Math. Annalen **63** (1907), 301–321.
9. B. Gambier, *Sur les équations différentielles du second ordre et du premier degré dont l'intégral général est à points critiques fixes*, Acta Math. **33** (1910), 1–55.
10. B. Grammaticos, A. Ramani, and V. Papageorgiou, *Do integrable mappings have the Painlevé property?*, Phys. Rev. Lett. **67** (1991), 1825–1828.
11. A. R. Its and V. Y. Novokshenov, *The isomonodromic deformation method in the theory of Painlevé equations*, Lecture Notes in Mathematics, vol. 1191, Springer-Verlag, Berlin-New York, 1986.
12. M. Jimbo and T. Miwa, *Monodromy preserving deformation of linear ordinary differential equations with rational coefficients.* II, Physica D **2** (1981), 407–448.
13. ———, *Monodromy preserving deformation of linear ordinary differential equations with rational coefficients.* III, Physica D **4** (1981), 26–46.
14. M. Jimbo, T. Miwa, and K. Ueno, *Monodromy preserving deformation of linear ordinary differential equations with rational coefficients.* I, Physica D **2** (1981), 306–352.
15. N. Joshi, D. Burtonclay, and R. Halburd, *Nonlinear nonautonomous discrete dynamical systems from a general discrete isomonodromy problem*, Lett. Math. Phys. **26** (1992), 123–131.
16. A. A. Kapaev and A. V. Kitaev, *Connection formulae for the first Painlevé transcendent in the complex domain*, Lett. Math. Phys. **27** (1993), 243–252.
17. K. Okamoto, *Isomonodromic deformation and Painlevé equations, and the Garnier system*, J. Fac. Sci. Univ. Tokyo Sect. IA Math. **33** (1986), 575–618.
18. P. Painlevé, *Sur les équations différentielles du second ordre et d'ordre supérieur dont l'intégrale générale est uniforme*, Acta Math. **25** (1902), 1–85.
19. A. Ramani, B. Grammaticos, and J. Hietarinta, *Discrete versions of the Painlevé equations*, Phys. Rev. Lett. **67** (1991), 1829–1832.
20. W. Wasow, *Asymptotic expansions for ordinary differential equations: Trends and problems*, vol. XIV, Interscience Publishers John Wiley & Sons, Inc., New York-London-Sydney 1965.

School of Mathematics, The University of New South Wales, Sydney NSW 2052, Australia.

Current address: Department of Pure Mathematics, University of Adelaide, Adelaide SA 5005, Australia.

E-mail address: clio@maths.unsw.edu.au

Centre de Recherches Mathématiques
CRM Proceedings and Lecture Notes
Volume **25**, 2000

Differential Identitites, Generalized Polynomials and Applications in Physics and Mathematics

G. Dattoli

ABSTRACT. We show that a wide class of differential identities can be derived by exploiting a method involving generalized forms of classical polynomials. We also discuss how the combined use of these identities and of exponential operational rules may offer an efficient tool to solve problems in classical and quantum mechanics.

1. Introduction

The Kampé de Fériet generalization of the Hermite polynomials reads[1]

$$H_n(x,y) = n! \sum_{r=0}^{[n/2]} \frac{y^r x^{n-2r}}{r!(n-2r)!}. \tag{1}$$

They can be derived from the generating function

$$\sum_{n=0}^{\infty} \frac{t^n}{n!} H_n(x,y) = e^{xt+yt^2}, \tag{2}$$

reduce to the ordinary $He_n(x)$ for $y = -1/2$ and can be exploited to state identities of the type

$$\left(\alpha \frac{d}{dx} + \beta x\right)^n = \sum_{s=0}^{n} \binom{n}{s} \alpha^s H_{n-s}\left(\beta x, \frac{1}{2}\alpha\beta\right) \frac{d^s}{dx^s}. \tag{3}$$

The proof of the validity of the above relation can be carried out by exploiting the method of the generating function and by noting that

$$\sum_{n=0}^{\infty} \frac{t^n}{n!} \left(\alpha \frac{d}{dx} + \beta x\right)^n = e^{(\alpha(d/dx)+\beta x)t} = e^{\beta x t + \alpha\beta t^2/2} e^{\alpha t (d/dx)}. \tag{4}$$

2000 *Mathematics Subject Classification.* Primary: 16R50; Secondary: 00A79.

The author expresses his sincere appreciation to Drs. S. Lorenzutta, A. Torre and C. Cesarano for a critical reading of the manuscript. It is also a pleasure to recognize the kind interest of Prof. D. Levi who stimulated the completion of the present work.

This is the final form of the paper.

[1]For a review, see e.g. [**7**]

The last term on the r.h.s. of (4) is a consequence of the Weyl decoupling identity, involving the two elements $\hat{A}$, $\widehat{B}$, of the ordinary Heisenberg algebra, namely

$$e^{\hat{A}+\widehat{B}} = e^{\hat{A}} e \widehat{B} e^{-k/2}, \quad [\hat{A}, \widehat{B}] = k, \ k \in \mathcal{C} \tag{5}$$

Equation (3) reduces to the Burchnall identity for $\beta = 2$, $\alpha = -1$ [**2**].

The interplay between differential relations and generalized special functions is particularly useful in application;, in this paper we will discuss generalizations of (3), which require further classes of polynomials. Before proceeding we will premise two decoupling theorems, which will be exploited in the following [**6**]

$$\exp(\hat{A} + \widehat{B}) = \exp\left[\frac{1 - e^{-m}}{m} \hat{A}\right] \cdot \exp(\widehat{B}), \quad [\hat{A}, \widehat{B}] = m\hat{A}, \ m \in \mathcal{C} \tag{6}$$

and

$$\begin{gathered} e^{\hat{A}+\widehat{B}} = e^{((m^2/12)-(m/2)\hat{A}^{1/2}+\hat{A})} e^{\widehat{B}}, \quad e^{\hat{A}+\widehat{B}} = e^{\widehat{B}} e^{(m^2/12)+(m/2)\hat{A}^{1/2}+\hat{A}}, \\ [\hat{A}, \widehat{B}] = m\hat{A}^{1/2}, \ m \in \mathcal{C}. \end{gathered} \tag{7}$$

We also note that, if $\hat{A}$ and $\widehat{B}$ satisfy the commutation brackets (3), the following Crofton-like rule [**3**] holds

$$e^{(\hat{A})^m} f(\widehat{B}) = f(\widehat{B} + mk(\hat{A})^{m-1}) e^{(\hat{A})^m}. \tag{8}$$

Along with the above identities we will exploit extended forms of Hermite polynomials, as e.g. those defined by the generating function[2]

$$e^{xt+yt^2+zt^3} = \sum_{n=0}^{\infty} \frac{t^n}{n!} H_n(x, y, z) \tag{9}$$

and linked to the $H_n(x, y)$ by

$$H_n(x, y, z) = n! \sum_{s=0}^{[n/3]} \frac{z^s H_{n-3s}(x, y)}{s!(n-3s)!}. \tag{10}$$

The above relations and the properties of generalized polynomials will be the key-points of the next sections.

2. Operational Identities and Generalized Polynomials

A direct consequence of (7) and of the definitions (9), (10) is the Theorem:

THEOREM 2.1. *The following differential identities hold for any natural number n and $\alpha \in \mathcal{C}$*

$$\begin{aligned} \left(x^2 + \alpha \frac{d}{dx}\right)^n &= \sum_{s=0}^{n} \binom{n}{s} \alpha^s H_{n-s}\left(x^2, \alpha x, \frac{\alpha^2}{3}\right) \frac{d^s}{dx^s}, \\ \left(x + \alpha \frac{d^2}{dx^2}\right)^n &= \sum_{s=0}^{n} \binom{n}{s} H_{n-s}\left(\alpha, \frac{d^2}{dx^2}, -\alpha, \frac{d}{dx}, \frac{\alpha}{3}\right) x^s. \end{aligned} \tag{11}$$

According to the second of (7), the last of (11) can also be cast in the form

$$\left(x + \alpha \frac{d^2}{dx^2}\right)^n = \sum_{s=0}^{n} \binom{n}{s} x^{n-s} H_s\left(\alpha \frac{d^2}{dx^2}, \alpha \frac{d}{dx}, \frac{\alpha}{3}\right). \tag{12}$$

[2] See e.g. [**8**]

Within the context of generalized polynomials, a role of noticeable importance is played by the Bell extension [**1**] of the ordinary Hermite, namely

$$\sum_{n=0}^{\infty} \frac{t^n}{n!} H_n(\{x\}) = e^{(\sum_{s=1}^{\infty} t^s x_s)}, \quad \{x\} = x_1, x_2, x_3, \dots. \tag{13}$$

This class of polynomials can be exploited, along with[3], to state the further differential identity

$$\left(x + \alpha x \frac{d}{dx}\right)^n = \sum_{s=0}^{n} \alpha^s \binom{n}{s} H_{n-s}(\{\beta\}) \left(x \frac{d}{dx}\right)^s, \quad \beta_n = \frac{\alpha^{n-1}}{n!} x. \tag{14}$$

It is clear that the above relations can be extended to more general structures. Consider e.g. a 3-dimensional complex algebra with an identity operator and with two elements $\hat{A}$, $\widehat{B}$ satisfying the commutation bracket in Eq. (5): in this case we get

$$(\hat{A} + \widehat{B})^n = \sum_{n=0}^{\infty} \binom{n}{s} H_{n-s}\left(\hat{A}, -\frac{k}{2}\right) \widehat{B}^s. \tag{15}$$

Let us now state some useful extensions of the above identities. We consider indeed the following generalization of the identity (3)

$$\Pi_{m,n}\left(x, \frac{d}{dx}; \alpha, \beta, \delta, \gamma\right) = \left(\beta x + \alpha \frac{d}{dx}\right)^m \left(\gamma x + \delta \frac{d}{dx}\right)^n. \tag{16}$$

By exploiting the method of the generating function we find

$$\sum_{m=0}^{\infty} \sum_{n=0}^{\infty} \frac{u^m}{m!} \frac{v^n}{n!} \Pi_{m,n}\left(x, \frac{d}{dx}; \alpha, \beta, \delta, \gamma\right) = e^{\beta x u + \alpha \beta u^2/2} e^{\alpha u (d/dx)} e^{\gamma x v + \gamma \delta v^2/2} e^{\delta v (d/dx)} \tag{17}$$

which on account of the Crofton relation yields

$$\sum_{n=0}^{\infty} \sum_{m=0}^{\infty} \frac{u^n}{n!} \frac{v^m}{m!} \Pi_{n,m}\left(x, \frac{d}{dx}; \alpha, \beta, \delta, \gamma\right) = e^{\beta x u + \alpha \beta u^2/2} e^{\gamma (x + \alpha u) v + \gamma \delta v^2/2} e^{\alpha u (d/dx)} e^{\delta v (d/dx)} \tag{18}$$

thus finally getting

$$\Pi_{m,n}\left(x, \frac{d}{dx}; \alpha, \beta, \delta, \gamma\right) = \sum_{r=0}^{m} \sum_{s=0}^{n} \binom{m}{r} \binom{n}{s} \alpha^r \delta^s H_{m-r,n-s}(x; \alpha, \beta, \delta, \gamma) \frac{d^{r+s}}{dx^{r+s}} \tag{19}$$

where $H_{m,n}(x; \alpha, \beta, \gamma, \delta)$ is a two index Hermite polynomial, defined as

$$H_{m,n}(x; \alpha, \beta, \delta, \gamma) = m!\, n! \sum_{r=0}^{\min=(m,n)} \frac{(\gamma\alpha)^q H_{m-q}(\beta x, \alpha\beta/2) H_{n-q}(\gamma x, \gamma\delta/2)}{q!(m-q)!(n-q)!}. \tag{20}$$

---

[3]See e.g. [**8**]

The outlined technique can be exploited to obtain further and more general relations as e.g.

$$\begin{aligned}(21)\quad &\left(x^2+\alpha\frac{d}{dx}\right)^m\left(x+\beta\frac{d}{dx}\right)^n\\ &=\sum_{r=0}^{m}\sum_{s=0}^{n}\binom{m}{r}\binom{n}{s}\alpha^r\beta^s H_{m-r,n-s}\left(x^2,\alpha x,\frac{\alpha^2}{3};x,\frac{1}{2}\beta\right)\frac{d^{r+s}}{dx^{r+s}},\\ &H_{m,n}\left(x^2,\alpha x,\frac{\alpha^2}{3};x,\frac{1}{2}\beta\right)\\ &=m!\,n!\sum_{q=0}^{\min(m,n)}\frac{\alpha^q H_{m-q}(x^2,\alpha x,(\alpha^2/3))H_{n-q}(x,\beta/2)}{q!(m-q)!(n-q)!}.\end{aligned}$$

The so far obtained results yield an idea of the flexibility of the method based on the use of generalized Hermite polynomials and of the wealth of identities one can obtain.

The previous identities, complemented with exponential operator expansion of the type [**1**]

$$(22)\qquad e^{\lambda\hat{A}\cdot\widehat{B}}=\sum_{n=0}^{\infty}\frac{1}{n!}\left[\frac{1-e^{-k\lambda}}{k}\right]^n\hat{A}^n\widehat{B}^n$$

which holds if $\hat{A}$, $\widehat{B}$ are elements of the Heisenberg algebra, may offer an efficient tool to solve evolution problems in classical and quantum mechanics, or to evaluate matrix elements in Fock spaces.

An elementary example is offered by the evolution operator of the quantum harmonic oscillator, which according to the identities (3) and (22) can be written as ($\hbar=m=1$)

$$\begin{aligned}(23)\quad &\widehat{U}(\tau)=e^{i/2\omega\tau}e^{-i/2\tau(-(d/dq)+\omega q)((d/dq)+\omega q)}=\\ &=e^{\omega\tau/2}\sum_{n=0}^{\infty}\frac{1}{n!}\left[\frac{e^{-i\omega\tau}-1}{2\omega}\right]^n\Pi_{n,n}\left(q,\frac{d}{dq};-1,\omega,1,\omega\right)\end{aligned}$$

A Gaussian initial state ($\Psi_0(q)=(1/\sqrt{\pi})e^{-(q^2/2)}$) ruled by (1) evolves according to

$$(24)\qquad \Psi(x,\tau)=e^{i/2\omega\tau}\sum_{n=0}^{\infty}\frac{1}{n!}\left[\frac{e^{-i\omega\tau}-1}{2\omega}\right]^n\Phi_n(q)$$

where

$$(25)\qquad \Phi_n(q)=\sum_{r=0}^{n}\sum_{s=0}^{n}\binom{n}{r}\binom{n}{s}(-1)^s H_{n-r,n-s}(q;-1,\omega,1,\omega)He_{r+s}(q)\Psi_0(q).$$

The solution provided by Eq. (24) is exact and, in spite of its apparently complicated aspect, easily manageable from the numerical point of view. In the case in which the problem to be dealt with is not so simple and anharmonic contributions of the type λq^4 must be included, one can combine the previous method with a symmetric split approximation of the evolution operator, namely

$$(26)\qquad \widehat{U}_a(\delta_\tau)\simeq e^{-i(\lambda/2)\delta_\tau q^4}\widehat{U}(\delta_\tau)e^{-i(\lambda/2)\delta_\tau q^4}+o(\delta_\tau^3)$$

where δ_τ is a suitable time interval step and $\widehat{U}(\delta_\tau)$ corresponds to the evolution operator (23), $o(\delta_\tau^3)$ represents the error of the method. According to (26) we can use an iterative procedure employing the Fourier transform as intermediate step. We can therefore write the evolution of an initial state $\Psi(q)$ as

$$\Psi(q,\tau) \simeq \widehat{U}_a(\delta_\tau)^M \Psi(q), \tag{27}$$

where M is the number of time steps δ_τ and at each step the evolution will be provided by

$$\Psi_s(q,\delta_\tau) = \frac{e^{-i(\lambda/2)\delta_\tau q^4} e^{i/2\omega\delta_\tau}}{\sqrt{2\pi}} \sum_{n=0}^{\infty} \frac{1}{n!}\left[\frac{e^{-i\omega\delta_\tau}-1}{2\omega}\right]^n \times \int_{-\infty}^{+\infty} \Pi_{n,n}(q,ik;-1,\omega,1,\omega)\widetilde{\Psi}_{s-1}(k,\delta_\tau)\cdot e^{ikq}dk \tag{28}$$

and

$$\widetilde{\Psi}_s(k,\delta_\tau) = \frac{1}{\sqrt{2\pi}}\int_{-\infty}^{+\infty} \Psi_s(q,\delta_\tau)\cdot e^{-ikq}dq \tag{29}$$

is the Fourier transform of the state wave function. More accurate symmetric decoupling schemes can also be used [4] they will not be discussed here, because, albeit more efficient from the numerical point of view, do not provide, within the present framework, significant conceptual differences.

3. Exponential Operators and Generalized Gauss Transforms

In the previous section we have seen that the combined use of exponential and differential operator identities offers a fairly useful tool to deal with different types of problems, both in pure and applied mathematics. We will now try to widen the scenario, by discussing further identities involving exponential operators and by presenting possible applications.

It has been shown that exponential operators of the type $e^{\lambda q(x)(d/dx)}$ once acting on a function $f(x)$ yield [**6**]

$$e^{\lambda q(x)(d/dx)} f(x) = f\big(F^{-1}(\lambda + F(x))\big), \quad F(x) = \int^x \frac{d\xi}{q(\xi)}, \tag{30}$$

with $F^{-1}(\cdots)$ denoting the inverse of $F(x)$. We can e.g. infer from (30) the following particular cases

$$e^{\lambda x(d/dx)} f(x) = f(e^\lambda x), \quad e^{\lambda x^2(d/dx)} f(x) = f\left(\frac{x}{1-\lambda x}\right), \quad x < \frac{1}{\lambda}. \tag{31}$$

Within the present framework a role of noticeable importance is played by the so called generalized Gauss transform [**4**], namely $(\lambda > 0)$

$$e^{\lambda[q(x)(d/dx)]^2} f(x) = \frac{1}{\sqrt{\pi}}\int_{-\infty}^{+\infty} e^{-\xi^2} f\big(F^{-1}(2\xi\sqrt{\lambda} + F(x)\big)d\xi, \tag{32}$$

[4] See e.g. [**8**]

which yields as trivial examples

$$
(33)\qquad
\begin{aligned}
e^{\lambda(\partial/\partial x)^2} f(x) &= \frac{1}{2\sqrt{\pi\lambda}} \int_{-\infty}^{+\infty} e^{[-((x-\xi)^2/4t)]} f(\xi)\, d\xi,\\
e^{\lambda(x(\partial/\partial x))^2} f(x) &= \frac{1}{\sqrt{\pi}} \int_{-\infty}^{+\infty} e^{-s^2} f(e^{2\sqrt{\lambda}s} x)\, ds.
\end{aligned}
$$

It is clear that the generalized Gauss transform can be viewed as the solution of the following "heat" equation

$$
(34)\qquad \frac{\partial}{\partial\lambda}\rho(x,\lambda) = \left(q(x)\frac{\partial}{\partial x}\right)^2 \rho(x,\lambda), \quad \rho(x,0) = f(x).
$$

Let us now discuss few examples of applications. We consider therefore the Fokker-Planck equation

$$
(35)\qquad \frac{\partial}{\partial t} f(v,t) = \left(\frac{\partial}{\partial v} v + \left(\frac{\partial}{\partial v}\right)^2\right) f(v,t), \quad f(v,0) = g(v),
$$

whose formal solution in terms of the corresponding evolution operator writes

$$
(36)\qquad f(v,t) = e^{t(v(\partial/\partial v) + (\partial/\partial v)^2)} g(v).
$$

By introducing the operators

$$
(37)\qquad \hat{A} = t\left(\frac{\partial}{\partial v}\right)^2, \quad \widehat{E} = t\left(v\frac{\partial}{\partial v}\right)
$$

by exploiting the decoupling rule (6) and the identities (30), (32) we end up with

$$
(38)\qquad f(v,t) = \frac{e^t}{2\sqrt{\pi S(t)}} \int_{-\infty}^{+\infty} e^{[-((v-\xi)^2/4S(t))]} g(e^t\xi)\, d\xi, \quad S(t) = \frac{1-e^{-2t}}{2}.
$$

The previous example was concerned with a particularly simple case, but provided a clear idea of how the various identities we have discussed in the paper can be combined to solve equations of the evolution type. Unfortunately the Fokker-Planck equations more commonly encountered in applications do not admit an exact solution. In these cases the most appropriate strategy is a combination of analytical and numerical techniques, as in the example of the quartic harmonic oscillator.

To better illustrate the previous point, we consider the Fokker-Planck equation ruling the evolution of the electron beam longitudinal phase space distributions in Storage-Rings [**9**]

$$
(39)\qquad
\begin{aligned}
\frac{\partial}{\partial t} f(x,v,t) &= \left[\left(v\frac{\partial}{\partial x} - p(x)\frac{\partial}{\partial x}\right) + \frac{2}{\xi}\left(v\frac{\partial}{\partial v} + \left(\frac{\partial}{\partial v}\right)^2\right) + \frac{2}{\xi}\right] f(x,v,t),\\
f(x,v,0) &= g(x,v).
\end{aligned}
$$

In formula (39) $p(x)$ denotes the radio-frequency potential. Without entering in the physical meaning of the equation, we note that by introducing the operators

$$
(40)\qquad \hat{L} = v\frac{\partial}{\partial x} - p(x)\frac{\partial}{\partial x} \quad \widehat{D} = \frac{2}{\xi}\left(v\frac{\partial}{\partial v} + \left(\frac{\partial}{\partial v}\right)^2\right),
$$

we can exploit the symmetric split approximation to write the evolution operator as

$$
(41)\qquad \widehat{U}(\delta_t) \simeq e^{1/2\delta_t \hat{L}} e^{\delta_t \widehat{D}} e^{1/2\delta_t \hat{L}} + o(\delta_t^3).
$$

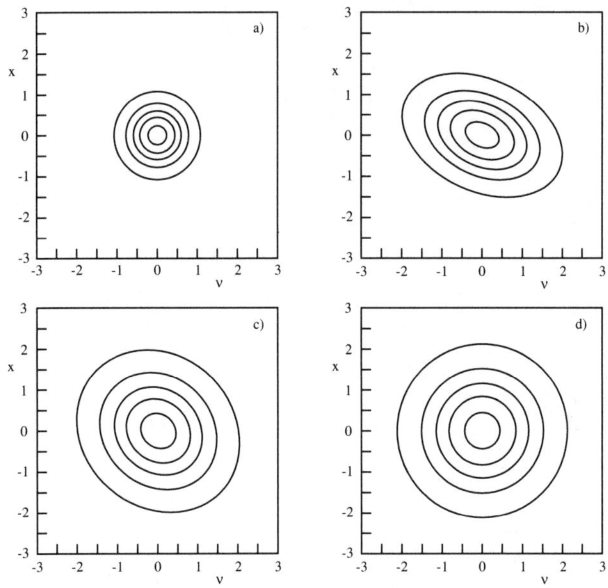

FIGURE 1. Longitudinal phase-spare contour plots $\sigma_\mu = 0.5$, $\sigma_x = 0.5$, $\xi = 1.1$ a) $\bar{t} = 0$, b) $\bar{t} = 1$, c) $\bar{t} = 2$, $\bar{t} = 3$.

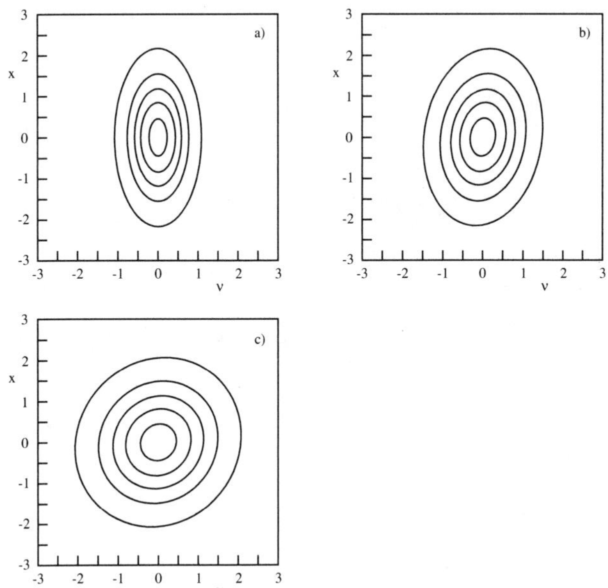

FIGURE 2. Same as Fig. 2 ($\sigma_\mu = 0.5$, $\sigma_x = 1$), $\xi = 1.1$ a) $\bar{t} = 0$, b) $\bar{t} = 0.1$, c) $\bar{t} = 0.5$.

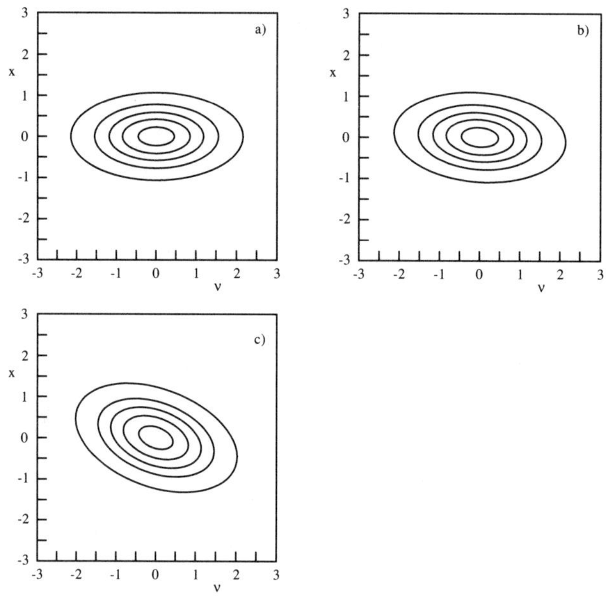

FIGURE 3. Same as Fig. 2 $\sigma_\mu = 1$, $\sigma_x = 0.5$, $\xi = 1.1$ a) $\bar{t} = 0$, b) $\bar{t} = 0.1$, c) $\bar{t} = 0.5$.

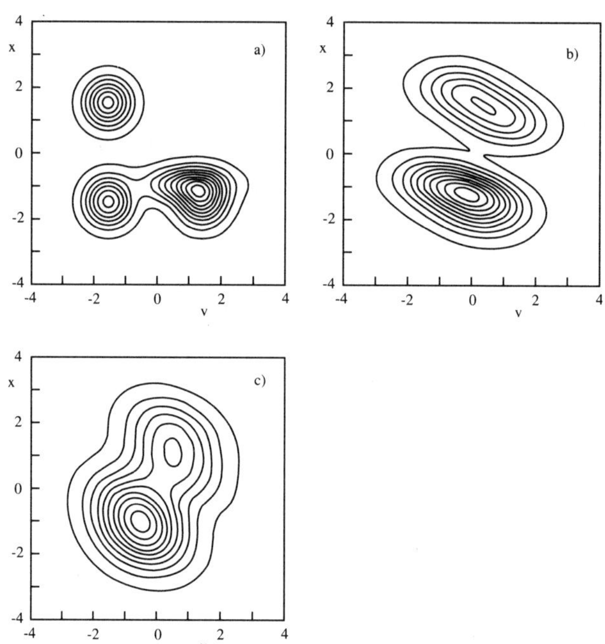

FIGURE 4. Phase-space contour plots for an initial nongaussian distribution a) $\bar{t} = 0.4$, b) $\bar{t} = 10.4$, c) $\bar{t} = 0$, $\bar{t} = 2$ the system reaches the stationary configuration.

Repeated application of the above operator may provided the evolution of the distribution f at different times and for different radio frequency potential as illustrated in Figs. 1–4. It is evident that when non linear terms are include in $p(x)$ the system relaxes towards a configuration significantly different from a gaussian.

Before concluding this paper it is worth noting that the concepts associated with the generalized shift operators can be exploited to derive solution for generalized shift equations as e.g.

$$\frac{f(e^a x) - f(x)}{a} = g(x). \tag{42}$$

We assume that $f(x)$ is unknown and $g(x)$ is known along with its Taylor expansion $\left(g(x) = \sum_{m=0}^{\infty} b_m x^m\right)$. By using the dilatation operator we can recast Eq. (42) in the form of a Jackson derivative

$$\frac{e^{axd/dx} - 1}{a} f(x) = g(x) \tag{43}$$

Which can be formally inverted yelding

$$f(e^\xi) = \frac{a}{e^{ad/d\xi} - 1} g(e^\xi) = \mathcal{D}_x^{-1} \sum_{n=0}^{\infty} \frac{a^n}{n!} B_n \left(\frac{d}{d\xi}\right)^n \sum_{m=0}^{\infty} b_m e^{m\xi}. \tag{44}$$

Where $\mathcal{D}_x^{-1}$ is the inverse of the derivative operator and B_n are Bernoulli numbers. By evaluating the derivatives, by performing the integration and by going back to the x-variable we find

$$f(x) = \sum_{m=0}^{\infty} \frac{a}{e^{ma} - 1} b_m x^m \tag{45}$$

which is essentially the Taylor expansion of $g(x)$, modified in such a way that the mth coefficient is multiplied by $(a/e^{ma} - 1)$. In a forthcoming investigation the condition of validity of solutions of the above type will be discussed more carefully.

In conclusion, the techniques outlined in the present paper and their combination with non trivial decoupling schemes may offer efficient computational means to treat a wide variety of problems.

References

1. E. T. Bell, *Exponential polynomials*, Ann. of Math. **35** (1934), 258–277.
2. J. L. Burchnall, *A note on the polynomials of Hermite*, Quart. J. Math., Oxford Ser. **12** (1941), 9–11.
3. M. W. Crofton, *Theorems in the calculus of operations*, Quart. J. Math. **16** (1879), 328–329.
4. G. Dattoli, L. Giannessi, M. Quattromini, and A. Torre, *Symmetric decomposition of exponential operators and evolution problems*, Physica D **111** (1998), 129–142.
5. G. Dattoli and L. Mezi, *The storage ring Fokker-Plank equations: Inclusion of the free electron laser amplifier contributions*, Phys. Rev. E (to appear).
6. G. Dattoli, P. L. Ottaviani, A. Torre, and L. Vazquez, *Evolution operator equations: Integration with algebraic and finite difference mthods, Applications to physical problems in classical and quantum mechanics and quantum field theory*, Riv. Nuovo Cimento Soc. Ital. Fis. (4) **20** (1997), no. 2.
7. G. Dattoli and A. Torre, *Theory and application of generalized Bessel functions*, ARACNE editrice ROME, 1996.
8. W. Louisell, *Quantum statisical properties of radiation*, Wiley, New York, 1972; G. Dattoli, S. Lorenzutta, M. Quattromini, and A. Torre , *Some operatorial identities and their applications*, Nuovo Cimento B (to appear).

9. M. Sands, *Stanford linear accelerator*, SLAC-121, 1970 (unpublished); A. W. Chao, *Physics of collective beam instabilities in high energy accelearators*, John Wiley & Sons Inc., New York, 1993.

ENEA—Dipartimento Innovazione, Divisione Fisica Applicata, Centro Ricerche di Frascati, C.P. 65, 00044 Frascati, Rome, Italy.

E-mail address: dattoli@frascati.enea.it

Centre de Recherches Mathématiques
CRM Proceedings and Lecture Notes
Volume **25**, 2000

Lattice Geometry of the Hirota Equation

Adam Doliwa

ABSTRACT. Geometric interpretation of the Hirota equation is presented as equation describing the Laplace sequence of two-dimensional quadrilateral lattices. Different forms of the equation are given together with their geometric interpretation: (i) the discrete coupled Volterra system for the coefficients of the Laplace equation; (ii) the gauge invariant form of the Hirota equation for projective invariants of the Laplace sequence; (iii) the discrete Toda system for the rotation coefficients and (iv) the original form of the Hirota equation for the τ-function of the quadrilateral lattice.

1. Introduction

The integrable discrete geometry deals with lattice submanifolds described by integrable equations. Among the integrable discrete (difference) equations an important role is played by the Hirota equation [**12**] which is the discrete analog of the two-dimensional Toda system [**17**]. Both the Toda system and the Hirota equation are important in the theory of integrable equations and in their applications. It turns out that the two-dimensional Toda system was studied in classical differential geometry and describes the so called Laplace sequences of two-dimensional conjugate nets [**3**].

The lattice geometric interpretation of the discrete Toda system was found in [**6**] and is based on the observation that the discrete analog of conjugate nets on a surface, which is given by two-dimensional lattices made of planar quadrilaterals [**18**], allows for definition of the corresponding Laplace sequence of discrete conjugate nets. Since then, multidimensional quadrilateral lattices [**10**] (multidimensional lattices made of planar quadrilaterals—discrete analogs of multidimensional conjugate nets) have been made one of central objects of the integrable discrete geometry, and many classical results of the theory of the conjugate nets and their reductions found discrete counterparts [**1, 2, 5, 7, 9, 11, 14, 16**].

The goal of the present article is to reinterpret and expand results obtained in [**6**] using new notions provided by the general theory of quadrilateral lattices.

2000 *Mathematics Subject Classification.* Primary: 35Q58; Secondary: 52C07, 53A20.

The author would like to thank the organizers of the SIDE III meeting for invitation and support.

This is the final form of the paper.

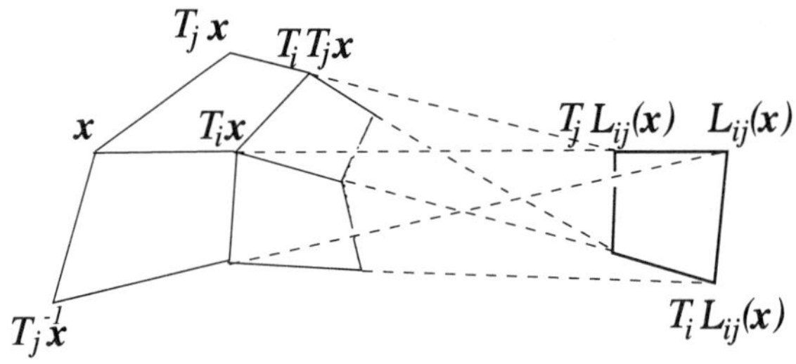

FIGURE 1. Laplace transformation of quadrilateral lattices.

Different formulations of the Hirota equations are reviewed and, for each of them, the geometric interpretation of the corresponding functions is given.

In Sections 2 and 3 we reformulate, using more convenient notation, results found in [**6**]. In Section 4 we present the discrete analog of the standard version of the Toda system as an equation governing Laplace transformations of rotation coefficients. Then in Section 5, basing on the geometric interpretation of the τ-function of the quadrilateral lattice given in [**9**], we show that the τ-functions of the Laplace sequence of quadrilateral lattices solve the original Hirota's form of the discrete Toda system—this last result fills out the missing point of paper [**6**].

2. Laplace Transformations of Quadrilateral Lattices

DEFINITION 2.1. A two-dimensional quadrilateral lattice is a mapping of the two-dimensional integer lattice into an M-dimensional projective space such that all the elementary quadrilaterals of the lattice are planar:

$$x\colon \mathbb{Z}^2 \to \mathbb{P}^M, \quad T_1T_2x \in \langle x, T_1x, T_2x\rangle.$$

In the above definition T_i, $i = 1,2$ denotes the shift operator along the i-th direction of the lattice.

In the nonhomogenous coordinates of the projective space a two dimensional quadrilateral lattice is represented by the mapping $\boldsymbol{x}\colon \mathbb{Z}^2 \to \mathbb{R}^M$ satisfying the Laplace equation [**6**]

$$\Delta_1\Delta_2\boldsymbol{x} = (T_1A_{12})\Delta_2\boldsymbol{x} + (T_2A_{21})\Delta_2\boldsymbol{x}, \tag{2.1}$$

which is equivalent to the planarity condition; here $\Delta_i = T_i - 1$, $i = 1,2$ is the partial difference operator, and the functions A_{12}, $A_{21}\colon \mathbb{Z}^2 \to \mathbb{R}$ define the position of the point $T_1T_2\boldsymbol{x}$ with respect to the points $\boldsymbol{x}$, $T_1\boldsymbol{x}$ and $T_2\boldsymbol{x}$.

Planarity of elementary quadrilaterals implies that the "tangent" lines containing opposite sides of an elementary quadrilateral intersect. The Laplace transformation $\mathcal{L}_{ij}$, $i \neq j$ of the lattice x is defined [**6,11**] as intersection of the line $\langle x, T_ix\rangle$ with the line $\langle T_j^{-1}x, T_iT_j^{-1}x\rangle$ (see Fig. 1).

Using elementary calculations one can show the following results [**6,11**].

PROPOSITION 2.1. *In the nonhomogeneous representation the Laplace transformation $\mathcal{L}_{ij}(\boldsymbol{x})$ of the quadrilateral lattice $\boldsymbol{x}$ is given by*

$$\mathcal{L}_{ij}(\boldsymbol{x}) = \boldsymbol{x} - \frac{1}{A_{ji}}\Delta_i\boldsymbol{x}. \tag{2.2}$$

PROPOSITION 2.2. *The Laplace transformations of quadrilateral lattices are again quadrilateral lattices and the corresponding transformations of the coefficients of the Laplace equation* (2.1) *of the lattice* $\mathcal{L}_{ij}(\boldsymbol{x})$ *read*

$$\mathcal{L}_{ij}(A_{ij}) = \frac{A_{ji}}{T_j A_{ji}}(T_i A_{ij} + 1) - 1, \tag{2.3}$$

$$\mathcal{L}_{ij}(A_{ji}) = T_j^{-1}\left(\frac{T_i \mathcal{L}_{ij}(A_{ij})}{\mathcal{L}_{ij}(A_{ij})}(A_{ji}+1)\right) - 1. \tag{2.4}$$

PROPOSITION 2.3. *Under the assumption that the transformed lattices are non-degenerate, i.e. their quadrilaterals do not degenerate to segments or points, we have*

$$\mathcal{L}_{ij} \circ \mathcal{L}_{ji} = \mathcal{L}_{ji} \circ \mathcal{L}_{ij} = id. \tag{2.5}$$

In this way given two-dimensional quadrilateral lattice $\boldsymbol{x}$ one can define a sequence of quadrilateral lattices

$$\boldsymbol{x}^{(l)} = \mathcal{L}_{12}^{l}(\boldsymbol{x}), \quad l \in \mathbb{N}, \quad \mathcal{L}_{12}^{-1} = \mathcal{L}_{21}.$$

In analogy to the Laplace sequence of conjugate nets, the above sequence can be called the Laplace sequence of quadrilateral lattices. Equations (2.3)-(2.4) can be then rewritten in the form

$$\frac{\Delta_2 A_{21}^{(l)}}{A_{21}^{(l)}} = \frac{T_1 A_{12}^{(l)} - A_{12}^{(l+1)}}{(T_1 A_{12}^{(l)} + 1)(A_{12}^{(l+1)} + 1)}, \tag{2.6}$$

$$\frac{\Delta_1 A_{12}^{(l)}}{A_{12}^{(l)}} = \frac{T_2 A_{12}^{(l)} - A_{21}^{(l-1)}}{(T_2 A_{21}^{(l)} + 1)(A_{21}^{(l-1)} + 1)}, \tag{2.7}$$

which is the discrete analog of the coupled Volterra system.

3. Projective Invariants of the Laplace Sequence

The planarity of elementary quadrilaterals of the quadrilateral lattice and the construction of the Laplace sequence is essentially of the projective nature [**6**]. It would be therefore interesting to know the pure projective-geometric version of the equation describing the Laplace sequence of quadrilateral lattices.

The basic numeric invariant of projective transformations is the so called cross-ratio of four collinear points, which is given in the affine representation as

$$\operatorname{cr}(a, b; c, d) = \left(\frac{c-a}{c-b}\right) : \left(\frac{d-a}{d-b}\right);$$

notice the simple identity

$$\operatorname{cr}(a, b; c, d) = \operatorname{cr}(b, a; d, c). \tag{3.1}$$

Define the function K_{ij} as the cross-ratio of x, $\mathcal{L}_{ij}(x)$, $T_i x$ and $T_j \mathcal{L}_{ij}(x)$. Elementary calculations show that

$$T_i \boldsymbol{x} - \mathcal{L}_{ij}(\boldsymbol{x}) = \frac{1 + A_{ji}}{A_{ji}} \Delta_i \boldsymbol{x},$$

$$T_j \mathcal{L}_{ij}(\boldsymbol{x}) - \boldsymbol{x} = -\frac{T_i A_{ij} + 1}{T_j A_{ji}} \Delta_i \boldsymbol{x},$$

$$T_j \mathcal{L}_{ij}(\boldsymbol{x}) - \mathcal{L}_{ij}(\boldsymbol{x}) = \left(\frac{1}{A_{ji}} - \frac{1 + T_i A_{ij}}{T_j A_{ji}}\right) \Delta_i \boldsymbol{x},$$

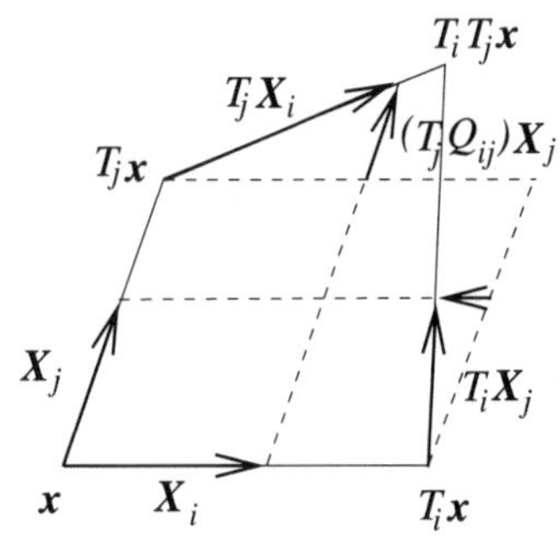

FIGURE 2. Definition of the rotation coefficients.

and, therefore,

$$K_{ij} = \mathrm{cr}\big(\boldsymbol{x}, \mathcal{L}_{ij}(\boldsymbol{x}); T_i\boldsymbol{x}, T_j\mathcal{L}_{ij}(\boldsymbol{x})\big) = \frac{A_{ji}(T_iA_{ij}+1) - T_jA_{ji}}{(1+T_iA_{ij})(1+A_{ji})}.$$

Equations (2.3) and (2.4) allow to find the Laplace transforms of the projective invariants [**6**]

$$\mathcal{L}_{ij}(K_{ij}) = T_j^{-1}\left(\frac{K_{ij}(T_iT_jK_{ij})}{(T_iK_{ij})(T_jK_{ij})}\frac{(T_iK_{ij}+1)(T_jK_{ij}+1)}{T_iK_{ji}+1} - 1\right), \tag{3.2}$$

$$\mathcal{L}_{ij}(K_{ji}) = K_{ij}; \tag{3.3}$$

notice that equation (3.3) is a simple consequence of Proposition 2.3 and property (3.1) of the cross-ratio.

Equations (3.2) and (3.3) can be rewritten in terms of the function $K = K_{12}$ in the following form

$$T_2\left(\frac{K^{(l+1)}+1}{K^{(l)}+1}\right)T_1\left(\frac{K^{(l-1)}+1}{K^{(l)}+1}\right) = \frac{(T_1T_2K^{(l)})K^{(l)}}{(T_1K^{(l)})(T_2K^{(l)})}, \tag{3.4}$$

known as the gauge invariant form of the Hirota equation.

4. Rotation Coefficients of the Quadrilateral Lattice

As it was shown in [**10**] it is convenient to reformulate the Laplace equation (2.1) as a first order system. We introduce the suitably scaled tangent vectors $\boldsymbol{X}_i$, $i = 1, 2$,

$$\Delta_i\boldsymbol{x} = (T_iH_i)\boldsymbol{X}_i, \tag{4.1}$$

in such a way that the j-th variation of $\boldsymbol{X}_i$ is proportional to $\boldsymbol{X}_j$ only (see Fig. 2)

$$\Delta_j\boldsymbol{X}_i = (T_jQ_{ij})\boldsymbol{X}_j, \quad i \neq j, \tag{4.2}$$

the coefficients Q_{ij} in equation (4.2) are called the rotation coefficients.

The scaling factors H_i in equation (4.1), called the Lamé coefficients, satisfy the linear equations

$$\Delta_iH_j = (T_iH_i)Q_{ij}, \quad i \neq j,$$

adjoint to (4.2); moreover

$$A_{ij} = \frac{\Delta_jH_i}{H_i}, \quad i \neq j.$$

The Laplace transformation of the Lamé coefficients and the scaled tangent vectors was found in [**11**] and is presented in the following

PROPOSITION 4.1. *The Lamé coefficients of the transformed lattice read*

$$\mathcal{L}_{ij}(H_i) = \frac{H_j}{Q_{ij}},$$

$$\mathcal{L}_{ij}(H_j) = T_j^{-1}\left(Q_{ij}\Delta_j\left(\frac{H_j}{Q_{ij}}\right)\right),$$

the tangent vectors of the new lattice are given by

$$\mathcal{L}_{ij}(\boldsymbol{X}_i) = -\Delta_i\boldsymbol{X}_i + \frac{\Delta_i Q_{ij}}{Q_{ij}}\boldsymbol{X}_i,$$

$$\mathcal{L}_{ij}(\boldsymbol{X}_j) = -\frac{1}{Q_{ij}}\boldsymbol{X}_i.$$

From above formulas follow transformation rules for the rotation coefficients.

PROPOSITION 4.2. *The rotation coefficients transform according to*

$$\mathcal{L}_{ij}(Q_{ij}) = T_j^{-1}\left(T_iQ_{ij} - \frac{Q_{ij}T_iT_jQ_{ij}}{T_jQ_{ij}}\left(1-(T_iQ_{ji})(T_jQ_{ij})\right)\right), \tag{4.3}$$

$$\mathcal{L}_{ij}(Q_{ji}) = \frac{1}{Q_{ij}}. \tag{4.4}$$

Equations (4.3) and (4.4) can be rewritten in terms of the function $Q = Q_{12}$ as

$$\Delta_2\frac{\Delta_1 Q^{(l)}}{Q^{(l)}} = T_1\left(\frac{T_2Q^{(l)}}{Q^{(l-1)}}\right) - \frac{T_2Q^{(l+1)}}{Q^{(l)}}. \tag{4.5}$$

5. Geometry of the τ-Function

To give the geometric meaning to the τ-function let us introduce [**9**] the vectors $\widetilde{\boldsymbol{X}}_i$ pointing in the negative directions

$$\tilde{\Delta}_i\boldsymbol{x} = (T_i^{-1}\widetilde{H}_i)\widetilde{\boldsymbol{X}}_i, \quad \text{or} \quad \Delta_i\boldsymbol{x} = \widetilde{H}_i(T_i\widetilde{\boldsymbol{X}}_i),$$

where $\tilde{\Delta}_i = T_i^{-1} - 1$ is the backward difference operator. The scaling factors $\widetilde{H}_i$ (the backward Lamé coefficients) are chosen in such a way that the $\tilde{\Delta}_i$ variation

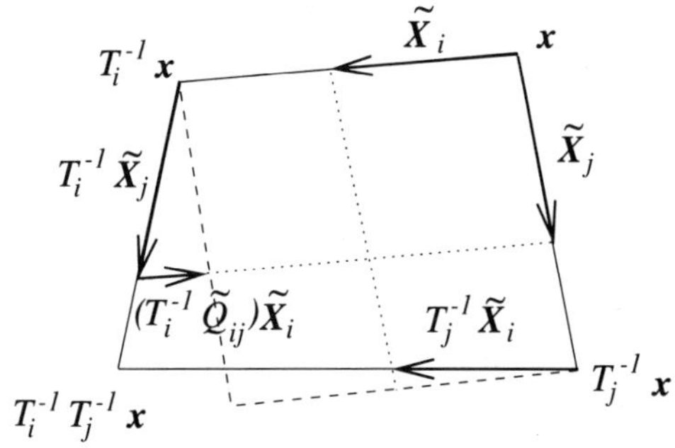

FIGURE 3. Definition of the backward data.

of $\widetilde{\boldsymbol{X}}_j$ is proportional to $\widetilde{\boldsymbol{X}}_i$ only (see Fig. 3). We define the backward rotation coefficients $\widetilde{Q}_{ij}$ as the corresponding proportionality factors

$$\tilde{\Delta}_i \widetilde{\boldsymbol{X}}_j = (T_i^{-1}\widetilde{Q}_{ij})\widetilde{\boldsymbol{X}}_i, \quad \text{or} \quad \Delta_i \widetilde{\boldsymbol{X}}_j = (T_i\widetilde{\boldsymbol{X}}_i)\widetilde{Q}_{ij}, \quad i \neq j. \tag{5.1}$$

The backward Lamé coefficients $\widetilde{H}_i$ satisfy then the following system of linear equations

$$\Delta_j \widetilde{H}_i = (T_j\widetilde{Q}_{ij})\widetilde{H}_j, \quad i \neq j,$$

adjoint to system (5.1)

Since the forward and backward data Q_{ij} and $\widetilde{Q}_{ij}$ describe the same lattice $\boldsymbol{x}$ but from different points of view, then one cannot expect that they are independent. Indeed, defining the functions $\rho_i \colon \mathbb{Z}^2 \to \mathbb{R}$, $i = 1, 2$, as the proportionality factors between $\boldsymbol{X}_i$ and $T_i\widetilde{\boldsymbol{X}}_i$ (both vectors are proportional to $\Delta_i \boldsymbol{x}$):

$$\boldsymbol{X}_i = -\rho_i(T_i\widetilde{\boldsymbol{X}}_i), \quad T_iH_i = -\frac{1}{\rho_i}\widetilde{H}_i, \quad i = 1, 2,$$

we have the following result [**9**].

PROPOSITION 5.1. *The forward and backward data of the lattice $\boldsymbol{x}$ are related through the following formulas*

$$\rho_j T_j \widetilde{Q}_{ij} = \rho_i T_i Q_{ji},$$

and the factors ρ_i are first potentials satisfying equations

$$\frac{T_j\rho_i}{\rho_i} = 1 - (T_iQ_{ji})(T_jQ_{ij}), \quad i \neq j. \tag{5.2}$$

The right hand side of equation (5.2) is symmetric with respect to the interchange of i and j, which implies the existence of a potential $\tau \colon \mathbb{Z}^2 \to \mathbb{R}$, such that

$$\rho_i = \frac{T_i\tau}{\tau};$$

therefore equation (5.2) defines the second potential τ:

$$\frac{(T_iT_j\tau)\tau}{(T_i\tau)(T_j\tau)} = 1 - (T_iQ_{ji})(T_jQ_{ij}), \quad i \neq j. \tag{5.3}$$

The potential τ connecting the forward and backward data is the *τ-function* of the quadrilateral lattice [**9**].

Let us find the Laplace transformation of the τ-function. Formulas (4.3) and (4.4) imply that

$$1 - \left(T_i\mathcal{L}_{ij}(Q_{ji})\right)\left(T_j\mathcal{L}_{ij}(Q_{ij})\right) = \frac{Q_{ij}T_iT_jQ_{ij}}{T_jQ_{ij}T_iQ_{ij}}\frac{\tau T_iT_j\tau}{T_i\tau T_j\tau}, \tag{5.4}$$

which, due to equation (5.3), allows for identification

$$\mathcal{L}_{ij}(\tau) = \tau Q_{ij}. \tag{5.5}$$

It should be mentioned here that the above formula was strongly suggested by the identification of the Schlesinger transformation of the theory of the multicomponent Kadomtsev–Petviashvili hierarchy [**4, 13, 15**] with the Laplace transformation of conjugate nets [**8**].

COROLLARY 5.2. *The geometric meaning of τ_{ij} as the Laplace transformation $\mathcal{L}_{ij}(\tau)$ of the τ-function applies for any dimension of the quadrilateral lattice.*

Finally, equation (5.3) rewritten in terms of the τ-function and its Laplace transformations take the following form

$$\tau^{(l)}T_1T_2\tau^{(l)} = (T_1\tau^{(l)})(T_2\tau^{(l)}) - (T_1\tau^{(l-1)})(T_2\tau^{(l+1)}), \tag{5.6}$$

which is the original Hirota's bilinear form of the discrete Toda system.

6. Conclusion

The geometric interpretation of the Hirota equation (integrable discrete analog of the Toda system) is given by the Laplace sequence of quadrilateral lattices, therefore various representations of the lattice give different versions of the equation. In the paper we presented four different forms of the Hirota equation: (i) the discrete coupled Volterra system (2.6)–(2.7) for the coefficients of the Laplace equations; (ii) the gauge invariant form of the Hirota equation (3.4) for projective invariants of the Laplace sequence; (iii) the discrete Toda system for the rotation coefficients (4.5), and (iv) the original form of the Hirota equation (5.6) for the τ-function of the quadrilateral lattice.

References

1. A. Bobenko and U. Pinkall, *Discrete isothermic surfaces*, J. Reine Angew. Math. **475** (1996), 187–208.
2. J. Cieśliński, A. Doliwa, and P. M. Santini, *The integrable discrete analogues of orthogonal coordinate systems are multidimensional circular lattices*, Phys. Lett. A **235** (1997), 480–488.
3. G. Darboux, *Leçons sur la théorie générale des surfaces* I, Gauthier-Villars, Paris, 1887; II, 1889; III, 1894; IV, 1896.
4. E. Date, M. Kashiwara, M. Jimbo, and T. Miwa, *Transformation groups for soliton equations*, Nonlinear Integrable Systems—Classical Theory and Quantum Theory, Proc. of RIMS Symposium (M. Jimbo and T. Miwa, eds.), World Scientific, Singapore, 1983, pp. 39–119.
5. A. Doliwa, *Quadratic reductions of quadrilateral lattices*, J. Geom. Phys., `solv-int/9802011` (in press).
6. ______, *Geometric discretization of the Toda system*, Phys. Lett. A **234** (1997), 187–192.
7. A. Doliwa, S. V. Manakov, and P. M. Santini, *$\bar\partial$-reductions of the multidimensional quadrilateral lattice: the multidimensional circular lattice*, Comm. Math. Phys. **196** (1998), 1–18.
8. A. Doliwa, M. Mañas, L. Martínez Alonso, E. Medina, and P. M. Santini, *Multicomponent* KP *hierarchy and classical transformations of conjugate nets*, `solv-int/9813015`.
9. A. Doliwa and P. M. Santini, *The symmetric and Egorov reductions of the quadrilateral lattice*, preprint, 1999.
10. ______, *Multidimensional quadrilateral lattices are integrable*, Phys. Lett. A **233** (1997), 365–372.
11. A. Doliwa, P. M. Santini, and M. Mañas, *Transformations of quadrilateral lattices*, `solv-int/9712017`.
12. R. Hirota, *Discrete analogue of a generalized Toda equation*, J. Phys. Soc. Japan **50** (1981), 3785–3791.
13. V. G. Kac and J. van de Leur, *The n-component* KP *hierarchy and representation theory*, Important Developments in Soliton Theory (A. S. Fokas and V. E. Zakharov, eds.), Springer Ser. Nonlinear DYnam., Springer, Berlin, 1993, pp. 302–343.
14. B. G. Konopelchenko and W. K. Schief, *Three-dimensional integrable lattices in Euclidean spaces: Conjugacy and orthogonality*, preprint, 1997.
15. J. van de Leur, *Schlessinger–Bäcklund transformation for N-component* KP, J. Math. Phys. **39** (1998), 2833–2847.
16. M. Mañas, A. Doliwa, and P. M. Santini, *Darboux transformations for multidimensional quadrilateral lattices.* I, Phys. Lett. A **232** (1997), 99–105.

17. A. V. Mikhailov, *Integrability of a two-dimensional generalization of the Toda chain*, JETP Lett. **30** (1979), 414–418.
18. R. Sauer, *Differenzengeometrie*, Springer, Berlin, 1970.

Instytut Fizyki Teoretycznej, Uniwersytet Warszawski, Hoża 69, 00-681 Warszawa, Poland.

E-mail address: doliwa@fuw.edu.pl

Centre de Recherches Mathématiques
CRM Proceedings and Lecture Notes
Volume **25**, 2000

Integrable Discrete Geometry: The Quadrilateral Lattice, Its Transformations and Reductions

Adam Doliwa and Paolo Maria Santini

ABSTRACT. We review recent results on Integrable Discrete Geometry. It turns out that most of the known (continuous and/or discrete) integrable systems are particular symmetries of the quadrilateral lattice (QL), a multidimensional lattice characterized by the planarity of its elemetary quadrilaterals. Therefore the linear property of planarity seems to be a basic geometric property underlying integrability. We solve the initial value problem for the quadrilateral lattice and we present the geometric meaning of its τ-function, as the potential connecting its forward and backward data. We present the theory of transformations of the quadrilateral lattice, which is based on the discrete analogue of the theory of rectilinear congruences. In particular, we discuss the discrete analogues of the Laplace, Combescure, Lévy, radial and fundamental transformations and their interrelations. We also show how the sequence of Laplace transformations of a quadrilateral surface is described by the discrete Toda system in three dimensions. We finally show that these classical transformations are strictly related to the basic operators associated with the quantum field theoretical formulation of the multicomponent Kadomtsev–Petviashvilii hierarchy. We review the properties of quadrilateral hyperplane lattices, which play an interesting role in the reduction theory, when the introduction of additional geometric structures allows to establish a connection between point and hyperplane lattices. We present and fully characterize some geometrically distinguished reductions of the quadrilateral lattice, like the symmetric, circular and Egorov lattices; we review also basic geometric results of the theory of quadrilateral lattices in quadrics, and the corresponding analogue of the Ribaucour reduction of the fundamental transformation.

We finally remark that the equations characterizing the above lattices are relevant in physics, being integrable discretizations of equations arising in hydrodynamics and in quantum field theory.

1. Quadrilateral Lattices (QL)

1.1. The quadrilateral lattice. Our first goal is to build a lattice which is integrable by construction. For the sake of simplicity, we consider first a 2-dimensional lattice $\mathbf{x}\colon \mathbf{Z}^2 \to \mathbf{R}^M$. Given the three arbitrary points $\mathbf{x}(n_1, n_2)$, $\mathbf{x}(n_1+$

2000 *Mathematics Subject Classification.* Primary: 35Q58; Secondary: 52C07, 53A20.

The author would like to thank the organizers of the SIDE III meeting for invitation and support.

This is the final form of the paper.

$1, n_2)$, $\mathbf{x}(n_1, n_2+1)$, the forth point $\mathbf{x}(n_1+1, n_2+1)$ is fixed prescribing a rule (the one defining the lattice); since we want an *integrable lattice*, this *rule must be linear.* We shall impose the simplest linear rule: *planarity*; i.e., the forth point $\mathbf{x}(n_1+1, n_2+1)$ will belong to the plane generated by the first three points $\mathbf{x}(n_1, n_2)$, $\mathbf{x}(n_1+1, n_2)$ and $\mathbf{x}(n_1, n_2+1)$ or, in algebraic terms,

$$\Delta_1\Delta_2\mathbf{x} = (T_1A_{12})\Delta_1\mathbf{x} + (T_2A_{21})\Delta_2\mathbf{x}, \tag{1}$$

where T_i is the translation operator in the i direction and $\Delta_i = T_i - 1$ is the corresponding difference operator. Given two discrete initial curves (two sequences of points) $\{\mathbf{x}_1^{(0)}\}$, $\{\mathbf{x}_2^{(0)}\}$ and two arbitrary sets of functions $A_{12}(n_1, n_2)$, $A_{21}(n_1, n_2)$: $\mathbf{Z}^2 \to \mathbf{R}$, the planarity constraint allows one to construct uniquely a 2-dimensional lattice, which we have called *quadrilateral* (or *planar*). The quadrilateral surface was first proposed in [**27**] as the proper discrete analogue of a conjugate net on a surface, without any connection with integrability.

Indeed, at the moment, the above construction does not seem to have anything to do with integrability; to show its profound connection with the familiar integrability schemes it is necessary to generalize the picture to higher dimensions, imposing the planarity constraint on each two-dimensional (ij) surface of the lattice.

Definition 1.1 ([**12**]). An N dimensional lattice $\mathbf{x}\colon \mathbf{Z}^N \to \mathbf{R}^M$ is a quadrilateral lattice (QL) iff its elementary quadrilaterals $\{\mathbf{x}, T_i\mathbf{x}, T_j\mathbf{x}, T_iT_j\mathbf{x}\}$ are planar; i.e., iff

$$\Delta_i\Delta_j\mathbf{x} = (T_iA_{ij})\Delta_i\mathbf{x} + (T_jA_{ji})\Delta_j\mathbf{x}, \quad i \neq j,\ i,j = 1,\dots,N. \tag{2}$$

The planarity constraints (2) are compatible only for the special class of data A_{ij} satisfying the nonlinear system

$$\Delta_kA_{ij} = (T_jA_{jk})A_{ij} + (T_kA_{kj})A_{ik} - (T_kA_{ij})A_{ik}, \quad i \neq j \neq k \neq i. \tag{3}$$

These equations characterize the quadrilateral lattice and we refer to them as the QL equations. They were first derived in [**4**] as integrable discrete analogues of the Darboux equations for conjugate nets, but without any geometric characterization.

We remark that, if a quadrilateral lattice exists, it is integrable by construction, since it is built out of the linear constraints (2). Indeed, in the language of the theory of integrable systems, the planarity constraints correspond to the set of linear spectral problems (2) and the resulting QL and its characterizing equations (3) correspond to the integrable nonlinear equations arising as the compatibility condition for such spectral problems.

Actually, in connection with integrability, certain constructions based on the quadrilateral surfaces were already known: the integrable discrete analogue of isothermic surfaces [**3**] and the Laplace sequence of quadrilateral surfaces [**9**], which provides the geometric meaning for the Hirota equation (see also Section 2.2 and [**10**] for more details).

It is often convenient to reformulate equations (2) as a first order system [**12**]. To do so we introduce the suitably scaled tangent vectors $\mathbf{X}_i$, $i = 1,\dots,N$:

$$\Delta_i\mathbf{x} = (T_iH_i)\mathbf{X}_i, \tag{4}$$

in such a way that the j-th variation of $\mathbf{X}_i$ is proportional to $\mathbf{X}_j$ only:

$$\Delta_j\mathbf{X}_i = (T_jQ_{ij})\mathbf{X}_j, \quad i \neq j. \tag{5}$$

The compatibility condition for the system (5) gives the following new form of the QL equations

$$\Delta_k Q_{ij} = (T_k Q_{ik}) Q_{kj}, \quad i \neq j \neq k \neq i, \tag{6}$$

and the scaling factors H_i, called the Lamé coefficients, solve the linear equations

$$\Delta_i H_j = (T_i H_i) Q_{ij}, \quad i \neq j, \tag{7}$$

whose compatibility gives equations (6) again; moreover $A_{ij} = (\Delta_j H_i / H_i)$ $i \neq j$.

The solution of the initial value problem for the QL is contained in the following result [**12**].

THEOREM 1.2 (The Initial Value Problem for the QL). *Given N initial discrete curves $\{\mathbf{x}_i^{(0)}\}$, $i = 1, \ldots, N$, assigning the initial data $A_{ij}^{(0)}(n_i, n_j)$, $A_{ji}^{(0)}(n_i, n_j)$, $i, j = 1, \ldots, N$, $i \neq j$ (or, equivalently, the initial data $H_i^{(0)}(n_i)$, $Q_{ij}^{(0)}$, $Q_{ji}^{(0)}$), one constructs the initial $N(N-1)/2$ quadrilateral surfaces of the lattice applying equations* (2) *(or, equivalently, equations* (5)*). Then the quadrilateral lattice* $\mathbf{x}$ *follows uniquely from the planarity constraint.*

In the continuous limit:

$$\Delta_i \sim \varepsilon \frac{\partial}{\partial u_i} = \varepsilon \partial_i, \quad Q_{ij} \sim \varepsilon \beta_{ij}, \quad 0 < \varepsilon \lll 1, \tag{8}$$

the QL reduces to an N dimensional conjugate net in $\mathbf{R}^M$, characterized by the famous Darboux equations [**6**]

$$\partial_k \beta_{ij} = \beta_{ik} \beta_{kj}, \quad i \neq j \neq k \neq i. \tag{9}$$

1.2. The backward representation of the quadrilateral lattice and the geometric meaning of the τ function. In this Section we define the backward data $\widetilde{\mathbf{X}}_i$, $\widetilde{H}_i$, $\widetilde{Q}_{ij}$ of the quadrilateral lattice. It turns out that the relation between the forward data $\mathbf{X}_i$, H_i, Q_{ij} and the backward data is given in terms of the τ-function [**13**], which is one of central objects of the theory of integrable systems.

In the previous Section the quadrilateral lattice was built through a *forward construction*; it is of course possible to build the lattice also through a *backward construction*. The backward tangent vectors $\widetilde{\mathbf{X}}_i$ and the backward Lamé coefficients $\widetilde{H}_i$, $i = 1, \ldots, N$ are defined by the equations

$$\widetilde{\Delta}_i \mathbf{x} = (T_i^{-1} \widetilde{H}_i) \widetilde{\mathbf{X}}_i \quad \text{or} \quad \Delta_i \mathbf{x} = \widetilde{H}_i (T_i \widetilde{\mathbf{X}}_i); \tag{10}$$

in terms of the backward difference operator $\widetilde{\Delta}_i := 1 - T_i^{-1}$. The backward Lamé coefficients are again chosen in such a way that the $\widetilde{\Delta}_i$ variation of $\widetilde{\mathbf{X}}_j$ is proportional to $\widetilde{\mathbf{X}}_i$ only. We define the backward rotation coefficients $\widetilde{Q}_{ij}$ as the corresponding proportionality factors

$$\widetilde{\Delta}_i \widetilde{\mathbf{X}}_j = (T_i^{-1} \widetilde{Q}_{ij}) \widetilde{\mathbf{X}}_i \quad \text{or} \quad \Delta_i \widetilde{\mathbf{X}}_j = (T_i \widetilde{\mathbf{X}}_i) \widetilde{Q}_{ij}, \quad i \neq j. \tag{11}$$

Comparing equations (7) and (11) we see immediately that the new functions $\widetilde{Q}_{ij}$ satisfy the MQL equations (6) as well. Moreover the new scaling factors $\widetilde{H}_i$ satisfy the following system of linear equations

$$\Delta_j \widetilde{H}_i = (T_j \widetilde{Q}_{ij}) \widetilde{H}_j, \quad i \neq j, \tag{12}$$

whose compatibility condition gives again the QL equations (6).

An easy consequence of equations (10), (11) and (12) is the following

PROPOSITION 1.3. *The vector function* $\mathbf{x}\colon \mathbf{Z}^N \to \mathbf{R}^M$ *representing a quadrilateral lattice satisfies the backward Laplace equation*

$$\tilde{\Delta}_i\tilde{\Delta}_j\mathbf{x} = (T_i^{-1}\tilde{A}_{ij})\tilde{\Delta}_i\mathbf{x} + (T_j^{-1}\tilde{A}_{ji})\tilde{\Delta}_j\mathbf{x}, \quad i \neq j, \tag{13}$$

where, in the notation of this Section, $\tilde{A}_{ij} = (\tilde{\Delta}_j\widetilde{H}_i/\widetilde{H}_i)$.

The forward and backward rotation coefficients Q_{ij} and $\widetilde{Q}_{ij}$ describe the same lattice $\mathbf{x}$ from different points of view, therefore they are not independent. Indeed, defining the functions $\rho_i\colon \mathbf{Z}^N \to \mathbf{R}$ as the proportionality factors between $\mathbf{X}_i$ and $T_i\widetilde{\mathbf{X}}_i$ (both vectors are proportional to $\Delta_i\mathbf{x}$):

$$\mathbf{X}_i = -\rho_i(T_i\widetilde{\mathbf{X}}_i), \quad T_iH_i = -\frac{1}{\rho_i}\widetilde{H}_i, \quad i = 1, \ldots, N, \tag{14}$$

we have the following

PROPOSITION 1.4. *The forward and backward data of the lattice* $\mathbf{x}$ *are related through the following formulas*

$$\rho_jT_j\widetilde{Q}_{ij} = \rho_iT_iQ_{ji}, \tag{15}$$

and the factors ρ_i *are first potentials satisfying equations*

$$\frac{T_j\rho_i}{\rho_i} = 1 - (T_iQ_{ji})(T_jQ_{ij}), \quad i \neq j. \tag{16}$$

REMARK 1.5. Since Q_{ij} and $\widetilde{Q}_{ij}$ are both solutions of the QL equations (6), then equations (15), (16) describe a special symmetry transformation of equations (15).

The RHS of equation (16) is symmetric with respect to the interchange of i and j; this implies the existence of a potential $\tau\colon \mathbf{Z}^N \to \mathbf{R}$, such that

$$\rho_i = \frac{T_i\tau}{\tau}; \tag{17}$$

therefore equation (16) defines the second potential τ:

$$\frac{(T_iT_j\tau)\tau}{(T_i\tau)(T_j\tau)} = 1 - (T_iQ_{ji})(T_jQ_{ij}), \quad i \neq j. \tag{18}$$

The potential τ connecting the forward and backward data:

$$T_j(\tau\widetilde{Q}_{ij}) = T_i(\tau Q_{ji}), \tag{19}$$

$$T_i(\tau\widetilde{\mathbf{X}}_i) = \tau\mathbf{X}_i, \tag{20}$$

$$\tau\widetilde{H}_i = T_i(\tau H_i), \tag{21}$$

is the famous τ-*function* of the quadrilateral lattice. In terms of the τ function, the QL equations read as follows [**16**]:

$$(T_iT_j\tau)\tau = (T_i\tau)T_j\tau - (T_i\tau_{ji})T_j\tau_{ij}, \tag{22}$$

$$(T_k\tau_{ij})\tau = (T_k\tau)\tau_{ij} + (T_k\tau_{ik})\tau_{kj}, \tag{23}$$

where

$$\tau_{ij} = \tau Q_{ij}. \tag{24}$$

2. Transformations of Quadrilateral Lattices

We present the basic ideas and results of the theory of the Darboux type transformations of the multidimensional quadrilateral lattice. Our approach follows the spirit of the book [**17**], which summarizes the theory of transformations of (two dimensional) conjugate nets. We also name the transformations of the quadrilateral lattice according to the classical geometric terminology of the transformations of conjugate nets.

All the results of this section can be found in [**14**], to whom we also refer for proofs and more detailed explanation.

2.1. Transformations and congruences. To define the transformations we need another geometric object—the *congruence*—which serves as the link between two quadrilateral lattices.

DEFINITION 2.1. An N-dimensional *rectilinear congruence* (or, simply, congruence) is a mapping $\mathbf{l}\colon \mathbf{Z}^N \to \mathrm{Gr}_A(2, M+1)$ from the integer lattice to the space of lines in $\mathbf{R}^M$ such that every two neighboring lines $\mathbf{l}$ and $T_i\mathbf{l}$, $i = 1, \ldots, N$, are coplanar.

We remark that with any N-dimensional quadrilateral lattice we may naturally associate N congruences as follows.

DEFINITION 2.2. Given an N-dimensional quadrilateral lattice $\mathbf{x}$, its i-th *tangent congruence* $\mathbf{t}_i(\mathbf{x})$ consists of the lines passing through the points $\mathbf{x}$ of the lattice and directed along the tangent vectors $\Delta_i\mathbf{x}$.

Reversing that process, one can associate with any N-dimensional congruence N lattices which are, in general, quadrilateral lattices.

DEFINITION 2.3. The i-th *focal lattice* $\mathbf{y}_i(\mathbf{l})$ of a congruence $\mathbf{l}$ is the lattice constructed out of the intersection points of the lines $\mathbf{l}$ with $T_i^{-1}\mathbf{l}$.

There exists a simple (but basic for the theory of transformations) relation between quadrilateral lattices and congruences.

DEFINITION 2.4. A quadrilateral lattice and a congruence are called *conjugate* if there exists a one-to-one correspondence between points of the lattice and lines of the congruence such that lines pass through the corresponding points.

COROLLARY 2.5. *Focal lattices of congruences conjugate to quadrilateral lattices are quadrilateral lattices.*

The above introduced notions allow to define all the transformations of quadrilateral lattices.

DEFINITION 2.6. Two quadrilateral lattices are related by the *fundamental transformation* $\mathcal{F}$ *of Jonas* when they are conjugate to the same congruence, which is called the congruence of the transformation.

Usually one assumes that, in the fundamental transformation, the congruence is *transversal* to the lattice, i.e., nontangent to the lattice in corresponding points. Otherwise we get the following reductions of the fundamental transformation

- the congruence of the transformation is the i-th tangent congruence of the original lattice—the *Lévy transformation* $\mathcal{L}_i$,
- the congruence of the transformation is the i-th tangent congruence of the new lattice—the *adjoint Lévy transformation* $\mathcal{L}_i^*$,

- both lattices are focal lattices of the congruence of the transformation— the *Laplace transformation.*

In addition, there are two useful special transformations

- the *projective* or *radial transformation*, when all the lines of the congruence meet in a single point,
- the *Combescure transformation*, when the transformed lattice is parallel to the original one.

Let us present some formulas for the transformations, the details can be found in [**14**].

THEOREM 2.7. *The fundamental transformation* $\mathcal{F}(\mathbf{x})$ *of the quadrilateral lattice* $\mathbf{x}$ *is given by*

$$\mathcal{F}(\mathbf{x}) = \mathbf{x} - \frac{\phi}{\phi_C}\mathbf{x}_C, \tag{25}$$

where

(i) $\phi\colon \mathbf{Z}^N \to \mathbf{R}$ *is a new solution of the Laplace equation* (2) *of the lattice* $\mathbf{x}$

(ii) $\mathbf{x}_C$ *is the Combescure transformation vector, which is a solution of the equations*

$$\Delta_i \mathbf{x}_C = (T_i\sigma_i)\Delta_i\mathbf{x}, \tag{26}$$

where, due to the compatibility of the system (26), *the functions* σ_i *satisfy*

$$\Delta_j\sigma_i = A_{ij}(T_j\sigma_j - T_j\sigma_i), \quad i \neq j; \tag{27}$$

moreover

(iii) ϕ_C *is a solution, corresponding to* ϕ, *of the Laplace equation of the lattice* $\mathbf{x}_C$, *i.e.*

$$\Delta_i\phi_C = (T_i\sigma_i)\Delta_i\phi. \tag{28}$$

The Combescure transformation vector $\mathbf{x}_C$ is used to construct the congruence of the transformation, and the function ϕ serves to place points of the new lattice on the lines of the congruence. The definitions (25) and (26)–(27) of the fundamental and Combescure transformations were first presented in [**20**].

REMARK 2.8. Notice that, given $\mathbf{x}_C$ and ϕ, then equation (28) determines ϕ_C uniquely, up to a constant of integration.

COROLLARY 2.9. *One can stop the transformation of the lattice* $\mathbf{x}$ *at the intermediate level of construction of the vector* $\mathbf{x}_C$ *to obtain the Combescure transformation*

$$\mathcal{C}(\mathbf{x}) = \mathbf{x} + \mathbf{x}_C, \tag{29}$$

characterized by the property that the tangent lines to both lattices in corresponding points are parallel.

DEFINITION 2.10. The i-th *Lévy transform* $\mathcal{L}_i(\mathbf{x})$ of the quadrilateral lattice $\mathbf{x}$ is a quadrilateral lattice conjugate to the i-th tangent congruence of $\mathbf{x}$.

Since the congruence of the transformation is known, to construct the transformation we need only the solution ϕ of the Laplace equation of the lattice $\mathbf{x}$.

PROPOSITION 2.11. *The Lévy transform $\mathcal{L}_i(\mathbf{x})$ of the quadrilateral lattice $\mathbf{x}$ is given by*

$$\mathcal{L}_i(\mathbf{x}) = \mathbf{x} - \frac{\phi}{\Delta_i \phi}\Delta_i \mathbf{x}. \tag{30}$$

Another way to reduce the fundamental transformation is to fix the way we place the new lattice on the congruence.

DEFINITION 2.12. The i-th *adjoint Lévy transform* $\mathcal{L}_i^*(\mathbf{x})$ of the quadrilateral lattice $\mathbf{x}$ is the i-th focal lattice of a congruence conjugate to $\mathbf{x}$.

To find the adjoint Lévy transformation we need only the congruence of the transformation, which can be find via the Combescure transformation vector $\mathbf{x}_{\mathcal{C}}$.

PROPOSITION 2.13. *The adjoint Lévy transform of the lattice $\mathbf{x}$ is given by*

$$\mathcal{L}_i^*(\mathbf{x}) = \mathbf{x} - \frac{1}{\sigma_i}\mathbf{x}_{\mathcal{C}}, \tag{31}$$

where the functions σ_i are defined in Theorem 2.7.

The deepest reduction of the fundamental transformation is the Laplace transformation where both the congruence of the transformation and the way to place the new lattice on the congruence are fixed.

DEFINITION 2.14. The *Laplace transform* $\mathcal{L}_{ij}(\mathbf{x})$ of the quadrilateral lattice $\mathbf{x}$ is the j-th focal lattice of its i-th tangent congruence.

PROPOSITION 2.15. *The Laplace transformation of the quadrilateral lattice $\mathbf{x}$ is given by*

$$\mathcal{L}_{ij}(\mathbf{x}) = \mathbf{x} - \frac{1}{A_{ji}}\Delta_i \mathbf{x}. \tag{32}$$

COROLLARY 2.16. *The superpositions of Laplace transformations satisfy the following identities*

$$\mathcal{L}_{ij} \circ \mathcal{L}_{ji} = \mathrm{id}, \tag{33}$$

$$\mathcal{L}_{jk} \circ \mathcal{L}_{ij} = \mathcal{L}_{ik}, \tag{34}$$

$$\mathcal{L}_{ki} \circ \mathcal{L}_{ij} = \mathcal{L}_{kj}. \tag{35}$$

Finally, the notion of radial (or projective) transformation is related to the so called radial congruence, whose focal lattices are degenerated to a single point, which can be taken as the origin.

DEFINITION 2.17. The radial (or projective) transform $\mathcal{P}(\mathbf{x})$ of the quadrilateral lattice $\mathbf{x}$ is a quadrilateral lattice conjugate to the radial congruence of $\mathbf{x}$.

PROPOSITION 2.18. *The radial transform $\mathcal{P}(\mathbf{x})$ is given by*

$$\mathcal{P}(\mathbf{x}) = \frac{1}{\phi}\mathbf{x}, \tag{36}$$

where $\phi\colon \mathbf{Z}^N \to \mathbf{R}$ is a solution of the Laplace equation (2) *of the lattice $\mathbf{x}$.*

The degeneration of the fundamental transformation to the Lévy, adjoint Lévy, the Combescure, radial, and the Laplace transformations can be shown also on the analytic level. From another point of view, the fundamental transformation can be decomposed into the superposition of the Lévy and adjoint Lévy transformations, or into the superposition of two Combescure and one radial transformation. Moreover the fundamental transformation can be put into the scheme of the Laplace transformation (preceeded by building a new level on the original lattice).

2.2. The Hirota equation and the Laplace transformations of quadrilateral surfaces. The Laplace transformations of two dimensional conjugate nets generates the two dimensional Toda system [**7, 24**]. Analogously, one can associate [**9**] (see also [**10**] for more details and new interpretations) the integrable discrete version of the Toda system (the Hirota equation [**18**]) with the Laplace transformations of a two dimensional quadrilateral lattice.

Given a quadrilateral surface $\mathbf{x}\colon \mathbf{Z}^2 \to \mathbf{R}^M$, consider the sequence $\mathbf{x}^{(k)}$ of quadrilateral surfaces obtained from $\mathbf{x}$ by recursive application of the Laplace transformation $\mathcal{L}_{12}$:

$$\mathbf{x}^{(k)} = (\mathcal{L}_{12})^k(\mathbf{x}), \quad k \in \mathbf{Z}. \tag{37}$$

The above sequence, which is the proper analogue of the Laplace sequence of conjugate nets, is well defined due to formula (33). The coefficients $A_{12}^{(k)}$, $A_{21}^{(k)}$ of the Laplace equations of the sequence of lattices satisfy the following system

$$\frac{A_{12}^{(k+1)}+1}{T_1 A_{12}^{(k)}+1} = \frac{A_{21}^{(k)}}{T_2 A_{21}^{(k)}}, \tag{38}$$

$$\frac{A_{21}^{(k-1)}+1}{T_2 A_{21}^{(k)}+1} = \frac{A_{12}^{(k)}}{T_1 A_{12}^{(k)}}. \tag{39}$$

Notice that one can consider the parameter $k \in \mathbf{Z}$ along the Laplace sequence as the third discrete variable. Actually, the Laplace sequence of quadrilateral surfaces may be viewed as a deep reduction of the three dimensional quadrilateral lattice in the same sense like the Laplace transformations are reductions of the fundamental transformation. An elementary quadrilateral of such reduced (or degenerated) lattice with $T_3 = \mathcal{L}_{12}$ is shown below.

Define the function $K\colon \mathbf{Z}^2 \to \mathbf{R}$ as the cross-ratio of four collinear points: x, $\mathcal{L}_{12}(x)$, $T_1 x$ and $T_j\mathcal{L}_{12}(x)$. Simple derivation shows that

$$K = \frac{A_{21}(T_1 A_{12}+1) - T_2 A_{21}}{(1+T_1 A_{12})(1+A_{21}}, \tag{40}$$

and equations (38)–(39) imply that K satisfies the gauge invariant form of the Hirota equation

$$T_2\left(\frac{K^{(l+1)}+1}{K^{(l)}+1}\right) T_1\left(\frac{K^{(l-1)}+1}{K^{(l)}+1}\right) = \frac{(T_1T_2K^{(l)})K^{(l)}}{(T_1K^{(l)})(T_2K^{(l)})}. \tag{41}$$

2.3. Superposition of fundamental transformations. Consider $K \geq 1$ fundamental transformations $\mathcal{F}_k(\mathbf{x})$, $k = 1, \ldots, K$, of the quadrilateral lattice $\mathbf{x}$, which are built from

(i) K solutions ϕ^k, $k = 1, \ldots, K$ of the Laplace equation of the lattice $\mathbf{x}$;
(ii) K Combescure transformation vectors $\mathbf{x}_{\mathcal{C},k}$, where

$$\Delta_i \mathbf{x}_{\mathcal{C},k} = (T_i\sigma_{i,k})\Delta_i\mathbf{x}, \quad i = 1, \ldots, N,\ k = 1, \ldots, K, \tag{42}$$

and $\sigma_{i,k}$ satisfy equations

$$\Delta_j\sigma_{i,k} = A_{ij}(T_j\sigma_{j,k} - T_j\sigma_{i,k}), \quad i \neq j; \tag{43}$$

(iii) K functions $\phi^k_{C,k}$, which satisfy

$$\Delta_i \phi^k_{C,k} = (T_i \sigma_{i,k}) \Delta_i \phi^k. \tag{44}$$

We arrange functions ϕ^k in the K component vector $\mathbf{\Phi} = (\phi^1, \ldots, \phi^K)^T$; similarily, we arrange the Combescure transformation vectors $\mathbf{x}_{C,k}$ into the $M \times K$ matrix $\mathbf{X}_C = (\mathbf{x}_{C,1}, \ldots, \mathbf{x}_{C,K})$; moreover we introduce

(iv) the $K \times K$ matrix $\mathbf{\Phi}_C = (\mathbf{\Phi}_{C,1}, \ldots, \mathbf{\Phi}_{C,K})$, whose columns are the K component vectors $\mathbf{\Phi}_{C,k} = (\phi^1_{C,k}, \ldots, \phi^K_{C,k})^T$, which are the Combescure transforms of $\mathbf{\Phi}$

$$\Delta_i \mathbf{\Phi}_{C,k} = (T_i \sigma_{i,k}) \Delta_i \mathbf{\Phi}. \tag{45}$$

REMARK 2.19. The diagonal part of $\mathbf{\Phi}_C$ is fixed by the initial fundamental transformations. To find the off-diagonal part of $\mathbf{\Phi}_C$ we integrate equations (45) introducing $K(K-1)$ arbitrary constants.

One can show that the vectorial fundamental transformation $\mathbf{F}(\mathbf{x})$ of the quadrilateral lattice $\mathbf{x}$, which is defined as

$$\mathbf{F}(\mathbf{x}) = \mathbf{x} - \mathbf{X}_C \mathbf{\Phi}_C^{-1} \mathbf{\Phi}, \tag{46}$$

is again a quadrilateral lattice. Moreover, the vectorial transformation is the superposition of the fundamental transformations

$$\mathbf{F}(\mathbf{x}) = (\mathcal{F}_{k_1} \circ \mathcal{F}_{k_2} \circ \cdots \circ \mathcal{F}_{k_K})(\mathbf{x}), \quad k_i \neq k_j \quad \text{for} \quad i \neq j, \tag{47}$$

and does not depend on the order in which the transformations are taken. In applying the fundamental transformations at the intermediate stages, the transformation data should be suitably transformed as well.

To prove the superposition and permutability statements, it is important to notice the following basic fact.

THEOREM 2.20 (Permutability theorem). *Assume the following splitting of the data of the vectorial fundamental transformation*

$$\mathbf{\Phi} = \begin{pmatrix} \mathbf{\Phi}^{(1)} \\ \mathbf{\Phi}^{(2)} \end{pmatrix}, \quad \mathbf{X}_C = (\mathbf{X}_{C(1)}, \mathbf{X}_{C(2)}), \quad \mathbf{\Phi}_C = \begin{pmatrix} \mathbf{\Phi}^{(1)}_{C(1)} & \mathbf{\Phi}^{(1)}_{C(2)} \\ \mathbf{\Phi}^{(2)}_{C(1)} & \mathbf{\Phi}^{(2)}_{C(2)} \end{pmatrix}, \tag{48}$$

associated with the partition $K = K_1 + K_2$. Then the vectorial fundamental transformation $\mathbf{F}(\mathbf{x})$ is equivalent to the following superposition of vectorial fundamental transformations:

1. *Transformation $\mathbf{F}_{(1)}(\mathbf{x})$ with the data $\mathbf{\Phi}^{(1)}$, $\mathbf{X}_{C(1)}$, $\mathbf{\Phi}^{(1)}_{C(1)}$:*

$$\mathbf{F}_{(1)}(\mathbf{x}) = \mathbf{x} - \mathbf{X}_{C(1)} (\mathbf{\Phi}^{(1)}_{C(1)} t)^{-1} \mathbf{\Phi}^{(1)}. \tag{49}$$

2. *Application on the result obtained in point 1., transformation $\mathbf{F}_{(2)}$ with the data transformed by the transformation $\mathbf{F}_{(1)}$ as well*

$$\mathbf{F}_{(2)}(\mathbf{F}_{(1)}(\mathbf{x})) = \mathbf{F}_{(1)}(\mathbf{x}) - \mathbf{F}_{(1)}(\mathbf{X}_{C(2)}) \big(\mathbf{F}_{(1)}(\mathbf{\Phi}^{(2)}_{C(2)})\big)^{-1} \mathbf{F}_{(1)}(\mathbf{\Phi}^{(2)}), \tag{50}$$

where

$$\mathbf{F}_{(1)}(\mathbf{X}_{C(2)}) = \mathbf{X}_{C(2)} - \mathbf{X}_{C(1)} (\mathbf{\Phi}^{(1)}_{C(1)})^{-1} \mathbf{\Phi}^{(1)}_{C(2)}, \tag{51}$$

$$\mathbf{F}_{(1)}(\mathbf{\Phi}^{(2)}) = \mathbf{\Phi}^{(2)} - \mathbf{\Phi}^{(2)}_{C(1)} (\mathbf{\Phi}^{(1)}_{C(1)})^{-1} \mathbf{\Phi}^{(1)}, \tag{52}$$

$$\mathbf{F}_{(1)}(\mathbf{\Phi}^{(2)}_{C(2)}) = \mathbf{\Phi}^{(2)}_{C(2)} - \mathbf{\Phi}^{(2)}_{C(1)} (\mathbf{\Phi}^{(1)}_{C(1)})^{-1} \mathbf{\Phi}^{(1)}_{C(2)}. \tag{53}$$

Another important feature of the fundamental transformation is that its recursive application generates new dimensions of the quadrilateral lattice. In particular, when applied to the continuous limit of the quadrilateral lattice—the conjugate net—this procedure explains, on the geometric level, the power of using the Darboux type transformations in discretizing the integrable equations.

2.4. Vertex operators as transformations of conjugate nets. The continuous limit of the above theory reduces to the classical theory of finite transformations of N dimensional conjugate nets. It was recently shown in [**16**] that these transformations of conjugate nets are strictly related to the quantum field theoretical formulation of the multicomponent Kadomtsev–Petviashvilii hierarchy.

The $b-c$ system of quantum fields, which appears as the system of ghost fields in string theory, is constructed in terms of the anticommutation relations

$$\{b_i(z), c_j(z')\} = \delta_{ij}\delta(z-z'), \tag{54}$$

$$\{b_i(z), b_j(z')\} = \{c_i(z), c_j(z')\} = 0, \tag{55}$$

where $b_i(z)$ and $c_i(z)$, $i = 1, \ldots, N$, are free charged fermion fields defined on the unit circle S^1, and $\delta(z - z')$ is the Dirac distribution on S^1.

The Clifford algebra generated by the $b-c$ system admits a representation in terms of bosonic variables. In this representation the fields act on the Fock space $\mathcal{F}$ of complex-valued functions

$$\tau = \sum_{\boldsymbol{\ell}} \tau(\boldsymbol{\ell}, \mathbf{t})\boldsymbol{\lambda}^{\boldsymbol{\ell}},$$

with

$$\boldsymbol{\ell} := (\ell_1, \cdots, \ell_N) \in \mathbf{Z}^N, \tag{56}$$

$$\mathbf{t} := (\mathbf{t}_1, \ldots, \mathbf{t}_N) \in \mathbf{C}^{N\cdot\infty}, \quad \mathbf{t}_i := (t_{i,1}, t_{i,2}, \ldots) \in \mathbf{C}^\infty,$$

$$\boldsymbol{\lambda} := (\lambda_1, \ldots, \lambda_N) \in \mathbf{C}^N, \quad \boldsymbol{\lambda}^{\boldsymbol{\ell}} := \lambda_1^{\ell_1} \ldots \lambda_N^{\ell_N}. \tag{57}$$

The representation of the b-c generators takes the form [**8**]:

$$b_i(z)\tau(\boldsymbol{\ell}, \mathbf{t}) = (-1)^{\sum_{j>i}\ell_j} z^{\ell_i - 1} \exp\big(\xi(z, \mathbf{t}_i)\big)\tau\big(\boldsymbol{\ell} - \mathbf{e}_i, \mathbf{t} - [1/z]\mathbf{e}_i\big), \tag{58}$$

$$c_i(z)\tau(\boldsymbol{\ell}, \mathbf{t}) = (-1)^{\sum_{j>i}\ell_j} z^{-\ell_i} \exp\big(-\xi(z, \mathbf{t}_i)\big)\tau\big(\boldsymbol{\ell} + \mathbf{e}_i, \mathbf{t} + [1/z]\mathbf{e}_i\big), \tag{59}$$

where

$$\xi(z, \mathbf{t}_i) := \sum_{n=1}^{\infty} z^n t_{i,n}, \quad [1/z] := \left(\frac{1}{z}, \frac{1}{2z^2}, \frac{1}{3z^3}, \ldots\right),$$

and $\{\mathbf{e}_i\}_{i=1}^N$ are the canonical generators of $\mathbf{C}^N$.

The Fock space decomposes into a direct sum of charge sectors

$$\mathcal{F} = \bigoplus_{q\in\mathbf{Z}} \mathcal{F}_q, \quad \mathcal{F}_q = \{\tau \in \mathcal{F} : Q\tau = q\tau\},$$

where the total charge operator $Q := \sum_{i=1}^N Q_i$ is the sum of N commuting partial charges $Q_i = \lambda_i\partial/\partial\lambda_i$, $i = 1, \ldots, N$; they correspond to the N different flavours of fermions of the model.

The N-component KP hierarchy can be formulated as the following bilinear identity

$$\mathcal{B}(\tau \otimes \tau) = 0, \quad \tau \in \mathcal{F}_0, \tag{60}$$

where

$$\mathcal{B} := \int_{S^1} \frac{dz}{z} \sum_{i=1}^{N} b_i(z) \otimes c_i(z).$$

In terms of the components $\tau(\boldsymbol{\ell}, \mathbf{t})$ we have

$$\text{(61)} \quad \int_{S^1} dz \sum_{m=1}^{N} (-1)^{\sum_{j>m} \ell_j + \ell'_j} \exp\bigl(\xi(z, \mathbf{t}_m - \mathbf{t}'_m)\bigr) z^{\ell_m - \ell'_m - 2} \times \tau\bigl(\boldsymbol{\ell} - \mathbf{e}_m, \mathbf{t} - [1/z]\mathbf{e}_m\bigr)\tau\bigl(\boldsymbol{\ell}' + \mathbf{e}_m, \mathbf{t}' + [1/z]\mathbf{e}_m\bigr) = 0,$$

for any $\mathbf{t}$, $\mathbf{t}'$ and $\boldsymbol{\ell}$, $\boldsymbol{\ell}'$ such that $\ell_1 + \cdots + \ell_N - 1 = \ell'_1 + \cdots + \ell'_N + 1 = 0$.

Denote by

$$\text{(62)} \qquad \tau_{ij}(\boldsymbol{\ell}, \mathbf{t}) = \mathcal{S}_{ij}\tau(\boldsymbol{\ell}, \mathbf{t}) := \tau(\boldsymbol{\ell} + \mathbf{e}_i - \mathbf{e}_j, \mathbf{t});$$

the shift operators $\mathcal{S}_{ij}$ correspond to the so called Schlesinger transformations in monodromy theory [**8, 19**] and do not alter the neutral character of the assembly of fermions. Moreover, $\mathcal{S}_{ij}$ are obvious symmetries of (60).

The $N \times N$ matrix Baker function ψ and its adjoint ψ^* can be defined in terms of the τ function as

$$\text{(63)} \qquad \psi_{ij}(z, \mathbf{t}) = \varepsilon_{ij} z^{\delta_{ij}-1} \frac{\tau_{ij}(\mathbf{t} - [1/z]\mathbf{e}_j)}{\tau(\mathbf{t})} \exp(\xi(z, \mathbf{t}_j)),$$

$$\text{(64)} \qquad \psi^*_{ij}(z, \mathbf{t}) = \varepsilon_{ji} z^{\delta_{ij}-1} \exp\bigl(-\xi(z, \mathbf{t}_i)\bigr) \frac{\tau_{ij}(\mathbf{t} + [1/z]\mathbf{e}_i)}{\tau(\mathbf{t})},$$

where $\varepsilon_{ij} := \operatorname{sgn}(j - i)$, $j \neq i$ ($\varepsilon_{ii} := 1$).

The connection of the N component hierarchy with the conjugate nets is given in the following result [**16**]

PROPOSITION 2.21. *The solutions of the N-component* KP *hierarchy describe N-dimensional conjugate nets with coordinates $u_i = t_{i,1}$, $i = 1, \ldots, N$, while the remaning times $t_{i,k}$, for $k > 1$, describe integrable iso-conjugate deformations of the nets.*

In particular, the rotation coefficients are given by

$$\text{(65)} \qquad \beta_{ij}(\mathbf{t}) = \varepsilon_{ij} \frac{\tau_{ij}(\mathbf{t})}{\tau(\mathbf{t})}, \quad i \neq j,\ i, j = 1, \ldots, N,$$

the tangent vectors $\mathbf{X}_i$ are constructed from the rows of the matrix

$$\text{(66)} \qquad X(\mathbf{t}) = \int_{S^1} dz \psi(z, \mathbf{t}) f(z),$$

for some distribution matrix $f(z) \in M_{N\times M}(\mathbf{C})$; the Lamé coefficients are given by the entries of the row matrix

$$\text{(67)} \qquad H(\mathbf{t}) = \int_{S^1} dz g(z) \psi^*(z, \mathbf{t}), \quad i = 1, \cdots, N,$$

for some distribution row matrix $g(z) \in \mathbf{C}^N$.

As one can expect, various elements of the KP theory have their counterparts on the level of the conjugate nets.

THEOREM 2.22. (i) *The Schlesinger transformation $\mathcal{S}_{ij}$ gives rise to the Laplace transformation $\mathcal{L}_{ij}$ of the conjugate nets.*

(ii) *The action of the (vertex) operators $b_i(p)$ and $c_i(q)$ gives rise to the Lévy and the adjoint Lévy transformations $\mathcal{L}_i$ and $\mathcal{L}^*_i$. Correspondingly*

(iii) *the soliton vertex operator generated by the superposition $b_i(p)c_i(q)$ gives rise to the fundamental transformation.*

3. Quadrilateral Hyperplane Lattices

In this Section we introduce and study the properties of quadrilateral lattices in the dual space (hyperplane lattices) [**13**]. These considerations turn out to be relevant in the reduction theory of the quadrilateral lattices, when the introduction of the additional geometric structure will allow to establish a direct connection between point lattices and hyperplane lattices.

Generic hyperplanes can be represented by the covectors $\mathbf{a}^* \in (\mathbf{R}^M)^*$ and consist of points $\mathbf{x}$ satisfying the equation

$$\langle \mathbf{a}^* \mid \mathbf{x} \rangle = a_1^* x^1 + \cdots + a_M^* x^M = 1; \tag{68}$$

the equation of the hyperplane passing through $\mathbf{0}$ and parallel to that represented by $\mathbf{a}^*$ can be written as $\langle \mathbf{a}^* \mid \mathbf{x} \rangle = 0$ (the covector $\mathbf{0}^*$ represents the hyperplane at infinity).

The linear mappings between points and hyperplanes are called *correlations.* Given the sphere of radius R centered at the origin, it defines the corresponding correlation P (called *polarity* with respect to that sphere). The image $P(\mathbf{v})$ of a point $\mathbf{v} \in \mathbf{R}^M$ is called the *polar hyperplane* of $\mathbf{v}$, and consists of points $\mathbf{x}$ satisfying equation $\mathbf{v}^T\mathbf{x} = R^2$, i.e., $P(\mathbf{v}) = (\mathbf{v}^T/R^2)$.

The notions dual to the parallelism of two lines and to the planarity of four points are contained in the following

DEFINITION 3.1. (i) Two subspaces of codimension 2 are called "coparallel" if there exists a hyperplane passing through the origin and containing them.
(ii) Four hyperplanes are "coplanar" if the fourth hyperplane contains the intersection of the first three.

A *hyperplane lattice* is simply a lattice $\mathbf{y}^* \colon \mathbf{Z}^N \to (\mathbf{R}^M)^*$, $N \leq M$, in the space of hyperplanes. From Definition 3.1 we obtain the following geometric object dual to the quadrilateral lattice.

DEFINITION 3.2. The hyperplane lattice $\mathbf{y}^* \colon \mathbf{Z}^N \to (\mathbf{R}^M)^*$ is "quadrilateral" ("coplanar") if, for any $i, j = 1, \ldots, N$, $i \neq j$, the hyperplane $T_iT_j\mathbf{y}^*$ contains the subspace $\mathbf{y}^* \cap T_i\mathbf{y}^* \cap T_j\mathbf{y}^*$.

This property reduces again to the Laplace equation:

$$\Delta_i \Delta_j \mathbf{y}^* = (T_i A_{ij}^*)\Delta_i \mathbf{y}^* + (T_j A_{ji}^*)\Delta_j \mathbf{y}^*, \quad i \neq j. \tag{69}$$

A convenient way to construct quadrilateral hyperplane lattices makes use of systems of parallel quadrilateral point lattices. We recall that two quadrilateral lattices $\mathbf{x}$, $\mathbf{x}'$ are parallel (Combescure related) iff $\Delta_i \mathbf{x}' \sim \Delta_i \mathbf{x}$; in this case the scaled tangent vectors of the two lattices can be chosen to be equal, then the rotation coefficients Q_{ij}, Q'_{ij} of both lattices coincide and the Lamé coefficients H_i and H'_i are solutions of the same equation.

DEFINITION 3.3. Consider a system of M parallel and linearly independent point lattices $\mathbf{x}_{(k)}$, $k = 1, \ldots, M$. Denote by $\mathbf{y}^*_{(k)}$ the hyperplane lattice made out of the hyperplanes passing through the (corresponding) points of $\mathbf{x}_{(k)}$ and spanned by the vectors $\mathbf{x}_{(l)}$, $l \neq k$, i.e.,

$$\langle \mathbf{y}^*_{(k)} \mid \mathbf{x}_{(l)} \rangle = \delta_{kl}. \tag{70}$$

We call the system $\{\mathbf{y}^*_{(k)}\}$ of hyperplane lattices the "dual system" to the system of parallel point lattices $\{\mathbf{x}_{(k)}\}$.

If we arrange the parallel system $\mathbf{x}_{(k)}$, $k = 1, \ldots, M$ of point lattices as columns of the *matrix* $\boldsymbol{\Omega}$ *of the system*:

$$\boldsymbol{\Omega} = (\mathbf{x}_{(1)}, \ldots, \mathbf{x}_{(M)}), \tag{71}$$

then we have the following results.

1. The dual system $\{\mathbf{y}^*_{(k)}\}$ consists of coparallel quadrilateral hyperplane lattices and is given by the rows of the matrix $\boldsymbol{\Omega}^{-1}$.
2. The rows of the matrix $\boldsymbol{\Omega}$ define a system $\{\mathbf{x}^*_{(k)}\}$ of coparallel quadrilateral hyperplane lattices, which we call the "adjoint system" to $\mathbf{x}_{(k)}$, characterized by the property that the forward rotation coefficients of the system $\{\mathbf{x}_{(k)}\}$ are the backward rotation coefficients of the system $\{\mathbf{x}^*_{(k)}\}$: $Q_{ij} = \widetilde{Q}^*_{ij}$.

4. Reductions of the Quadrilateral Lattice

We now study some geometrically distinguished reductions of the QL which possess additional geometric properties that, once imposed on the initial surfaces, "propagate" everywhere through the construction of the lattice. Since the quadrilateral lattice is integrable, these reductions will inherit its integrability properties.

4.1. The symmetric lattice. Forward and backward data allow one to introduce the first basic reduction of the quadrilateral lattice [**13**].

DEFINITION 4.1. A quadrilateral lattice $\mathbf{x}$ is *symmetric* iff its forward rotation coefficients are also its backward rotation coefficients:

$$\widetilde{Q}_{ij} = Q_{ij}. \tag{72}$$

The symmetric lattice admits the following different characterizations [**13**].

1. A quadrilateral lattice is symmetric iff, for a given set of rotation coefficients Q_{ij}, there exists a τ-function of the lattice such that
$$T_i(\tau Q_{ji}) = T_j(\tau Q_{ij}), \quad i \neq j, \tag{73}$$
or equivalently, in terms of the corresponding potentials ρ_i,
$$\rho_i T_i Q_{ji} = \rho_j T_j Q_{ij}. \tag{74}$$
2. A quadrilateral lattice is symmetric iff, for a given set of forward tangent vectors $\mathbf{X}_i$ of the lattice, there exists a complementary set of the backward tangent vectors $\widetilde{\mathbf{X}}_i$ such that the parallelograms $P(T_i\widetilde{\mathbf{X}}_i, T_j\widetilde{\mathbf{X}}_j)$ and $P(\Delta_i\mathbf{X}_j, \Delta_j\mathbf{X}_i)$ are similar.
3. A quadrilateral lattice is symmetric iff, for different indices i, j, k, its rotation coefficients satisfy the following constraint
$$(T_iQ_{ji})(T_jQ_{kj})(T_kQ_{ik}) = (T_jQ_{ij})(T_iQ_{ki})(T_kQ_{jk}). \tag{75}$$

We remark that, unlike the others, the characterization (75) is *local.* After the work [**13**] was completed, we were told that the algebraic formulation (74) of a symmetric lattice was already known to Schief [**29**].

As we have anticipated, the constraints discussed in this paper allow one to establish a connection between quadrilateral point lattices and their duals, the quadrilateral hyperplane lattices. The following proposition describes this connection in the case of the symmetry constraint.

PROPOSITION 4.2. *Given a system of parallel quadrilateral lattices* $\{\mathbf{x}_{(k)}\}_{k=1}^{M}$ *and the associated matrix* $\mathbf{\Omega}$, *then the following properties are equivalent.*

(i) *The matrix* $\mathbf{\Omega}$ *of the system is symmetric:*

$$\mathbf{\Omega} = \mathbf{\Omega}^T. \tag{76}$$

(ii) *The polar hyperplane* $P(\mathbf{x}_{(k)})$ *of the point lattice* $\mathbf{x}_{(k)}$ *coincides with the hyperplane lattice* $\mathbf{x}^*_{(k)}$:

$$P(\mathbf{x}_{(k)}) = \mathbf{x}^*_{(k)}, \quad k = 1, \dots, M. \tag{77}$$

(iii) *The lattices* $\mathbf{x}_{(k)}$, $k = 1, \dots, M$ *are symmetric. Furthermore the associated tangent vectors* $\mathbf{X}_i$ *and* $\mathbf{X}^*_i$ *are related in the following way*

$$\mathbf{X}_i^T = \rho_i (T_i \mathbf{X}^*_i), \quad i = 1, \dots, N. \tag{78}$$

COROLLARY 4.3. *A quadrilateral lattice* $\mathbf{x}$ *is symmetric iff it is adjoint to its own polar.*

In the continuous limit (8) the symmetric quadrilateral lattice reduces to a *symmetric conjugate net*, for which the rotation coefficients β_{ij} satisfying the Darboux equations (9) are symmetric

$$\beta_{ij} = \beta_{ji}. \tag{79}$$

4.2. The circular lattice. The discrete analogue of an N-dimensional orthogonal system of coordinates is the circular lattice [**5**].

DEFINITION 4.4. An N-dimensional lattice is circular if and only if its elementary quadrilaterals are inscribed in circles.

This notion was first proposed in [**23, 25**] for $N = 2$, $M = 3$, as a discrete analogue of surfaces parametrized by curvature lines (see also [**3**]); later in [**2**] for $N = M = 3$ and, finally, for arbitrary $N \leq M$ in [**5**]; subsequently a convenient set of equations characterizing the circular lattices in $\mathbf{E}^3$ was found in [**20**]. A geometric proof of the integrability of the circular lattice was first given in [**5**] while the analytic proof of its integrability was given in [**15**] through the $\bar{\partial}$ method.

An elementary characterization of circular quadrilaterals states that, if a circular quadrilateral is convex, then the sum of its opposite angles is π; when the quadrilateral is skew, then its opposite angles are equal. Other convenient characterizations are the following.

1. A quadrilateral lattice is circular if and only if:

$$\cos\angle(\mathbf{X}_i, T_i\mathbf{X}_j) + \cos\angle(\mathbf{X}_j, T_j\mathbf{X}_i) = 0, \tag{80}$$

2. A quadrilateral lattice is circular if and only if [**15**]:

$$\mathbf{X}_i \cdot T_i\mathbf{X}_j + \mathbf{X}_j \cdot T_j\mathbf{X}_i = 0, \quad i \neq j. \tag{81}$$

3. A quadrilateral lattice $\mathbf{x}$ is circular iff the scalars

$$v_i := (T_i\mathbf{x} + \mathbf{x}) \cdot \mathbf{X}_i, \quad i = 1, \ldots, N, \tag{82}$$

solve the linear system (5) or, equivalently, iff the function $|\mathbf{x}|^2$ (the square of the norm of $\mathbf{x}$) satisfies the Laplace equation (2) of $\mathbf{x}$. This characterization was found in [**20**] and explained geometrically in [**11**].

In addition, the circularity constraint (81) implies the following formula [**15**]:

$$\frac{T_i|\mathbf{X}_j|^2}{|\mathbf{X}_j|^2} = 1 - (T_iQ_{ji})(T_iQ_{ji}); \tag{83}$$

which, compared with equations (16)–(18), allows to fix, without loss of generality, the backward formulation of the circular lattice in the following way:

$$|\mathbf{X}_i|^2 = \rho_i = \frac{T_i\tau}{\tau} \implies |T_i\widetilde{\mathbf{X}}_i|^2 = \frac{1}{\rho_i} = \frac{\tau}{T_i\tau}. \tag{84}$$

A distinguished sub-class of circular lattices corresponds to the particular case in which the lattice points $\mathbf{x}$ belong to the sphere of radius R: $|\mathbf{x}| = R$. In this case there exists, like for the symmetric reduction, an elegant relation between point lattices and hyperplane lattices [**13**].

PROPOSITION 4.5. *Given a system of parallel quadrilateral lattices $\{\mathbf{x}_{(k)}\}_{k=1}^M$ and the associated matrix $\mathbf{\Omega}$ of the system, then the following properties are equivalent.*

(i) *The matrix $\mathbf{\Omega}/R$ is orthogonal:*

$$\mathbf{\Omega}\mathbf{\Omega}^T = \mathbf{\Omega}^T\mathbf{\Omega} = R^2 I, \quad \mathbf{\Omega}^T = R^2\mathbf{\Omega}^{-1}. \tag{85}$$

(ii) *The polar hyperplane $P(\mathbf{x}_{(k)})$ coincides with the dual hyperplane $R^2\mathbf{y}^*_{(k)}$:*

$$P(\mathbf{x}_{(k)}) = R^2\mathbf{y}^*_{(k)}, k = 1, \ldots, M. \tag{86}$$

(iii) *The quadrilateral lattices $\mathbf{x}_{(k)}/R, k = 1, \ldots, M$ form an orthonormal basis:*

$$\mathbf{x}_{(i)} \cdot \mathbf{x}_{(j)} = R^2\delta_{ij}, i, j = 1, \ldots, M. \tag{87}$$

In addition, the associated tangent vectors $\mathbf{X}_i$, $\mathbf{X}^*_i$, $i = 1, \ldots, N$, are related by the following formulas

$$\mathbf{X}_i = \frac{\rho_i}{2R^2}T_i(\mathbf{\Omega}\mathbf{X}_i^{*T}) = -\frac{\rho_i}{2R^2}\mathbf{\Omega}(T_i\mathbf{X}_i^{*T}), \quad i = 1, \ldots, N, \tag{88}$$

$$T_i\mathbf{X}^*_i = -\frac{2}{\rho_i}\mathbf{X}_i^T\mathbf{\Omega}, \quad i = 1, \ldots, N, \tag{89}$$

with

$$|\mathbf{X}_i|^2 = \rho_i, \quad T_i|\mathbf{X}^*_i|^2 = \frac{4R^2}{\rho_i}, \tag{90}$$

and satisfy the circularity constraint (81) *and its adjoint*

$$C^{\circ *}_{ij} := \mathbf{X}^*_i \cdot T_i^{-1}\mathbf{X}^*_j + \mathbf{X}^*_j \cdot T_j^{-1}\mathbf{X}^*_i = 0. \tag{91}$$

In the continuous limit, equations (81) become the orthogonality conditions

$$\mathbf{X}_i \cdot \mathbf{X}_j = 0, \quad i \neq j, \tag{92}$$

and the circular lattice reduces to an orthogonal conjugate net.

4.3. The Egorov lattice. In the previous sections we introduced two integrable reductions of the quadrilateral lattice: the symmetric and the circular lattice. Imposing both properties on the initial surfaces of the lattice, they propagate through the lattice in the unique construction determined by the planarity condition and the corresponding lattice will be an integrable deeper reduction of the quadrilateral lattice. This lattice is called a *Egorov lattice* since, in the continuous limit, it reduces to a Egorov system of coordinates on submanifolds.

DEFINITION 4.6 ([**28**]). A quadrilateral lattice is a Egorov lattice iff the internal angles corresponding to the vertices $T_i\mathbf{x}$ and $T_j\mathbf{x}$ are right angles; i.e.:

$$\mathbf{X}_i \cdot T_i\mathbf{X}_j = 0, \quad i \neq j. \tag{93}$$

A Egorov lattice exhibits the following properties can be characterized as follows.

1. A Egorov lattice is a circular and symmetric.
2. If $N = M$, then a Egorov lattice is a circular lattice invariant under translations along the main diagonal [**13, 29**], i.e., such that:

$$T\mathbf{X}_i = \mathbf{X}_i \implies TQ_{ij} = Q_{ij}, \tag{94}$$

   where T is the total shift: $T := \prod_{i=1}^{N} T_i$.

We remark that this invariance property implies that the lattice is effectively $N-1$ dimensional, since it depends on the differences of the variables n_i:

$$\mathbf{x} = \mathbf{x}(n_1 - n_2, n_2 - n_3, \ldots, n_{N-1} - n_N). \tag{95}$$

We have been recently informed of a work on the finite gap formulation of the circular and Egorov lattices [**21**].

The continuous limits of the above equations:

$$\partial_i\beta_{jk} = \beta_{ji}\beta_{ik}, \quad \beta_{ij} = \beta_{ji}, \quad \mathbf{X}_i \cdot \mathbf{X}_j = 0, \quad i \neq j, \tag{96}$$

characterize submanifolds parametrized by Egorov systems of conjugate coordinates (Egorov nets). At last, the invariance property (94) reduces to

$$\sum_{\ell=1}^{N} \partial_\ell \beta_{ij} = 0, \quad \sum_{\ell=1}^{N} \partial_\ell \mathbf{X}_i = 0, \tag{97}$$

implying that $\beta_{ij} = \beta_{ij}(u_1 - u_2, \ldots, u_{N-1} - u_n)$. For $N = 3$, we recover a classical characterization of the Egorov net [**1, 6**].

4.4. Quadratic reductions and the Ribaucour transformation. Quadrilateral lattices in quadrics generalize (in a sense which will be explained in Proposition 4.7) the circular lattices. Their continuous analogs—the so called *quadratic conjugate nets* (see [**17**] and references therein) were the subject of intensive study by Ribaucour, Darboux, Bianchi and Eisenhart. The particular type of fundamental transformation of conjugate nets which preserves the quadratic constraint is called the Ribaucour transformation. In this section we present the basic geometric results concerning quadrilateral lattices in quadrics (see [**11**] for details); in particular, we give the analogue of the Ribaucour transformation.

Consider the quadrilateral lattice $\mathbf{x}$ in the quadric $\mathcal{Q}$ defined by the equation

$$\mathbf{x}^T Q\mathbf{x} + \mathbf{a}^T\mathbf{x} + c = 0, \tag{98}$$

where Q is a symmetric matrix, $\mathbf{a}$ is a constant vector, c is a number.

As it was shown in [**11**], such a constraint is compatible with the geometric integrability scheme of the quadrilateral lattice. This integrability statement can be generalized to quadrilateral lattices in spaces obtained by intersection of many quadric hypersurfaces like, for example, lattices in the Grassmann manifolds and in Segré or Veronese varieties.

To show a simple example, consider a quadrilateral lattice in the standard M dimensional sphere S^M. The intersection of the plane of any elementary quadrilateral of the lattice $\mathbf{x}\colon \mathbf{Z}^N \to S^M$ with the sphere S^M is a circle. In the stereographic projection $S^M \to \mathbf{E}^M \cup \{\infty\}$ circles of the sphere S^M are mapped into circles (or straight lines, i.e., circles passing through the infinity point) of $\mathbf{E}^M$, therefore we have:

PROPOSITION 4.7. *Quadrilateral lattices in the sphere S^M are mapped during the stereographic projection into multidimensional circular lattices in $\mathbf{E}^M$; conversely, any circular lattice in $\mathbf{E}^M$ can be obtained in this way.*

Theorem 2.7 states that, in order to construct the fundamental transformation of the lattice $\mathbf{x}$, we need three new ingredients: ϕ, $\mathbf{x}_C$ and ϕ_C. In looking for the Ribaucour reduction of the fundamental transformation, we can use the following additional information:

(i) the initial lattice $\mathbf{x}$ satisfies the quadratic constraint (98),
(ii) the final lattice $\mathcal{R}(\mathbf{x})$ should satisfy the same constraint as well.

This should allow to reduce the number of the necessary data and, indeed, to find the Ribaucour transformation, it is enough to know the Combescure transformation vector $\mathbf{x}_C$ only.

PROPOSITION 4.8. *The Ribaucour reduction $\mathcal{R}(\mathbf{x})$ of the fundamental transformation of the quadrilateral lattice $\mathbf{x}$ subjected to quadratic constraint (98) is determined by the Combescure transformation vector $\mathbf{x}_C$, provided that $\mathbf{x}_C$ is not annihilated by the bilinear form Q of the constraint*

$$\mathbf{x}_C^t Q \mathbf{x}_C \neq 0. \tag{99}$$

The functions ϕ and ϕ_C entering in formula (25) are then given by

$$\phi = 2\mathbf{x}^t Q \mathbf{x}_C + \mathbf{a}^t \mathbf{x}_C, \tag{100}$$

$$\phi_C = \mathbf{x}_C^t Q \mathbf{x}_C. \tag{101}$$

As it was shown in [**11**], the Ribaucour transformation for circular lattices in $\mathbf{E}^M$, defined algebraically in [**20**], can be obtained combining the stereographic projection with the Ribaucour transformation of the quadrilateral lattices in S^M.

To present the Ribaucour reduction of the vectorial fundamental transformation we use results and notation of Section 2, together with equations (100) and (101), to obtain

$$\phi^k = 2\mathbf{x}^t Q \mathbf{x}_{C,k} + \mathbf{a}^t \mathbf{x}_{C,k}, \tag{102}$$

$$\phi^k_{C,k} = \mathbf{x}^t_{C,k} Q \mathbf{x}_{C,k}. \tag{103}$$

Equations (45) and (102) lead to

$$\Delta_i(\phi^k_{C,\ell} + \phi^\ell_{C,k}) = (T_i\mathbf{x}^t_{C,k} + \mathbf{x}^t_{C,k}) Q(\Delta_i \mathbf{x}_\ell) + (T_i \mathbf{x}^t_{C,\ell} + \mathbf{x}^t_{C,\ell}) Q(\Delta_i \mathbf{x}_k), \tag{104}$$

which implies that

$$\phi^k_{C,\ell} + \phi^\ell_{C,k} = 2\mathbf{x}^t_{C,k} Q \mathbf{x}_{C,\ell}; \tag{105}$$

the constant of integration was found from the condition $(\mathcal{R}_k \circ \mathcal{R}_\ell)(\mathbf{x}) \subset \mathcal{Q}$.

PROPOSITION 4.9. *The vectorial Ribaucour transformation* $\mathbf{R}$, *i.e., the reduction of the vectorial fundamental transformation* (46) *compatible with the quadratic constraint* (98), *is given by the following constraints*

$$\mathbf{\Phi}^T = 2\mathbf{x}^t Q \mathbf{X}_C + \mathbf{a}^T \mathbf{X}_C, \tag{106}$$

$$\mathbf{\Phi}_C + \mathbf{\Phi}_C^T = 2\mathbf{X}_C^T Q \mathbf{X}_C. \tag{107}$$

One can also consider the superposition of several Ribaucour transformations and show the corresponding permutability theorem.

THEOREM 4.10. *Assume the following splitting of the data of the vectorial Ribaucour transformation*

$$\mathbf{\Phi} = \begin{pmatrix} \mathbf{\Phi}^{(1)} \\ \mathbf{\Phi}^{(2)} \end{pmatrix}, \quad \mathbf{X}_C = (\mathbf{X}_{C(1)}, \mathbf{X}_{C(2)}), \quad \mathbf{\Phi}_C = \begin{pmatrix} \mathbf{\Phi}^{(1)}_{C(1)} & \mathbf{\Phi}^{(1)}_{C(2)} \\ \mathbf{\Phi}^{(2)}_{C(1)} & \mathbf{\Phi}^{(2)}_{C(2)} \end{pmatrix}, \tag{108}$$

associated with the partition $K = K_1 + K_2$. *Then the vectorial Ribaucour transformation* $\mathbf{R}(\mathbf{x})$ *is equivalent to the following superposition of vectorial Ribaucour transformations:*

1. *Transformation* $\mathbf{R}_{(1)}(\mathbf{x})$ *with the data* $\mathbf{\Phi}^{(1)}$, $\mathbf{X}_{C(1)}$, $\mathbf{\Phi}^{(1)}_{C(1)}$.
2. *Application on the result obtained in point* 1. *of the transformation* $\mathbf{R}_{(2)}$ *with the data* $\mathbf{R}_{(1)}(\mathbf{X}_{C(2)})$, $\mathbf{R}_{(1)}(\mathbf{\Phi}^{(2)}_{C(2)})$, $\mathbf{R}_{(1)}(\mathbf{\Phi}^{(2)})$ *given by* $\mathcal{R}$*-analogs of formulas* (51)–(53).

Combining the stereographic projection with the vectorial Ribaucour transformation of the quadrilateral lattices in S^M, one can obtain the vectorial Ribaucour transformation for circular lattices in $\mathbf{E}^M$ and prove its permutability property. This result was also found independently in [**22**], directly on the level of circular lattices in $\mathbf{E}^M$.

References

1. L. Bianchi, *Lezioni di Geometria Differenziale*, 3rd ed., Zanichelli, Bologna, 1924.
2. A. Bobenko, *Discrete conformal maps and surfaces*, *Symmetries and Integrability of Difference Equations* (P. Clarkson and F. Nijhoff, eds.), Cambridge Univ. Press, 1999, pp. 97–108.
3. A. Bobenko and U. Pinkall, *Discrete isothermic surfaces*, J. Reine Angew. Math. **475** (1996) 187–208.
4. L. V. Bogdanov and B. G. Konopelchenko, *Lattice and q-difference Darboux–Zakharov–Manakov systems via $\bar\partial$ method*, J. Phys. A **28** (1995), L173–L178.
5. J. Cieśliński, A. Doliwa, and P. M. Santini, *The integrable discrete analogues of orthogonal coordinate systems are multidimensional circular lattices*, Phys. Lett. A **235** (1997), 480–488.
6. G. Darboux, *Leçons sur les systèmes orthogonaux et les coordonnées curvilignes. Principes de géométrie analytique*, 2e éd., Gauthier-Villars, Paris, 1910.
7. ———, *Leçons sur la théorie générale des surfaces* I–IV, Gauthier-Villars, Paris, 1887–1896.
8. E. Date, M. Kashiwara, M. Jimbo, and T. Miwa, *Transformation groups for soliton equations*, Nonlinear Integrable Systems—Classical Theory and Quantum Theory, Proc. RIMS Symposium (M. Jimbo and T. Miwa, eds.) (Kyoto, 1981), World Scientific, 1983, Singapore, pp. 39–119.
9. A. Doliwa, *Geometric discretisation of the Toda system*, Phys. Lett. A **234** (1997), 187–192.
10. ———, *Lattice geometry of the Hirota equation*, this volume.
11. ———, *Quadratic reductions of quadrilateral lattices*, J. Geom. Phys. **30** (1999), 169–186.
12. A. Doliwa and P. M. Santini, *Multidimensional quadrilateral lattices are integrable*, Phys. Lett. A **233** (1997), 365–372.

13. ______, *The symmetric and Egorov reductions of the quadrilateral lattice*, solv-int/9907012.
14. A. Doliwa, P. M. Santini, and M. Mañas, *Transformations of quadrilateral lattices*, J. Math. Phys. (in press), solv-int/9712017.
15. A. Doliwa, S. V. Manakov, and P. M. Santini, *$\bar\partial$-reductions of the multidimensional quadrilateral lattice: The multidimensional circular lattice*, Comm. Math. Phys. **196** (1998), 1–18.
16. A. Doliwa, M. Mañas, L. Martínez Alonso, E. Medina, and P. M. Santini, *Charged free Fermions, vertex operators and transformation theory of conjugate nets*, J. Phys. A **32** (1999), 1197–1216.
17. L. P. Eisenhart, *Transformations of surfaces*, Princeton Univ. Press, Princeton, N.J., 1923.
18. R. Hirota, *Discrete analogue of a generalized Toda equation*, J. Phys. Soc. Japan **50** (1981), 3785–3791.
19. J. van de Leur, *Schlesinger–Bäcklund transformation for N-component KP*, J. Math. Phys. **39** (1998), 2833–2847.
20. B. G. Konopelchenko and W. K. Schief, *Three-dimensional integrable lattices in Euclidean spaces: Conjugacy and orthogonality*, Proc. Roy. Soc. London A **454** (1998), 3075–3104.
21. I. M. Krichever, seminar given at the Dept. of Mathematics, University of Rome "La Sapienza", June 1998.
22. Q. P. Liu and M. Mañas, *Superposition formulae for the discrete Ribaucour transformations of circular lattices*, Phys. Lett. A **249** (1998), 424–430.
23. R. R. Martin, J. de Pont, and T. J. Sharrock, *Cyclic surfaces in computer aided design*, *The Mathematics of Surfaces* (J. A. Gregory, ed.), Oxford, Clarendon Press, 1986, pp. 253–268,
24. A. V. Mikhailov, *Integrability of a two-dimensional generalization of the Toda chain*, JETP Lett. **30** (1979), 414–418.
25. A. W. Nutbourne, *The solution of a frame matching equation*, The Mathematics of Surfaces (J. A. Gregory, ed.), Oxford, Clarendon Press, 1986, pp. 233–252,
26. D. Pedoe, *Geometry, a comprehensive course*, Dover Publ., New York, 1988.
27. R. Sauer, *Differenzengeometrie*, Springer, Berlin, 1970.
28. W. K. Schief, talk given at the Workshop: Nonlinear Systems, Solitons and Geometry, Oberwolfach, October 1997.
29. W. K. Schief, private communication.

Instytut Fizyki Teoretycznej, Uniwersytet Warszawski, Hoża 69, 00–681 Warszawa, Poland.

E-mail address: doliwa@fuw.edu.pl

Current address: Istituto Nazionale di Fisica Nucleare, Sezione di Roma and Dipartimento di Fisica, Università di Roma "La Sapienza", P.le Aldo Moro 2, I–00185 Roma, Italy.

E-mail address: Paolo.Santini@roma1.infn.it

Centre de Recherches Mathématiques
CRM Proceedings and Lecture Notes
Volume **25**, 2000

On the Kleinian Construction of Abelian Functions of Canonical Algebraic Curves

J. C. Eilbeck, V. Z. Enolskii, and D. V. Leykin

1. Introduction

The discovery of classical and quantum completely integrable systems led to an increase of interest in the theory of Abelian functions in theoretical physics and applied mathematics. This area was considered traditionally as a field of pure mathematics. This new trend makes it necessary to reconsider classical results in the area from the point of view of modern applications.

In this paper we consider an arbitrary algebraic curve V of genus g and construct the field of meromorphic functions on its Jacobi variety $\mathrm{Jac}(V)$ in terms of Kleinian $\wp$-functions,

$$\wp_{ij}(\mathbf{u}) = -\frac{\partial^2}{\partial u_i \partial u_j} \ln \sigma(\mathbf{u}), \quad i, j = 1, \ldots, g,$$

where the vector $\mathbf{u} \in \mathrm{Jac}(V)$ and σ is the Kleinian σ-function. The effective construction of the σ-function plays the principal role in our approach. It is defined on the universal space of Jacobians, which is the fibration with the base given by the space of moduli, $M(V)$ of the curve V of dimension $d \leq 3g-3$ and a fibre generated by Jacobi variety $\mathrm{Jac}(V)$. The Kleinian σ function represents a natural generalization of the Weierstrass elliptic function to the case of an arbitrary algebraic curve, and has the following properties:

- The σ-function is automorphic with respect to the action of the symplectic group $\mathrm{Sp}(2g, \mathbb{Z})$.
- The σ-function is an entire function on the universal space and is expanded in a power series whose entries are monomials
$$u_1^{\alpha_1} \cdots u_g^{\alpha_g} \lambda_1^{\beta_1} \cdots \lambda_d^{\beta_d},$$
where $\mathbf{u} \in \mathrm{Jac}(V)$, $\lambda \in M(V)$, $\alpha \in (\mathbb{Z}_+)^g$, $\beta \in (\mathbb{Z}_+)^d$, $(\mathbb{Z}_+) = 0 \cup \mathbb{N}$.

2000 *Mathematics Subject Classification.* Primary: 14H55, 14H40; Secondary: 35Q53, 58F07.

The research described in this publication was supported in part by grants from the Civil Research Development Foundation, CRDF grant no. UM1-325, the INTAS grant no. 96-770 and the Engineering and Physical Sciences Research Committee, grant GR/L06119 (JCE and VZE).

This is the final form of the paper.

- The addition formula for the σ function inherits the form of the addition theorem for the Weierstrass σ function and is written as

$$\frac{\sigma(\mathbf{u}+\mathbf{v})\sigma(\mathbf{u}-\mathbf{v})}{\sigma(\mathbf{u})^2\sigma(\mathbf{v})^2} = \text{polynomial in } \wp_{i,j}.$$

This construction began with Weierstrass [**27, 28**] and Klein [**15**] and was well documented in [**1, 2**]; recently V. M. Buchstaber and two of the authors reviewed the hyperelliptic case[**4–6**].

The principal result of [**4, 5**] is the discovery of a $(g+2)\times(g+2)$ matrix H of a rank ≤ 4 whose entries are $\wp$-functions and the moduli of the curve V and which dictates all the theory: the 4×4 minors give the embedding of the Kummer variety $\mathrm{Kum}(V)$ in the projective space, 3×3 minors describe in the same way the Jacobi variety $\mathrm{Jac}(V)$, and the associated KdV hierarchy is constructed in terms of 2×2 minors of the matrix H.

This paper seeks to develop the analogous matrix realization of the Kleinian construction of Abelian functions for an arbitrary algebraic curve, including the case of a singular curve. The paper is based on the recent results of [**7**], where the construction of the Kleinian σ functions was given for a nonhyperelliptic curve, and contains a set of explicit formulae to realize the approach developed in [**7**].

The relation to problems of integrable discrete mappings, to which this Workshop is devoted, lies in the intention to find a uniformization in terms of Kleinian functions for the integrable discrete map given by the fundamental conformally invariant double cross ratio

$$\frac{(z_{n,m}-z_{n+1,m})(z_{n,m+1}-z_{n+1,m+1})}{(z_{n,m}-z_{n,m+1})(z_{n+1,m}-z_{n+1,m+1})} = \frac{p^2}{q^2},$$

which was introduced in [**18**]. The map can be then interpreted as a discrete Schwarzian KdV equation, because the double cross ratio constitutes the discrete analogue of the Schwarzian derivative (see e.g. [**12**]). It was already shown by F.Nijhoff and two of the authors [**11, 19**] in the case of the lattice KdV system that by imposing a periodicity condition of the form

$$z_{n+2g+2,m} = z_{n,m} \quad \forall m\in\mathbb{Z}, \quad n=1,\dots,2g+2, \; g\in\mathbb{N}, \tag{1.1}$$

the map is uniformized by an Abelian functions of the hyperelliptic algebraic curve $V=(y,x)$ of genus g. The argument $\mathbf{u}_{n,m}$ of the Abelian functions depends on discrete indices n,m and arises as an image of the Abel map of the kind "divisor+point". The extension of this result to integrable discrete systems, which are associated with general algebraic curves, e.g. discrete mappings associated with trigonal curves [**20**], represents one of the motivations to undertake the investigation described below.

The paper starts from an example of a hyperelliptic curve of genus two for which we give the basic formulae of the theory and their application to completely integrable systems. We concentrate further on the construction of the principle objects—Kleinian σ functions for a wide class of algebraic curves (the so called canonical curves). With this purpose we derive the canonical Abelian differentials using the Weierstrass gap theorem as the main working tool. We obtain as the result the Kleinian formula (3.28), which is a generating one for deriving the complete set relations between $\wp$ functions and their derivatives. The principle result of the paper is the explicit solution of the Jacobi inversion problem, which is an alternative to that given by M. Noether [**21**]. We consider as a main example the case of a non singular trigonal curve, for which we give the complete set of formulae, analogous to those given in [**1, 2**] for hyperelliptic curve. The paper is completed by a short discussion on the application of our approach to completely integrable equations and of the further perspectives of the development of the theory.

2. Kleinian Functions of the Genus Two Hyperelliptic Curve

In this section we consider the simplest example—the Kleinian functions of an algebraic curve of genus two, and demonstrate how these functions work in completely integrable systems.

We consider the Riemann surface of a curve $V(y,x)$ of genus $g=2$,

$$\begin{aligned} y^2 &= 4x^5 + \lambda_4 x^4 + \lambda_3 x^2 + \lambda_2 x^2 + \lambda_1 x + \lambda_0, \\ &= 4\prod_{k=1}^{5}(x-a_k), \quad a_i \neq a_j \in \mathbb{C}, \end{aligned} \tag{2.1}$$

equipped with a homology basis $(\mathfrak{a}_1,\mathfrak{a}_2;\mathfrak{b}_1,\mathfrak{b}_2)\in H_1(V,\mathbb{Z})$.

Introduce the canonical basis in the space of holomorphic differentials $\mathrm{d}\mathbf{u}^T = (\mathrm{d}u_1,\mathrm{d}u_2)\in\mathcal{H}^1(V,\mathbb{C})$ as follows

$$\mathrm{d}u_1 = \frac{\mathrm{d}x}{y}, \quad \mathrm{d}u_2 = \frac{x\mathrm{d}x}{y}.$$

The associated canonical meromorphic differentials of the second kind $\mathrm{d}\mathbf{r}^T = (\mathrm{d}r_1, \mathrm{d}r_2)$ have the form

$$\mathrm{d}r_1 = \frac{\lambda_3 x + 2\lambda_4 x^2 + 12x^3}{4y}dx, \quad \mathrm{d}r_2 = \frac{x^2}{y}dx. \tag{2.2}$$

The 2×2 matrices of their periods,

$$2\omega = \left(\oint_{\mathfrak{a}_k}\mathrm{d}u_l\right)_{k,l=1,2}, \quad 2\omega' = \left(\oint_{\mathfrak{b}_k}\mathrm{d}u_l\right)_{k,l=1,2},$$
$$2\eta = \left(-\oint_{\mathfrak{a}_k}\mathrm{d}r_l\right)_{k,l=1,2}, \quad 2\eta' = \left(-\oint_{\mathfrak{b}_k}\mathrm{d}r_l\right)_{k,l=1,2},$$

satisfy the equations,

$$\omega'\omega^T - \omega\omega'^T = 0, \quad \eta'\omega^T - \eta\omega'^T = -\frac{\imath\pi}{2}1_2, \quad \eta'\eta^T - \eta\eta'^T = 0,$$

which generalizes the Legendre relations between complete elliptic integrals to the case $g=2$.

The fundamental σ function in this case is a natural generalization of the Weierstrass elliptic σ function and is defined as follows

$$\sigma(\mathbf{u}) = \frac{\pi}{\sqrt{\det(2\omega)}}\frac{\varepsilon}{\sqrt[4]{\prod_{1\le i<j\le 5}(a_i-a_j)}} \times \exp\left\{\mathbf{u}^T\eta(2\omega)^{-1}\mathbf{u}\right\}\theta[\varepsilon]\big((2\omega)^{-1}\mathbf{u}\mid\omega'\omega^{-1}\big),$$

where $\varepsilon^8=1$ and $\theta[\varepsilon](\mathbf{v}\mid\tau)$ is the θ function with an odd characteristic $[\varepsilon] = \left[\begin{smallmatrix}\varepsilon_1 & \varepsilon_2\\ \varepsilon_1' & \varepsilon_2'\end{smallmatrix}\right]$,

$$\theta[\varepsilon](\mathbf{v}\mid\tau) = \sum_{\mathbf{m}\in\mathbb{Z}^2}\exp\imath\pi\big\{(\mathbf{m}+\varepsilon)^T\tau(\mathbf{m}+\varepsilon) + 2(\mathbf{v}+\varepsilon')^T\tau(\mathbf{m}+\varepsilon)\big\}.$$

The expansion of the σ function near $\mathbf{u} = 0$ has the form

$$\sigma_1(\mathbf{u}) = u_1 + \frac{1}{24}\lambda_2 u_1^3 - \frac{1}{3}u_2^3 + o(\mathbf{u}^5), \tag{2.3}$$

and the further terms can be computed with the help of bilinear differential equation [**2**]. We denote

$$\wp_{11}(\mathbf{u}) = -\frac{\partial^2}{\partial u_1^2}\ln\sigma(\mathbf{u}), \quad \wp_{12}(\mathbf{u}) = -\frac{\partial^2}{\partial u_1 \partial u_2}\ln\sigma(\mathbf{u}),$$
$$\wp_{22}(\mathbf{u}) = -\frac{\partial^2}{\partial u_2^2}\ln\sigma(\mathbf{u}).$$

The multi-index symbols $\wp_{i,j,k}$ etc. are defined as logarithmic derivatives by the variable u_i, u_j, u_k on the corresponding indices i, j, k etc.

The equations of the Jacobi inversion problem, [1]

$$u_1 = \int_{a_1}^{x_1} du_1 + \int_{a_2}^{x_2} du_1, \quad u_2 = \int_{a_1}^{x_1} du_2 + \int_{a_2}^{x_2} du_2,$$

are equivalent to an algebraic equation

$$\mathcal{P}(x, \mathbf{u}) = x^2 - \wp_{22}(\mathbf{u})x - \wp_{12}(\mathbf{u}) = 0, \tag{2.4}$$

that is, the pair (x_1, x_2) is the pair of roots of (2.4). So we have

$$\wp_{22}(\mathbf{u}) = x_1 + x_2, \quad \wp_{12}(\mathbf{u}) = -x_1x_2. \tag{2.5}$$

The corresponding y_i is expressed as

$$y_i = \wp_{222}(\mathbf{u})x_i + \wp_{122}(\mathbf{u}), \quad i = 1, 2. \tag{2.6}$$

There is the following expression for the function $\wp_{11}(\mathbf{u})$ in terms of x_1, x_2 and $y_1 4$, y_2:

$$\wp_{11}(\mathbf{u}) = \frac{F(x_1, x_2) - 2y_1y_2}{4(x_1 - x_2)^2}, \tag{2.7}$$

where

$$F(x_1, x_2) = \sum_{r=0}^{2} x_1^r x_2^r \big[2\lambda_{2r} + \lambda_{2r+1}(x_1 + x_2)\big]. \tag{2.8}$$

All the possible pairwise products of the $\wp_{ijk}$ functions are expressed as follows in terms of $\wp_{22}$, $\wp_{12}$, $\wp_{11}$ and constants λ_s of the defining equation (2.1). We give here only basis equations

$$\wp_{222}^2 = 4\wp_{22}^3 + 4\wp_{12}\wp_{22} + \lambda_4\wp_{22}^2 + 4\wp_{11} + \lambda_3\wp_{22} + \lambda_2,$$
$$\wp_{222}\wp_{122} = 4\wp_{12}\wp_{22}^2 + 2\wp_{12}^2 - 2\wp_{11}\wp_{22} + \lambda_4\wp_{12}\wp_{22} + \frac{1}{2}\lambda_3\wp_{12} + \frac{1}{2}\lambda_1,$$
$$\wp_{122}^2 = 4\wp_{22}\wp_{12}^2 - 4\wp_{11}\wp_{12} + \lambda_4\wp_{12}^2 - \lambda_0.$$

All such expressions may be rewritten in the form of an *extended cubic relation* as follows. For arbitrary $\mathbf{l}, \mathbf{k} \in \mathbb{C}^4$ the following formula is valid [**2**]

$$\mathbf{l}^T \pi \pi^T \mathbf{k} = -\frac{1}{4}\det\begin{pmatrix} H & \mathbf{l} \\ \mathbf{k}^T & 0 \end{pmatrix}, \tag{2.9}$$

[1] In what follows we shall denote the integral bounds by the second coordinate of the curve $V = V(y, x)$.

where $\pi^T = (\wp_{222}, -\wp_{221}, \wp_{211}, -\wp_{111})$ and H is the 4×4 matrix:

$$H = \begin{pmatrix} \lambda_0 & \frac{1}{2}\lambda_1 & -2\wp_{11} & -2\wp_{12} \\ \frac{1}{2}\lambda_1 & \lambda_2 + 4\wp_{11} & \frac{1}{2}\lambda_3 + 2\wp_{12} & -2\wp_{22} \\ -2\wp_{11} & \frac{1}{2}\lambda_3 + 2\wp_{12} & \lambda_4 + 4\wp_{22} & 2 \\ -2\wp_{12} & -2\wp_{22} & 2 & 0 \end{pmatrix}.$$

The vector π satisfies the equation $H\pi = 0$, and so the functions $\wp_{22}$, $\wp_{12}$ and $\wp_{11}$ are related by the equation

$$\det H = 0. \tag{2.10}$$

The equation (2.10) defines the quartic Kummer surface $\mathbb{K}$ in $\mathbb{C}^3$ [**14**].

The $\wp_{ijkl}$ functions are expressed as follows

$$\wp_{2222} = 6\wp_{22}^2 + \frac{1}{2}\lambda_3 + \lambda_4\wp_{22} + 4\wp_{12}, \tag{2.11}$$

$$\wp_{2221} = 6\wp_{22}\wp_{12} + \lambda_4\wp_{12} - 2\wp_{11}, \tag{2.12}$$

$$\wp_{2211} = 2\wp_{22}\wp_{11} + 4\wp_{12}^2 + \frac{1}{2}\lambda_3\wp_{12}, \tag{2.13}$$

$$\wp_{2111} = 6\wp_{12}\wp_{11} + \lambda_2\wp_{12} - \frac{1}{2}\lambda_1\wp_{22} - \lambda_0, \tag{2.14}$$

$$\wp_{1111} = 6\wp_{11}^2 - 3\lambda_0\wp_{22} + \lambda_1\wp_{12} + \lambda_2\wp_{11} - \frac{1}{2}\lambda_0\lambda_4 + \frac{1}{8}\lambda_1\lambda_3. \tag{2.15}$$

These equations can be identified with completely integrable partial differential equations and dynamical systems, which are solved in terms of Abelian functions of hyperelliptic curve of genus two.

To demonstrate this statement we consider the stationary *Veselov–Novikov* equation,

$$\begin{gathered} \alpha u_{xxx} + \beta u_{yyy} - 3\alpha(vu)_x - 3\beta(wu)_y = 0, \\ w_x = u_y, \quad v_y = u_x, \end{gathered} \tag{2.16}$$

where $\alpha, \beta \neq 0$ are constants. Then the following proposition is valid

PROPOSITION 2.1. *The stationary flow of the Veselov–Novikov equation is satisfied if we set*

$$\begin{aligned} u(x,y) &= 2\wp_{12}(y,x), \\ v(x,y) &= 2\wp_{22}(y,x) + \frac{1}{3}\left(\lambda_4 - \frac{\beta}{2\alpha}\lambda_1\right), \\ w(x,y) &= 2\wp_{11}(y,x) + \frac{1}{3}\left(\lambda_2 - \frac{2\alpha}{\beta}\right). \end{aligned} \tag{2.17}$$

PROOF. A straightforward substitution of (2.17) into (2.16) and use of the relations (2.12), (2.14) and $H\pi = 0$. □

Because the Kleinian functions $\wp_{22}$, $\wp_{12}$, $\wp_{11}$ are the coordinates of the Kummer surface, the stationary Veselov–Novikov equation, being associated with a hyperelliptic curve of genus two, describes the Kummer surface. The link between the stationary Veselov–Novikov equation and the Kummer surface was recently discussed by Ferapontov [**13**].

It can be also shown, that the equations (2.11)–(2.15) describe hierarchies of KdV and "sine-Gordon" equations associated with the hyperelliptic curve of genus two.

In what follows we develop the Kleinian construction of Abelian functions for a certain class of nonhyperelliptic algebraic curves.

3. Kleinian Construction of Abelian Functions

3.1. The curve. Let V be an algebraic curve given by an irreducible equation

$$f(x,y)=0, \quad f(x,y)=a_0(x)y^n-\sum_{k=1}^{n}a_k(x)y^{n-k}, \tag{3.1}$$

where $a_k(x)$ are polynomials in x and $a_0(x)$ and $a_k(x)$, $k=1\dots,n$ have no common factors. The curve is called *singular* if $a_0(x)\not\equiv 1$ and *nonsingular* otherwise. In other words, a singular curve is the curve which has points $(y=\infty, x\neq\infty)$.

DEFINITION 3.1. The order N of an arbitrary *rational function* $\psi(x,y)$ *on the curve* V is the number N of common solutions $(x_1,y_1),\dots,(x_N,y_N)$ of the equations

$$f(x,y)=0, \quad \psi(x,y)=0. \tag{3.2}$$

We shall call the positive integers (n,s) *orders* of the curve V because, clearly x is a function of order n and y is a function of order s.

The definition given above permit us to formulate the *Weierstrass gap theorem*, which serves as a principal working tool in what follows.

THEOREM 3.1 (Weierstrass Lückensatz). *For V there exists a number g of positive integers $n_1,\dots n_g$ ("gap sequence") such that for each number n_i the equations*

$$f(x,y)=0, \quad \psi(x,y)=0, \quad \mathrm{ord}_V\psi(x,y)=n_i, \tag{3.3}$$

have no solutions and

- *all exceptional integers ("gaps") belong to the interval $[1,2g-1]$*

$$0<n_1<n_2<\cdots<n_g<2g,$$

- *if the numbers n_i and n_j—are permitted ("nongaps") then the numbers pn_i+qn_j, where p, $q\in\mathbb{N}$ are also nongaps.*

The following corollaries are valid

COROLLARY 3.1.1. *Consider the gap sequence with the number of gaps g. Let n be the lowest from the nongaps. Denote by s_i, $i=1,\dots,n-1$ the lowest nongaps such that $s_i(\mathrm{mod}\, n)=i<n$. Then the following equality is valid*

$$\sum_{i=1}^{n-1}\left[\frac{s_i}{n}\right]=g. \tag{3.4}$$

COROLLARY 3.1.2. *Suppose that the condition of Corollary 3.1.1 are satisfied. Let $s=\min\{s_1,\dots,s_{n-1}\}$. Then the following equality is valid*

$$\frac{(n-1)(s-1)}{2}-g=\delta, \quad \textit{where} \quad \delta\geq 0. \tag{3.5}$$

The Weierstrass gap theorem introduced two important positive integers g and δ which are the principal geometrical invariants of the curve.

DEFINITION 3.2. The nonnegative integer δ is the number of *multiple points* of the curve, which can not exceed $(n-1)(s-1)/2$. The number of gaps g, which is the difference between the maximal number of multiple points of a curve of orders (n, s) and number δ of multiple points, which it actually has, is called the *genus* [2] of the curve.

To describe the number δ analytically we compute the discriminant (in y) $D_y(x)$ of the curve (3.1),

$$D_y(x) = \prod_{i=1}^{n-1} d_i^{\varepsilon_i}(x) r(x), \tag{3.6}$$

where $\varepsilon \in \mathbb{N} \geq 2$. Zeros of $d_i(x)$ $i = 1, \ldots, n-1$, where

$$d_i(x) = \prod_{j=1}^{\delta_j} (x - \kappa_{ij}), \quad \delta_j \in \mathbb{N}, \ \kappa_{ij} \in \mathbb{C},$$

are *multiple* points, whence zeros of the polynomial $r(x)$ are the branching points of the curve. Let us write the number

$$\delta = \sum_{i=1}^{n-1} \delta_i(\varepsilon_i - 1) \in \mathbb{Z}_+, \tag{3.7}$$

where we denote $\mathbb{Z}_+ = \mathbb{N} \oplus 0$, as *number of multiple points.* The curve is called *nondegenerate* if $\delta = 0$ and *degenerate* otherwise.

Introduce on V the structure of one-dimensional compact complex manifold—*the Riemann surface of algebraic curve* by introducing a local parametrization of the point $(y, x) = \big(y(\xi), x(\xi)\big)$ in the vicinity of a point (a, b), where ξ is the *local coordinate*:

$$(y,x) = \begin{cases} (a+\xi, b+\xi) & \text{if } (a,b) \text{ is an regular point,} \\ (a+\xi, b+\xi^m) & \text{if } (a,b) \text{ is an branching point,} \\ \left(\frac{1}{\xi^s}, \frac{1}{\xi^n}\right) & \text{if } (a=\infty, b=\infty) \text{ is branching point at infinity,} \\ (a+\xi^m, b+\xi) & \text{if } (a,b) \text{ is an multiple point of multiplicity } m. \end{cases}$$

We will employ further the same notation for the plane curve and the Riemann surface—V. Combining notions introduced so far, we come up with

DEFINITION 3.3. The algebraic curve V of orders (n, s) is called *nonsingular nondegenerate canonical* iff $a_0(x) = 1$, $\delta = 0$ and (n, s) are coprime.

It is clearly that the Weierstrass gap sequence of the nonsingular nondegenerate canonical curve V of orders (n, s) is generated by two coprime integers n and s. Alternatively, the Weierstrass gap sequence of a nonsingular degenerate canonical curve V of orders (n, s) is generated by the integers n and $s_1 = s, \ldots, s_k$, where $1 < k \leq n-1$.

The canonical algebraic curve has a branching point at infinity, where the coordinates x and y are given as $x = 1/(\xi^n) = \infty$, $y = 1/(\xi^s) = \infty$, $2 < n < s$, where ξ is the local coordinate. The polynomial $f(x, y)$ defining the canonical curve can be written in the form

$$f(x, y) = y^n - x^s + \text{terms of lower order.}$$

[2] In classical literature—*Geschlecht* or *deficiency*

Let us fix in the form of a proposition the important property of the nonsingular curve

PROPOSITION 3.2. *The order of a monomial* $x^P y^Q$, P, $Q \in \mathbb{N}$ *on nonsingular nondegenerate curve of orders* (n, s) *equals*

$$\operatorname{ord}_V x^P y^Q = nP + sQ. \tag{3.8}$$

EXAMPLE 3.1. Consider all canonical curves of genus $g = 4$ and construct the associated Weierstrass gap sequences.

- *Nonsingular canonical hyperelliptic curve*

$$\overline{0}, 1, \overline{2}, 3, \overline{4}, 5, \overline{6}, 7, \overline{8, 9, \ldots}.$$

  The sequence is generated by the orders 2 and 9.
- *Nonsingular canonical trigonal curve*

$$\overline{0}, 1, 2, \overline{3}, 4, \overline{5, 6}, 7, \overline{8, 9, \ldots}.$$

  The sequence is generated by the orders 3 and 5.
- *Singular canonical trigonal curve*

$$\overline{0}, 1, 2, \overline{3}, 4, 5, \overline{6, 7, 8, \ldots}.$$

  The sequence is generated by the orders 3, 7 and 8.
- *Singular canonical four-sheeted covering*

$$\overline{0}, 1, 2, 3, \overline{4, 5, 6}, 7, \overline{8, 9, \ldots}.$$

  The sequence is generated by the orders 4, 5 and 6.

3.2. The differentials and integrals. All the construction is based on the explicit realization of the *fundamental second order* 2*-differential.*

DEFINITION 3.4. The 2-differential $\mathrm{d}\Omega(y, x; w, z)$ on $V \times V$ is called *a fundamental second order* 2*-differential* if it is symmetric $\mathrm{d}\Omega(x, y; z, w) = \mathrm{d}\Omega(z, w; x, y)$ and has the only pole of the second order along the diagonal, $x = z$ in the vicinity of which it can be given as

$$\mathrm{d}\Omega(y, x; w, z) = \frac{\mathrm{d}\xi \mathrm{d}\eta}{(\xi - \eta)^2} + O(1)\, \mathrm{d}\xi\, \mathrm{d}\eta, \tag{3.9}$$

where ξ, η—are local coordinates of the points x and z respectively. We shall look for a realization of $\mathrm{d}\Omega(y, x; w, z)$ in the form [**1**]

$$\mathrm{d}\Omega(y, x; w, z) = \frac{F(y, x; w, z)\, \mathrm{d}x\, \mathrm{d}z}{(x - z)^2 f_y(x, y) f_w(z, w)}, \tag{3.10}$$

where $F(y, x; w, z)$ is a polynomial of its variables.

Holomorphic differentials can be represented locally at *every* point $(X, Y) \in V$ in the form $\mathrm{d}u = h(\xi)\mathrm{d}\xi$, where h—is the holomorphic function and ξ—is the local parameter in the vicinity of the point (X, Y). For the algebraic curve of genus g there exist exactly g independent holomorphic differentials, which can be written in the form

$$\mathrm{d}u_i = \frac{x^{P_i} y^{Q_i}\, \mathrm{d}x}{f_y(x, y)}, \quad i = 1 \ldots, g, \tag{3.11}$$

where $\mathrm{ord}_V x^{P_i} y^{Q_i}$, $i = 1, \dots, g$ are exactly g first nongaps of the Weierstrass gap sequence. Introduce the g-vector

$$\mathcal{U}^T = (\dots, \mathcal{U}_{i,j}, \dots) = (\mathcal{U}_1, \dots, \mathcal{U}_g), \tag{3.12}$$

whose components $\mathcal{U}_{i,j} = x^{P_i} y^{Q_i}$ are ordered by increasing order of the monomials:

$$\mathcal{U}_1 = \mathcal{U}_{0,0} = 1, \quad \mathcal{U}_2 = \mathcal{U}_{1,0} = x, \dots .$$

DEFINITION 3.5. Let $\mathcal{P}(x)$ be a polynomial in x of order $\mathrm{ord}_x \mathcal{P}(x) = n$ and $F_k(x)$ polynomials of orders k, such that the equality

$$\frac{\mathcal{P}(x) - \mathcal{P}(z)}{x - z} = \sum_{k=0}^{n} z^{n-k} F_k(x),$$

is valid. Then $D_z^k \mathcal{P}(z) = F_k(x)$ is the *umbral derivative of order* k of the polynomial $\mathcal{P}(z)$.

DEFINITION 3.6. The rational function $\Psi(x, y)$ on the nonsingular curve V is called an *entire rational function*, if $\Psi(x, y) \to \infty$ iff $x \to \infty$. Entire rational functions generate a ring, which we denote by $\mathcal{O}(V)$.

Let

$$\psi(x, y)^T = \big(1, \psi_1(x, y), \dots, \psi_{n-1}(x, y)\big),$$

be a basis in $\mathcal{O}(V)$, where

$$\psi_i(x, y) = \left(\frac{D_y^{n-i} f(x, y)}{d_i(x)}\right), \quad i = 0, \dots, n-1, \tag{3.13}$$

where D_y^k—is the umbral derivative in y of order k, and $f(x, y)$ is the polynomial defining the curve.

Then it follows from the definition of the umbral derivative and the existence of such functions

$$\phi(x, y)^T = \big(\phi_0(x, y), \phi_1(x, y), \dots, \phi_{n-1}(x, y)\big), \tag{3.14}$$

that

$$\psi(x, y)^T \phi(x, Y) = \frac{f(x, Y) - f(x, y)}{Y - y}. \tag{3.15}$$

The functions ϕ_i have, clearly, the form

$$\phi_i(x, y) = d_i(x) y^{n-i-1}, \quad i = 0, \dots, n-1.$$

The property (3.15) allows us to construct *A canonical differential of the third kind* with unique poles of the first order in the points (x_1, y_1) and (x_2, y_2) as follows [**1**]

$$\mathrm{d}\Pi_{x_1, x_2}(x) = \frac{\mathrm{d}x}{f_y(x, y)} \left\{ \frac{\psi(x_1, y_1)^T \phi(x, y)}{x - x_1} - \frac{\psi(x_2, y_2)^T \phi(x, y)}{x - x_2} \right\}. \tag{3.16}$$

To construct explicitly the fundamental second order 2-differential $\mathrm{d}\Omega(x, y; z, w)$ we consider on $V \times V$ the 2-differential

$$\mathrm{d}z \frac{\mathrm{d}x}{f_y(x, y)} \frac{\partial}{\partial z} \frac{\psi^T(z, w) \phi(x, y)}{x - z}.$$

It follows from the definition of the umbral derivatives, that this differential has a pole of the second order along the diagonal, where it is represented in the form (3.9). But being holomorphic in (x, y) away from the diagonal, this differential has nevertheless poles in the variables (z,w) at $z=\infty$. Restore the symmetry by setting

$$\begin{aligned}(3.17)\quad \mathrm{d}\Omega(x,y;z,w) = \mathrm{d}z\frac{\mathrm{d}x}{f_y(x,y)}\frac{\partial}{\partial z}\frac{\psi^T(z,w)\phi(x,y)}{x-z} \\ +\boldsymbol{\mathcal{R}}^T(z,w)\boldsymbol{\mathcal{U}}(x,y)\frac{\mathrm{d}x\mathrm{d}z}{f_y(x,y)f_w(z,w)},\end{aligned}$$

where the g-vector $\boldsymbol{\mathcal{U}}(x,y)$ is defined by the holomorphic differentials and the g-vector $\boldsymbol{\mathcal{R}}^T=(\mathcal{R}_1,\dots,\mathcal{R}_g)$ is found from the equality

$$\begin{aligned}(3.18)\quad \frac{\partial f(z,w)}{\partial w}\frac{\partial}{\partial z}\frac{\psi^T(z,w)\phi(x,y)}{x-z} - \frac{\partial f(x,y)}{\partial y}\frac{\partial}{\partial x}\frac{\psi^T(x,y)\phi(z,w)}{z-x} \\ = \boldsymbol{\mathcal{R}}^T(x,y)\boldsymbol{\mathcal{U}}(z,w) - \boldsymbol{\mathcal{U}}^T(x,y)\boldsymbol{\mathcal{R}}(z,w).\end{aligned}$$

The canonical differentials of the second kind $\mathrm{d}\mathbf{r}^T=(\mathrm{d}r_1,\dots,\mathrm{d}r_g)$ *associated with the holomorphic differentials* (3.11) are then given by the formula

$$(3.19)\qquad \mathrm{d}r_i = \frac{\mathcal{R}_i(x,y)\,\mathrm{d}x}{f_y(x,y)},\quad i=1\dots,g,$$

where the correspondence

$$\mathcal{R}_i(x,y)\leftrightarrow\mathcal{U}_i(x,y),\quad i=1,\dots,g,$$

of the polynomials $\mathcal{R}(x,y)_i$ with monomials $\mathcal{U}_i(x,y)$ being ordered according to the Weierstrass gap sequence as was pointed in (3.12) is supposed.

The Weierstrass gap sequence gives complete information about the curve. Consider the Weierstrass gap sequence and prescribe by the lowest nongap entries (n,s), $n<s$ orders $\mathrm{ord}_V x=n$ and $\mathrm{ord}_V y=s$. Write the curve in the form (3.1).

Define the functions (3.13). $d_0=1$, and $d_1(x),\dots,d_{n-1}(x)$ polynomials in x, such that $\mathrm{ord}_V\psi_i(x,y)=s_i$, $i=1,\dots,n-1$ are the lowest nongaps, and that $s_i(\mathrm{mod}\ n)<i<n$. The functions

$$\psi_0(x,y)=1,\quad \psi_1(x,y),\dots,\psi_{n-1}(x,y),$$

give a basis in $\mathcal{O}(V)$. Therefore for every pair $i,k=0,\dots,n-1$ the following equalities are valid

$$(3.20)\qquad \psi_i(x,y)\psi_k(x,y)=\sum_{j=0}^{n-1}\alpha_{ik}^{(j)}(x)\psi_j(x,y),$$

where $\alpha_{ik}^{(j)}(x)$ are polynomials in x. By equating powers of y in equations (3.20) one can find the form of the polynomials $a_k(x)$ entering the equation of the curve (3.1) and also the property, which the polynomials $d_i(x)$ possess

$$(3.21)\qquad d_{i-1}(x)d_{i+1}(x)\ \text{divisible by}\ d_i^2(x).$$

As result we come to the following proposition:

PROPOSITION 3.3. *The curve* V *of orders* (n,s) *is given by the formula*

$$(3.22)\qquad a_0(x)y^n-\sum_{k=0}^{n-1}a_k(x)\prod_{i=2}^{n-1}\tilde{d}_i^{n-k}(x)y^{k-1}=0,$$

where $a_k(x)$ are polynomials of orders $\mathrm{ord}_x a_k(x) \le s$, and the polynomials $\tilde{d}_i(x)$ are defined by the equation

$$\tilde{d}_{i+1} = \frac{d_{i-1}(x)d_{i+1}(x)}{d_i^2(x)}, \quad i = 1\dots n-1, \tag{3.23}$$

where the polynomials d_i posses the property (3.21).

This construction can be inverted: the Weierstrass gap sequence is uniquely defined by the curve and can be algorithmically constructed by the curve.

3.3. Kleinian formula. Let $(\mathfrak{a}, \mathfrak{b})$ be a basis of cycles in $\mathrm{H}_1(V, \mathbb{Z})$ with the intersection matrix $\mathfrak{a}_i \circ \mathfrak{a}_k = 0$ and $\mathfrak{b}_i \circ \mathfrak{b}_k = 0$, $\mathfrak{a}_i \circ \mathfrak{b}_k = -\mathfrak{b}_k \circ \mathfrak{a}_i = 1$. $g \times g$ matrices of their periods $\mathfrak{a}$, $\mathfrak{b}$-periods,

$$2\omega = \left(\oint_{\mathfrak{a}_k} du_l \right)_{k,l=1,\dots,g}, \quad 2\omega' = \left(\oint_{\mathfrak{b}_k} du_l \right)_{k,l=1,\dots,g},$$

are nondegenerate and the matrix $\tau = \omega'\omega^{-1}$ is symmetric and has positive imaginary part. Under the action of the transformation $(2\omega)^{-1}$ the vector $d\mathbf{u} = (du_1, \dots, du_g)^T$ becomes the vector of normalised holomorphic differentials $d\mathbf{v} = (dv_1, \dots, dv_g)^T$, namely the vector $H^1(V, \mathbb{C})$ satisfying the conditions $\oint_{\mathfrak{a}_k} dv_k = \delta_{kl}, k$, $l = 1 \dots, g$. Introduce also the period matrices of the canonical differentials of the second kind

$$(2\eta, 2\eta') = \left(\{ -\oint_{\mathfrak{a}_i} dr_j \}, \{ -\oint_{\mathfrak{b}_i} dr_j \} \right)_{i,j=1,\dots,n}. \tag{3.24}$$

Denote by $\mathrm{Jac}(V)$ the *Jacobian* of the curve V, which is the factor $\mathbb{C}^g/\Gamma$, where $\Gamma = 2\omega \oplus 2\omega'$ is the lattice, generated by the periods of the holomorphic differentials.

Divisors on Riemann surfaces are given by formal sums of analytic points $\mathcal{D} = \sum_i^n m_i(y_i, x_i)$, and $\deg \mathcal{D} = \sum_i^n m_i$. The *effective divisor* is such that $m_i > 0 \forall i$. Let $\mathcal{D}$ be a divisor of degree 0, $\mathcal{D} = \mathcal{X} - \mathcal{Z}$, with $\mathcal{X}$ and $\mathcal{Z}$ the effective divisors $\deg \mathcal{X} = \deg \mathcal{Z} = n$ given by $\mathcal{X} = \{(y_1, x_1), \dots, (y_n, x_n)\}$ and $\mathcal{Z} = \{(w_1, z_1), \dots, (w_n, z_n)\} \in (V)^n$, where $(V)^n$ is the n-th symmetric power of V.

The *Abel map* $\mathfrak{A}$: $(V)^n \to \mathrm{Jac}(V)$ puts into correspondence the divisor $\mathcal{D}$, with fixed $\mathcal{Z}$, and the point $\mathbf{u} \in \mathrm{Jac}(V)$, according to the rule

$$\mathbf{u} = \int_{\mathcal{Z}}^{\mathcal{X}} d\mathbf{u}, \quad \text{or} \quad u_i = \sum_{k=1}^{n} \int_{z_k}^{x_k} du_i, \quad i = 1, \dots, g. \tag{3.25}$$

The *Jacobi inversion problem* is formulated as the problem of inversion of the Abel map.

The standard θ function $\theta(\mathbf{v} \mid \tau)$ is defined by its Fourier series,

$$\theta(\mathbf{v} \mid \tau) = \sum_{\mathbf{m} \in \mathbb{Z}^g} \exp \pi i \{ \mathbf{m}^T \tau \mathbf{m} + 2\mathbf{v}^T \mathbf{m} \}. \tag{3.26}$$

The θ function possesses the following periodicity properties $\forall k \in 1, \dots, g$:

$$\theta(v_1, \dots, v_k + 1, \dots, v_g \mid \tau) = \theta(\mathbf{v} \mid \tau),$$
$$\theta(v_1 + \tau_{1k}, \dots, v_k + \tau_{kk}, \dots, v_g + \tau_{gk} \mid \tau) = \mathrm{e}^{-i\pi\tau_{kk} - 2\pi i v_k} \theta(\mathbf{v} \mid \tau).$$

Let $\mathbf{w}^T = (w_1, \dots, w_g) \in \mathrm{Jac}(V)$ be some fixed vector. The function,

$$\Theta(x) = \theta\left(\int_{x_0}^{x} d\mathbf{v} - \mathbf{w} \mid \tau \right), \quad x \in V,$$

is called the *Riemann θ function.* The Riemann θ function $\Theta(x)$ is either identically 0, or it has exactly g zeros $x_1,\ldots,x_g \in V$, for which the *Riemann vanishing theorem* says that

$$\sum_{k=1}^{g}\int_{x_0}^{x_i} \mathrm{d}\mathbf{v} = \mathbf{w} + \mathbf{K}_{x_0},$$

where $\mathbf{K}_{x_0}^T = (K_1,\ldots,K_g)$ is the vector of Riemann constants with respect to the base point x_0.

DEFINITION 3.7. T*he fundamental Abelian σ function* is defined by the formula

$$\sigma(\mathbf{z}) = \frac{1}{\sqrt[4]{D(V)}}\sqrt{\frac{\pi^g}{\det(\omega)}}\exp\left\{\frac{1}{2}\mathbf{u}^T\eta\omega^{-1}\mathbf{u}\right\}\theta\big(\mathbf{u}-\mathbf{K}_{x_0}\mid\omega'\omega^{-1}\big), \tag{3.27}$$

$\mathbf{K}_{x_0}$ is the vector of Riemann constants with the base point x_0, the function $D(V) = D_x\big(D_y f(x,y)\big)$ is the discriminant of the defining equation $f(x,y)=0$ of the curve V.

The fundamental σ function *is invariant with respect to the action of the symplectic group* $\mathrm{Sp}(2g,\mathbb{Z})$ and therefore inherits the corresponding property of the σ function of the Weierstrass theory of elliptic functions.

Kleinian $\wp$ functions are defined as logarithmic derivatives of the fundamental σ function

$$\wp_{ij}(\mathbf{u}) = -\frac{\partial^2\ln\sigma(\mathbf{u})}{\partial u_i\partial u_j},\quad \wp_{ijk}(\mathbf{u}) = -\frac{\partial^3\ln\sigma(\mathbf{u})}{\partial u_i\partial u_i\partial u_k}\cdots,$$
$$i,j,k,\dots = 1,\ldots,g.$$

The $\wp$ functions have the following periodicity properties

$$\wp_{\mathbf{I}}\big(\mathbf{u}+2\boldsymbol{\Omega}(\mathbf{m},\mathbf{m}')\big) = \wp_{\mathbf{I}}(\mathbf{u}), \mathbf{I} = \{i,j\},\{i,j,k\},\ldots \quad i,j,k,=1,\ldots,g,$$

where $\boldsymbol{\Omega}(\mathbf{m},\mathbf{m}') = \omega\mathbf{m}+\omega'\mathbf{m}'$.

For the remaining results, the principal role is played by the following theorem

THEOREM 3.4 (Klein). *Let $\big(y(x_0),x_0\big)$, (y,x) be arbitrary distinct points on V and let $\big\{(y_1,x_1),\ldots,(y_g,x_g)\big\}$ be arbitrary sets of distinct points $\in (V)^g$. Then the following relation is valid for every $r=1,\ldots,g$*

$$\sum_{i,j=1}^{g}\wp_{ij}\left(\int_{x_0}^{x}\mathrm{d}\mathbf{u} - \sum_{k=1}^{g}\int_{x_0}^{x_k}\mathrm{d}\mathbf{u}\right)\mathcal{U}_i(x,y)\mathcal{U}_j(x_r,y_r) = \frac{F(x,y;x_r,y_r)}{(x-x_r)^2}, \tag{3.28}$$

where monomials $\mathcal{U}_i(x,y)$ a re defined by the holomorphic differentials and $F(x,y;$ $z,w)$ is the polynomial entering into the definition of the fundamental 2*-differential* (3.9).

PROOF. The proof is based on the application of the Riemann vanishing theorem to compute the integral over the boundary of the fundamental domain $\partial\Delta$,

$$\oint_{\partial\Delta}\int_z^x \mathrm{d}\Omega(x,y;z,w)\mathrm{d}\,\log\Theta(x),$$

where $\Theta(x)$ is the Riemann θ function and $\mathrm{d}\Omega(x,y;z,w)$ is the fundamental 2-differential. □

The Kleinian formula (3.28) gives an additional theorem of a *"point + divisor"* kind. In the case of $g = 1$ it is exactly the addition theorem for the Weierstrass $\wp$ function (function $\wp_{11}(u)$ in our notation).

Expanding the formula (3.28) at $x = 1/\xi^n = \infty$ we obtain from the condition of vanishing of the principal part of the poles the complete set of relations between $\wp$ functions and their derivatives. In particular, the first $n-1$ equations are polynomial in x, y (written instead x_r, y_r, $r = 1, \ldots, g$) with coefficients depending on Kleinian $\wp$ functions. Their derivatives represent an over-determined system of algebraic equations whose zeros give the solution of Jacobi inversion problem. This statement will be exemplified further by the considering of the trigonal curve.

We summarise the results of this section in the form of a proposition

PROPOSITION 3.5. *Consider a nonsingular* ($a_0 \equiv 1$) *nondegenerate* ($d_1 \equiv 1, \ldots, d_{n-1} \equiv 1$) *canonical algebraic curve of orders* (n, s) *and genus* g. *Construct the associated Weierstrass gap sequence with* g *gaps. Then*

- *the* g *independent canonical holomorphic differentials* (3.11) *are defined by monomials* $\mathcal{U}_{i,j} = x^{P_i} y^{Q_j}$ *whose orders* $nP_i + sQ_j$ *are exactly the first* g *nongaps in the Weierstrass gap sequence.*
- *The associated differentials of the second kind are given by* (3.19) *with the polynomials*

$$\text{(3.29)}\quad \mathcal{R}_{ij}(x,y) = \sum_{l=0}^{j-1} \sum_{k=\alpha_{n-l-1}}^{\alpha_n+1} \mathfrak{R}_{i,j,k,l} x^k y^l + y^j \sum_{k=i+1}^{\alpha_n+1} \mathfrak{R}_{i,j,k,j} x^k + \sum_{l=j+1}^{n-1} \sum_{k=0}^{\alpha_n+1} \mathfrak{R}_{i,j,k,l} x^k y^l,$$

where $\alpha_k = \mathrm{ord}_x a_k(x)$, $k = 0, \ldots, n$, $\mathfrak{R}_{i,j,k,l} = \tilde{\mathfrak{R}}_{i,j,k,l} - \tilde{\mathfrak{R}}_{k,l,i,j}$ *and*

$$\text{(3.30)}\quad \tilde{\mathfrak{R}}_{i,j,k,l} = \chi\binom{l}{j} \sum_{p=0}^{n-1-l} \sum_{q=0}^{\alpha_{n-1-j-p}} \chi\binom{\alpha_{n-1-l-p}}{i+k-q+2} C\binom{n-l-1, l-j, p}{i+k-q-2, q, i},$$

and

$$C\binom{i,j,k}{l,r,s} = (-1)^{\delta_{0,i}} 2^{-\delta_{0,j}} a_{i-k,l} a_{i+k+j,r} \times \begin{cases} (-1)^{\delta_{0,i}\delta_{0,j}} (n-i-j)(r-s-1), & \text{if } k=0 \\ (-1)^{\delta_{i,k}} \big[k(l+r-2s-2) + j(r-s-1)\big], & \text{if } k>0 \end{cases},$$

where $\delta_{i,k}$ *is the Kronecker symbol and* $\chi\binom{l}{k} = 1$ *if* $l > k$ *and* 0 *otherwise.*

PROOF. The proof of the first part of the theorem is based on the Corollary 3.1.1. The formula (3.29) is derived by direct transformations of the right hand side of the condition (3.18) to the form of the right hand side. □

The formula (3.29) generalizes the formula given by Abel and Weierstrass for the polynomial $\mathcal{R}_i$ in the case of a hyperelliptic curve (4.2). In the next sections we consider as examples hyperelliptic and trigonal curves.

4. Hyperelliptic Curve

The hyperelliptic curve V of genus g, given by the formula

$$f(x,y) = y^2 - p(x), \quad p(x) = 4x^{2g+1} + \sum_{i=0}^{2g} \lambda_i x^i, \tag{4.1}$$

is the canonical hyperelliptic curve of order $(2, 2g+1)$. Suppose, that the curve is nonsingular, i.e. the discriminant $D_y\big(f(x,y)\big) \equiv p(x)$ has no multiple zeros.

The sets of canonical holomorphic differentials and associated second kind differentials have the form [**1**]

$$\mathrm{d}u_k = \frac{x^{k-1}\mathrm{d}x}{y}, \quad \mathrm{d}r_i = \frac{\mathrm{d}x}{4y} \sum_{k=i}^{2g+1-i} (k+1-i)\lambda_{k+1+i} x^k, \tag{4.2}$$

where $i = 1, \ldots, g$. The differential of the third kind with first order poles at the points $x_{1,2}$ is defined as

$$\mathrm{d}\Pi_{x_1,x_2}(x) = \frac{\mathrm{d}x}{2y}\left(\frac{y+y_1}{x-x_1} - \frac{y+y_2}{x-x_2}\right). \tag{4.3}$$

The 2-differential of the second kind $\mathrm{d}\Omega(x,y;z,w)$ with the unique pole along the diagonal is given by (3.10) with the polynomial

$$F(x,y;z,w) = 2yw + \sum_{k=0}^{g} x^k z^k (2\lambda_{2k} + (x+z)\lambda_{2k+1}). \tag{4.4}$$

The Abel pre-image of the point $\mathbf{u} \in \mathrm{Jac}(V)$ is given by the set $\big\{(y_1, x_1), \ldots, (y_g, x_g)\big\} \in (V)^g$, where $\{x_1, \ldots, x_g\}$ are the zeros of the *Bolza equation* [**1, 3**]

$$x^g - x^{g-1}\wp_{g,g}(\mathbf{u}) - x^{g-2}\wp_{g,g-1}(\mathbf{u}) - \cdots - \wp_{g,1}(\mathbf{u}) = 0, \tag{4.5}$$

and $\{y_1, \ldots, y_g\}$ are given by

$$y_k = -\left.\frac{\partial \mathcal{P}(x;\mathbf{u})}{\partial u_g}\right|_{x=x_k}, \quad k = 1, \ldots, g. \tag{4.6}$$

5. Trigonal Curve

Consider the trigonal curve. Set $d_0 = 1$, then we have

$$\begin{aligned} \psi_0(x,y) &= 1, \\ \psi_1(x,y) &= \frac{y - a_1(x)}{d_1(x)}, \\ \psi_2(x,y) &= \frac{y^2 - a_1(x)y - a_2(x)}{d_2(x)}. \end{aligned} \tag{5.1}$$

Write the equations

$$\begin{aligned} \psi_1^2 &= \alpha_{11}^{(0)}\psi_0 + \alpha_{11}^{(1)}\psi_1 + \alpha_{11}^{(2)}\psi_2, \\ \psi_1\psi_2 &= \alpha_{12}^{(0)}\psi_0 + \alpha_{12}^{(1)}\psi_1 + \alpha_{12}^{(2)}\psi_2, \\ \psi_2^2 &= \alpha_{22}^{(0)}\psi_0 + \alpha_{22}^{(1)}\psi_1 + \alpha_{22}^{(2)}\psi_2. \end{aligned}$$

By equating powers of y and eliminating $\alpha_{ij}^{(k)}$ we arrive at the equation of the curve

$$y^3 - r(x)d_1(x)y^2 - p(x)d_2(x)y - q(x)\frac{d_2^2(x)}{d_1(x)} = 0, \tag{5.2}$$

where d_2 is divisible by d_1^2. We restrict ourselves to the case of the nonsingular curve $d_1 = d_2 = 1$ of the form

$$f(x,y) = 0, \quad f(x,y) = y^3 - p(x)y - q(x), \tag{5.3}$$

$$p(x) = \sum_{k=0}^{g-[(g+1)/3]} p_k x^k, \quad q(x) = x^{g+1} + \sum_{k=0}^{g-1} q_k x^k, \tag{5.4}$$

with the coefficients p_i, $q_j \in \mathbb{C}$, the fraction $(g+1)/3$ being nonreducible, and the discriminant $D_y\big(f(x,y)\big) = 27q^2(x) - 4p^3(x)$ is assumed to be without multiple zeros.

Further we shall distinguish two cases $s = 3K+1$ and $s = 3K+2$, which we shall denote in what follows as (I) and (II) respectively. For example, the curves of lowest genera of order (3,4) and (3,5) correspondingly yield the following Weierstrass gap sequences, in which we over-line the nongap numbers

$$\bar{0}\; 1,\; 2,\; \overline{3,\; 4},\; 5,\; \overline{6,\; 7,\; \ldots} \qquad \text{case } (3,4),$$
$$\bar{0}\; 1,\; 2,\; \bar{3}, 4, \overline{5,\; 6},\; 7,\; \overline{8,\; 9,\; \ldots} \qquad \text{case } (3,5).$$

The holomorphic differentials are given by the formula (3.11) with the monomials

$$\mathcal{U}_{i,0}, i = 0,\ldots,\left[\frac{s_1}{3}\right], \quad \mathcal{U}_{i,1}, i = 0,\ldots,\left[\frac{s_2}{3}\right],$$

where s_i, $i = 1, 2$ are the lowest nongaps such that $s_i(\mod 3) = i < 3$ (see Corollary 3.1.1). The associated second kind differentials are given by the formula (3.19) with the polynomials $\mathcal{R}_i(x,y)$ given by (3.29). In particular,

$$\mathcal{R}_g(x,y) = \begin{cases} -x^{2K} & \text{for (I)} \\ -x^K y & \text{for (II)} \end{cases}, \tag{5.5}$$

$$\mathcal{R}_{g-1}(x,y) = \begin{cases} -2yx^K & \text{for (I)} \\ -2x^{2K+1} - p_{2K+1}x^K y & \text{for (II)} \end{cases}. \tag{5.6}$$

To define the fundamental 2-differential we introduce the following polynomials

$$\mathcal{P}(x,z) = \sum_{l=0}^{2K+\varepsilon} (p_{2l} + xp_{2l+1})x^l z^l,$$

$$\mathcal{Q}(x,z) = \sum_{l=0}^{K} q_l\big(lx + (3-l)z\big)z^{l-1} + z^{K+1}\sum_{l=-1}^{s-K-2} q_{K+l+2}\big(\beta(l)x + (3-\beta(l)z)\big)z^{[l/2]},$$

where $\varepsilon = 0$ for the case (I) and $\varepsilon = 1$ for the case (II) and in the summand the positive integers $\beta(l) = K - 3\big[(l-1)/2)\big] - 1$. Note, that the polynomials, $\mathcal{P}(x,z)$, $\mathcal{Q}(x,z)$ have the property

$$\mathcal{P}(x,x) = p(x), \quad \mathcal{Q}(x,x) = 3q(x).$$

The 2-differential of the second kind $d\Omega(x,y;z,w)$ with the unique pole along the diagonal is given by (3.10) with the polynomial $F(x,y;z,w)$ being defined by the formula

$$F(x,y;z,w) = 3y^2w^2 + 2yw\big(\mathcal{P}(x,z) + \mathcal{P}(z,x)\big) - w^2p(x) - y^2p(z) \\ + w\mathcal{Q}(x,z) + y\mathcal{Q}(z,x) - \mathcal{P}(x,z)\mathcal{P}(z,x).$$

Denote $\pi_{i,j}(\mathbf{u}) = \wp_{i,j}(\mathbf{u}) - \wp_{i,j+1,j+1}(\mathbf{u})$. Then the solution of the Jacobi inversion problem reduces to the solution of algebraic equations

$$w\sum_{i=0}^{K-1} \pi_{K+2i+2,g}z^i + \sum_{i=0}^{K-1}(\pi_{i+1,g} + z^g\pi_{K+2i+1,g})z^i = \mathcal{R}_g,$$

$$\sum_{i=0}^{K-1}(\pi_{i+1,g-1} + z^g\pi_{K+2i+1,g-1})z^i + w\sum_{i=0}^{K-1}\pi_{K+2i+2,g-1}z^i = \mathcal{R}_{g-1},$$

where $\mathcal{R}_g$ and $\mathcal{R}_{g-1}$ are monomials given in (5.6). Each pair of equations is reduced after elimination of w to an algebraic equation of degree g, whose coefficients expressed in terms of the functions $\pi_{i,j}(\mathbf{u})$ give the solution of the Jacobi inversion problem. These equations generalizes the Bolza equation (4.5) to the case of trigonal curve.

In particular the first and the last symmetric functions of the arguments $x_1, \ldots, x_g$ are

$$s_1 = -\frac{1}{2}\wp_{ggg} + \frac{3}{2}\wp_{g-1,g}, \text{ for (I)} \quad \text{and} \quad +\frac{1}{2}p_{2K+1}\wp_{g,g}, \text{ for (II)},$$
$$s_g = (-1)^g\frac{\varepsilon}{2}\big[(\wp_{1,g}\wp_{K+2,g-1} - \wp_{1,g-1}\wp_{K+2,g}) + (\wp_{1,g,g}\wp_{K+2,g} - \wp_{1,g}\wp_{K+2,g,g})\big],$$

where in the second formula $\varepsilon = 1$ for the case (I) and -1 for the case (II). In particular for the curve of order (3,4) the symmetric functions are given as

$$s_1 = \frac{3}{2}\wp_{23} - \frac{1}{2}\wp_{333},$$
$$s_2 = -\frac{1}{2}\wp_{33}(\wp_{22} - \wp_{233}) + \frac{1}{2}\wp_{23}(\wp_{23} - \wp_{333}) - \wp_{13},$$
$$s_3 = \frac{1}{2}\wp_{33}(\wp_{12} - \wp_{133}) - \frac{1}{2}\wp_{13}(\wp_{23} - \wp_{333}).$$

In the case of the curve of order (3,5)

$$s_1 = -\frac{1}{2}\wp_{444} + \frac{3}{2}\wp_{34} + \frac{1}{2}p_3\wp_{44},$$
$$s_2 = -\frac{1}{2}\wp_{34}(\wp_{34} - \wp_{444}) + \frac{1}{2}\wp_{244} - \frac{1}{2}\wp_{23}, \\ -\frac{1}{2}\wp_{44}(\wp_{33} - \wp_{344}) + \frac{1}{2}p_3\wp_{24},$$
$$s_3 = \frac{1}{2}\wp_{24}(\wp_{33} - \wp_{344}) - \frac{1}{2}\wp_{34}(\wp_{23} - \wp_{244}) + \frac{1}{2}\wp_{13} \\ + \frac{1}{2}\wp_{144} + \frac{1}{2}p_3\wp_{14}$$
$$s_4 = \frac{1}{2}\wp_{34}(\wp_{13} - \wp_{144}) - \frac{1}{2}\wp_{14}(\wp_{33} - \wp_{344}).$$

6. Discussion

It was mentioned in the introduction that in [**4, 5**] the matrix

$$H = \{h_{i,k}\}_{i=1,\dots,g+1,j=1,\dots,g+1},$$

was found for the genus g hyperelliptic curve (4.1) of rank ≤ 4 with the entries

$$h_{ik} = 4\wp_{i-1,k-1} - 2\wp_{k,i-2} - 2\wp_{i,k-2} + \frac{1}{2}\big(\delta_{ik}(\lambda_{2i-2} + \lambda_{2k-2}) + \delta_{k,i+1}\lambda_{2i-1} + \delta_{i,k+1}\lambda_{2k-1}\big).$$

It was shown there, that the following relations are valid

$$\wp_{ggi}\wp_{ggk} = -\det H\begin{bmatrix} i, g+1 g+2 \\ k, g+1, g+2\end{bmatrix} = 0, \quad i,k = 1,\dots,g,$$

$$\det H\begin{bmatrix} i,j,g+1,g+2 \\ k,l,g+1,g+2\end{bmatrix} = 0, \quad i,j,k,l = 1,\dots,g,$$

$$\wp_{gggi} = -\det H\begin{bmatrix} i, g+1 \\ g+1, g+2\end{bmatrix} + \det H\begin{bmatrix} i,g,g+2 \\ g,g,g+2\end{bmatrix} - \det H\left[i-1, g+2 g+1, g+2\right], \quad i = 1,\dots,g,$$

where $H\left[\begin{smallmatrix} i_1\dots i_m \\ j_1, j_n\end{smallmatrix}\right] = \{h_{i_k,j_l}\}_{k=1,\dots,m;\, l=1,\dots,n}$ denotes an $m\times n$ sub-matrix. The first group of these relations describe the Jacobi variety $\mathrm{Jac}(V)$ as algebraic variety, the second group describes the Kummer variety, $\mathrm{Kum}(V) = \mathrm{Jac}(V)/\mathbf{u} \to -\mathbf{u}$ and the third group describes the KdV hierarchy.

The techniques developed in this paper is aimed to enlarge this result to the case of arbitrary algebraic curve. We mention here only, that for the trigonal curve of order (3,4) the condition of vanishing the next principal part (after those, which give the solution of Jacobi inversion problem) in the expansion of the Kleinian formula leads to the equation

$$\wp_{3333} = 6\wp_{33}^2 - 3\wp_{22} + 4p_2\wp_{33},$$

which after double differentiation by u_3 becomes the Boussinesq equation with respect to the function $u(x,t) = 2\wp_{g,g}(\mathrm{const}, t, x)$. Because the Kleinian function appears to be natural coordinates of the Boussinesq hierarchy, then the approach presented here creates insights to contribute to an algebro-geometrical description of solutions of the Boussinesq equation, [**8–10, 16, 17, 22, 26**], and the trigonal generalizations of the Neumann system [**23–25**]. We also believe that our results will shed some new light on the integrable discrete maps associated with non hyperelliptic curves [**20**].

References

1. H. F. Baker, *Abels theorem and the allied theory including the theory of theta functions*, Cambridge Univ. Press, Cambridge, 1897.
2. ———, *Multiply periodic functions*, Cambridge Univ. Press, Cambridge, 1907.
3. O. Bolza, *On the first and second derivatives of hyperelliptic σ-functions*, Amer. J. Math. **17** (1895), 11–36.
4. V. M. Buchstaber, V. Z. Enolskii, and D. V. Leykin, *Hyperelliptic Kleinian functions and applications*, Solitons, Geometry, and Topology: On the Crossroad (College Part, 1997), Amer. Math. Soc. Transl. Ser. 2, vol. 179, Amer. Math. Soc., Providence, RI, 1997, pp. 1–33.
5. ———, *Kleinian functions, hyperelliptic Jacobians and applications*, Rev. Math. and Mathematical Physics, vol. 10, no. 2, Gordon and Breach, London, 1997.

6. ———, *Recursive family of polynomials generated by Sylvester's identity and addition theorem for hyperelliptic Kleinian functions*, Funct. Anal. Appl. **31** (1997), no. 4, 19–32.
7. ———, *σ-functions of (n, s) curves*, Russian Math. Surveys **54** (1999), no. 3, 155–156.
8. R. Dickson, F. Gesztesy, and K. Unterkofler, *A new approach to the Boussinesq hierarchy*, Math. Nachr. **98** (1999), 51–108.
9. ———, *Algebro-geometric solutions of the Boussinesq hierarchy*, Rev. Math. Phys. **11** (1999), 823–879.
10. B. A. Dubrovin, *Theta functions and nonlinear equations*, Russian Math. Surveys **36** (1981), 11–80.
11. V. Z. Enolskii, D. V. Leykin, and F. W. Nijhoff, *Lattice* KdV *type equations and Abel theorem*, 1999 (in preparation).
12. L. D. Faddeev and L. A. Takhtajan, *Liouville model on the lattice*, Field Theory, Quantum Gravity and Strings (Meudon/Paris, 1984/1985), Lecture Notes in Phys. Springer vol. 246, Springer, Berlin-New York, 1986, pp. 166–179.
13. E. V. Ferapontov, *Stationary Veselov–Novikov equation and isothermally asymptotic surfaces in projective differential geometry*, preprint, 1998.
14. R. W. H. T. Hudson, *Kummer's quartic surface*, Cambridge Univ. Press, Cambrigde, 1990.
15. F. Klein, *Über hyperelliptische Sigmafunctionen*, Math. Ann. **32** (1888), 351–380.
16. I. M. Krichever, *Algebraic-geometric construction of the Zakharov–Shabat equations and their periodic solutions*, Dokl. Akad. Nauk. UzSSR **227** (1976), 291–294.
17. V. B. Matveev and A. O. Smirnov, *Symmetry reductions of Riemann θ-function and some of their applications to the nonlinear Schrödinger equation and the Boussinesq equation*, Amer. Math. Soc. Transl. Ser. 2, **157** (1993), 227–237.
18. F. W. Nijhoff and H. W. Capel, *The discrete Korteweg de Vries equation*, Acta Appl. Math. **39** (1995), 133–158.
19. F. W. Nijhoff and V. Z. Enolskii, *Integrable mappings of* KdV *type and hyperelliptic addition theorems*, Symmetries and Integrabilty of Difference Equations (SIDE II) (P. Clarkson and F. Nijhoff, eds.) (Canterbery, 1996), Lecture Note Series, vol. 255, Cambridge Univ. Press, Cambridge, pp. 64–158.
20. F. W. Nijhoff, V. G. Papageorgiou, H. W. Capel, and G. R. W. Quispel, *The lattice Gel'fand–Dikii hierarchy*, Invent. Problems **6** (1992), 597–621.
21. M. Noether, *Zum Umkerhproblem in der Theorie der Abelschen Functionen*, Math. Anal. **28** (1887), 354–380.
22. E. Previato, *Monodromy of Boussinesq elliptic operators*, Acta Appl. Math. **36** (1994), 49–55.
23. R. J. Schilling, *Generalizations of the Neumann system—A curve theoretical approach.* I, Comm. Pure Appl. Math. **40** (1987), 455–522.
24. ———, *Generalizations of the Neumann system—A curve theoretical approach part* II, Comm. Pure Appl. Math. **42** (1989), 409–442.
25. ———, *Generalizations of the Neumann system—A curve theoretical approach.* III, Comm. Pure Appl. Math. **45** (1992), 775–820.
26. A. O. Smirnov, *On some set of elliptic solutions of Boussinesq equation*, Teoret. Mat. Fiz. **109** (1996), no. 3, 347–356.
27. K. Weierstrass, *Zur Theorie der Abelschen Functionen*, J. Reine Angew. Math. **47** (1854), 289–306.
28. ———, *Gesammelte Werke*, vol. 4, Teübner, 1902.

Department of Mathematics, Heriot-Watt University, Edinburgh, UK EH14 4AS.
E-mail address: J.C.Eilbeck@ma.hw.ac.uk

Theoretical Physics Division, NASU Institute of Magnetism, 36–b Vernadsky str., Kiev-680, 252142, UK.
E-mail address: vze@imag.kiev.ua

Theoretical Physics Division, NASU Institute of Magnetism, 36–b Vernadsky str., Kiev-680, 252142, UK.
E-mail address: dile@imag.kiev.ua

Centre de Recherches Mathématiques
CRM Proceedings and Lecture Notes
Volume **25**, 2000

Discrete Equations and the Singular Manifold Method

P. G. Estévez and P. A. Clarkson

Abstract. The Painlevé expansion for the second Painlevé equation (PII) and fourth Painlevé equation (PIV) have two branches. The singular manifold method therefore requires two singular manifolds. The double singular manifold method is used to derive Miura transformations from PII and PIV to modified Painlevé type equations for which auto-Bäcklund transformations are obtained. These auto-Bäcklund transformations can be used to obtain discrete equations.

1. Introduction

A very important subject related to the study of discrete equations is that concerning the discretization of Painlevé equations (cf. [**2, 16, 27, 30**]). Recently there has been substantial interest in the discrete Painlevé equations (dP1–dPVI), which arise in a variety of physical applications (and indeed dPI and dPII were first discovered in physical situations [**5, 17, 28**]). In the continuous limit, the discrete Painlevé equations yield their corresponding continuous Painlevé equation. This is not all as they have a variety of other properties in common with their continuous counterparts. For example they have Lax pairs, bilinear representations, Bäcklund transformations and particular solutions for certain parameter values, expressible in terms of rational functions or discrete special functions (see, for example, [**14**] and the references therein).

The discrete equations, though, are much richer than the continuous equations. They have a host of properties which are lost in the continuous limit. Perhaps the most fundamental difference is that the continuous Painlevé equations have a unique canonical form whereas this is not true for the discrete equations: for any discrete Painlevé equation there exists a number of different forms.

It is well known that continuous Painlevé equations satisfy the Painlevé property (cf. [**8, 22**]) which means that all their solutions $y(x)$ can be locally expressed as

$$y(x) = \sum_{j=0}^{\infty} a_j (x - x_0)^{j-\alpha}, \tag{1.1}$$

2000 *Mathematics Subject Classification.* Primary: 58F07; Secondary: 39A10.
This is the final form of the paper.

in a neighborhood of the movable singularity $\phi \equiv x - x_0 = 0$, where x_0 is an arbitrary constant, α is a positive integer and a_j, $j = 0, 1, \ldots$, are constants to be determined. More precisely, the singular manifold method (SMM) [**36**] consists of truncation of the series (1.1) to the constant level

$$y(x) = \sum_{j=0}^{\alpha} a_j (x - x_0)^{j-\alpha}.$$

This condition seems to be very restrictive, nevertheless it provides much information about a given equation (cf. [**7, 36**]).

In the present paper we shall apply the SMM to the second and fourth Painlevé equations, namely PII and PIV, which are usually written as

$$\text{(1.2)} \qquad \text{PII}: \qquad y'' = 2y^3 + xy + \ell,$$

$$\text{(1.3)} \qquad \text{PIV}: \qquad y'' = \frac{(y')^2}{2y} + \frac{3}{2}y^3 + 4xy^2 + 2(x^2 - 2\ell)y + \frac{\mu}{y},$$

where $' \equiv \mathrm{d}/\mathrm{d}x$ and ℓ and μ are constants (cf. [**22**]).

1.1. Two Painlevé branches and the singular manifold method. These equations (1.2), (1.3) have the property that the leading analysis of the expansion (1.1) provides $\alpha = 1$ and $a_0 = \pm 1$, where the $\pm$ sign in a_0 indicates that there are two possible Painlevé branches.

Equations with two Painlevé branches and how to extend the SMM to them have been studied in previous papers (cf. [**9, 10**]). According to these, when an equation, such as (1.2) or (1.3), has two Painlevé branches, then the truncation of (1.1) should be made for both branches simultaneously. This means that we should work with truncated solutions $\widetilde{y}$ of the following form [**9**]

$$\text{(1.4)} \qquad \widetilde{y} = y + \frac{g'}{g} - \frac{h'}{h},$$

where $y(x)$ is also a solution of the equation, and $g(x) = 0$ is the singularity for the + expansion and $h(x) = 0$ the singularity for the − expansion. Henceforth we shall call g and h *singular manifolds* [**36**].

The SMM usually requires that each coefficient in the different powers of the singular manifold that arises from substitution of the truncated expansion in the differential equation be equal to zero. Due to the nonlinearity of PII and PIV, the substitution of (1.4) in (1.2) and (1.3) respectively provides terms that mix the powers in h and g. To solve this problem we use a *decoupling ansatz* [**9**] given by

$$\text{(1.5)} \qquad \frac{g'}{g}\frac{h'}{h} = A\frac{g'}{g} + B\frac{h'}{h},$$

where A and B are functions of y, and g and h to be determined from the truncation itself. Furthermore, taking the derivative of (1.5) with respect to x we have

$$\text{(1.6)} \qquad A' = A(r - A - B), \qquad B' = B(v - A - B),$$

where $v = g''/g'$ and $r = h''/h'$.

Within this framework, the objective of this paper is to prove that for PII and PIV the improved version of the SMM including two singular manifolds provides the following results:

(i) Modified versions of PII and PIV (namely mPII and mPIV).

(ii) Miura transformations between PII and mPII and between PIV and mPIV.
(iii) Bäcklund transformations for mPII and mPIV.
(iv) By regarding the parameter ℓ as a discrete variable, discrete equations can be derived associated with mPII and mPIV.
(v) Solutions of these discrete equations allow us to construct solutions of PII and PIV as a "linear" superposition of solutions of the discrete equations.

2. Painlevé II

Substituting (1.4) into PII (1.2) and using (1.5) to decouple the crossed terms we get:

$$A = y + \frac{1}{2}v, \tag{2.1}$$

$$B = -y + \frac{1}{2}r, \tag{2.2}$$

$$0 = v' + v^2 - 6y^2 - x + 6A(2y + B - A), \tag{2.3}$$

$$0 = r' + r^2 - 6y^2 - x - 6B(2y + B - A). \tag{2.4}$$

The decoupling ansatz (1.6), once used in (2.1)–(2.2), gives

$$AB = k,$$

where k is a constant. By substituting (2.1)–(2.2) into (2.3)–(2.4), the following singular manifold equations are obtained

$$v' - \frac{1}{2}v^2 + 6k - x = 0, \qquad r' - \frac{1}{2}r^2 + 6k - x = 0. \tag{2.5}$$

2.1. A Miura transformation for modified PII. If we return to equation (1.6), substituting A and B as given by (2.1) and (2.2), respectively, and using (2.5), we have

$$y' - y^2 - \frac{1}{2}x + v' + 2k = 0, \quad -y' - y^2 - \frac{1}{2}x + r' + 2k = 0, \tag{2.6}$$

which means that the expression $y' - y^2 - \frac{1}{2}x$ depends only on the singular manifold g whilst $-y' - y^2 - \frac{1}{2}x$ depends only on h. Hence it is useful to define the following functions m' and n' as follows

$$2m' = y' - y^2 - \frac{1}{2}x, \quad 2n' = -y' - y^2 - \frac{1}{2}x. \tag{2.7}$$

With the aid of (1.2), these equations (2.7) can be integrated to give

$$\begin{aligned} 2m &= (y')^2 - \left(y^2 + \frac{1}{2}x\right)^2 - (2\ell - 1)y, \\ 2n &= (y')^2 - \left(y^2 + \frac{1}{2}x\right)^2 - (2\ell + 1)y, \end{aligned} \tag{2.8}$$

respectively. In order to identify the equations that m and n satisfy, we take the derivative of (2.7), which gives

$$2m'' = -4ym' + \left(\ell - \frac{1}{2}\right), \quad 2n'' = 4yn' - \left(\ell - \frac{1}{2}\right). \tag{2.9}$$

Now if we take y' from (2.7) and y from (2.9) and substitute them in (2.8), then we obtain the following equation for m and n

$$\frac{(M_\ell'')^2}{M_\ell'} + 4(M_\ell')^2 + 2(xM_\ell' - M_\ell) - \frac{(2\ell-1)^2}{16M_\ell'} = 0, \tag{2.10}$$

where $m = M_\ell$ and $n = M_{\ell+1}$. It is easy to prove that this equation has the Painlevé property. In fact, equation (2.10) is the potential version of the equation 34 of the Gambier classification [**13**] (see also [**11, 22, 30**]), which is commonly referred to as P34. Indeed, if we set $M_\ell' = Q_\ell$, then (2.10) becomes

$$Q_\ell'' - \frac{(Q_\ell')^2}{2Q_\ell} + 4Q_\ell^2 + xQ_\ell + \frac{(\ell-1/2)^2}{8Q_\ell} = 0, \tag{2.11}$$

which is precisely P34. Equations (2.7) and (2.9) can be written as

$$\begin{aligned} 2Q_\ell &= y' - y^2 - \tfrac{1}{2}x, & 2Q_{\ell+1} &= -y' - y^2 - \tfrac{1}{2}x, \\ y &= \frac{-2Q_\ell' + \ell - 1/2}{4Q_\ell}, & y &= \frac{-2Q_{\ell+1}' + \ell + 1/2}{4Q_{\ell+1}}. \end{aligned}$$

These can be interpreted as a Miura transformation between PII (1.2) and P34 (2.11) [**11**]. According to Ramani and Grammaticos [**30**], P34 (2.11) may be also be thought of as a modified PII (mPII), since the relationship between their solutions is analogous to that between solutions of the Korteweg-de Vries and modified Korteweg-de Vries equations (see also [**11**]).

2.2. Auto-Bäcklund transformations for modified PII. By subtracting the two equations of (2.8), we have

$$y(x) = M_\ell(x) - M_{\ell+1}(x), \tag{2.12}$$

which combined with (2.9) gives

$$M_{\ell+1} = M_\ell + \frac{4M_\ell'' - (2\ell-1)}{8M_\ell'}, \qquad M_\ell = M_{\ell+1} + \frac{4M_{\ell+1}'' + (2\ell+1)}{8M_{\ell+1}'}, \tag{2.13}$$

which are auto-Bäcklund transformations for potential mPII.

2.3. Linear superposition for PII. Suppose we have two solutions M_ℓ and $M_{\ell+1}$ of potential mP34 (2.10), related by the Bäcklund transformations (2.13), then we can construct a solution of PII (1.2) by using (2.12).

2.4. Discrete equations for potential P34. If we let $\ell \to \ell - 1$ in the second of equations (2.13) then

$$M_{\ell+1} = M_\ell + \frac{4M_\ell'' - (2\ell-1)}{8M_\ell'}, \qquad M_{\ell-1} = M_\ell + \frac{4M_\ell'' + (2\ell-1)}{8M_\ell'}. \tag{2.14}$$

Adding and subtracting these two equations gives

$$M_\ell' = -\frac{2\ell-1}{4(M_{\ell+1} - M_{\ell-1})}, \qquad M_\ell'' = -\frac{(2\ell-1)(M_{\ell+1} + M_{\ell-1} - 2M_\ell)}{4(M_{\ell+1} - M_{\ell-1})}. \tag{2.15}$$

By substituting (2.15) in (2.10), the result is the nonautonomous discrete equation

$$(M_{\ell-1} - M_{\ell+1})\left\{\left(\ell - \frac{1}{2}\right)\left[(M_{\ell-1} - M_\ell)(M_{\ell+1} - M_\ell) + \frac{1}{2}x\right] - M_\ell(M_{\ell-1} - M_{\ell+1})\right\} + \left(\ell - \frac{1}{2}\right)^2 = 0,$$

where the parameter ℓ can be interpreted as the discrete variable.

3. Painlevé IV

In an analogous way as we studied PII in the previous section, in this section we study PIV (1.3) by using the truncated expansion (1.4) together with the decoupling ansatz (1.5). The result is

$$A = y + x + \frac{1}{2}v, \tag{3.1}$$

$$B = -y - x + \frac{1}{2}r, \tag{3.2}$$

$$\begin{aligned} 0 = {} & (24x + 6v)(A - y) - 4x^2 + 8\ell + v^2 + 2(v' + y') \\ & - 18y^2 - 16A^2 + 36yA + 8AB - 2rA, \end{aligned} \tag{3.3}$$

$$\begin{aligned} 0 = {} & (-24x + 6r)(B + y) - 4x^2 + 8\ell + r^2 + 2(r' - y') \\ & - 18y^2 - 16B^2 - 36yB + 8AB - 2vB. \end{aligned} \tag{3.4}$$

Equations (3.1)–(3.2) combined with (1.6) gives $AB = k$, with k a constant. Then substituting (3.1)–3.2 into (3.3)–(3.4), for the singular manifolds we have the equations

$$v' - \frac{1}{2}v^2 + 6k + 2x^2 + 8\ell - 2 = 0, \quad r' - \frac{1}{2}r^2 + 6k + 2x^2 + 8\ell + 2 = 0. \tag{3.5}$$

3.1. A Miura transformation for modified PIV. The substitution of (2.1)–(2.2) into (1.6) gives

$$y' - y^2 - 2xy + v' + 2(k + 2\ell) = 0, \quad -y' - y^2 - 2xy + r' + 2(k + 2\ell) = 0. \tag{3.6}$$

It is reasonable therefore to define new functions m and n in the following form

$$2m' = y' - y^2 - 2xy, \quad 2n' = -y' - y^2 - 2xy. \tag{3.7}$$

These equations can be integrated, taking (1.3) into account, as

$$\begin{aligned} 2m &= \frac{(y')^2 + 2\mu - (y^2 + 2xy)^2}{4y} + (2\ell + 1)y, \\ 2n &= \frac{(y')^2 + 2\mu - (y^2 + 2xy)^2}{4y} + (2\ell - 1)y. \end{aligned} \tag{3.8}$$

Furthermore, the derivation of (3.7) provides

$$\begin{aligned} 2m'' &= \frac{2(m')^2 + \mu}{y} - 2(m' + 2\ell + 1)y, \\ 2n'' &= -\frac{2(n')^2 + \mu}{y} + 2(n' + 2\ell - 1)y. \end{aligned} \tag{3.9}$$

Taking y' as defined by (3.7) and substituting in (3.8) we have

$$\begin{aligned} 4(m-xm') &= \frac{2(m')^2+\mu}{y} + 2(m'+2\ell+1)y, \\ 4(n-xn') &= \frac{2(n')^2+\mu}{y} + 2(n'+2\ell-1)y. \end{aligned} \tag{3.10}$$

Adding and subtracting (3.9) and (3.10) the result is

$$\begin{aligned} y &= \frac{2(m-xm')-m''}{2(m'+2\ell+1)}, \quad \frac{1}{y} = \frac{2(m-xm')+m''}{2(m')^2+\mu}, \\ y &= \frac{2(n-xn')+n''}{2(n'+2\ell-1)}, \quad \frac{1}{y} = \frac{2(n-xn')-n''}{2(n')^2+\mu}. \end{aligned} \tag{3.11}$$

We can eliminated y by multiplication of the left equation by the right one. Thus we then have

$$(M_\ell'')^2 - 4(M_\ell - xM_\ell')^2 + 2(M_\ell' + 2\ell + 1)\left[2(M_\ell')^2 + \mu\right] = 0, \tag{3.12}$$

where $m = M_\ell$ and $n = M_{\ell-1}$. Equation (3.12) is of Painlevé type, and it can be considered as the modified version of PIV (mPIV). The Miura transformation that relates PII and mPII are (3.9), which can now be written as

$$2M_\ell' = y' - y^2 - 2xy, \quad 2M_{\ell-1}' = -y' - y^2 - 2xy. \tag{3.13}$$

3.2. Auto-Bäcklund transformations. Subtracting the two equations of (3.8), and with the aid of (3.13), we have

$$y = M_\ell - M_{\ell-1}, \tag{3.14}$$

then combined with the left part of (3.11) yields to

$$\begin{aligned} M_{\ell-1} &= M_\ell + \frac{M_\ell'' - 2(M_\ell - xM_\ell')}{2(M_\ell' + 2\ell + 1)}, \\ M_\ell &= M_{\ell-1} + \frac{M_{\ell-1}'' + 2(M_{\ell-1} - xM_{\ell-1}')}{2(M_{\ell-1}' + 2\ell - 1)}, \end{aligned} \tag{3.15}$$

which are auto-Bäcklund transformation for mPIV (3.12).

3.3. Linear superposition for PIV. If we have two solutions M_ℓ and $M_{\ell+1}$ of potential mPIV (3.12), related by the Bäcklund transformations (3.15), then we can construct a solution of PIV (1.3) by using (3.14).

3.4. Discrete equations for mPIV. As we did in the previous section, the Bäcklund transformations (3.15) can be written as

$$\begin{aligned} M_{\ell-1} &= M_\ell + \frac{M_\ell'' - 2(M_\ell - xM_\ell')}{2(M_\ell' + 2\ell + 1)}, \\ M_{\ell+1} &= M_\ell + \frac{M_\ell'' + 2(M_\ell - xM_\ell')}{2(M_\ell' + 2\ell + 1)}. \end{aligned} \tag{3.16}$$

Then by addition and subtraction, it is easy to show that

$$\begin{aligned} M_\ell' &= \frac{2M_\ell - (2\ell+1)(M_{\ell+1} - M_{\ell-1})}{M_{\ell+1} - M_{\ell-1} + 2x}, \\ M_\ell'' &= \frac{(M_\ell' + 2\ell + 1)(M_{\ell+1} + M_{\ell-1} - 2M_\ell)}{M_{\ell+1} - M_{\ell-1} + 2x}, \end{aligned} \tag{3.17}$$

whose substitution in (3.13) yields the discrete equation

$$(M_{\ell+1} - M_{\ell-1} + 2x)(M_\ell - M_{\ell-1})(M_\ell - M_{\ell+1})\left[M_\ell + 2\left(\ell + \frac{1}{2}\right)x\right]$$
$$+ \left\{2\left[M_\ell - \left(\ell + \frac{1}{2}\right)\right](M_{\ell+1} - M_{\ell-1})\right\}^2 + \frac{1}{2}x(M_{\ell+1} - M_{\ell-1} + 2x)^2 = 0,$$

where ℓ is the discrete variable.

4. Conclusions

In this paper we have derived Miura transformations, modified equations and associated discrete equations for the second and fourth Painlevé equations. Recently there have been several studies of the derivation of discrete equations, and in particular the discrete Painlevé equations, from Bäcklund transformations of the (continuous) Painlevé equations [**12, 14–16, 20, 21, 27, 33, 34**].

Hierarchies of solutions of PII and PIV are well-known (cf. [**1,3,6,18,19,25,26, 29, 35**]). Since there is an explicit relationship between PII and PIV and discrete equations, then these Hierarchies of solutions of PII and PIV also satisfy difference equations in addition to the ordinary differential equations. This is analogous to the situation for the classical special functions, such as Bessel, hypergeometric, Legendre, Weber–Hermite and Whittaker functions, which satisfy both an ordinary differential equation and a recurrence relation, which is a discrete equation. This provides further evidence that the Painlevé equations may be thought of as nonlinear special functions and that there is a deep relationship between the classical special functions, the Painlevé equations and the discrete Painlevé equations (see, for example, [**32**]).

References

1. H. Airault *Rational solutions of Painlevé equations*, Stud. Appl. Math. **61** (1979), 31–53.
2. A. P. Bassom and P. A. Clarkson, *New exact solutions of the discrete fourth Painlevé equation*, Phys. Lett. A **194** (1994), 358–370.
3. A. P. Bassom, P. A. Clarkson, and A. C. Hicks, *Bäcklund transformations and solution hierarchies for the fourth Painlevé equation*, Stud. Appl. Math. **95** (1995), 1–71.
4. ———, *On the application of solution hierarchies of the fourth Painlevé equation to several physically significant nonlinear partial differential equations*, Adv. Differential Equations **1** (1996), 175–198.
5. E. Brézin and V. A. Kazakov, *Exactly solvable field theories of closed strings*, Phys. Lett. B **236** (1990), 144–150.
6. P. A. Clarkson, *New exact solutions of the Boussinesq equation*, European J. Appl. Math. **1** (1990), 279–300.
7. P. A. Clarkson, N. Joshi, and A. Pickering *Bäcklund transformations for the second Painlevé hierarchy: A modified truncation approach*, Inverse Problems **15** (1999), 175–187.
8. H. Davis, *Introduction to nonlinear differential and integral equations*, Dover, New York, 1962.
9. P. G. Estévez and P. R. Gordoa, *Darboux transformations via Painlevé analysis*, Inverse Problems **13** (1997), 939–957.
10. P. G. Estévez, P. R. Gordoa, L. Martinez-Alonso, and E. Medina-Reus, *Modified singular manifold expansion: Application to the Boussinesq and Mikhailov-Shabat systems*, J. Phys. A **26** (1993), 1915–1925.
11. A. S. Fokas and M. J. Ablowitz, *On a unified approach to transformations and elementary solutions of Painlevé equations*, J. Math. Phys. **23** (1982), 2033–2042.
12. A. S. Fokas, B. Grammaticos, and A. Ramani, *From continuous to discrete Painlevé equations*, J. Math. Anal. Appl. **180** (1993), 342–360.

13. B. Gambier, *Sur les équations différentielles du second ordre et du premeir degre dont l'intégrale générale est à points critiques fixés*, Acta Math. **33** (1910), 1–55.
14. B. Grammaticos, F. W. Nijhoff, and A. Ramani, *The Painlevé property, one century later* (R. Conte, ed.), CRM Series in Mathematical Physics, Springer-Verlag, Berlin, 1999, pp. 413–516.
15. B. Grammaticos and A. Ramani, *Applications of analytic and geometric methods to nonlinear differential equations* (P. A. Clarkson, ed.), NATO ASI Series C, Kluwer, Dordrecht, 1993, pp. 299–313.
16. ———, *From continuous Painlevé* IV *to the asymmetric discrete Painlevé*. I, J. Phys. A **31** (1998), 5787–5798.
17. D. J. Gross and A. A. Migdal, *Nonperturbative two-dimensional quantum gravity*, Phys. Rev. Lett. **64** (1990), 127–130.
18. V. I. Gromak, *One-parameter systems of solutions of Painlevé's equations*, Differential Equations **14** (1978), 1510–1513.
19. ———, *The Painlevé property, one century later* (R. Conte, ed.), CRM Series in Mathematical Physics, Springer-Verlag, Berlin, 1999, pp. 687–734.
20. V. I. Gromak and V. V. Tsegel'nik, *Functional relations between solutions to P-type equations*, Differential Equations **30** (1994), 1037–1043.
21. ———, *Some properties of the Bäcklund transformations of the Painlevé equations*, Differential Equations **32** (1996), 1024–1029.
22. E. L. Ince, *Ordinary differential equations*, Dover, New York, 1956.
23. K. Kajiwara and Y. Ohta, *Determinant structure of the rational solutions for the Painlevé* II *equation*, J. Math. Phys. **37** (1996), 4693–4704.
24. ———, *Determinant structure of the rational solutions for the Painlevé* IV *equation*, J. Phys. A **31** (1998), 2431–2446.
25. N. A. Lukashevich, *Elementary solutions of certain Painlevé equations*, Differential Equations **1** (1965), 561–564.
26. Y. Murata, *Rational solutions of the second and the fourth Painlevé equations*, Funkcial. Ekvac. **28** (1985), 1–32.
27. F. W. Nijhoff, J. Satsuma, K. Kajiwara, B. Grammaticos, and A. Ramani, *A study of the alternative discrete Painlevé* II. *Equation*, Inverse Problems **12** (1996), 697–716.
28. V. Periwal and D. Shevitz, *Unitary-matrix models as exactly solvable string theories*, Phys. Rev. Lett. **64** (1990), 1326–1329.
29. K. Okamoto, *Studies on the Painlevé equations* III. *Second and fourth Painlevé equations,* PII *and* PIV, Math. Ann. **275** (1986), 221–225.
30. A. Ramani and B. Grammaticos, *Miura transforms for discrete Painlevé equations*, J. Phys. A **25** (1992), L633–L637.
31. ———, *Discrete Painlevé equations: Coalescences, limits and degeneracies*, Phys. A **228** (1996), 160–171.
32. K. M. Tamizhmani, A. Ramani, B. Grammaticos, and K. Kajiwara, *Coalescence cascades and special-function solutions for the continuous and discrete Painlevé equations*, J. Phys. A **31** (1998), 5799–5810.
33. V. V. Tsegel'nik, *On a relationship between solutions of the third Painlevé equation*, Theoret. and Math. Phys. **102**, 265–266 (1995).
34. ———, *Analytic properties of solutions of a nonlinear P-type equation*, Differential Equations **32** (1996), 1433–1435.
35. H. Umemura and H. Watanabe, *Solutions of the second and fourth Painlevé equations* . I, Nagoya Math. J. **148** (1997), 151–198.
36. J. Weiss, *The Painlevé property for partial differential equations.* II. *Bäcklund transformation, Lax pairs, and the Schwarzian derivative*, J. Math. Phys. **24** (1983), 1405–1413.

AREA DE FISICA TEÓRICA, FACULTAD DE CIENCIAS, UNIVERSIDAD DE SALAMANCA, 37008 SALAMANCA, SPAIN.

E-mail address: `pilar@sonia.usal.es`

INSTITUTE OF MATHEMATICS AND STATISTICS, UNIVERSITY OF KENT AT CANTERBURY, CANTERBURY, KENT, CT2 7NF, UK.

E-mail address: `P.A.Clarkson@ukc.ac.uk`

Centre de Recherches Mathématiques
CRM Proceedings and Lecture Notes
Volume **25**, 2000

Discrete Versions of Some Algebraic Integrable Systems Related to Generalized Jacobians

Yu. Fedorov

ABSTRACT. It is known that the integrable ellipsoidal billiard can be regarded as the limit of a geodesic flow on an ellipsoid, when one of its semi-axis tends to zero. This phenomenon can be used to construct discrete analogs of other algebraic integrable systems. Trajectories of such discrete systems are defined as limit points of certain asymptotic trajectories of the continuous counterparts. The latter are linearized on generalized Jacobians of singular spectral curves, whereas the former are described as shifts on reductions of such Jacobians. We illustrate this approach by producing a discretization of the classical Euler top problem on the group SO(3) and give its geometric model of motion.

1. Introduction. Ellipsoidal Billiards

It is known that one of the classical discrete integrable systems, the ellipsoidal billiard in $\mathbb{R}^n$, can be regarded as the limit of a geodesic flow on an n-dimensional ellipsoid (quadric) when one of its semi-axis tends to zero. In this sense, the billiard is a discrete version of the geodesic problem.

Let us briefly recall algebro-geometric background of the limit procedure (see [**3**]). Consider a family of confocal quadrics in $\mathbb{R}^{n+1}$

$$\widetilde{Q}(s) = \left\{ \frac{x_1^2}{D_1 - s} + \cdots + \frac{x_{n+1}^2}{D_{n+1} - s} = 1 \right\}, \quad s \in \mathbb{R}$$

with positive and distinct $D_1, \ldots, D_{n+1}$. The geodesic flow on the ellipsoid $\widetilde{Q} = \widetilde{Q}(0)$ is integrable due to the following fundamental geometric property (the *Chasles theorem*): tangent lines to a fixed geodesic are also tangent to other $n-1$ fixed confocal quadrics $\widetilde{Q}(c_1), \ldots, \widetilde{Q}(c_{n-1})$.

Let $\lambda_1, \ldots, \lambda_n$ be elliptic coordinates on $\widetilde{Q}$. It is known that after the change of time $t \to t'$ by the formula $dt = \lambda_1 \ldots \lambda_n \, dt'$, the geodesic flow is linearized on an unramified covering of the Jacobi variety of the genus n hyperelliptic curve

$$\Gamma = \left\{ w^2 = -\lambda(\lambda - D_1) \ldots (\lambda - D_{n+1})(\mu - c_1) \ldots (\mu - c_{n-1}) \right\},$$

2000 *Mathematics Subject Classification.* 58F07, 34A20, 14K99.
This is the final form of the paper.

[7,16], which is also the spectral curve of a Lax representation of the problem (see [10]).

Now let $D_{n+1} \to 0$. As a result, $\widetilde{Q}$ transforms to the interior of an $(n-1)$-dimensional ellipsoid Q, the geodesic flow to a straight line motion inside Q, whereas the points of intersection of the geodesics with the (hyper) plane $\{x_{n+1} = 0\}$ are mapped into impact points on Q. The curve Γ becomes singular. Its regularization is the genus $n-1$ curve

$$\Gamma' = \{w^2 = R(\lambda)\}, \quad R(\lambda) = (\lambda - D_1) \dots (\lambda - D_n)(\mu - c_1) \dots (\mu - c_{n-1}),$$

with the pair of points $E_\pm = \left(0, \pm\sqrt{R(0)}\right)$ arising from the double point on Γ. The variables λ_i then become elliptic coordinates in $\mathbb{R}^n = (x_1, \dots, x_n)$ and form a divisor of n points $P_i = (\lambda_i, w_i)$ on Γ'.

It turns out that n independent holomorphic differentials on the original curve Γ survive the regularization procedure and give rise to a basis of holomorphic differentials $\omega = (\omega_1, \dots, \omega_{n-1})$ on Γ' and one differential of third kind $\Omega_{E_\pm}$ having a pair of simple poles at E_-, E_+. As a result, the Jacobian of Γ transforms to the generalized Jacobian variety $\operatorname{Jac}(\Gamma', E_\pm)$, an algebraic extension of $\operatorname{Jac}(\Gamma')$ by $\mathbb{C}^*$.

Let $z \in \mathbb{C}^{n-1}$, $Z \in \mathbb{C}$ be coordinates on the universal coverings of $\operatorname{Jac}(\Gamma')$ and $\mathbb{C}^*$ respectively, such that $\{e^Z\} = \mathbb{C}^*$. The straight line motion inside the ellipsoid Q is described as a special straight line flow on the generalized Jacobian with $z = \text{const.}$ and Z depending linearly on t'.

An important fact is that as t' and Z vary along real axis from $-\infty$ to ∞, the mass point inside the ellipsoid Q moves from one impact point on Q to the next one, whereas one of the points P_i on Γ', say P_n, evolves from the pole E_- of $\Omega_{E_\pm}$ to E_+ (notice that the infinite "time" t' between any subsequent impacts corresponds to a finite interval in the original time t). It follows that for the given constants of motion, the impact points on Q are characterized by the divisor $P_1, \dots, P_{n-1}$ on Γ', i.e., by a point in the customary Jacobian of Γ'. Passage from one impact point to the next one corresponds to the shift in $\operatorname{Jac}(\Gamma')$ by the vector $q = \int_{E_-}^{E_+} \omega \subset \mathbb{C}^{n-1}$, which coincides with the vector of b-periods of the *normalized* differential $\Omega_{E_\pm}$ on Γ'.

Thus a discrete billiard trajectory can be obtained as a sequence of limit points of special geodesic trajectories corresponding to the singular spectral curve ($D_{n+1} = 0$). In this sense the trajectories can be viewed as *asymptotic*.

It seems natural that this phenomenon can be applied to produce discrete analogs of other algebraic integrable systems that have multidimensional generalizations. To obtain their asymptotic trajectories with limit points, in some problems, like in the geodesic flow, it is possible to choose a critical value of a constant parameter, in other cases we have to take a special value of a constant of motion. However, in both situations we deal with singular spectral curves and, as a consequence, with generalized Jacobians. According to above considerations, the sequence of limit points forming a trajectory of the discrete analog will automatically be described as translations on reductions of such Jacobians.

In given paper we construct a discrete analog of the classical problem of the Euler top moving in space by considering special asymptotic solutions of its 4-dimensional generalization, the Frahm–Manakov top on the Lie group SO(4). This result can be extended to N-dimensional case.

In §2 we describe both problems from the algebraic geometrical point of view. In §3 we consider a class of asymptotic trajectories on SO(4) leading to discrete trajectories on SO(3) and give their explicit solutions in terms of generalized theta-functions. Then it is necessary to solve "the inverse problem", i.e., to find the mapping generating the discrete trajectories. This is done in §4, where we introduce purely algebraic coordinates on the affine part of a generalized hyperelliptic Jacobian (a modification of the Jacobi–Mumford construction) and describe their transformation under a shift depending on parameters (the Bäcklund–Darboux transformation). In §5 we present a transparent geometric model of motion of the discrete Euler top and discuss its "equations of motion".

2. The Euler Top on SO (3) and SO (4)

Recall that the Euler–Frahm equations describing free motion of an m-dimensional rigid body have the Hamiltonian form

$$\dot{M} = [M, \Omega], \quad \Omega = \frac{\partial H(M)}{\partial M}, \tag{2.1}$$

where $H(M)$ is the Hamiltonian, $\Omega \in \mathrm{so}(m)$ is the angular velocity, $M \in \mathrm{so}^*(m)$ the angular momentum of the body [**5**]. These equations are known to be integrable provided the operators M and Ω are related as follows: $[M, b] = [\Omega, a]$, where a, b are constant diagonal matrices $a = \mathrm{diag}(a_1, \dots, a_m)$, $b = \mathrm{diag}(b_1, \dots, b_m)$ and all the eigenvalues of a and b are distinct.

Let us consider the case $b = a^2$. Then $\Omega = aM + Ma$, and equations (2.1) take the simple form

$$\dot{M}_{ij} = (a_j - a_i) \sum_{k=1}^{m} M_{ik} M_{kj}, \quad i, j = 1, \dots, m. \tag{2.2}$$

The integrability of (2.2) follows from the Lax representation with a rational spectral parameter found in [**8**] or the following hyperelliptic Lax pair [**2**]

$$\begin{gathered}
\dot{L}(s) = \big[L(s), A(s)\big], \quad s \in \mathbb{C}, \quad L(s), A(s) \in \mathrm{so}(m), \\
L(s)_{ij} = \frac{\sqrt{\Phi(s)}}{\sqrt{(s-a_i)(s-a_j)}} M_{ij}, \quad A(s)_{ij} = \sqrt{(s-a_i)(s-a_j)} M_{ij}, \\
\Phi(s) = (s-a_1)(s-a_2)\dots(s-a_m), \quad i, j = 1, \dots, m.
\end{gathered}$$

Writing out the characteristic polynomial for $L(s)$, we obtain the following families of first integrals of the problem

$$\begin{gathered}
\mathcal{I}_k(s, M) = \sum_I \frac{\Phi(s)}{(s-a_{i_1})\dots(s-a_{i_k})} |M|_I^I = \sum_{\nu=0}^{m-k} s^\nu H_{k\nu}(M), \\
k = 2, 4, \dots, 2[m/2],
\end{gathered} \tag{2.3}$$

where $|M|_I^I$ are diagonal minors of order k corresponding to the multi-indices $I = \{i_1 \dots i_k\} \subset \{1 \dots m\}$, $1 \le i_1 < \dots < i_k \le m$. The polynomials $H_{k\nu}(M)$, $k = 2, 4, \dots \le m$, $\mu = 0, 1, \dots, m$ form a complete set of independent first integrals of the system in involution. The leading coefficients $H_{k,m-k}(M)$ provide a complete set of Casimir functions of the coalgebra $\mathrm{so}^*(m)$.

In the classical case $m = 3$ the problem of motion of the Euler top in space was first solved by Jacobi [**5**]. We give a different method of finding the rotation matrix of the top by applying separation of variables (see also [**3**]).

The Euler equations (2.2) can now be written in vector form

$$\dot{\mathbf{M}} = \mathbf{M} \times \omega, \quad \omega = a\mathbf{M}, \quad \mathbf{M}, \omega \in \mathbb{C}^3, \quad a = \operatorname{diag}(a_1, a_2, a_3),$$

and from (2.3) we have the family of quadratic integrals

$$\mathcal{I}_2(s, M) = \sum_{i=1}^{3} (s - a_i)\mathbf{M}_i^2 = s(\mathbf{M}, \mathbf{M}) - (a\mathbf{M}, \mathbf{M}), \quad s \in \mathbb{C}. \tag{2.4}$$

Let us fix constants of motion by putting $\mathcal{I}_2(s, M) = c_0(s - c)$, c_0, $c = \text{const}$. To determine the position of the body in space, i.e., on the group SO(3), we introduce unit vectors α, β, γ forming a fixed orthogonal frame. Their components in the moving frame forming a 3×3 rotation matrix $\mathcal{R}$ of the body, satisfy the Poisson equations

$$\dot{\alpha} = \alpha \times \omega, \quad \dot{\beta} = \beta \times \omega, \quad \dot{\gamma} = \gamma \times \omega. \tag{2.5}$$

Suppose α is directed along the angular momentum $\mathbf{M}$, and β, γ are orthogonal to it:

$$\alpha = \mathbf{M}/\sqrt{c_0}, \quad (\mathbf{M}, \beta) = 0, \quad (\mathbf{M}, \gamma) = 0. \tag{2.6}$$

Then, using equations (2.5) for γ, we express $\mathbf{M}$ in terms of γ and $\dot{\gamma}$:

$$\mathbf{M} = a^{-1} \frac{\dot{\gamma} \times a^{-1}\gamma}{(\gamma, a^{-1}\gamma)}. \tag{2.7}$$

Let λ_1, λ_2 be spheroconic coordinates on the Poisson sphere $S^2 = \{(\gamma, \gamma) = 1\}$ such that

$$\gamma_i^2 = \frac{(a_i - \lambda_1)(a_i - \lambda_2)}{(a_i - a_j)(a_i - a_k)}, \quad (i, j, k) = (1, 2, 3), \tag{2.8}$$
$$\sum_{i=1}^{3} \gamma_i^2/a_i = \frac{\lambda_1\lambda_2}{a_1a_2a_3}, \quad 2\dot{\gamma}_i/\gamma_i = \frac{\dot{\lambda}_1}{a_i - \lambda_1} + \frac{\dot{\lambda}_2}{a_i - \lambda_2}.$$

Substituting these relations into (2.7), then to the integrals (2.4), after calculations we arrive at the equations

$$\frac{d\lambda_1}{2\sqrt{R(\lambda_1)}} + \frac{d\lambda_2}{2\sqrt{R(\lambda_2)}} = -\sqrt{c_0}\, dt, \quad \frac{\lambda_1 d\lambda_1}{2\sqrt{R(\lambda_1)}} + \frac{\lambda_2 d\lambda_2}{2\sqrt{R(\lambda_2)}} = 0, \tag{2.9}$$
$$R(\lambda) = -(\lambda - a_1)(\lambda - a_2)(\lambda - a_3)(\lambda - c).$$

Here the variables λ_1, λ_2 must be regarded as coordinates of points $P_1 = (\lambda_1, w_1)$, $P_2 = (\lambda_2, w_2)$ on the elliptic curve $\mathcal{C} = \{w^2 = R(\lambda)\}$. The equations include one holomorphic differential and one differential of 3rd kind with a pair of simple poles at the infinite points ∞_+, ∞_- on $\mathcal{C}$. Thus, (2.9) defines an Abel–Jacobi mapping of the symmetric product $(\mathcal{C} \setminus \{\infty_+, \infty_-\})^2$ to the generalized Jacobian, a group extension of $\operatorname{Jac}(\mathcal{C})$ by $\mathbb{C}^*$ denoted as $\operatorname{Jac}(\mathcal{C}, \infty_\pm)$.

By using (2.9), (2.7), we express $\dot{\gamma}$ and $\mathbf{M}$, α in terms of λ_1, λ_2, c. Then, applying $\beta = \gamma \times \alpha$, we get

$$\beta_i = \frac{\sqrt{\Sigma_i(\lambda_1)(\lambda_2 - a_i)} - \sqrt{\Sigma_i(\lambda_2)(\lambda_1 - a_i)}}{\sqrt{(a_i - a_j)(a_i - a_k)(\lambda_1 - \lambda_2)}}, \quad \Sigma_i(\lambda) = \frac{R(\lambda)}{\lambda - a_i}. \tag{2.10}$$

Let us fix a canonical basis of cycles A, B on $\mathcal{C}$ and the corresponding normalized differential of first kind $\bar{\omega}$, as well as the normalized differential of third kind $\Omega_{\infty_\pm}$ having a pair of simple poles at ∞_-, ∞_+. Let z, Z be corresponding coordinates on the universal coverings of $\mathrm{Jac}(\mathcal{C})$ and its extension $\mathbb{C}^*$ respectively.

By comparing expressions (2.8), (2.10) with theta-functional expressions for standard root functions (see [**3**]), the components of γ, β, and α can be expressed as the following quotients of customary and generalized theta-functions related to $\mathrm{Jac}(\mathcal{C}, \infty_\pm)$:

$$\begin{gathered}\gamma_i, \beta_i = \sqrt{\pm 1}\,\kappa_i \frac{e^{-Z/2}\theta[\Delta + \eta_i](z - q/2) \pm e^{Z/2}\theta[\Delta + \eta_i](z + q/2)}{\theta[\Delta](z)}, \\ \alpha_i = d_i \frac{\theta[\Delta + \eta_i](z)}{\theta[\Delta](z)}, \quad \kappa_i, d_i = \text{const.}, \quad i = 1, 2, 3, \quad q = \int_{\infty_-}^{\infty_+} \bar{\omega},\end{gathered} \tag{2.11}$$

where $\theta[\Delta + \eta_i](z), \theta[\Delta](z)$ are customary theta-functions related to the curve $\mathcal{C}$ with appropriate half-integer theta-characteristics Δ, η_i; the factors κ_i, d_i depend on the modulus of $\mathcal{C}$ only, and

$$z = \frac{2\sqrt{c_0}}{\xi} t + z_0, \quad Z = \frac{2\sqrt{c_0}\,\Xi}{\xi} t + Z_0, \quad z_0, Z_0 = \text{const.}, \tag{2.12}$$

ξ, Ξ being A-periods of the two differentials in (2.9) respectively. Expressions (2.11), (2.12) give a complete solution to the Euler top problem.

It follows from (2.11) that the components of γ, β are meromorphic functions on the 8-fold unramified covering of $\mathrm{Jac}(\mathcal{C}, \infty_\pm)$ obtained by doubling its three periods, whereas those of $\mathbf{M}$ and α are meromorphic on the corresponding 4-fold covering of $\mathrm{Jac}(\mathcal{C})$ itself.

Now we concentrate on the case $m = 4$. The families (2.3) take the form

$$\begin{gathered}\mathcal{I}_2(s, M) = \sum_{i<j} \frac{\Phi(s)}{(s - a_i)(s - a_j)} M_{ij}^2 = s^2 H_{22}(M) + s H_{21}(M) + H_{20}(M), \\ M \in \mathrm{so}(4), \quad \Phi(s) = (s - a_1) \dots (s - a_4), \\ H_{22} = \sum_{i<j}^{4} M_{ij}^2, \quad \mathcal{I}_4(M) = (M_{12}M_{34} - M_{13}M_{24} + M_{23}M_{14})^2.\end{gathered} \tag{2.13}$$

They give a set of 4 independent integrals of the system on the coalgebra $\mathrm{so}^*(4)$. Real generic invariant manifolds are 2-dimensional tori, on which the functions $H_{21}(M)$, $H_{20}(M)$ generate 2 commuting Hamiltonian flows with respect to the Lie–Poisson bracket on the coalgebra.

Let us fix constants of motion by putting

$$\mathcal{I}_2(s, M) = c_0(s - c_1)(s - c_2), \quad \mathcal{I}_4(M) = 0, \quad c_0, c_1, c_2 = \text{const.} \tag{2.14}$$

The latter choice corresponds to special solutions lying on *singular* rank 2 orbits of the coadjoint representation of the group SO(4) on the coalgebra (rank $M = 2$). Following [**1, 10**] we shall refer to them as *rank* 2 *solutions.*

To describe the motion of the top in space, we choose a fixed basis $\{\beta, \gamma, \alpha, \delta\}$ of unit orthogonal vectors in $\mathbb{R}^4$ such that α, δ span the 2-dimensional kernel of M: $M\alpha = M\delta = 0$ and $M = \sqrt{c_0}\gamma \wedge \beta$. Introduce spheroconic coordinates on $S^3 = \big\{(\gamma, \gamma) = 1\big\}$ by putting

$$\gamma_i^2 = \frac{(a_i - \lambda_1)(a_i - \lambda_2)(a_i - \lambda_3)}{(a_i - a_j)(a_i - a_k)(a_i - \lambda_l)}, \quad (i, j, k, l) = (1, 2, 3, 4). \tag{2.15}$$

Using the Poisson equations $\dot{\gamma} = -\Omega\gamma$, $\Omega = \partial H(M)/\partial M \in \mathrm{so}(4)$ and the condition $\mathcal{I}_4(M) = 0$, we find

$$M_{ij} = \frac{a_i \dot{\gamma}_i \gamma_j - a_j \dot{\gamma}_j \gamma_i}{(\gamma, a^{-1}\gamma)}, \quad \beta_i = \frac{a_i \dot{\gamma}_i}{\sqrt{c_0}(\gamma, a^{-1}\gamma)}, \tag{2.16}$$

and then represent M_{ij} in terms of $\lambda, \dot{\lambda}$. Substituting the latter expressions to the integrals (2.14), we obtain the following result.

THEOREM 2.1. *Let $M(t)$ be a rank 2 solution of the system defined by Hamiltonian*

$$H = \frac{f_2}{2} H_{22}(M) + \frac{f_1}{2} H_{21}(M) + \frac{f_0}{2} H_{20}(M), \quad f_0, f_1, f_2 = \text{const.},$$

and the constants of motion in (2.14). *Then the evolution of λ_1, λ_2, λ_3 is described by the generalized Abel–Jacobi equations*

$$\begin{gathered} \sum_{k=1}^{3} \frac{d\lambda_k}{2\sqrt{R_6(\lambda_k)}} = \sqrt{c_0} f_0 \, dt, \quad \sum_{k=1}^{3} \frac{\lambda_k d\lambda_k}{2\sqrt{R_6(\lambda_k)}} = \sqrt{c_0} f_1 \, dt, \\ \sum_{k=1}^{3} \frac{\lambda_k^2 d\lambda_k}{2\sqrt{R_6(\lambda_k)}} = \sqrt{c_0} f_2 \, dt, \quad R_6(\lambda) = -\Phi(\lambda)(\lambda - c_1)(\lambda - c_2). \end{gathered} \tag{2.17}$$

These equations include two holomorphic differentials and one differential of third kind with a pair of simple poles at the infinite points ∞_+, ∞_- on the genus 2 hyperelliptic curve $\Gamma = \{w^2 = R_6(\lambda)\}$. It follows that the motion on the group SO(4) in the rank 2 case is linealized on the generalized Jacobian $\mathrm{Jac}(\Gamma, \infty_\pm)$.

Applying Theorem 2.1, relations (2.17) and (2.16), we express the components of M, β in terms of λ only:

$$\begin{aligned} \beta_i &= \frac{\sqrt{U(\lambda, c_1)}\sqrt{U(\lambda, c_2)}\sqrt{U(\lambda, a_i)}}{\sqrt{(a_i - a_j)(a_i - a_k)(a_i - a_l)}} \\ &\qquad \times \sum_{s=1}^{3} \frac{\sqrt{R_6(\lambda_s)}}{\prod_{p \neq s}(\lambda_s - \lambda_p) \cdot (\lambda_s - c_1)(\lambda_s - c_2)(\lambda_s - a_i)}, \\ M_{ij} &= \sqrt{c_0} \frac{\sqrt{U(\lambda, a_i)}\sqrt{U(\lambda, a_j)}}{(a_i - a_j)} \\ &\qquad \times \sum_{s=1}^{3} \frac{\sqrt{R_6(\lambda_s)}}{\prod_{p \neq s}(\lambda_s - \lambda_p) \cdot (\lambda_s - a_i)(\lambda_s - a_j)}, \\ & U(\lambda, s) = (s - \lambda_1)(s - \lambda_2)(s - \lambda_3), \quad (i, j, k, l) = (1, 2, 3, 4). \end{aligned} \tag{2.18}$$

Then we deduce analogous expressions for α, δ.

Comparing them, as well as (2.18), with standard root functions in [**3**], one can show that the components of β, γ are meromorphic functions on an unramified covering of $\mathrm{Jac}(\Gamma, \infty_\pm)$, whereas those of α, δ and of the matrix M are meromorphic on a covering of the customary Jacobian of Γ.

REMARK 2.2. Notice that in the classical case $m = 3$, by fixing the angular momentum M with respect to the basis α, β, γ (condition (2.6)), we made a Marsden–Weinstein reduction of the Euler top slow on the cotangent bundle $T^* \mathrm{SO}(3)$ to a flow on the group $\mathrm{SO}(3)$ itself. Similarly, for $m = 4$, under the condition $\mathcal{I}_4(M) = 0$ and the above choice of M in the fixed frame α, β, γ, δ, the rank 2 trajectories on $T^* \mathrm{SO}(4)$ reduce to a flow on a 5-dimensional subvariety of $\mathrm{SO}(4)$. In both cases the components of β, γ play the role of excessive coordinates on the reduced phase spaces. As noticed in [**1, 10**], the reduced systems admit the Lax pairs

$$\dot{L}(\lambda) = \big[L(\lambda), A(\lambda)\big], \quad L(\lambda) = \sum_{i=1}^{m} \frac{\Phi(\lambda)}{\lambda - a_i} \begin{pmatrix} \beta_i\gamma_i & \gamma_i^2 \\ -\beta_i^2 & -\beta_i\gamma_i \end{pmatrix}, \tag{2.19}$$

$$A(\lambda) = \sum_{i=1}^{m} \sqrt{c_0} \frac{\Phi(\lambda)}{(\lambda - a_i)^2} \begin{pmatrix} \beta_i\gamma_i & \gamma_i^2 \\ -\beta_i^2 & -\beta_i\gamma_i \end{pmatrix} + \begin{pmatrix} 0 & \lambda \\ \lambda & 0 \end{pmatrix}, \quad m = 3, 4,$$

$$\Phi(\lambda) = (\lambda - a_1) \dots (\lambda - a_m).$$

For $m = 3$, the scalar equations defined by (2.19) are equivalent to the Poisson equations (2.5) with $\alpha = \gamma \times \beta$, $\omega = \sqrt{c_0} a(\gamma \times \beta)$. In both cases the characteristic equation $\big|L(\lambda) - \mu I\big| = 0$ has the form $\mu^2 - \Phi(\lambda)\mathcal{I}_2(\lambda, M) = 0$ which gives rise to the (hyper)elliptic spectral cuves involved in the Abel mappings (2.9), (2.17).

3. Asymptotic Trajectories on SO (4) and Discrete Trajectories on SO (3)

Following the idea described in §1, in order to obtain a discrete Euler top motion on the group $\mathrm{SO}(3)$, we must apply a limit procedure to the top on $\mathrm{SO}(4)$. Let us consider its asymptotic rank 2 solutions by putting $c_2 = a_4$. Then the spectral curve Γ becomes singular. Its regularization is the elliptic curve $\Gamma' = \{w^2 = R_4(\lambda)\}$, $R_4(\lambda) = -(\lambda - a_1)(\lambda - a_2)(\lambda - a_3)(\lambda - c_1)$ having a pair of distinguished points E_-, E_+ with $\lambda(E_\pm) = a_4$. The Abelian differentials in (2.17) transform to one holomorphic differential and two differentials of third kind having simple poles at $E_\pm$ and $\infty_\pm$ on Γ' respectively. Equations (2.17) give rise to the following generalized Abel–Jacobi mapping

$$\begin{gathered} \sum_{k=1}^{3} \int_{P_0}^{P_k} \frac{d\lambda}{2\sqrt{R_4(\lambda)}} = (f_1 - a_4 f_0)\, dt, \quad \sum_{k=1}^{3} \int_{P_0}^{P_k} \frac{\lambda d\lambda}{2\sqrt{R_4(\lambda)}} = (f_2 - a_4 f_1)\, dt, \\ \sum_{k=1}^{3} \int_{P_0}^{P_k} \frac{d\lambda}{2(\lambda - a_4)\sqrt{R_4(\lambda)}} = f_0\, dt, \qquad P_k = (\lambda_k, w_k) \in \Gamma', \end{gathered} \tag{3.1}$$

where we choose the basepoint P_0 to be the Weierstrass point $(a_3, 0)$. Thus the asymptotic rank 2 trajectories are linearized on an unramified covering of the generalized Jacobian $\mathrm{Jac}(\Gamma', \infty_\pm, E_\pm)$, the extension of $\mathrm{Jac}(\Gamma') = \Gamma'$ by $\mathbb{C}^* \times \mathbb{C}^*$.

Notice that for real motion of the top there are 2 possibilities: $R_4(a_4) < 0$ and $R_4(a_4) > 0$. Below we consider the case $R_4(a_4) > 0$, which corresponds to a real asymprotic motion on SO(4). Then the poles of the second and third differentials in (3.1) have imaginary and real residues respectively.

Let $\bar{\omega}_1$, $\Omega_{\infty_\pm}$, $\Omega_{E_\pm}$ be the *normalized* differentials on Γ' and z_1, Z_1, Z_2 the corresponding coordinates on the universal covering $\mathbb{C}^3$ of $\mathrm{Jac}(\Gamma', \infty_\pm, E_\pm)$. According to (3.1), they change linearly in time t. Then, in the real case, z_1, Z_1 vary along the imaginary axis, whereas Z_2 evolves along the real axis.

There is a transparent relation between the flow on $\mathrm{Jac}(\Gamma', \infty_\pm, E_\pm)$ and motion of the top in the 4-dimensional space. Namely, according to the mapping (3.1), as t and Z_2 vary from $-\infty$ to ∞ along the real axis, one of the points of the divisor on Γ', say P_3, must evolve from the pole E_- to E_+ (or vic e versa). Substituting the limit value $\lambda_3 = a_4$ into (2.15), (2.18) and analogous expressions for α, δ with $c_2 = a_4$, we get the following result.

THEOREM 3.1. *As t and Z_2 tend to $\pm\infty$, the components M_{i4}, δ_i, $i = 1, 2, 3$ and $\alpha_4, \beta_4, \gamma_4$, tend to zero, whereas $\delta_4 \to \mp 1$ respectively. This describes an asymptotic trajectory on the group* SO(4)*, which, for $t \to -\infty$ and $t \to +\infty$, tends to different connected components* $\mathrm{SO}_-(3)$, $\mathrm{SO}_+(3)$ *of the same subgroup* O(3) *given by conditions $\delta_4 = \pm 1$. The limit motion of the top is thus a quasiperiodic rotation in the 3-dimensional space spanned by α, β, γ.*

Now consider special rank 2 asymptotic solutions of the Frahm–Manakov system on SO(4) with $c_2 = a_4$ defined by the Hamiltonian $H = \mathcal{I}_2(c_2, M)/2$. According to Theorem 2.1, in this case $f_1 = a_4 f_0$, $f_2 = a_4 f_1$. In view of the equations (3.1), this corresponds to a special flow on $\mathrm{Jac}(\Gamma', \infty_\pm, E_\pm)$ with $z_1 = \text{const.}$, $Z_1 = \text{const.}$ As t and $Z_2 \to \pm\infty$, the trajectories on SO(4) have limit *points* on $\mathrm{SO}_-(3)$ and $\mathrm{SO}_+(3)$ respectively. Indeed, for $Z_2 = \pm\infty$, we have $P_3 = E_\pm$, and the mapping (3.1) leads to

$$\text{(3.2)} \quad \int_{P_0}^{P_1} \bar{\omega} + \int_{P_0}^{P_2} \bar{\omega} = z_1 - \int_{P_0}^{E_\pm} \bar{\omega}, \quad \int_{P_0}^{P_1} \Omega_{\infty_\pm} + \int_{P_0}^{P_2} \Omega_{\infty_\pm} = Z_1 - \int_{P_0}^{E_\pm} \Omega_{\infty_\pm}.$$

Hence, fixed z_1, Z_1 define a fixed divisor (P_1, P_2) on Γ' and therefore a fixed position of the top. On the other hand, a divisor of 2 points on Γ' defines a point on the generalized Jacobian $\mathrm{Jac}(\Gamma', \infty_\pm)$, a reduction of $\mathrm{Jac}(\Gamma', \infty_\pm, E_\pm)$. According to (3.2), limit points on $\mathrm{SO}_-(3)$ and $\mathrm{SO}_+(3)$ joined by one and the same asymptotic trajectory correspond to points on $\mathrm{Jac}(\Gamma', \infty_\pm)$ which are different by the vector

$$\text{(3.3)} \qquad (e, S)^T = \int_{E_-}^{E_+} (\bar{\omega}, \Omega_{\infty_\pm})^T.$$

Let us assume an endpoint of such an asymptotic trajectory to be the start point of another asymptotic trajectory with the same Hamiltonian. By induction, we obtain an infinite sequence of points on $\mathrm{O}(3) \subset \mathrm{SO}(4)$. In view of Theorem 3.1, they are characterized by the first 3 nonzero components of $\alpha, \beta, \gamma \in \mathbb{R}^4$, and the sign of δ_4. By disregarding the sign, we get a sequence of points on the same subgroup SO(3). *This will be regarded as a trajectory of the discrete Euler top.*

We define its angular momentum vector and columns of rotation matrix as follows: for $Z_2 = \pm\infty$, $\mathbf{M} = (M_{23}, M_{31}, M_{12})^T$ and α, β, $\gamma \in \mathbb{R}^3$ ($\mathbb{C}^3$) are just truncations of the above 4-dimensional vectors. Now we identify Γ' with the elliptic curve $\mathcal{C}$ in §2 by putting $c_1 = c$. Then we can also indentify $\text{Jac}(\Gamma', \infty_\pm)$ specified by the mapping (3.2) with the generalized Jacobian for the classical Euler top, by assuming that $z_1 - q/2 = z$, $Z_1 - S/2 = Z$. In view of (3.3), we conclude that *for given constants of motion, trajectories of the discrete Euler top are described by theta-functional expressions* (2.11) *with*

$$z = z_0 + eN, \quad Z = Z_0 + SN, \quad z_0, Z_0 = \text{const.}, \quad N \in \mathbb{Z}.$$

Notice that the shift vector $(e, S)^T$ on $\text{Jac}(\mathcal{C}, \infty_\pm)$ depends on the parameter $a_4 = c_2$ and is generally not parallel to the velocity vector $(\dot{z}, \dot{Z})$ of the continuous system defined in (1.Z).

REMARK 3.2. As follows from the solutions (2.11), at limit points of the special asymptotic trajectories on SO(4) the components M_{12}, M_{23}, M_{13} of the angular momentum are generally nonzero, but, on the other hand, the angular velocity $\Omega \in \text{so}(4)$ must vanish. There is no contradiction: such asymptotic trajectories always have zero energy: $\mathcal{I}_2(c_2, M) \equiv 0$.

4. Two-Point Discrete Transformation of the Generalized Jacobi–Mumford System

Let $X_1, Y_1, \ldots, X_s, Y_s$ be s distinct pairs of points on a genus g Riemann surface Γ, $\omega_1, \ldots, \omega_g$ be independent holomorphic differentials and $\Omega_{X_l Y_l}$, $l = 1, \ldots, s$, be the normalized differentials of third kind on Γ having pairs of simple poles at the points X_l, Y_l.

Let $\mathcal{A}(P) = \int_{P_0}^{P} (\omega, \Omega_{XY})^T \in \mathbb{C}^n$, $n = g + s$ denote the Abel mapping of a point $P \in \Gamma$ with a basepoint $P_0 \in \widehat{\Gamma} = \Gamma \setminus \{X_1, Y_1, \ldots, X_s, Y_s\}$, $\Lambda \subset \mathbb{C}^n$ be the rank $2g + s$ lattice of periods of the mapping. The quotient variety $\mathbb{C}^{g+s}/\Lambda$ is known to be the generalized Jacobian $\text{Jac}(\Gamma, X, Y)$, a noncompact algebraic group.

Now suppose that Γ is a smooth hyperelliptic curve given by equation $\mu^2 = R(\lambda)$, where $R(\lambda)$ is a *monic* polynomial of degree $2g + 2$. To a generic divisor of points $P_1 = (\lambda_1, \mu_1), \ldots, P_n = (\lambda_n, \mu_n)$ on Γ we associate the point $\mathcal{A}(P_1) + \cdots + \mathcal{A}(P_n)$ in $\text{Jac}(\Gamma, X, Y)$ and the polynomials

$$U(\lambda) = (\lambda - \lambda_1) \ldots (\lambda - \lambda_n) = \lambda^n + u_1 \lambda^{n-1} + \cdots + u_n,$$
$$V(\lambda) = \sum_{k=1}^{n} \prod_{j \neq k} \mu_k \frac{\lambda - \lambda_k}{\lambda_k - \lambda_j} = v_1 \lambda^{n-1} + \cdots + v_n,$$

of order n, $n - 1$ respectively and the unique polynomial $W(\lambda)$ of order $g + \kappa$, where $\kappa = 1$ for $s = 1$, $\kappa = s - 2$ for $s > 1$, such that

$$W(\lambda)U(\lambda) + V^2(\lambda) = R(\lambda). \tag{4.1}$$

Under this condition, the coefficients of $U(\lambda)$, $V(\lambda)$, $W(\lambda)$ form coordinates on the affine part $\text{Jac}(\Gamma, X, Y) \setminus \widetilde{\Theta}$, where $\widetilde{\Theta}$ is the generalized theta-divisor defined e.g., in [**3,5**]. In addition, any translationally invariant flow on $\text{Jac}(\Gamma, X, Y)$ can be described as a system of differential equations on the coefficients. This generalizes the Jacobi–Mumford construction giving a purely algebraic description of the affine parts of customary Jacobians of hyperelliptic curves and providing so called *master systems* linearized on such Jacobians [**12, 14**].

Consider a translation in $\operatorname{Jac}(\Gamma, X, Y)$ by the fixed vector $\mathcal{A}(P)-\mathcal{A}(Q)$, $P=(\lambda_P,\mu_P)$, $Q=(\lambda_Q,\mu_Q)\in\widehat{\Gamma}$. In terms of divisors on $\widehat{\Gamma}$, the translation is described in form

$$\mathcal{A}(P_1)+\cdots+\mathcal{A}(P_n)+\mathcal{A}(P)=\mathcal{A}(Q_1)+\cdots+\mathcal{A}(Q_n)+\mathcal{A}(Q) \pmod{\Lambda}. \tag{4.2}$$

THEOREM 4.1, [**13**]. *Relation (4.2) holds if and only if there exists a meromorphic function $F(P)$ on Γ such that*

1. *it has poles at $P_1,\ldots,P_n,P$ and zeros at $Q_1,\ldots,Q_n,Q$,*
2. *for $l=1,\ldots,s$, the function $F(P)$ the same finite values at X_l, Y_l.*

Let $\widetilde{U}(\lambda)$, $\widetilde{V}(\lambda)$, $\widetilde{W}(\lambda)$ be the polynomials associated to the divisor $Q_1,\ldots,Q_n$. Below we describe the transformation $\mathcal{B}:(U(\lambda),V(\lambda),W(\lambda))\to(\widetilde{U}(\lambda),\widetilde{V}(\lambda),\widetilde{W}(\lambda))$ generated by the translation. The latter depends on two points on Γ. It is natural to refer to $\mathcal{B}$ as a 2-point Bäcklund–Darboux transformation.

In given paper we restrict ourselves to the case $s=1$, $X_1=\infty_-$, $Y_1=\infty_+$. Then the polynomials U,V,W have degree $g+1$, g, $g+1$ respectively and the above meromorphic function can be chosen in form

$$F(\lambda,\mu)=\frac{\mu+\mathcal{V}(\lambda|P_1,\ldots,P_n,P,Q)}{U(\lambda)(\lambda-\lambda_P)},$$

where the degree $g+2$ polynomial $\mathcal{V}(\lambda)$ is such that $\mathcal{V}(\lambda_i)=\mu_i$, $i=1,\ldots,n$, $\mathcal{V}(\lambda_P)=\mu_P$, $\mathcal{V}(\lambda_Q)=-\mu_Q$. Then

$$F(\lambda,\mu)=\frac{\mu+V(\lambda)+U(\lambda)(A\lambda+B)}{U(\lambda)(\lambda-\lambda_P)}, \tag{4.3}$$

$$A\lambda+B=\frac{1}{\lambda_P-\lambda_Q}\left(\frac{\lambda-\lambda_Q}{U(\lambda_P)}(\mu_P-V(\lambda_P))-\frac{\lambda-\lambda_P}{U(\lambda_Q)}(-\mu_Q-V(\lambda_Q))\right).$$

Notice that $F(\infty_-)=F(\infty_+)=A$.

The numerator in (4.3) has simple zeros at $\iota P_1,\ldots,\iota P_n$, ιP and $Q_1,\ldots,Q_n$, Q, where ι denotes the hyperelliptic involution on Γ. Applying the condition (4.1), we obtain

$$\begin{aligned}
\widetilde{U}&=-\frac{2\mathcal{Z}V+W-\mathcal{Z}^2U}{A^2\Psi},\qquad \Psi=(\lambda-\lambda_P)(\lambda-\lambda_Q),\\
\widetilde{V}&=\frac{2\mathcal{Y}\mathcal{Z}V+\mathcal{Y}W-\mathcal{Z}^2\mathcal{Y}U+(\mathcal{Z}A^2U-A^2V)\Psi}{A^2\Psi}=-V+\mathcal{Z}U-\mathcal{Y}\widetilde{U},\\
\widetilde{W}&=\frac{-(\mathcal{Y}\mathcal{Z}-A^2\Psi)^2U}{A^2\Psi}+\frac{2\mathcal{Y}^2\mathcal{Z}V-2\mathcal{Y}A^2\Psi V+\mathcal{Y}^2W}{A^2\Psi},
\end{aligned} \tag{4.4}$$

where

$$\widetilde{U}=\lambda^{g+1}+\widetilde{u}_1\lambda^g+\cdots+\widetilde{u}_{g+1},\quad \widetilde{W}=\lambda^{g+1}+\widetilde{w}_1\lambda^g+\cdots+\widetilde{w}_{g+1},$$
$$\widetilde{V}=\widetilde{v}_1\lambda^g+\cdots+\widetilde{v}_{g+1},\quad \mathcal{Z}=A\lambda+B,\quad \mathcal{Y}=A\lambda+\widetilde{B},$$

and the coefficient $\widetilde{B}$ must be chosen in such a way that the second expression in (4.4) is a polynomial of degree g. Expanding this expression in λ and imposing the condition, we find

$$\mathcal{Z}=A\lambda+\frac{1}{2}A(\widetilde{u}_1-u_1-\lambda_P-\lambda_Q),\quad \mathcal{Y}=A\lambda+\frac{1}{2}A(u_1-\widetilde{u}_1-\lambda_P-\lambda_Q). \tag{4.5}$$

The transformation (4.4) can be packed into the matrix form

$$(4.6)\qquad \widetilde{L}(\lambda)M(\lambda)=M(\lambda)L(\lambda),\quad L(\lambda)=\begin{pmatrix} V(\lambda) & U(\lambda)\\ W(\lambda) & -V(\lambda)\end{pmatrix},$$
$$\widetilde{L}(\lambda)=\begin{pmatrix} \widetilde{V}(\lambda) & \widetilde{U}(\lambda)\\ \widetilde{W}(\lambda) & -\widetilde{V}(\lambda)\end{pmatrix},\quad M(\lambda)=\begin{pmatrix} -\mathcal{Y} & 1\\ \mathcal{Y}\mathcal{Z}-A^2\Psi & -\mathcal{Z}\end{pmatrix},$$

where, in view of (4.5),

$$\mathcal{Y}\mathcal{Z}-A^2\Psi=\frac{1}{4}A^2(\lambda_Q-\lambda_P+u_1-\widetilde{u}_1)(\lambda_Q-\lambda_P+\widetilde{u}_1-u_1).$$

The coefficient A can be extracted from (4.3), (4.1) or by comparing terms in (4.6). In the latter case,

$$A=\frac{-2(v_1+\widetilde{v}_1)}{(u_1-\widetilde{u}_1)(u_1+\widetilde{u}_1+\lambda_Q+\lambda_P+2)},$$

and the matrix $M(\lambda)$ depends on the original and new coordinates in a symmetric way. This describes the transformation $\mathcal{B}$ in an implicit form.

Now we put $g=1, n=2$ and identify matrix $L(\lambda)$ in (4.6) with that of the Lax pair (2.19) for the Euler top flow on SO(3). Then we find

$$(4.7)\qquad \begin{gathered} u_1=(\gamma,a\gamma)-\operatorname{tr}a,\quad w_1=(\beta,a\beta)-\operatorname{tr}a,\quad v_1=\sqrt{-1}(\beta,a\gamma),\\ u_2=\det a\,(\gamma,a^{-1}\gamma),\quad w_2=\det a(\beta,a^{-1}\beta),\quad v_2=\sqrt{-1}\det a\,(\beta,a^{-1}\gamma),\\ a=\operatorname{diag}(a_1,a_2,a_3).\end{gathered}$$

This also leads to identifying the families of curves Γ with the elliptic spectral curves $\mathcal{C}$. As a result, (4.6) describes a *correspondence* $(\beta,\gamma)\to(\widetilde{\beta},\widetilde{\gamma})$ on the group SO(3), where $\widetilde{\beta},\widetilde{\gamma}$ are related to the coefficients of $\widetilde{U}$, $\widetilde{V}$, $\widetilde{W}$ in the same manner as in (4.7). The correspondence is an exact mapping modulo the action of the finite group of reflections $(\beta_i,\gamma_i)\to(-\beta_i,-\gamma_i)$ for any i.

Now we recall that in the shift vector (3.3) for the discrete Euler top the points P, Q are involutive: $P=\iota Q=E_+$, $\lambda_Q=\lambda_P=a_4$. Then the transformation (4.6) takes the following particular form

$$(4.8)\qquad L(\lambda|\widetilde{\beta},\widetilde{\gamma})M(\lambda)=M(\lambda)L(\lambda|\beta,\gamma),$$
$$L(\lambda|\beta,\gamma)=\sum_{i=1}^{3}\frac{\Phi(\lambda)}{\lambda-a_i}\begin{pmatrix}\sqrt{-1}\beta_i\gamma_i & \gamma_i^2\\ \beta_i^2 & -\sqrt{-1}\beta_i\gamma_i\end{pmatrix},$$
$$M(\lambda)=\begin{pmatrix}-A(\lambda-a_4)+Ad/2 & 1\\ -A^2d^2/4 & -A(\lambda-a_4)-Ad/2\end{pmatrix},\quad d=(\widetilde{\gamma},a\widetilde{\gamma})-(\gamma,a\gamma).$$

The factor A is found from (4.3) by taking the limit $Q\to\iota P$:

$$(4.9)\qquad A=\frac{\partial}{\partial\lambda}\left(\frac{\mu-V(\lambda)}{U(\lambda)}\right)\bigg|_{\lambda=a_4}.$$

Now putting in (4.8) $\lambda = a_1,\ a_3,\ a_3$ subsequently, and calculating the matrices $ML(a_i|\beta,\gamma)M^{-1}$, we find

$$\widetilde{\gamma}_i^2 d = -\frac{(2\beta_i - 2\sqrt{-1}A(a_i - a_4)\gamma_i - \sqrt{-1}A\,d\gamma_i)^2}{4A^2(a_i - a_4)^2},$$

$$\widetilde{\beta}_i^2 = \frac{(-4(a_i - a_4)\beta_i + 2d\beta_i - \sqrt{-1}Ad^2\gamma_i)^2}{16(a_i - a_4)^2},$$

$$\sqrt{-1}\widetilde{\beta}_i\widetilde{\gamma}_i = \frac{(-4(a_i - a_4)\beta_i + 2d\beta_i - \sqrt{-1}A\,d^2\gamma_i)}{8A(a_i - a_4)^2} \times (2\beta_i - 2\sqrt{-1}A(a_i - a_4)\gamma_i - \sqrt{-1}A\,d\gamma_i).$$

Notice that in the real case, under the condition $R_4(a_4) > 0$ imposed in §3, (4.9) implies that A is pure imaginary. Putting $A = \sqrt{-1}\hat{A}$, $\hat{A} \in \mathbb{R}$, from the last expressions we extract

$$\widetilde{\beta}_i + \beta_i = \frac{2d\beta_i + \hat{A}d^2\gamma_i}{4(a_i - a_4)}, \quad \widetilde{\gamma}_i + \gamma_i = \frac{-2\beta_i - \hat{A}d\gamma_i}{2\hat{A}(a_i - a_4)},$$
$$d = (\widetilde{\gamma}, a\widetilde{\gamma}) - (\gamma, a\gamma) = (\beta, a\beta) - (\widetilde{\beta}, a\widetilde{\beta}), \quad i = 1,2,3, \tag{4.10}$$

which gives an exact mapping in an implicit form generating the trajectories of the discrete Euler top. Another mapping having a similar form was found in [**8**]

5. A Geometric Model and Equations of Motion

The classical Euler top motion in $\mathbb{R}^3$ admits the remarkable Poinsot kinematical model, which can be found in many books on analytical dynamics. Here we present a "discrete" version of the model arising as a limit of the following kinematical model of rank 2 motions of the Frahm–Manakov top on SO(4). Namely, let $x_1, \ldots, x_4$ be coordinates in the moving frame whose origin O coincides with the fixed point, c_0, c_1, c_2 be constants specified in (2.14). We attach to the top two confocal cones $K_p = \{(X, (a - c_pI)^{-1}X) = 0\}$, $p = 1, 2$, which are always real for the real motion. Let Π be the fixed 2-plane in $\mathbb{R}^4$ representing the kernel of M: $\Pi = \operatorname{span}(\beta, \gamma)$.

THEOREM 5.1, [**4**]. *Let $M(t)$, $\alpha(t)$, $\beta(t)$, $\gamma(t)$, $\delta(t)$ be a rank 2 solution of the top on* SO(4) *with the constants of motion c_0, c_1, c_2 and the Hamiltonian*

$$H = \frac{1}{2}\mathcal{I}_2(c_p, M) + \frac{\rho}{2}H_{22}(M), \quad p = 1 \text{ or } 2, \quad \rho = \text{const.},$$

$\mathcal{I}_2(s, M)$, $H_{22}(M)$ being defined in (2.13). *Then, during the motion, the attached cones K_1, K_2 remain tangent to Π. At the points of the tangency lines $\{K_1 \cap \Pi\}$, $\{K_2 \cap \Pi\}$, normal vectors of K_1, K_2 are parallel to α and δ respectively. In addition, the cone K_p rolls without sliding over the plane Π, which, in turn, rotates with constant angular velocity $\rho\sqrt{c_0}$ about its origin.*

We recall that the rolling without sliding means that instant velocities of the points of $\{K_p \cap \Pi\}$ lying on K_p and on Π coincide. Notice that the rank 2 trajectory on SO(4) is uniquely defined by the above construction.

Let us choose the moving frame where the operator a is diagonal. Consider a limit of this model corresponding to $H = \frac{1}{2}\mathcal{I}_2(c_2, M)$ and $c_2 = a_4$. Then the plane Π does not rotate. The cone K_2 degenerates to the interior of the 2-dimensional

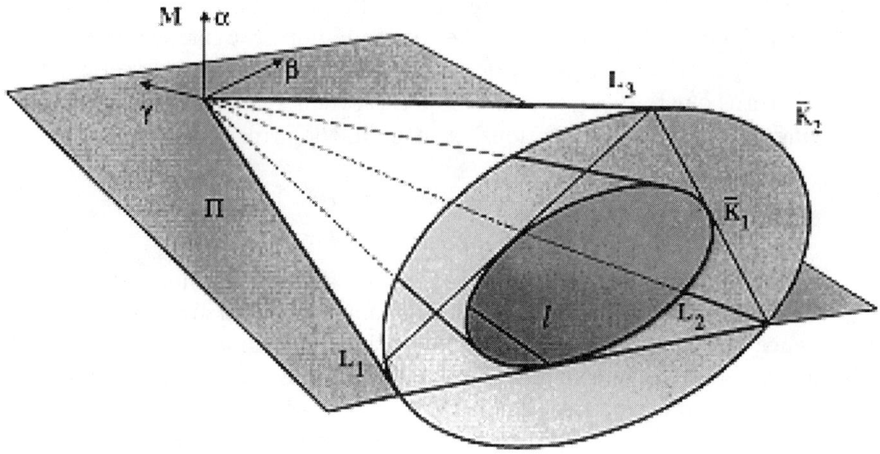

FIGURE 1

cone $\overline{K}_2 = \{x_1^2/(a_1-c_2)+x_2^2/(a_2-c_2)+x_3^2/(a_3-c_2)=0\}\cap\{x_4=0\}$, which is real in view of the condition $R(a_4)>0$ imposed in §3. According to Theorem 3.1, for $t=\pm\infty$ the hyperplane $\{x_4=0\}$ coincides with the fixed 3-dimensional subspace $\mathbb{R}^3$ spanned by α, β, γ. Hence, at these limit positions the cones $\overline{K}_2$ and $\overline{K}_1 = \{x_1^2/(a_1-c_1)+x_2^2/(a_2-c_1)+x_3^2/(a_3-c_1)=0\}\cap\{x_4=0\}$ lie in $\mathbb{R}^3$ and are confocal as well. The tangency line $K_1\cap\Pi$ coincides with $\ell=\overline{K}_1\cap\Pi$, whereas the line $K_2\cap\Pi$ blows to the sector in Π between the intersection lines $\{L_1,L_2\}=\overline{K}_2\cap\Pi$.

In view of the rolling condition and the degeneracy of K_2, when t varies from $-\infty$ to ∞, the tangency line $K_2\cap\Pi$ remains fixed in Π and coincides with L_1 or L_2. For definiteness we suppose $K_2\cap\Pi=L_2$. At the end of the asymptotic trajectory the cones $\overline{K}_1$, $\overline{K}_2$ again appear in $\mathbb{R}^3$ in a different position with $\overline{K}_1$ being tangent to Π and $\overline{K}_2$ intersecting Π along L_2 and another line L_3. Thus, the new configuration in $\mathbb{R}^3$ is obtained from the previous one by rotating the both 2-dimensional cones about the intersection line L_2 up to a position in which $\overline{K}_1$ is again tangent to Π. The next position is obtained by rotating about the line L_3. Applying this procedure recursively by rotating about L_N, we get a sequence of positions of the cones in $\mathbb{R}^3$. In view of the definition in §3, this provides a trajectory of the discrete Euler top on SO(3), which, for the given constants c_1, c_2, is described by theta-functional expressions (2.11). The resulting geometric model is shown in Fig. 1.

It is seen easily that in a frame attached to the top the sequence of 2-planes Π has the same property as one-dimensional billiard trajectories inside an ellipse: it has an elliptic caustic. In this sense our geometric model can be viewed as an elliptic billiard in a 2-dimensional projective space.

For $c_1\to a_4$, the cones $\overline{K}_1$, $\overline{K}_2$ confluent, and according to the model, the tangent cone $\overline{K}_1$ is rolling without sliding over the fixed plane Π providing the continuous motion of the top with the Hamiltonian $\mathcal{I}_2(c_1,M)/2=c_0\big((\mathbf{M},\mathbf{M})/2-c(\mathbf{M},a\mathbf{M})/2\big)$.

Let $\{\mathcal{R}_N \in \mathrm{SO}(3)\}$ be a sequence of rotation matrices of the discrete Euler top. The matrix $\hat{\Omega}_N = \mathcal{R}_N^T \mathcal{R}_{N-1} \in \mathrm{SO}(3)$ represents the discrete analog of the angular velocity in the body and describes a finite rotation about the intersection line L_N. Let $M_N \in \mathrm{so}(3)$ be the momentum operator corresponding to the vector $\mathbf{M}_N$ in the body. It turns out that *for any constants* c_0, c, the sequences $\{\hat{\Omega}_N, M_N\}$ satisfy the following discrete "Euler equations" written in the principal axes of the top

$$
\begin{gathered}
M_{N+1} = \hat{\Omega}_N M_N \hat{\Omega}_N^T, \quad M_N = \hat{\Omega}_N^T J - J \hat{\Omega}_N, \\
J^{-1} = \operatorname{diag}(a_1 - a_4, a_2 - a_4, a_3 - a_4),
\end{gathered} \tag{5.1}
$$

where J plays the role of tensor of inertia of the top and Ω_N is uniquely determined by M_N under the extra condition: the rotation axis of Ω_N belongs to the cone $\overline{K}_2$. In the absence of free parameter a_4, these equations were first written in [**11, 15**].

Given the above discrete model, equations (5.1) have a clear geometric interpretation. The first one expresses conservation of the angular momentum is space. J is the operator of duality with respect to the cone $\overline{K}_2$. Operator M_N uniquely defines the position of the plane $\Pi = \operatorname{span}(L_{N-1}, L_N)$ in the body. Then the second equation is equivalent to the "elastic reflection" condition: a normal vector of $\overline{K}_2$ at a point of L_N has equal angular distances from the planes $\operatorname{span}(L_{N-1}, L_N)$, $\operatorname{span}(L_N, L_{N+1})$.

I am deeply indebted to A. Bobenko, Yu. Suris, P. Vanhaecke, and V. Kuznetsov for discussions and valuable remarks.

References

1. M. Adams, J. Harnad, and J. Hurtubise, *Darboux coordinates on coadjoint orbits of Lie algebras*, Lett. Math. Phys. **40** (1997), 41–57.
2. Yu. Fedorov, *Integrable systems, hyperelliptic Lax pairs and confocal quadrics*, Amer. Math. Soc. Transl. **168** (1995), no. 2, 173–199.
3. ______, *Classical integrable systems and billiards related to generalized Jacobians*, Acta Appl. Math. **55** (1999), no. 2, 1–52.
4. Yu. Fedorov and V. Kozlov, *A memoir on integrable systems*, Springer-Verlag, Berlin, 1998.
5. L. Gagnon, J. Harnad, J. Hurtubise, and P. Winternitz, *Abelian integrals and the reduction method for an integrable Hamiltonian system*, J. Math. Phys. **26** (1885), 1605–1612.
6. C. G. J. Jacobi, *Sur la rotation d'un corps*, Gesamelte Werke **2** (1884), 139–172.
7. H. Knörrer, *Geodesics on the ellipsoid*, Invent. Math. **59** (1980), 119–143.
8. V. Kuznetsov, private communication.
9. A. Manakov, *Note on the integration of Euler's eqations of the dynamics of an n-dimensional rigid body*, Functional Anal. Appl. **10** (1976), 328–329.
10. J. Moser, *Geometry of quadrics and spectral theory*, The Chern Symposium 1979, Proc. Internat. Sympos. (Berkeley, 1979), Springer, New York-Berlin, 1979, 1980, pp. 147–188.
11. J. Moser and A. Veselov, *Discrete versions of some classical integrable systems and factorization of matrix polynomials*, Comm. Math. Phys. **139** (1991), 217–243.
12. D. Mumford, *Tata lectures on theta II*, Progr. Math., Birkhäuser, Boston, 1984.
13. J.-P. Serre, *Groupes algébriques et corps de classes*, Hermann, Paris, 1959.
14. P. Vanhaecke, *Linearizing two-dimensional integrable systems and the construction of action-angle variables*, Math. Z. **211** (1992), 265–313.
15. A. P. Veselov, *Integrable systems with discrete time and difference operators*, Functional Anal. Appl. **22** (1988), 1–13. (Russian)
16. K. Weierstrass, *Über die geodätischen Linien auf dem dreiachsigen Ellipsoid*, Mathematische Werke. I. Abhandlungen 1, Mayer & Muller, Berlin, 1894, pp. 257–266.

Department of Mathematics and Mechanics, Moscow Lomonosov University, Moscow 119899, Russia

E-mail address: fedorov@mech.math.msu.su

Centre de Recherches Mathématiques
CRM Proceedings and Lecture Notes
Volume **25**, 2000

New Potential Symmetries

Maria Luz Gandarias

ABSTRACT. In this paper new classes of symmetries for partial differential equations which can be written in the form of conservation laws are introduced. These symmetries, called *nonclassical potential* symmetries are realized as nonclassical symmetries of an associated system and yield solutions which are neither nonclassical solutions of the original equation, nor solutions arising from potential symmetries.

1. Introduction

Local symmetries admitted by a PDE are useful for finding invariant solutions. These solutions are obtained by using group invariants to reduce the number of independent variables. The fundamental basis of the technique is that, when a differential equation is invariant under a Lie group of transformations, a reduction transformation exists. The machinery of Lie group theory provides a systematic method to search for these special group-invariant solutions. For PDE's with two independent variables, a single group reduction transforms the PDE into ODE's, which are generally easier to solve than the original PDE. Most of the required theory and description of the method can be found in [**4, 9, 10**].

Local symmetries admitted by a nonlinear PDE are also useful to discover whether or not the equation can be linearized by an invertible mapping and construct an explicit linearization when one exists. A nonlinear scalar PDE is linearizable by an invertible contact (point) transformation if and only if it admits an infinite-parameter Lie group of contact transformations satisfying specific criteria [**3, 4**].

An obvious limitation of group-theoretic methods based on local symmetries, in their utility for particular PDE's, is that many of these equations does not have local symmetries. It turns out that PDE's can admit nonlocal symmetries whose infinitesimal generators depend on the integrals of the dependent variables in some specific manner. It also happens that if a nonlinear scalar PDE does not admit an infinite-parameter Lie group of contact transformations is not linearizable by an

2000 *Mathematics Subject Classification.* Primary: 22E05; Secondary: 35Q53.
This is the final form of the paper.

invertible contact transformation. However most of the interesting linearizations involve noninvertible transformations, such linearizations can be found by embedding given nonlinear PDE's in auxiliary systems of PDE's [**4**].

In [**3, 4**] Bluman introduced a method to find a new class of symmetries for a PDE. By writing a given PDE, denoted by $R\{x,t,u\}$ in the form of a conservation law, a related system denoted by $S\{x,t,u,v\}$ with potentials as additionals dependent variables is obtained. If $u(x,t)$, $v(x,t)$ satisfies $S\{x,t,u,v\}$, then $u(x,t)$ solves $R\{x,t,u\}$ and $v(x,t)$ solves an integrated related equation $T\{x,t,v\}$. Any Lie group of point transformations admitted by $S\{x,t,u,v\}$ induces a symmetry for $R\{x,t,u\}$; when at least one of the generators of the group depends explicitly of the potential, then the corresponding symmetry is neither a point nor a Lie-Bäcklund symmetry. These symmetries of $R\{x,t,u\}$ are called *potential* symmetries.

The nature of potential symmetries allows one to extend the uses of point symmetries to such nonlocal symmetries. In particular:

(1) Invariant solutions of $S\{x,t,u,v\}$, respectively $T\{x,t,v\}$, yield solutions of $R\{x,t,u\}$ which are not invariant solutions for any local symmetry admitted by $R\{x,t,u\}$.
(2) If $R\{x,t,u\}$ admits a potential symmetry leading to the linearization of $S\{x,t,u,v\}$, respectively $T\{x,t,v\}$, then $R\{x,t,u\}$ is linearized by a noninvertible mapping.

Suppose $S\{x,t,u,v\}$ admits a local Lie group of transformations with infinitesimal generator

$$X_S = p(x,t,u,v)\frac{\partial}{\partial x} + q(x,t,u,v)\frac{\partial}{\partial t} + r(x,t,u,v)\frac{\partial}{\partial u} + s(x,t,u,v)\frac{\partial}{\partial v}, \tag{1.1}$$

this group maps any solution of $S\{x,t,u,v\}$ to another solution of $S\{x,t,u,v\}$ and hence induces a mapping of any solution of $R\{x,t,u\}$ to another solution of $R\{x,t,u\}$. Thus (1.1) defines a symmetry group of $R\{x,t,u\}$. If

$$\left(\frac{\partial p}{\partial v}\right)^2 + \left(\frac{\partial q}{\partial v}\right)^2 + \left(\frac{\partial r}{\partial v}\right)^2 \neq 0, \tag{1.2}$$

then (1.1) yields a nonlocal symmetry of $R\{x,t,u\}$, such nonlocal symmetry is called a *potential* symmetry of $R\{x,t,u\}$, otherwise X_S projects onto a point symmetry of $R\{x,t,u\}$.

Suppose

$$X_T = p^T(x,t,v)\frac{\partial}{\partial x} + q^T(x,t,v)\frac{\partial}{\partial t} + s^T(x,t,v)\frac{\partial}{\partial v}, \tag{1.3}$$

defines a point symmetry of the related integrated equation $T\{x,t,v\}$. Then X_T yields a nonlocal potential symmetry of $R\{x,t,u\}$ if and only if

$$X_S = X_T + r(x,t,u,v)\frac{\partial}{\partial u}, \tag{1.4}$$

yields a nonlocal potential symmetry of $R\{x,t,u\}$.

Motivated by the fact that symmetry reductions for many PDE's are known that are not obtained using the classical Lie group method, there have been several generalizations of the classical Lie group method for symmetry reductions. Bluman and Cole developed the nonclassical method to study the symmetry reductions of the heat equation; Clarkson and Mansfield [**5**] presented an algorithm for calculating the determining equations associated with the nonclassical symmetries.

The basic idea of the nonclassical method is that the PDE is augmented with the invariance surface condition

$$pu_x + qu_t - r = 0, \tag{1.5}$$

which is associated with the vector field

$$X_R = p(x,t,u)\frac{\partial}{\partial x} + q(x,t,u)\frac{\partial}{\partial t} + r(x,t,u)\frac{\partial}{\partial u}. \tag{1.6}$$

By requiring that both the Burgers equation

$$u_t + uu_x - u_{xx} = 0, \tag{1.7}$$

and (1.5) are invariant under the transformation with infinitesimal generator (1.6) one obtains an overdetermined, nonlinear system of equations for the infinitesimals $p(x,t,u)$, $q(x,t,u)$, $r(x,t,u)$. The number of determining equations arising in the nonclassical method is smaller than for the classical method, consequently the set of solutions is larger than for the classical method as in this method one requires only the subset of solutions of (1.7) and (1.5) to be invariant under the infinitesimal generator (1.6). However, the associated vector fields do not form a vector space.

The determining equations by applying the nonclassical method to the Burgers equation were first derived by Ames. This set of determining equations were partially solved by Pucci [**11**], Pucci also obtained by using the nonclassical method due to Bluman some new solutions of the Burgers equation which cannot been obtained by the direct method of Clarkson and Kruskal. In [**1**] Arrigo *et al.* formulate a criterion for determining when a solution obtained from a symmetry reduction of any equation calculated by Bluman and Cole's method is recoverable by Clarkson and Kruskal approach and obtained some new solutions for the nonclassical determining equations of the Burgers equation as well as some new solutions for the Burgers equation.

Knowing that an associated system $S\{x,t,u,v\}$ to the Boussinesq equation has the same classical symmetries as the Boussinesq equation, Clarkson proposed as an open problem if an auxiliary system $S\{x,t,u,v\}$ of the Boussinesq equation does posses more or less nonclassical symmetries than the equation itself. Bluman says [**2**] that the ansatz to generate nonclassical solutions of $S\{x,t,u,v\}$ could yield solutions of $R\{x,t,u\}$ which are neither nonclassical solutions of $R\{x,t,u\}$ nor solutions arising from potential symmetries. However as far as we know none of these new symmetries have been obtained.

The aim of this work is to obtain new symmetries that we will call *nonclassical potential* symmetries for the Burgers equation.

2. Potential Symmetries and Nonclassical Potential Symmetries

Let $R\{x,t,u\}$ be Burgers equation (1.7). In order to find the potential symmetries of (1.7) we write the equation in the form of a conservation law

$$u_t + \left(\frac{u^2}{2} - u_x\right)_x = 0.$$

From this form the associated auxiliary system $S\{x,t,u,v\}$ is given by

$$v_x = u, \quad v_t = u_x - \frac{1}{2}u^2. \tag{2.1}$$

If $\big(u(x), v(x)\big)$ satisfies (2.1) then $u(x)$ solves the Burgers equation and $v(x)$ solves an integrated Burgers equation

$$v_t + \frac{v_x^2}{2} - v_{xx}. \tag{2.2}$$

Bluman [**2**] derived that (2.1) admits an infinite-parameter Lie group of point symmetries. The basic idea for obtaining *nonclassical potential* symmetries is that the potential system $S\{x, t, u, v\}$ is augmented with the invariance surface condition

$$pv_x + qv_t - s = 0 \tag{2.3}$$

which is associated with the vector field (1.3).

By requiring that both (2.1) and (2.3) are invariant under the transformation with infinitesimal generator (1.3) one obtains an overdetermined, nonlinear system of equations for the infinitesimals.

Then X_S yields a nonclassical potential symmetry of (1.7) and only if X_S is a nonclassical symmetry of (2.1) and (1.2) is satisfied. This new symmetry is a potential symmetry of (1.7) which do not arise from a Lie symmetry of (2.1) but from a nonclassical symmetry of (2.1). In the case $q \neq 0$, without loss of generality, we may set $q(x, t, u, v) = 1$. The nonclassical method applied to (2.1) give rise to two determining equations for the infinitesimals $p(x, t, u, v)$, $r(x, t, u, v)$ and $s(x, t, u, v)$. If we require that,

$$\frac{\partial p}{\partial u} = \frac{\partial s}{\partial u} = 0,$$

then $p = p(x, t, v)$, $s = s(x, t, v)$ and

$$r = -\frac{\partial p}{\partial v}u^2 + \left(\frac{\partial s}{\partial v} - \frac{\partial p}{\partial x}\right)u + \frac{\partial s}{\partial x}, \tag{2.4}$$

where $p(x, t, v)$ and $s(x, t, v)$ are related by the following conditions:

$$2\frac{\partial^2 p}{\partial v^2} + \frac{\partial p}{\partial v} = 0,$$
$$-2\frac{\partial^2 s}{\partial v^2} - \frac{\partial s}{\partial v} + 4\frac{\partial^2 p}{\partial x} + 4p\frac{\partial p}{\partial v} = 0,$$
$$-\frac{\partial s}{\partial x} - 2\frac{\partial^2 s}{\partial x} - 2\frac{\partial p}{\partial}s + \frac{\partial^2 p}{\partial x^2} + 2p\frac{\partial p}{\partial x} + \frac{\partial p}{\partial t} = 0,$$
$$\frac{\partial^2 s}{\partial x^2} + \frac{\partial s}{\partial t} + 2\frac{\partial p}{\partial x}s = 0.$$

Solving these equations we obtain

$$p = p(x, t), \quad s = \mathrm{s}_1(x, t)e^{v/2} + \mathrm{s}_2(x, t) \tag{2.5}$$

where p, s_1, and s_2 are related by

$$\frac{\partial \mathrm{s}_2}{\partial x} + \frac{\partial^2 \mathrm{p}}{\partial x^2} - 2\mathrm{p}\frac{\partial \mathrm{p}}{\partial x} - \frac{\partial \mathrm{p}}{\partial t} = 0, \tag{2.6}$$

$$-\frac{\partial^2 \mathrm{s}_1}{\partial x^2} + \frac{\partial \mathrm{s}_1}{\partial t} + 2\frac{\partial \mathrm{p}}{\partial x}\mathrm{s}_1 = 0, \tag{2.7}$$

$$-\frac{\partial \mathrm{s}_1}{\partial t} + 2\frac{\partial \mathrm{p}}{\partial x}\mathrm{s}_2 + \frac{\partial^2 \mathrm{s}_2}{\partial x^2} = 0. \tag{2.8}$$

Although the previous equations are too complicated to be solved in general, some solutions can be obtained. Choosing $p = p(x)$, $\mathrm{s}_1 = \mathrm{s}_1(x)$ and $\mathrm{s}_2 = 0$ we can distinguish the following cases: For $p = \tan(x)$ solving the invariant surface condition we obtain the nonclassical symmetry reduction

$$(2.9)\quad z = t - \log\big(\sin(x)\big), \quad v = 2\log\left(-\frac{4}{2\mathrm{k}_1\log\sin x + \mathrm{k}_1 x^2 + 2\mathrm{k}_2 x + 2H(z)}\right),$$

where $H(z)$ satisfies

$$H'' + H' - k_1 = 0,$$

from which we obtain that an exact solution for (2.2) is

$$v = -2\big(\log(-2\mathrm{k}_4\sin x - \mathrm{k}_1 e^t x^2 - 2\mathrm{k}_2 e^t x - \big(2\mathrm{k}_1 t + 2k\big)e^t\big) - t - 2\log 2\big)$$

and by (2.1) a new exact solution of the Burgers equation is

$$u = -\frac{2(-2\mathrm{k}_4\cos x - 2\mathrm{k}_1 e^t x - 2\mathrm{k}_2 e^t)}{-2\mathrm{k}_4\sin x - \mathrm{k}_1 e^t x^2 - 2\mathrm{k}_2 e^t x - (2\mathrm{k}_1 t + 2k)e^t}.$$

For $p = -1/x$ setting $c_2 = 0$, leads to the similarity variable and to the implicit solution ansatz

$$z = \frac{x^2}{2} + t, \quad H(z) - \frac{k_2 x^4}{4} - \frac{k_1 x}{3} + 2e^{-v/2} = 0,$$

where $H(z)$ satisfies

$$H'' - 3k_2 = 0,$$

from which we obtain the exact solution

$$v = 2\log\left(-\frac{48}{3\mathrm{k}_2 x^4 + (36\mathrm{k}_2 t - 12\mathrm{k}_3)x^2 + 8\mathrm{k}_1 x + 36\mathrm{k}_2 t^2 - 24\mathrm{k}_3 t + 24\mathrm{k}_4}\right),$$

and by (2.1) a new exact solution of the Burgers equation is

$$u = -\frac{2(12\mathrm{k}_2 x^3 + 2(36\mathrm{k}_2 t - 12\mathrm{k}_3)x + 8\mathrm{k}_1)}{3\mathrm{k}_2 x^4 + (36\mathrm{k}_2 t - 12\mathrm{k}_3)x^2 + 8\mathrm{k}_1 x + 36\mathrm{k}_2 t^2 - 24\mathrm{k}_3 t + 24\mathrm{k}_4}.$$

3. Concluding Remarks

In this work we have introduced new classes of symmetries for the Burgers equation. If the Burgers equation is written in the form of a conservation law, then a related system (2.1) may be obtained. The ansatz to generate nonclassical solutions of the associated system (2.1) yields solutions of (1.7) which are neither nonclassical solutions of (1.7) nor solutions arising from potential symmetries.

References

1. D. J. Arrigo, P. Broadbridge, and J. M. Hill, *Nonclassical symmetry solutions and the methods of Bluman–Cole and Clarkson–Kruskal*, J. Math. Phys **34**, (1993), 4692–4703.
2. G. W. Bluman, *Potential symmetries and linearization*, Applications of Analytic and Geometric Methods to Nonlinear Differential Equations (Exeter, 1992), NATO Adv. Sci. Inst. Ser. C Math. Phys. Sci., vol. 413, Kluwer Acad. Publ., Dordrecht, 1993.
3. G. W. Bluman and S. Kumei, *On the remarkable nonlinear diffusion equation* $(a(u+b)^{-2}u_x)_x - u_t = 0$, J. Math. Phys. **21** (1980), 1019–1023.
4. ———, *Symmetries and differential Equations*, Appl. Math. Sci., vol. 81, Springer-Verlag, New York-Berlin, 1989.
5. P. A. Clarkson and E. L. Mansfield, *Algorithms for the nonclassical method of symmetry reductions*, SIAM J. Appl. Math **55** (1994), 1693–1719.
6. M. L. Gandarias, *Classical point symmetries of a porous medium equation*, J. Phys A **29** (1996), 607–633.

7. ———, *Potential symmetries of a porous medium equation*, J. Phys A **29** (1996), 5919–5934.
8. M. L. Gandarias, J. L. Romero, and J. M. Diaz, *Nonclassical symmetry reductions of a porous medium equation with convection* 1998, J. Phys. A **32** (1999), 1–13.
9. P. J. Olver,*Applications of Lie groups to differential equations*, Grad. Texts in Math., vol. 107, Springer-Verlag, New York-Berlin, 1986.
10. L. V. Ovsiannikov, *Group analysis of differential equations*, Academic Press, New York, 1982,
11. E. Pucci, *Similarity reductions of partial differential equations*, J. Phys. A **25** (1992), 2631–2640.
12. P. Rosenau, *Fast and superfast diffusion processes*, Phys. Rev. Lett. **7** (1995), 1056–1059.
13. A. M. Vinogradov, *Symmetries of partial differential equations. Conservation laws—applications—algorithms*, Reprint of Acta Appl. Math. **15** (1989), no. 1–2, and **16** (1989); no. 1 and no. 2 (A. M. Vinogradov, ed.), Kluwer Academic Publishers, Dordrecht, 1989.

Departamento de Matematicas, Universidad de Cádiz, P.O.Box 40, 11510 Puerto Real, Cádiz, Spain.

E-mail address: mlgand@merlin.uca.es

Centre de Recherches Mathématiques
CRM Proceedings and Lecture Notes
Volume **25**, 2000

Symmetry Classification of Systems of Differential-Difference Equations

D. Gómez-Ullate Oteiza, S. Lafortune, and P. Winternitz

ABSTRACT. The Lie point symmetries of a coupled system of two nonlinear differential-difference equations are investigated. The equations arise from a model describing the interaction of two long molecular chains. We have classified all symmetry algebras containing an $\mathrm{sl}(2,\mathbb{R})$ subalgebra. Examples are given for which the symmetry algebra is 10-dimensional and infinite-dimensional.

1. Introduction

We have performed a symmetry analysis of a system of two coupled differential-difference equations of the form

$$\begin{aligned} E_1 &= \ddot{u}_n - F_n(t, u_{n-1}, u_n, u_{n+1}, v_{n-1}, v_n, v_{n+1}) = 0, \\ E_2 &= \ddot{v}_n - G_n(t, u_{n-1}, u_n, u_{n+1}, v_{n-1}, v_n, v_{n+1}) = 0, \end{aligned} \tag{1.1}$$

where the dots denote time derivatives. The discrete variable n plays the role of a space variable which labels positions along a one-dimensional lattice. The variables u_n and v_n represent deviations from equilibrium positions of two different types of atoms U and V. Each atom at site n interacts with its nearest neighbours on the two molecular chains, at sites $n-1$, n and $n+1$. The functions F_n and G_n represent the interactions between different atoms. Due to the lack of a realistic model, the interactions are a priori unspecified. Our aim is to classify equations of the type (1.1) according to the Lie point symmetries that they allow.

We do not restrict to two-body forces, nor do we impose translational invariance for the chain. We do however assume there is no dissipation, i.e. system (1.1) does not involve first derivatives with respect to time.

Such differential-difference equations typically arise when modeling phenomena in molecular physics, biophysics, or simply coupled oscillations in classical mechanics [**1**, **9**].

2000 *Mathematics Subject Classification.* Primary: 22E70; Secondary: 39A10.

The authors thank D.Levi and M.A.Rodríguez for helpful discussions. The research of S.L. and P.W. was partly supported by NSERC of Canada and FCAR du Québec. D.G.U's work was partly supported by DGES grant PB95-0401. He would like to express his gratitude to the Centre de Recherches Mathématiques for their kind hospitality.

This is the final form of the paper.

A recent article [**5**] was devoted to a similar problem, but was concerned with a single species, i.e. one dependent variable $u_n(t)$. Several different treatments of Lie symmetries of difference and differential-difference equations exist in the literature [**3, 5–8**]. The approach adopted here is the "intrinsic method" [**5–7**]: we shall consider only symmetries acting on the continuous variables t, u_n and v_n. Transformations of the discrete variable n must then be studied separately.

The Lie algebra of the symmetry group is realized by vector fields of the form

$$\widehat{X} = \tau(t, u_n, v_n)\partial_t + \phi_n(t, u_n, v_n)\partial_{u_n} + \psi_n(t, u_n, v_n)\partial_{v_n}. \tag{1.2}$$

The algorithm for finding the functions τ, ϕ_n and ψ_n in (1.2) is to construct the appropriate prolongation $\mathrm{pr}\,\widehat{X}$ of $\widehat{X}$ and to impose that it should annihilate the system of equations (1.1) on their solution set

$$\mathrm{pr}\,\widehat{X}E_1|_{E_1=E_2=0} = 0, \quad \mathrm{pr}\,\widehat{X}E_2|_{E_1=E_2=0} = 0. \tag{1.3}$$

Our classification will be up to conjugacy under a group of "allowed transformations" [**2**]. These are fiber preserving locally invertible point transformations

$$u_n = \Omega_n(\tilde{u}_n, \tilde{v}_n, \tilde{t}), \quad v_n = \Gamma_n(\tilde{u}_n, \tilde{v}_n, \tilde{t}), \quad t = t(\tilde{t}), \tag{1.4}$$

that preserve the form of equations (1.1), but not necessarily the functions F_n and G_n.

2. Determining Equations and the Classification Group

To find the Lie point symmetries of the system (1.1) we write the second prolongation of the vector field (1.2) in the form [**5–7**]:

$$\begin{aligned}\mathrm{pr}^{(2)}\,\widehat{X} = \tau(t, u_n, v_n)\partial_t &+ \sum_{k=n-1}^{n+1} \phi_k(t, u_k, v_k)\partial_{u_k} \\ &+ \sum_{k=n-1}^{n+1} \psi_k(t, u_k, v_k)\partial_{v_k} + \phi_n^{tt}\partial_{\ddot{u}_n} + \psi_n^{tt}\partial_{\ddot{u}_n},\end{aligned} \tag{2.1}$$

with

$$\begin{aligned}\phi_n^{tt} &= D_t^2\phi_n - (D_t^2\tau)\dot{u}_n - 2(D_t\tau)\ddot{u}_n, \\ \psi_n^{tt} &= D_t^2\psi_n - (D_t^2\tau)\dot{v}_n - 2(D_t\tau)\ddot{v}_n,\end{aligned} \tag{2.2}$$

where D_t is the total time derivative. The result is that an element $\widehat{X}$ of the symmetry algebra must have the form

$$\begin{aligned}\widehat{X}_n = \tau(t)\partial_t &+ \left[\left(\frac{\dot{\tau}}{2} + a_n\right)u_n + b_n v_n + \lambda_n(t)\right]\partial_{u_n} \\ &+ \left[c_n u_n + \left(\frac{\dot{\tau}}{2} + d_n\right)v_n + \mu_n(t)\right]\partial_{v_n},\end{aligned} \tag{2.3}$$

where the dots denote time derivatives. The functions $\tau(t)$, $\lambda_n(t)$, $\mu_n(t)$, a_n, b_n, c_n and d_n satisfy the two remaining determining equations, namely

$$
\begin{aligned}
&\frac{\dddot{\tau}}{2}u_n + \ddot{\lambda}_n + \left(a_n - \frac{3}{2}\dot{\tau}\right)F_n + b_nG_n - \tau F_{n,t} \\
&\quad - \sum_{k=n-1}^{n+1} F_{n,u_k}\left[\left(\frac{\dot{\tau}}{2} + a_k\right)u_k + b_kv_k + \lambda_k(t)\right] \\
&\quad - \sum_{k=n-1}^{n+1} F_{n,v_k}\left[\left(\frac{\dot{\tau}}{2} + d_k\right)v_k + c_ku_k + \mu_k(t)\right] = 0,
\end{aligned} \tag{2.4}
$$

$$
\begin{aligned}
&\frac{\dddot{\tau}}{2}v_n + \ddot{\mu}_n + \left(d_n - \frac{3}{2}\dot{\tau}\right)G_n + c_nF_n - \tau G_{n,t} \\
&\quad - \sum_{k=n-1}^{n+1} G_{n,u_k}\left[\left(\frac{\dot{\tau}}{2} + a_k\right)u_k + b_kv_k + \lambda_k(t)\right] \\
&\quad - \sum_{k=n-1}^{n+1} G_{n,v_k}\left[\left(\frac{\dot{\tau}}{2} + d_k\right)v_k + c_ku_k + \mu_k(t)\right] = 0.
\end{aligned} \tag{2.5}
$$

Substituting (1.4) into eq. (1.1) and requiring that the form of these two equations be preserved, we find that the allowed transformations are

$$
\begin{aligned}
&\begin{pmatrix} u_n(t) \\ v_n(t) \end{pmatrix} = \begin{pmatrix} Q_n & R_n \\ S_n & T_n \end{pmatrix} \dot{\tilde{t}}^{-1/2} \begin{pmatrix} \tilde{u}_n(\tilde{t}) \\ \tilde{v}_n(\tilde{t}) \end{pmatrix} + \begin{pmatrix} \alpha_n(t) \\ \beta_n(t) \end{pmatrix}, \\
&\tilde{t} = \tilde{t}(t), \quad \frac{d\tilde{t}}{dt} \neq 0.
\end{aligned} \tag{2.6}
$$

A short-hand notation for the symmetry generator X_n of eq. (2.3) is

$$
\left\{\tau(t), A_n, \begin{pmatrix} \lambda_n(t) \\ \mu_n(t) \end{pmatrix}\right\}, \quad A_n = \begin{pmatrix} a_n & b_n \\ c_n & d_n \end{pmatrix}. \tag{2.7}
$$

Under an allowed transformation (2.6) it transforms into one of the same form with

$$
\begin{aligned}
\tilde{\tau}(\tilde{t}) &= \tau(t(\tilde{t}))\dot{\tilde{t}}, \\
\tilde{A}_n &= {M_n}^{-1}A_nM_n, \\
\begin{pmatrix} \tilde{\lambda}_n(\tilde{t}) \\ \tilde{\mu}_n(\tilde{t}) \end{pmatrix} &= {M_n}^{-1}\dot{\tilde{t}}^{1/2}\left[\left(A_n + \frac{\dot{\tau}}{2}\right)\begin{pmatrix} \alpha_n \\ \beta_n \end{pmatrix} - \tau\begin{pmatrix} \dot{\alpha}_n \\ \dot{\beta}_n \end{pmatrix} + \begin{pmatrix} \lambda_n \\ \mu_n \end{pmatrix}\right].
\end{aligned}
$$

This transformation will be used to simplify the symmetry generator (2.3). Then we shall find interactions F_n and G_n compatible with such a symmetry, by solving the remaining determining equations (2.4) and (2.5).

3. Classification of Symmetry Algebras

There are 10 equivalence classes of one-dimensional symmetry algebras under allowed transformations [**3**]. The interactions that allow higher dimensional algebras become more and more specified, and the arbitrary functions involved depend on the joint invariants of the group action. For a certain dimension of the symmetry algebra (in some cases infinity), F_n and G_n are completely specified functions.

These are the most interesting cases of the classification. The interested reader is referred to [**3**] for a more thorough discussion.

We shall concentrate on $\mathrm{sl}(2,\mathbb{R})$ symmetry algebras and show that up to equivalence there are 4 different classes. The statement is:

Precisely 4 *classes of* $\mathrm{sl}(2,\mathbb{R})$ *algebras can be realized by vector fields of the form* (2.3). *Any such* $\mathrm{sl}(2,\mathbb{R})$ *algebra can be taken by an allowed transformation* (2.6) *into one of the following algebras*:

$$\mathrm{sl}(2,\mathbb{R})_1: \quad \begin{aligned} X_1 &= v_n \partial_{u_n}, \\ X_2 &= \frac{1}{2}(u_n \partial_{u_n} - v_n \partial_{v_n}), \\ X_3 &= u_n \partial_{v_n}, \end{aligned} \tag{3.1}$$

$$\mathrm{sl}(2,\mathbb{R})_2: \quad \begin{aligned} X_1 &= \partial_t, \\ X_2 &= t\partial_t + \frac{1}{2}(u_n \partial_{u_n} + v_n \partial_{v_n}), \\ X_3 &= t^2\partial_t + t(u_n \partial_{u_n} + v_n \partial_{v_n}), \end{aligned} \tag{3.2}$$

$$\mathrm{sl}(2,\mathbb{R})_3: \quad \begin{aligned} X_1 &= \partial_t + v_n \partial_{u_n}, \\ X_2 &= t\partial_t + u_n \partial_{u_n}, \\ X_3 &= t^2\partial_t + t u_n \partial_{u_n} + (t v_n - u_n)\partial_{v_n}, \end{aligned} \tag{3.3}$$

$$\mathrm{sl}(2,\mathbb{R})_4: \quad \begin{aligned} X_1 &= \partial_t + v_n \partial_{u_n}, \\ X_2 &= t\partial_t + (u_n + t)\partial_{u_n} + \partial_{v_n}, \\ X_3 &= t^2\partial_t + (t u_n + 2t^2)\partial_{u_n} + (t v_n - u_n + 2t)\partial_{v_n}. \end{aligned} \tag{3.4}$$

We shall focus on those symmetry algebras which contain $\mathrm{sl}(2,\mathbb{R})_1$ as a subalgebra. The classification can be summed up in the following way: in addition to $\mathrm{sl}(2,\mathbb{R})_1$ of eq. (3.1) we have a further algebra L_C (the "complementary" algebra). The structure of each symmetry algebra is

$$L = \mathrm{sl}(2,\mathbb{R})_1 \dot{+} L_C, \quad [\mathrm{sl}(2,\mathbb{R})_1, L_C] \subseteq L_C, \quad [L_C, L_C] \subseteq L_C. \tag{3.5}$$

The symbol $\dot{+}$ denotes a direct sum of vector spaces. Moreover, eq.(3.5) shows that L is either a direct sum or a semidirect one [**4**]. The algebra L_C is also a representation space for $\mathrm{sl}(2,\mathbb{R})_1$. Irreducible representations in this case can be of dimension 1 or 2. All higher dimensional representations are completely reducible into sums of 1 and 2 dimensional representations. The symmetry algebras can be classified into 4 series, according to the structure of the Lie algebra L_C.

- **Series A:** L_C is solvable and each element is an $\mathrm{sl}(2,\mathbb{R})_1$ singlet. There exist 3 Lie algebras in this family for which the interactions are completely specified. Two of them are infinite dimensional, and the other one has dimension 7.
- **Series B:** L_C is solvable and contains precisely one $\mathrm{sl}(2,\mathbb{R})_1$ doublet. The algebras are either 5 or 6 dimensional.
- **Series C:** L_C contains two $\mathrm{sl}(2,\mathbb{R})_1$ doublets. There are two completely specified interactions in this family, both having symmetry algebras of dimensions 9 and 10.

Series D: L_C contains $\mathrm{sl}(2,\mathbb{R})_2$ and (possibly) further elements. There are 4 such algebras of dimensions 6, 7, 8 and 10, the last two having completely specified interactions.

Due to limited space, all the symmetry algebras with their corresponding invariant interactions cannot be listed in this work. We will show two examples of the most interesting cases, and refer the reader to [**3**] for the complete classification.

EXAMPLE 1. The following dynamical equations:

$$\begin{aligned}\ddot{u}_n &= u_{n+1}\,\frac{u_{n-1}v_n - u_n v_{n-1}}{u_{n+1}v_{n-1} - u_{n-1}v_{n+1}}\,h_n + u_n k_n,\\ \ddot{v}_n &= v_{n+1}\,\frac{u_{n-1}v_n - u_n v_{n-1}}{u_{n+1}v_{n-1} - u_{n-1}v_{n+1}}\,h_n + v_n k_n,\end{aligned} \tag{3.6}$$

with h_n and k_n constant (depending on n) allow an infinite dimensional symmetry algebra

$$A_2(\infty) \sim \mathrm{sl}(2,\mathbb{R})_1 \dot{+} \left\{T(b_n), V(a_{k,n}),\;\; k \in \mathbb{Z}^{>}\right\}, \tag{3.7}$$

with

$$V(a_n) = a_n(u_n\partial_{u_n} + v_n\partial_{v_n}), \quad T(b_n) = \partial_t + b_n(u_n\partial_{u_n} + v_n\partial_{v_n}).$$

The algebra belongs to the Series A discussed above, the subindex refers to the classification in [**3**] and the number in brackets is the dimension of the algebra.

EXAMPLE 2. The following system:

$$\begin{aligned}\ddot{u}_n &= (\xi_{n+1} - \xi_n - \xi_{n-1})^{-2}\big[(u_{n+1} - u_{n-1})p_n + (u_n - u_{n+1})q_n\big],\\ \ddot{v}_n &= (\xi_{n+1} - \xi_n - \xi_{n-1})^{-2}\big[(v_{n+1} - v_{n-1})p_n + (v_n - v_{n+1})q_n\big],\end{aligned} \tag{3.8}$$

with p_n and q_n constants (depending on n) and

$$\begin{aligned}\xi_n &= u_{n+1}v_{n-1} - u_{n-1}v_{n+1},\\ \xi_{n+1} &= u_{n+1}v_n - u_n v_{n+1},\\ \xi_{n-1} &= u_{n-1}v_n - u_n v_{n-1},\end{aligned}$$

has a 10-dimensional symmetry algebra given by

$$C_{10}(10) \sim \mathrm{sl}(2,\mathbb{R})_1 \oplus \mathrm{sl}(2,\mathbb{R})_2 \dot{+} \left\{Y_u(1), Y_v(1), Y_u(t), Y_v(t)\right\}, \tag{3.9}$$

where $Y_u(t) = t\partial_{u_n}$, $Y_v(t) = t\partial_{v_n}$, $Y_u(1) = \partial_{u_n}$ and $Y_v(1) = \partial_{v_n}$.

References

1. A. Campa, A. Giansanti, A. Tenenbaum, D. Levi, and O. Ragnisco, *Quasisolitons at room temperature*, Phys. Rev. B **48** (1993), 10168–10182.
2. J. P.Gazeau and P. Winternitz, *Symmetries of variable coefficient Korteweg-de Vries equations*, J. Math. Phys. **33** (1992), 4087–4102.
3. D. Gómez-Ullate, S. Lafortune, and P.Winternitz, *Symmetries of discrete dynamical systems involving two species*, J. Math. Phys. **40** (1999), 2782–2804.
4. N. Jacobson, *Lie algebras*, Dover, New York, 1979.
5. D. Levi and P. Winternitz, *Symmetries of discrete dynamical systems*, J. Math. Phys. **37** (1996), 5551–5576.
6. ———, *Continuous symmetries of discrete equations*, Phys. Lett. A **152** (1991), 335–338.
7. ———, *Symmetries and conditional symmetries of differential-difference equations*, J. Math. Phys. **34** (1993), 3713–3730.
8. D. Levi, L. Vinet, and P. Winternitz, *Lie group formalism for difference equations*, J. Phys. A. **30** (1997), 633–649.
9. S. Pneumaticos, N. Flytzanis, and M. Remoissenent, *Soliton dynamics of nonlinear diatomic lattices*, Phys. Rev. B **33** (1986), 2308–2320.

Departamento de Física Teórica II, Facultad de Ciencias Fisicas, 28040 Universidad Complutense, Madrid, Spain.
E-mail address: dgu@eucmos.sim.ucm.es

Centre de Recherches Mathématiques, Univ. de Montréal, C.P. 6128, Succ. Centre-ville, Montréal, Qc, H3C 3J7, Canada.
E-mail address: lafortus@crm.umontreal.ca

Centre de Recherches Mathématiques, Univ. de Montréal, C.P. 6128, Succ. Centre-ville, Montréal, Qc, H3C 3J7, Canada.
E-mail address: wintern@crm.umontreal.ca

Centre de Recherches Mathématiques
CRM Proceedings and Lecture Notes
Volume **25**, 2000

Integrable Boundary Conditions for Nonlinear Lattices

I. T. Habibullin and A. N. Vil'danov

ABSTRACT. Integrable boundary conditions in $1+1$ and $2+1$ dimensions are discussed from the higher symmetries point of view. Boundary conditions consistent with the discrete Landau–Lifshitz model and infinite 2D Toda lattice are represented.

1. Introduction

The inverse scattering transform method is a powerful tool for solving the Cauchy problem for nonlinear integrable equations. However the method is not sufficiently effective in application to the initial boundary value problems on a half line or a finite interval when both the boundary condition and the initial data are chosen arbitrary. At the same time there is a special kind of boundary conditions, called integrable, which are completely consistent with the integrability property of the equation. Then the inverse scattering method is effectively applied to the initial boundary value problem when these boundary conditions are imposed. Several definitions of integrable boundary conditions are known in the literature (see, for instances, [**5, 10**]). In essence they are more or less equivalent. We will concentrate ourselves on one of them based on the symmetry approach.

During the last decade the classes of integrable boundary conditions have been studied for a large number of physically interesting equations in dimension 1+1 like the sine-Gordon equation [**2, 4, 10**], affine Toda lattices [**3**], the KdV equation [**1**] etc.

In dimensions higher than $1+1$ the problem is less studied. The main difficulty appearing in multi-dimensionality is connected with the presence of nonlocal variables in higher symmetries and conserved quantities. Recently [**6**] it has been shown that the symmetry approach can effectively be used to find integrable boundary conditions in $1+2$ dimensions.

In the present paper a boundary condition is found consistent with the Toda lattice by using the symmetry method. The boundary 2D lattice found reduces to the well-known boundary affine Toda field equations [**3**] by imposing the periodicity or similar other closure constraints.

2000 *Mathematics Subject Classification.* Primary: 35; Secondary: 46.
Supported by RFBR, grant # 98-01-00576.
This is the final form of the paper.

Another model dealt with is the discrete version of the famous Landau–Lifshitz equation. Some integrable boundary conditions are represented for this chain. The boundary conditions found are differential constraints consistent with the nearest symmetry of the continuous L-L equation. An interesting fact is that under this constraint the symmetry turns into the Krichever-Novikov equation.

2. Boundary Condition for the Discrete L-L Model

Let us consider an integrable lattice of the form

$$u_x = f\big(u(n-1), u(n), u(n+1)\big), \quad -\infty < n < +\infty, \tag{2.1}$$

where $u = u(n, x)$ is the unknown function and lower index denotes the derivative. Impose the following boundary condition at the point $n = k$

$$u(k) = F\big(u(k+1), \ldots, u(k+m)\big). \tag{2.2}$$

Under this constraint the lattice (2.1) turns into a semi-infinite one, defined for all $n \geq k+1$. Let us given one more lattice of similar form

$$u_t(n) = g\big(u(n-s), u(n-s+1), \ldots, u(n+s)\big), \tag{2.3}$$

which is a symmetry of the lattice (2.1), i.e. two flows defined by chains (2.1) and (2.3) commute: $\big(u_x(n)\big)_t = \big(u_t(n)\big)_x$. It is clear that the constraint (2.1) and its differential consequences reduce the infinite chain (2.3) to a semi-infinite chain defined for $n \geq k+1$. The boundary condition (2.2) is called consistent with the symmetry (2.3), if these two semi-infinite lattices obtained are commuting.

If the boundary condition is consistent with an infinite set of symmetries of the lattice then it is called consistent with the integrability.

The commutativity of two semi infinite lattices mentioned above might be checked directly but in some cases it is more convenient to use the so called associated system. Let us rewrite the pair of equations (2.1), (2.3) as a system of partial differential equations. To this end let us introduce another set of dynamical variables, representing these chains and consisting of $u = u(k), v = u(k+1)$ and their x-derivatives u_x, v_x, u_{xx}, $v_{xx}, \ldots$. One expresses $u(n)$ for $n < k$ and $n > k+1$ through the new variables by using the equation (2.1) and its differential consequences. Substitute now the expressions obtained instead of the variables $u(n)$ into the chain (2.3). As a result one gets a system of the form

$$u_t = g_1(u, v, u_x, v_x, \ldots, u_s), \quad v_t = g_2(u, v, u_x, v_x, \ldots, v_s), \tag{2.4}$$

where $u_s = \partial^s u/\partial x^s$, $v_s = \partial^s v/\partial x^s$. The boundary condition (2.2) reads as

$$u = G(v, u_x, v_x, \ldots, u_m). \tag{2.5}$$

The following important statement is a consequence of these transformations [5].

PROPOSITION 2.1. *The boundary condition* (2.2) *is consistent with the symmetry* (2.3) *if and only if the constraint* (2.5) *is consistent with the associated system* (2.4).

It is clear that among the symmetries consistent with the integrable boundary condition one can find that of the smallest order. We refer to such a symmetry as a trial one. Our hypothesis (checked on numerous examples) claims that for the integrable equation given all trial symmetries connected with the certain type boundary conditions have one and the same order and this order can be pointed out a priori by rather simple preliminary analysis.

Let us consider an illustrative example. The following integrable discrete analogue of the well known Landau–Lifshits equation was found in [**9**] years ago:

$$u_{nx} = v_n, \quad v_{nx} = -f_n, \tag{2.6}$$

where the function f_n is defined as

$$f_n = (v_n^2 + P(u_n))\left(\frac{1}{u_{n+1} - u_n} - \frac{1}{u_n - u_{n-1}}\right) + \frac{P'(u_n)}{2},$$

and $P(u) = au^4 + bu^3 + cu^2 + du + e$ is an arbitrary polynomial of forth order with constant coefficients. The lattice (2.6) admits higher symmetries, the simplest one is of the form

$$u_{nt} = h_n, \quad v_{nt} = D_x(h_n), \tag{2.7}$$

where

$$h_n = (v_n^2 + P(u_n))\left(\frac{1}{u_{n+1} - u_n} + \frac{1}{u_n - u_{n-1}}\right).$$

Let us introduce new variables $q = u_{-1}$ and $u = u_0$ and then by excluding the explicit n-dependence, as it was discussed above, bring the symmetry (2.7) to the form of associated system

$$u_t = u_{xx} - 2\frac{u_x^2 + P(u)}{u - q} + \frac{P'(u)}{2}, \quad -q_t = q_{xx} + 2\frac{q_x^2 + P(q)}{u - q} + \frac{P'(q)}{2}, \tag{2.8}$$

which is an integrable generalization of the Landau–Lifshits model. In the particular case, if the polynomial $P(u)$ subject to the additional constraint $P(-u) = P(u)$ then the system (2.8) coincides with the L-L model taken under the stereographic projection. The next symmetry of the chain (2.6) in terms of the variables q, u is at the same time a symmetry of the system (2.8). We take it in the form of the associated system

$$\begin{aligned} u_\tau &= 2u_{xxx} + u_x P''(u) - \frac{12u_x}{u - q}\left(u_{xx} + \frac{P'(u)}{2}\right) + \frac{12u_x}{(u - q)^2}(u_x^2 + P(u)), \\ q_\tau &= 2q_{xxx} + q_x P''(q) + \frac{12q_x}{u - q}\left(q_{xx} + \frac{P'(q)}{2}\right) + \frac{12q_x}{(u - q)^2}(q_x^2 + P(q)). \end{aligned} \tag{2.9}$$

To find the boundary condition of the form

$$u_{-1} = F(u_0), \tag{2.10}$$

consistent with the integrability property of the Landau–Lifshits chain one has to answer the question when the constraint of the form $q = H(u)$ is compatible with the trial symmetry (2.9). Only three choices are possible:

(1) $q = c$,
(2) $q = -u + c$,
(3) $(c_1 q + c_2)(c_1 u + c_2) = -1$.

Here c, c_1, and c_2 are arbitrary constants.

3. Higher Dimensions

In dimensions higher than $1+1$ the phenomenon of integrable boundary conditions is less studied. The classical generalized Toda lattices corresponding to infinite series of Lie algebras of finite grows can be interpreted as finite reductions of infinite 2D Toda lattice

$$q_{xy}(n) = e^{q(n+1)-q(n)} - e^{q(n)-q(n-1)}, \tag{3.1}$$

with integrable cutting off conditions at two fixed points $n=0$, $n=N$. In [**6**] the question was examined when the boundary condition of the form

$$q(1) = F\big(q(0), q_x(0), q_y(0), q(-1)\big), \tag{3.2}$$

is consistent with the integrability property of the Toda lattice (3.1). In higher dimensions the main obstacle is the existence of nonlocal variables which higher symmetries depend on. For instance, the following two equations

$$q_{t_1}(n) = b_1(n) + b_1(n-1) + q_x(n)^2, \tag{3.3}$$

and

$$\begin{aligned} q_t(n) = b_2(n-2) + b_2(n-1) + b_2(n) + b_1(n)\big[2q_x(n) + q_x(n+1)\big] \\ + b_1(n-1)[2q_x(n) + q_x(n-1)] + q_x(n)^3, \end{aligned} \tag{3.4}$$

are two symmetries of the Toda lattice (3.1). They depend on two nonlocal variables $b_1(n)$ and $b_2(n)$ which are introduced as solutions to the equations [**8**]:

$$q_{xx}(n) = b_1(n) - b_1(n-1), \tag{3.5}$$

$$b_{1,y}(n) = c(n)\big[q_x(n+1) - q_x(n)\big], \tag{3.6}$$

$$b_{1,x}(n) = b_1(n)\big[q_x(n+1) - q_x(n)\big] + b_2(n) - b_2(n-1), \tag{3.7}$$

$$b_{2,y}(n) = c(n)b_1(n+1) - c(n+1)b_1(n), \tag{3.8}$$

where $c(n)$ is described by the equation $c(n) = e^{q(n+1)-q(n)}$. Other nonlocal variables b_j, $j>1$ satisfy similar equations:

$$\begin{aligned} b_{j,x}(n) &= b_j(n)\big[q_x(n+j) - q_x(n)\big] + b_{j+1}(n) - b_{j+1}(n-1), \\ b_{j+1,y}(n) &= c(n)b_j(n+1) - c(n+j)b_j(n). \end{aligned}$$

Let us pass from the standard set of local dynamical variables $q(n)$, $q_x(n)$, $q_{xx}(n)$, $\dots$, $q_y(n)$, $q_{yy}(n), \dots$ for all $n = 0, \pm 1, \pm 2, \dots$ to the set consisting of variables u, v and their all x- and y-derivatives by setting $u = e^{q(1)}$, $v = e^{-q(0)}$. For example, $q(-1)$ may be expressed from the Toda equation written in the form $e^{-q(-1)} = e^{q(1)-2q(0)} - q_{xy}(0)e^{-q(0)}$, and so on. In terms of these new variables the symmetries (3.3) and (3.4) become (see, also [**8**])

$$u_{t_1} = u_{xx} + 2ru, \quad v_{t_1} = -v_{xx} - 2rv, \tag{3.9}$$

$$\begin{aligned} u_t &= u_{xxx} + 3ru_x + 3su, \\ v_t &= v_{xxx} + 3rv_x - 3sv + 3r_xv, \end{aligned} \tag{3.10}$$

where nonlocalities $r = b_1(0)$ and $s = b_2(0) + r(\log u)_x$ obviously satisfy the equations $r_y = (uv)_x$, $s_y = (u_xv)_x$. The boundary condition (3.2) takes the form

$$u = \widetilde{F}(v, v_x, v_y, v_{xy}). \tag{3.11}$$

A very useful consequence of the change of variables is the following statement:

PROPOSITION 3.1. *The boundary condition* (3.2) *is compatible with the symmetry* (3.4) (*or* (3.3)) *if and only if the constraint* (3.11) *is consistent with the system* (3.10) (*or* (3.9)).

REMARK 3.2. The connection between the "new" and "old" dynamical variables has some singularities at points $u = 0$ and $v = 0$ because of the formulae $s = b_2(0) + r(\log u)_x$ and $s = b_2(-1) - r(\log v)_x + b_{1,x}(0)$. That is why these cases should be checked directly without passing to the associated system.

In [**6**] two constraints has been classified

$$u = \widetilde{F}(v, v_x, v_y, v_{xy}), \tag{3.12}$$

and

$$u_y = \widetilde{G}(u, v, v_x). \tag{3.13}$$

In both cases the equation (3.10) was taken as the trial symmetry because the other one (3.9) does not admit any constraint of the above forms, except two degenerate ones $u = 0$ and $v = 0$ (see Remark above), for it is skew-symmetric in the highest order derivatives.

The consistency condition with the t-dynamics allows one to extract the following five choices for the function $\widetilde{F}$, here we give also the additional constraint the nonlocalities have to satisfy to:

(i) $u = 0$, $s_y = 0$;
(ii) $u = a$, $s = 0$;
(iii) $u = av$, $s = \frac{1}{2}r_x$;
(iv) $u = \frac{v_{xy}}{(a-v^2)} + \frac{vv_xv_y}{(a-v^2)^2}$, $s = r_x - \frac{v_xv_{xx}}{a-v^2} - \frac{vv_x^3}{(a-v^2)^2}$;

here a is an arbitrary constant,

(v) $u = -\frac{v_{xy}}{v^2} + \frac{v_xv_y}{v^3}$, $r = -\frac{v_{xx}}{v} + \frac{v_x^2}{v^2} + b$,

and b is an arbitrary function of x.

Rewritten in terms of the basic variables these choices read:

(1) $e^{q(1)} = 0$, $\quad b_{2,y}(0) = 0$;
(2) $q(1) = \text{const.}$, $\quad b_2(0) = 0$;
(3) $q(1) = -q(0) + \text{const.}$, $\quad b_2(0) = \frac{b_{1,x}(0)}{2} + b_1(0)q_x(0)$;
(4) $ae^{q(1)} = e^{-q(-1)} + \frac{aq_x(0)q_y(0)}{ae^{q(0)}} - e^{-q(0)}$,
$\quad b_2(0) = b_{1,x}(0) - b_1(0)q_x(1) + \frac{aq_x(0)^3}{(ae^{q(0)} - e^{-q(0)})^2} - \frac{q_x(0)q_{xx}(0)e^{-q(0)}}{ae^{q(0)} - e^{-q(0)}}$;
(5) $e^{-q(-1)} = 0$, $b_{1,y}(-1) = 0$.

All these boundary conditions are known as integrable [**7**]. Among constraints of the form (3.13) the only is consistent with the trial symmetry (3.10): $\widetilde{G} = av_x$, where $a \neq 0$ is an arbitrary constant (if $a = 0$ then the constraint $u_y = 0$ is reduced by integration to the form $u = \text{const.}$). Here nonlocalities should satisfy the constraint $s = u_x - uu_x/a$.

Turning back to the original variables yields the following boundary condition

$$q_y(1) = -ae^{-q(1)-q(0)}q_x(0),$$

which is not reduced to any of standard ones. So the following system of hyperbolic equations

$$(3.14)\qquad \begin{aligned} q_x(0) &= -a_1 e^{q(1)+q(0)} q_y(1), \\ q_{xy}(j) &= e^{q(j+1)-q(j)} - e^{q(j)-q(j-1)}, \quad 1 \geq j \geq N, \\ q_y(N+1) &= -a_N e^{-q(N+1)-q(N)} q_x(N), \end{aligned}$$

is an integrable reduction of the 2D Toda lattice. Similarly one can reduce the Toda lattice (3.1) by imposing different kind closure constraints at the ends:

$$(3.15)\qquad \begin{aligned} q_x(0) &= -a_1 e^{q(1)+q(0)} q_y(1), \\ q_{xy}(j) &= e^{q(j+1)-q(j)} - e^{q(j)-q(j-1)}, \quad 1 \geq j \geq N, \\ q(N+1) &= \widetilde{F}\big(q(N), q_x(N), q_y(N), q(N-1)\big), \end{aligned}$$

where $\widetilde{F}$ is one of the boundary conditions (1)–(5) above. Undoubtedly the systems (3.14), (3.15) are integrable, but it is not clear what is the algebraic structure they are related to.

4. Higher Dimensions—Boundary Conditions of the Other Kind

Now let us represent the Toda lattice as an infinite system of hyperbolic equations

$$(4.1)\qquad q_{tt}(n) - q_{zz}(n) = e^{q(n+1)-q(n)} - e^{q(n)-q(n-1)}, \quad -\infty < n < +\infty,$$

where the new independent variables are introduced as follows $t = x + y$ $z = y - x$. Consider this system on the half-line $z > 0$ imposing along the border $z = 0$ a boundary condition of the form

$$(4.2)\qquad q_z(n) = H_n\big(q(n+1), q(n), q(n-1)\big), \quad -\infty < n < +\infty.$$

The problem is now to extract from the class of boundary conditions (4.2) those consistent with the integrability. To solve the problem we will use the symmetry approach. First choose up the trial symmetry and rewrite it as an associated system. Notice that under the reflection type transformation $x \to y$, $y \to x$ the symmetries of the Toda lattice turn into symmetries, for the lattice itself is invariant under this transformation. So the symmetry (3.3), (3.5) produces a new one: $q_{t_2}(n) = p_1(n) + p_1(n-1) + q_y(n)^2$, $p_1(n) = p_1(n-1) + q_{yy}(n)$. The sum of this symmetry and its counterpart (3.3), (3.5) is again a symmetry but now it is invariant under the x, y reflection $x \to y$, $y \to x$:

$$(4.3)\qquad \begin{aligned} q_\tau(n) &= h(n) + h(n-1) + q_x(n)^2 + q_y(n)^2, \\ h(n) &= h(n-1) + q_{xx}(n) + q_{yy}(n), \end{aligned}$$

here $h(n) = p_1(n) + b_1(n)$. Notice that the symmetry (4.3) can be reduced to the famous Davey-Stewardson equation. In terms of the new independent variables t, z it reads

$$(4.4)\qquad \begin{aligned} q_\tau(n) &= 4q_{tt}(n) - 2c(n) + 2c(n-1) + 2h(n-1) + 2q_t(n)^2 + 2q_z(n)^2, \\ h(n) &= h(n-1) + 4q_{tt}(n) - 2c(n) + 2c(n-1). \end{aligned}$$

This symmetry will be taken as a trial one. The dynamical set of variables serving the system (4.4) consists of the local variables $q(n)$, $q_z(n)$ and their t-derivatives for all n and nonlocal variables $h(n)$ and their z-derivatives. Let us put $v(n) = q_z(n)$, $g(n) = h_z(n)$ and represent the symmetry (4.4) as an associated system

$$
\begin{aligned}
q_\tau(n) &= 4q_{tt}(n) - 2c(n) + 2c(n-1) + 2h(n-1) + 2q_t(n)^2 + 2v(n)^2, \\
v_\tau(n) &= 4v_{tt} - 2c(n)\big(v(n+1) - v(n)\big) + 2c(n-1)\big(v(n) - v(n-1)\big) \\
&\qquad + 2g(n-1) + 4q_t(n)v_t(n) + 4v(n)\big(q_{tt}(n) - c(n) + c(n-1)\big), \\
h(n) &= h(n-1) + 4q_{tt}(n) - 2c(n) + 2c(n-1), \\
g(n) &= g(n-1) + 4v_{tt} - 2c(n)\big(v(n+1) - v(n)\big) \\
&\qquad + 2c(n-1)\big(v(n) - v(n-1)\big).
\end{aligned} \tag{4.5}
$$

The boundary condition (4.2) for the Toda lattice (4.1) is consistent with the trial symmetry (4.4) if and only if the differential constraint

$$v(n) = H_n\big(q(n-1), q(n), q(n+1)\big) \tag{4.6}$$

is consistent with the system of equations (4.5). Direct computations show that only choice of H is (remind that $v = q_z$)

$$q_z(n)|_{z=0} = c_n e^{(q(n+1)-q(n))/2} - c_{n-1} e^{(q(n)-q(n-1))/2}, \tag{4.7}$$

where $c_n^2 = 1$ for all n, either $c_n = 0$ for all n. Combining the periodicity closure constraint $q(n) = q(n+N)$ with the boundary condition (4.7) one gets a finite system of hyperbolic equations on a half-plane

$$
\begin{aligned}
q_{tt}(n) - q_{zz}(n) &= e^{q(n+1)-q(n)} - e^{q(n)-q(n-1)}, \quad 0 < z < +\infty, \\
q_z(n)|_{z=0} &= c_n e^{(q(n+1)-q(n))/2} - c_{n-1} e^{(q(n)-q(n-1))/2}, \\
q(0) &= q(N), \quad q(N+1) = q(1),
\end{aligned} \tag{4.8}
$$

found recently by E. Corrigan *et al.* (see [**3**]). Imposing the closure conditions of the form (1)–(5) in addition to the boundary condition (4.7) leads again to integrable boundary value problem from [**3**]. This kind of problems for finite Toda systems have a beautiful interpretation in the field theory.

References

1. V. Adler, B. Gürel, M. Gürses, and I. Habibullin, *Boundary conditions for integrable equations*, J. Phys. A **30** (1997), 3505–3513.
2. R. Bikbaev and V. Tarasov, *Nonhomogeneous boundary problems on a half-line and a segment*, Algebra i Analiz **3** (1991), 78–93.
3. E. Corrigan, P. E. Dorey, R. H. Rietdijk, and R. Sasaki, *Affine Toda field theory on half-line*, Phys. Lett. B **333** (1994), 83–103.
4. S. Ghoshal and A. B. Zamolodchikov, *Boundary S matrix and boundary state in two-dimensional integrable quantum field theory*, Internat. J. Modern Phys. A **9** (1994), 3841–3886.
5. B. Gürel, M. Gürses, and I. Habibullin, *Boundary value problems, compatible with symmetry algebra*, J. Math. Phys. **36** (1995), 6809–6825.
6. B. Gürel and I. Habibullin, *Boundary conditions for two-dimensional integrable chains*, Phys. Lett. A. **233** (1997), 68–72.
7. M. A. Olshanetsky and A. M. Perelomov, *Classical integrable finite-dimensional systems related to Lie algebras*, Phys. Rep. **71** (1981), 313–400.
8. A. B. Shabat, *Higher symmetries of two-dimensional lattices*, Phys. Lett. A **200** (1995) 121–133.

9. A. B. Shabat and R. I. Yamilov, *Symmetries of nonlinear chains*, Algebra i Analiz **2** (1990), no. 2, 377–400.
10. E. K. Sklyanin, *Boundary conditions for integrable systems*, Functional Anal. Appl. **21** (1987), 86–87.

INSTITUTE OF MATHEMATICS OF RUSSIAN ACADEMY OF SCIENCES, CHERNYSHEVSKAYA STR. 112, UFA 450000, RUSSIA.

E-mail address: `ihabib@imat.rb.ru`

Centre de Recherches Mathématiques
CRM Proceedings and Lecture Notes
Volume **25**, 2000

Modular Invariants and Generalized Halphen Systems

J. Harnad and J. McKay

ABSTRACT. Generalized Halphen systems are solved in terms of functions that uniformize genus zero Riemann surfaces, with automorphism groups that are commensurable with the modular group. Rational maps relating these functions imply subgroup relations between their automorphism groups and symmetrization relations between the associated differential systems.

1. Introduction. Halphen Systems and Modular Invariants

1.1. Darboux-Halphen equations. The Darboux-Halphen differential system:

$$\begin{aligned} w_1' &= w_1(w_2 + w_3) - w_2 w_3, \\ w_2' &= w_2(w_3 + w_1) - w_3 w_1, \\ w_3' &= w_3(w_1 + w_2) - w_1 w_2, \end{aligned} \tag{1.1}$$

originally appeared in the work of Darboux [**Da**] on orthogonal coordinate systems. It was subsequently solved by Halphen [**Ha**], who related it to the hypergeometric equation of Legendre type

$$\lambda(1-\lambda)\frac{d^2y}{d\lambda^2} + (1-2\lambda)\frac{dy}{d\lambda} - \frac{1}{4}y = 0, \tag{1.2}$$

and also generalized it to a 3-parameter family of systems admitting a similar relation to the general hypergeometric equation [**Ha, Br**]. The general solution to (1.1) may be expressed [**Ha**] in terms of the *elliptic modular function*; that is, the square of the elliptic modulus, viewed as a function of the ratio τ of the periods of the Jacobi elliptic functions. More recently, the system (1.1) has found applications in mathematical physics in relation to magnetic monopole dynamics [**AH**], self-dual Einstein equations [**GP, Hi, To**] and topological field theory [**Du**].

1991 *Mathematics Subject Classification.* Primary: 58F07, 35Qxx; Secondary: 34L05.

Talk presented by J. Harnad at the SIDE III international meeting, Sabaudia, May, 1998. Research supported in part by the Natural Sciences and Engineering Research Council of Canada and the Fonds FCAR du Québec.

This is the final form of the paper.

To relate the Darboux-Halphen system to the hypergeometric equation (1.2), we first form the ratio of two linearly independent solutions of the latter

$$\tau(\lambda) := \frac{y_1}{y_2}, \tag{1.3}$$

and note that the inverse function $\lambda(\tau)$ satisfies the Schwarzian equation [**GS, H**]

$$\{\lambda, \tau\} + \lambda^2 - \lambda + \frac{1}{2}\lambda^2(1-\lambda)^2\lambda'^2 = 0, \tag{1.4}$$

where the Schwarzian derivative is defined as

$$\{f, \tau\} := \frac{f'''}{f'} - \frac{3}{2}\left(\frac{f''}{f'}\right)^2, \quad \left(f' := \frac{df}{d\tau}\right). \tag{1.5}$$

In view of the invariance properties of the Schwarzian derivative, the general solution of (1.4) is obtained by composing a particular one with the Möbius (linear fractional) transformations

$$\tau \to \frac{a\tau + b}{c\tau + d}, \quad \begin{pmatrix} a & b \\ c & d \end{pmatrix} \in \mathrm{SL}(2, \mathbf{C}). \tag{1.6}$$

The solutions to the Darboux-Halphen system are then given [**Br, Ha**] by setting

$$w_1 := \frac{1}{2}\frac{d}{d\tau}\ln\frac{\lambda'}{\lambda}, \quad w_2 := \frac{1}{2}\frac{d}{d\tau}\ln\frac{\lambda'}{(\lambda-1)}, \quad w_3 := \frac{1}{2}\frac{d}{d\tau}\ln\frac{\lambda'}{\lambda(\lambda-1)}, \tag{1.7}$$

where $\lambda(\tau)$ is a solution of (1.4). A particular solution is provided [**H, GS**] by the ellliptic modular function

$$\lambda(\tau) = k^2(\tau), \tag{1.8}$$

(with τ is interpreted as the ratio of two elliptic periods), whose automorphism group is the principal congruence subgroup $\Gamma(2) \subset \Gamma$ of the full modular group $\Gamma := \mathrm{PSL}(2, \mathbf{Z})$.

An explicit representation of $\lambda(\tau)$ may be given in terms of null ϑ-functions [**WW**]

$$\lambda(\tau) = \frac{\vartheta_2^4(\tau)}{\vartheta_3^4(\tau)} = 1 - \frac{\vartheta_4^4(\tau)}{\vartheta_3^4(\tau)}. \tag{1.9}$$

Substituting this in (1.7) and using the differential identities satisfied by the null theta functions leads to the explicit formulae

$$w_1 = 2\frac{d}{d\tau}\ln\vartheta_4, \quad w_2 = 2\frac{d}{d\tau}\ln\vartheta_2, \quad w_3 = 2\frac{d}{d\tau}\ln\vartheta_3. \tag{1.10}$$

1.2. Symmetrization under $\mathbf{S_3 = \Gamma/\Gamma(2)}$. The Chazy equation. The modular transformations

$$\tau \mapsto \tau + 1, \quad -\frac{1}{\tau}, \tag{1.11}$$

do not leave λ invariant, but generate the *group of anharmonic ratios* [**H**]

$$\lambda \mapsto \lambda, \quad \frac{1}{\lambda}, \quad 1-\lambda, \quad \frac{1}{1-\lambda}, \quad \frac{\lambda}{\lambda-1}, \quad \frac{\lambda-1}{\lambda}. \tag{1.12}$$

The symmetric invariant for this group is given by Klein's J–function

$$J = \frac{4(\lambda^2 - \lambda + 1)^3}{27\lambda^2(\lambda-1)^2} = \frac{(\vartheta_2^8 + \vartheta_3^8 + \vartheta_4^8)^3}{54\vartheta_2^8\vartheta_3^8\vartheta_4^8}, \tag{1.13}$$

which has the full modular group Γ as automorphism group, and satisfies the Schwarzian equation

$$\{J,\tau\} + \frac{36J^2 - 41J + 32}{72J^2(J-1)^2} J'^2 = 0. \tag{1.14}$$

In a similar manner to the above, this may be associated with the hypergeometric equation

$$J(1-J)\frac{d^2y}{dJ^2} + \left(\frac{2}{3} - \frac{7}{6}J\right)\frac{dy}{dJ} - \frac{1}{144}y = 0. \tag{1.15}$$

In terms of the Halphen variables (1.7), it follows that the corresponding elementary symmetric polynomials

$$\sigma_1 := w_1 + w_2 + w_3, \quad \sigma_2 := w_1w_2 + w_2w_3 + w_3w_1, \quad \sigma_3 := w_1w_2w_3, \tag{1.16}$$

satisfy the symmetrized system

$$\sigma_1' = \sigma_2 \quad \sigma_2' = 6\sigma_3 \quad \sigma_3' = 4\sigma_1\sigma_3 - \sigma_2^2, \tag{1.17}$$

which reduces to the Chazy equation [**Ch**]

$$W''' = 2WW'' - 3W'^2 \tag{1.18}$$

for

$$W := 2\sigma_1 = \frac{1}{2}\frac{d}{d\tau}\ln\frac{J'^6}{J^4(J-1)^3}. \tag{1.19}$$

More generally, it is easy to see that, forming the ratio, as in (1.3), of a pair (y_1, y_2) of linearly independent solutions of the general hypergeometric equation

$$f(1-f)\frac{d^2y}{df^2} + (c - (a+b+1)f)\frac{dy}{df} - aby = 0, \tag{1.20}$$

and *assuming* that the inverse function $f = f(\tau)$ is well defined, this similarly provides solutions to the Schwarzian equation

$$\{f,\tau\} + \frac{1}{2}\left(\frac{1-\lambda^2}{f^2} + \frac{1-\mu^2}{(f-1)^2} + \frac{\lambda^2+\mu^2-\nu^2-1}{f(f-1)}\right)f'^2 = 0, \tag{1.21}$$

where (λ, μ, ν) are the relative Frobenius exponents for (1.20) at $(0, 1, \infty)$, given in terms of the parameters (a, b, c) by

$$\lambda := 1 - c, \quad \mu := c - a - b, \quad \nu := b - a.$$

Introducing the general *Halphen-Brioschi variables* [**Br**]

$$W_1 := \frac{1}{2}\frac{d}{d\tau}\ln\frac{f'}{f}, \quad W_2 := \frac{1}{2}\frac{d}{d\tau}\ln\frac{f'}{(f-1)}, \quad W_3 := \frac{1}{2}\frac{d}{d\tau}\ln\frac{f'}{f(f-1)}, \tag{1.22}$$

these are similarly seen to satisfy the *general Halphen system*:

$$\begin{aligned} W_1' &= W_1(W_2+W_3) - W_2W_3 + X(\lambda,\mu,\nu), \\ W_2' &= W_2(W_3+W_1) - W_3W_1 + X(\lambda,\mu,\nu), \\ W_3' &= W_3(W_1+W_2) - W_1W_2 + X(\lambda,\mu,\nu), \end{aligned} \tag{1.23}$$

where

$$\begin{aligned} X(\lambda,\mu,\nu) := {} & \mu^2W_1^2 + \lambda^2W_2^2 + \nu^2W_3^2 + (\nu^2-\lambda^2-\mu^2)W_1W_2 \\ & + (\lambda^2-\mu^2-\nu^2)W_3W_1 + (\mu^2-\lambda^2-\nu^2)W_2W_3. \end{aligned} \tag{1.24}$$

This procedure, although formally identical to the two cases treated above, can really only be viewed as providing global solutions to such generalized systems if certain additional conditions, relating to functional inversion and modularity of the resulting functions, are satisfied. We therefore end this introductory section by posing the following questions:

(1) When is the functional inversion $\tau(f) \to f(\tau)$ well-defined?
(2) Are there other cases in which the resulting function $f(\tau)$ is a modular function, with known properties, analogous to λ and J?
(3) Are there any further generalizations of the above systems admitting solutions in terms of modular functions?

These questions will be addressed in the following sections.

2. Modular Solutions of General Halphen Systems

2.1. Replicable functions, Hauptmoduls and Schwarzian equations. A sufficient condition for the existence of a well-defined inverse function $f(\tau)$ is that the (projectivized) monodromy group of the hypergeometric equation in question be a Fuchsian group of the first type. Essentially, this means that the possibly infinite multivaluedness of the functions $\tau(f)$ defined by the ratio of two solutions can be characterized by subdividing its image in the τ-plane into fundamental domains (each having a finite number of sides), which are permuted amongst themselves by the action of the monodromy group, and into which each branch of the function is mapped in a single-valued way. This characterization may be applied not only to hypergeometric equations, but more generally, to Fuchsian differential equations having an arbitrary number of regular singular points, the number coinciding generally with the number of vertices in a fundamental domain. In the case of hypergeometric equations, there are three such singular points, at $(0, 1, \infty)$, and the domains are necessarily triangular.

In this section, a class of modular functions $f(\tau)$ is presented that provide instances of globally defined inverses of the multivalued functions $\tau(f)$ given by ratios of solutions of the hypergeometric equation (1.20) for certain particular parameter values (a, b, c). By "globally", we here understand an open domain of definition with natural boundary, in the interior of which the functions are holomorphic, and beyond which they do not admit analytic continuation. As in the case of the modular functions λ and J, this domain in the first instance consists of the upper half of the complex τ plane, with the real axis as boundary, but upon application of the Möbius transformations (1.6), it may become the interior of any disc. In subsequent sections, examples of functions within the same general class will be given that provide solutions to further generalizations of Halphen systems.

The particular class of functions to be considered here are the *replicable functions* [**CN, FMN**], which arose originally in the context of "modular moonshine". These functions provide generalizations of the J-function and are similarly defined in the upper half-plane by a normalized q-series [**CN, FMN**]

$$F(q) = \frac{1}{q} + \sum_{n=1}^{\infty} a_n q^n, \quad q := e^{2i\pi\tau} \tag{2.1}$$

satisfying suitable invariance properties under generalized Hecke operators. To relate these series to the functions obtained from solutions of the hypergeometric equations, an affine transformation must be applied

$$F(\tau) = af(\tau) + b, \tag{2.2}$$

with the constants (a, b) chosen so that the values of f at the vertices of the fundamental domains are $(0, 1, \infty)$. The main properties of these functions that are of importance in what follows are:

(1) They are *uniformizing functions* for genus zero Riemann surfaces $\mathfrak{H}/\mathfrak{G}_f$ formed by factoring the upper half-plane $\mathfrak{H}$ by the automorphism group $\mathfrak{G}_f$ of the function. Such functions are referred to as *Hauptmoduls.*

(2) The automorphism group $\mathfrak{G}_f$ is a subgroup of $\mathrm{PGL}(2, \mathbf{Q})$ that is *commensurable* with the modular group Γ; that is, the intersection $\mathfrak{G}_f \cap \Gamma$ is of finite index in both.

(3) In view of the form of the q-series (2.1), each such function has a *cusp* at $\tau = i\infty$.

(4) The automorphism group $\mathfrak{G}_f$ in each case contains a subgroup of type

$$\Gamma_0(N) := \left\{ \begin{pmatrix} a & b \\ c & d \end{pmatrix} \in \mathrm{SL}(2, \mathbf{Z}),\ c \equiv 0 \mod N \right\}, \tag{2.3}$$

with the smallest such N referred to as the *level* of the function (or group).

(5) Finally, for the cases to be considered here, the coefficients a_n are all integers (although this is not part of the properties shared by all replicable functions).

A list of such replicable functions, together with many of their properties, is provided in [**CN, FMN**].

The fact that each such function is the generator of the field of meromorphic functions on a genus zero Riemann surface implies that they all satisfy a Schwarzian equation of the same general form as (1.21)

$$\{f, \tau\} + 2R(f)f'^2 = 0, \tag{2.4}$$

where $R(f)$ is *some* rational function. If the fundamental domains are triangular, with vertices mapping to $(0, 1, \infty)$ in the f-plane, $R(f)$ will have the form appearing in eq. (1.21) and the functions may be associated to solutions of a corresponding hypergeometric equation. The angles at the vertices of the fundamental domain are determined in terms of the hypergeometric parameters as $(\lambda\pi, \mu\pi, \nu\pi)$. In general, $R(f)$ could have any number of poles at arbitrary locations, but in the case of the replicable functions arising in [**CN, FMN**], this number never exceeds 26. The remainder of this section concerns only the triangular case; in the following sections, cases with higher numbers of vertices and the corresponding generalized Halphen systems will be discussed.

2.2. Triangular replicable functions. Table 1, which is taken from [**HM**], lists all cases, up to equivalence under Möbius transformations (1.6) and affine transformations of f, of replicable triangular functions with integer a_n's. These are the modular functions whose automorphism groups are the arithmetic triangular groups of noncompact type classified in [**Ta**].

The first column identifies the groups according to the notation of [**CN, FMN**], with the integer indicating the level. The second and third columns give the associated hypergeometric parameters and angles at the vertices $(0, 1, \infty)$, respectively. The fourth column gives the generators $(\rho_0\ , \rho_1)$ of the automorphism group fixing a pair of finite vertices mapping to the points 0 and 1 in the f-plane. The third generator, stabilizing the point $i\infty$, is

$$\rho_\infty = \begin{bmatrix} 1 & 1 \\ 0 & 1 \end{bmatrix}, \tag{2.5}$$

and the three generators together satisfy the consistency relation

$$\rho_\infty \rho_1 \rho_0 = \mathbf{I}. \tag{2.6}$$

The fifth column in the table indicates the relation between the function F normalized as in (2.1) and f normalized as in the hypergeometric equation, so that the vertices are mapped to $(0, 1, \infty)$. The last column gives explicit expressions for the modular function f in terms of null theta functions or the Dedekind eta function $\eta(\tau)$. In this table, the case 1A is just the J function, which therefore determines solutions to the symmetrized Halphen system (1.17), while the case 4C is essentially the function λ (more precisely, composed with the transformation $\tau \to 2\tau$, $\lambda \to 1/\lambda$ as indicated in the sixth column), and hence determines solutions to the original Darboux-Halphen system (1.1). The other cases provide seven further examples of globally defined modular solutions to the general Halphen system (1.23).

Although these cases are distinct, it is important to note that, just as the functions λ and J are related by the rational map defined in (1.13), so all these replicable functions are linked by algebraic relations, some of which may similarly be expressed in terms of explicit rational maps. This implies that the general Halphen-Brioschi variables for the various cases are also algebraically related, just as in the Halphen-Chazy case, where the ones for the J case are essentially the elementary symmetric invariants of those for the λ case. A further illustrative example of such algebraic relations is given in the next subsection; the full set of rational maps relating various triangular cases is given in [**HM**], to which the reader is referred for details.

2.3. Rational transformations. If $f(\tau)$ satisfies a Schwarzian equation of the form (1.21) (where $R(f)$ is any rational function) and there is a map

$$f = Q(g), \tag{2.7}$$

relating it to another function $g(\tau)$, where $Q(g)$ satisfies the Schwarzian equation

$$\{Q, g\} + 2R\big(Q(g)\big)Q'^2 = 2\widetilde{R}(g), \tag{2.8}$$

for some function $\widetilde{R}(g)$, then g satisfies the transformed Schwarzian equation

$$\{g, \tau\} + 2\widetilde{R}(g)g'^2 = 0. \tag{2.9}$$

Also, if $y(f)$ satisfies the second order linear equation

$$\frac{d^2y}{df^2} + R(f)y = 0, \tag{2.10}$$

then $\widetilde{y}(g) := (Q')^{-1/2}y\big(Q(g)\big)$ satisfies the transformed equation:

$$\frac{d^2\widetilde{y}}{dg^2} + \widetilde{R}(g)\widetilde{y} = 0. \tag{2.11}$$

Similar transformations may be applied to second order equations in which first order derivative terms are also present. For the case of hypergeometric equations, such transformations, with $Q(g)$ rational functions of degree up to four, were studied by Goursat [**Go**] in his thesis. They imply corresponding *symmetrizations* of the associated general Halphen systems, such as the ones relating (1.1) and (1.17). These are discussed in detail in [**HM**]; here, we present just one illustrative example taken from [**HM**].

EXAMPLE 2.1. $3B \mapsto 1A$

In this case the relevant identity relating the associated hypergeometric functions is

$$F\left(\frac{1}{12},\frac{1}{12};\frac{2}{3};\frac{x(x+8)^3}{64(x-1)^3}\right) = (1-x)^{1/4} F\left(\frac{1}{3},\frac{1}{3};\frac{2}{3};x\right). \tag{2.12}$$

The corresponding rational map $Q(g)$ is therefore defined by

$$f = \frac{g(g+8)^3}{64(g-1)^3}, \tag{2.13}$$

where f denotes the Hauptmodul of case $1A$ (i.e., the modular function J), and g is the Hauptmodul of case $3B$. The relationship between the corresponding automorphism groups is given by a symmetrization quotient which may also be expressed as a quotient of finite groups:

$$\mathfrak{G}_f/\mathfrak{G}_g = \Gamma/\Gamma_0(3) = \mathbf{S}^{9B}_{1A}/\mathbf{S}^{9B}_{3B} = \mathbf{A}_4/\mathbf{Z}_3. \tag{2.14}$$

(The subgroup $\mathfrak{G}_g$ for this case is not normal in $\mathfrak{G}_f$, so the corresponding field extension is non-Galois. The expressions $\mathbf{S}^{9B}_{1A}$, $\mathbf{S}^{9B}_{3B}$ denote the finite groups obtained by quotienting the respective automorphism group of $1A$ and $3B$ by the largest subgroup that is normal in both, which in this case also corresponds to a replicable function, denoted by $9B$, appearing in Table 2 below.) We may associate the following polynomial invariants to this symmetrization:

$$\begin{aligned} \Sigma_1 &:= 3w_1 + 2w_2 + w_3, \\ \Sigma_2 &:= (w_1 - w_3)(9w_1 - 8w_2 - w_3), \\ \Sigma_3 &:= (w_1 - w_3)(27w_1^2 - 36w_1w_2 + 8w_2^2 - 18w_1w_3 + 2w_2w_3 - w_3^2), \end{aligned} \tag{2.15}$$

where (w_1, w_2, w_3) are the Halphen-Brioschi variables for the case $3B$, in terms of which the corresponding variables for the case $1A$, denoted (W_1, W_2, W_3) are determined by

$$3W_1 + 2W_2 + W_3 = \Sigma_1, \quad W_1 - W_3 = -\frac{\Sigma_2^2}{\Sigma_3}, \quad W_2 - W_3 = -\frac{\Sigma_3}{\Sigma_2}. \tag{2.16}$$

3. Generalized Halphen Systems

3.1. Fuchsian equations, monodromy and automorphism groups. Up to projective equivalence (i.e., multiplication of solutions by a common function that does not alter the location of the singular points), second order Fuchsian equations may be expressed in the form (2.10), where $R(f)$ is a rational function of the form

$$R(f) = \frac{N(f)}{(D(f))^2}, \quad D(f) = \prod_{i=1}^{n}(f - a_i), \tag{3.1}$$

and the numerator is a polynomial $N(f)$ of degree $\leq 2n-2$. The ratio of two linearly independent solutions

$$\tau(f) := \frac{y_1}{y_2}, \tag{3.2}$$

then satisfies the Schwarzian differential equation [**GS, H**]

$$\{\tau, f\} = 2R(f), \tag{3.3}$$

and the inverse function $f = f(\tau)$ (if well defined) satisfies eq. (2.4) with $R(f)$ given by (3.1). The image of the monodromy representation for eq. (2.10) determines a subgroup $\mathfrak{G}_f \subset \mathrm{PGL}(2, \mathbf{C})$ that acts on the ratio τ of solutions by linear fractional transformations (1.6) leaving the inverse function $f(\tau)$ invariant.

3.2. Constrained dynamical systems and flows on the SL(2, C) manifold. Following Ohyama [**Oh**], we may associate to any such Fuchsian system a dynamical system in $n+1$ variables, subject to $n-2$ independent quadratic constraints. Define the following $n+1$ variables, which serve to generalize the Halphen-Brioschi variables,

$$X_0 := \frac{1}{2}\frac{d}{d\tau}\ln f', \quad X_i := \frac{1}{2}\frac{d}{d\tau}\ln\frac{f'}{(f-a_i)^2}, \quad i = 1, \dots, n. \tag{3.4}$$

Equivalently, we may use the linear combinations

$$u := X_0 = \frac{1}{2}\frac{f''}{f'}, \quad v_i := \frac{1}{2}(X_0 - X_i) = \frac{1}{2}\frac{f'}{f - a_i}. \tag{3.5}$$

These satisfy the set of quadratic constraints

$$(a_i - a_j)v_i v_j + (a_j - a_k)v_j v_k + (a_k - a_i)v_k v_i = 0, \ 1 \le i, \ j, k \le n \tag{3.6}$$

(of which $n-2$ are independent), and the differential equations:

$$v_i' = -2v_i^2 + 2uv_i, \ i = 1, \dots n, \tag{3.7a}$$

$$u' = u^2 - \sum_{i,j=1}^{n} r_{ij}v_i v_j, \tag{3.7b}$$

where the coefficients r_{ij} defining the quadratic form

$$r(v) := \sum_{i,j=1}^{n} r_{ij}v_i v_j, \tag{3.8}$$

appearing in (3.7) are obtained by expressing $R(f)$ in the form

$$R(f) = \frac{1}{4}\sum_{i,j=1}^{n}\frac{r_{ij}}{(f-a_i)(f-a_j)}. \tag{3.9}$$

(This leaves a residual ambiguity in their definition, which just consists of adding any linear combination of the vanishing quadratic forms appearing in (3.6). The definition may be made unique by fixing an ordering for the singular points, and requiring that all coefficients r_{ij} vanish, except when $i = j$, $j \pm 1$.) The quadratic form $r(v)$ encodes all the relevant information about the monodromy of the associated Fuchsian system and, in the cases where the associated group is modular, about the geometry of the fundamental domains. In particular, the angles at the vertices are $\{\alpha_i\pi\}_{i=1,n}$ where

$$r_{ii} = 1 - \alpha_i^2. \tag{3.10}$$

Equivalently, these systems may be viewed as unconstrained dynamical systems on the $SL(2, \mathbf{C})$ group manifold. To see this, let

$$g(\tau) := \begin{pmatrix} A & B \\ C & D \end{pmatrix} \in \mathrm{SL}(2, \mathbf{C}), \tag{3.11}$$

denote an integral curve in $\mathrm{SL}(2, \mathbf{C})$ for the equation

$$g' = \begin{pmatrix} 0 & \gamma \\ -1 & 0 \end{pmatrix} g, \tag{3.12}$$

where

$$\gamma := -\frac{1}{4C^2} \sum_{i,j=1}^{n} \frac{r_{ij}}{(Ca_i + D)(Ca_j + D)} = -\frac{1}{C^2} R\left(-\frac{D}{C}\right). \tag{3.13}$$

Defining $f(\tau)$ as

$$f := -\frac{D}{C}, \tag{3.14}$$

it follows that this satisfies (2.4), and that

$$u := \frac{A}{C}, \quad v_{a_i} := \frac{1}{2C(Ca_i + D)}, \; i = 1, \dots, n, \tag{3.15}$$

satisfy the system (3.7), (3.7) and the constraints (3.6). Ohyama's variables (3.4) are recovered by applying g as a linear fractional transformation to $\{\infty, a_1, \dots, a_n\}$.

$$X_0 = \frac{A}{C}, \quad X_i = \frac{Aa_i + B}{Ca_i + D}, \; i = 1, \dots, n. \tag{3.16}$$

3.3. Examples with four vertices. Table 2, which is also taken from [**HM**], lists all cases of replicable functions having integer q-series coefficients, whose fundamental domains have four vertices, and which are related by a rational map of degree ≤ 4 to one of the triangular cases. These functions, composed with the Möbius transformation (1.6), therefore provide the general solutions to generalized Halphen systems of the type (3.6)–(3.7), with $n = 3$.

The first column again gives the label of the associated automorphism group $\mathfrak{G}_f$ for each function $f(\tau)$ in the notation of [**CN,FMN**], while the second specifies the location of the finite singular points (a_1, a_2, a_3) of the associated Fuchsian equation (2.10). The third column gives the automorphism group generators stabilizing three finite vertices mapping to (a_1, a_2, a_3), and the fourth gives the quadratic form $r(v)$ appearing in eq. (3.7). The fifth column lists the affine relations connecting the normalized q-series of the form (2.1) with the functions $f(\tau)$ mapping the vertices to the singular points (a_1, a_2, a_3) of (2.10), and the last column gives explicit expressions for $f(\tau)$ in terms of the Dedekind η-function.

There are also algebraic relations connecting these dynamical systems with triangular cases, following from the existence of rational maps satisfying (2.7)–(2.11). The following example is an illustration of such rational maps and the implied relations between the generalized Halphen-Brioschi variables. (Further details and a list of the other cases may be found in [**HM**].)

EXAMPLE 3.1. $8E \mapsto 4C$. The rational map (2.7) relating these two cases is given by

$$f = \frac{(g+1)^2}{4g}, \tag{3.17}$$

where f denotes the Hauptmodul for the case $4C$ and g the one for $8E$. The automorphism group for $4C$ is $\Gamma_0(4)$, and that for $8E$ is $\Gamma_0(8)$, a normal subgroup, and therefore in this case the function field generated by g is a Galois extension of the one generated by f. The map is of degree two and the quotient group characterizing the symmetrization is

$$\mathbf{S}_{4C}^{8E} = \Gamma_0(4)/\Gamma_0(8) = \mathbf{Z}_2, \tag{3.18}$$

whose action is generated by

$$\tau \mapsto \frac{-\tau}{4\tau - 1}, \quad g \mapsto \frac{1}{g}. \tag{3.19}$$

The effect of this on the generalized Halphen-Brioschi variables is

$$(u, v_{-1}, v_0, v_1) \mapsto (u - 2v_0, v_1 - v_0, -v_0, v_1 - v_0), \tag{3.20}$$

and the associated polynomial invariants are

$$\Sigma_1 := u + v_{-1} - v_0 - v_1, \quad \Sigma_1' := 4v_1 - 2v_0, \quad \Sigma_2 := 4v_0^2. \tag{3.21}$$

The Halphen-Brioschi variables for the $4C$ case are determined in terms of these by

$$W_1 = \Sigma_1, \quad W_1 - W_3 = \frac{\Sigma_2}{\Sigma_1'}, \quad W_2 - W_3 = \Sigma_1'. \tag{3.22}$$

3.4. Two examples with 26 vertices. In the list of replicable functions with integer q-series coefficients, [**CN,FMN**], there are three cases, denoted $72e$, $96a$ and $144^{\sim}e$, for which the fundamental domains have 26 vertices, the maximal number occurring. We give here the corresponding generalized Halphen-Brioschi variables, and the quadratic forms determining the associated constrained dynamical system for the cases $72e$ and $96a$. (The case $144^{\sim}e$ is related to $72e$ by multiplication of f by $e^{(i\pi/12)}$.)

EXAMPLE 3.2. $72e$

The modular function for this case may be expressed as follows in terms of the Dedekind η-function

$$f = \frac{\eta(24\tau)\eta(36\tau)}{\eta(12\tau)\eta(72\tau)}. \tag{3.23}$$

The fundamental domain has 25 finite vertices, and the rational function $R(f)$ to whose poles these are mapped is

$$R(f) = \frac{1}{4f^2}\left(1 + \frac{2^7 3^3 f^{12}(f^{12}+1)^2}{(f^{24} - 34f^{12} + 1)^2}\right). \tag{3.24}$$

The poles are located at the origin, and at the vertices of two regular dodecagons centered at the origin, at radial distances $(\sqrt{2} \pm 1)^{1/3}$, forming angles that are multiples of $\pi/12$ with the axes:

$$\begin{aligned} &a_0 := 0, \quad a_m := e^{((m-1)\pi i/6)}(\sqrt{2}-1)^{1/3}, \\ &a_{12+m} := e^{((m-1)\pi i/6)}(\sqrt{2}+1)^{1/3}, \qquad m = 1, \ldots, 12. \end{aligned} \tag{3.25}$$

The quadratic form $r(v)$ defining the dynamical system in this case is

$$\sum_{i,j=1}^{n} r_{ij}v_iv_j = v_0^2 + 3/4\sum_{m=1}^{24} v_m^2 - 3/8\sum_{m=1}^{11}(1-e^{(m\pi i/6)})\big((2-\sqrt{2})v_mv_{m+1} + (2+\sqrt{2})v_{12+m}v_{13+m}\big). \tag{3.26}$$

EXAMPLE 3.3. $96a$

For this case, the modular function may be expressed as

$$f = \frac{\eta^2(48\tau)}{\eta(24\tau)\eta(96\tau)}, \tag{3.27}$$

and the rational function $R(f)$ is

$$R(f) = \frac{1}{4f^2}\left(1 + \frac{2^{10}3^3f^{24}}{(f^{24}-2^6)^2}\right). \tag{3.28}$$

The poles are therefore located at the origin, and at the vertices of a regular 24-gon centered at the origin, at radial distance $2^{1/4}$, forming angles that are multiples of $\pi/24$ with the axes:

$$a_0 := 0, \quad a_m := e^{((m-1)\pi i/12)}2^{1/4}, \quad m = 1, \dots, 24. \tag{3.29}$$

The quadratic form $r(v)$ defining the dynamical system for this case is

$$\sum_{i,j=1}^{n} r_{ij}v_iv_j = v_0^2 + 3/4\sum_{m=1}^{24} v_m^2 - 3/4\sum_{m=1}^{23}(1-e^{(m\pi i/12)})v_mv_{m+1}. \tag{3.30}$$

References

[AH] M. F. Atiyah and N. J. Hitchin, *The geometry and dynamics of magnetic monopoles*, Princeton Univ. Press, Princeton, N.J., 1988.

[Br] M. Brioschi, *Sur un système d'équations différentielles*, C. R. Acad. Sci. Paris Sér. I Math. **92** (1881), 1389–1393.

[Ch] J. Chazy, *Sur les équations différentielles dont l'intégrale générale possède une coupure essentielle mobile*, C. R. Acad. Sci. Paris Sér. I Math. **150** (1910), 456–458.

[CN] J. Conway and S. P. Norton, *Monstrous moonshine*, Bull. London Math. Soc. **11** (1979), 308–339.

[Da] G. Darboux, *Leçons sur les systèmes othogonaux*, 2nd ed., Gauthiers-Villars, Paris, 1910.

[Du] B. A. Dubrovin, *Geometry of 2D topological field theories*, Lecture Notes in Math. **1620**, Springer-Verlag, Berlin, Heidelberg, New York (1996).

[FMN] D. Ford, J. McKay, and S. Norton, *More on replicable functions*, Comm. Algebra **22** (1994), 5175–5193.

[GP] G. W. Gibbons and C. N. Pope, *The positive action conjecture and asymptotically Euclidean metrics in quantum gravity*, Comm. Math. Phys. **66** (1979), 267–290.

[Go] E. Goursat, *Sur l'équation différentielle linéaire qui admet pour intégrale la série hypergéométrique*, Gauthier-Villars, Paris, 1881.

[GS] J. Gerretsen and G. Sansone, *Lectures on the theory of functions of a complex variable. II. Geometric theory.* Walters-Noordhoff Publ., Gröningen, 1969.

[H] E. Hille, *Ordinary differential equations in the complex domain*, Section 7.3, Chapter 10, Wiley, New York, 1976; *Analytic function theory*, vol. II, Chelsea, New York, 1987.

[Ha] G.-H. Halphen, *Sur des fonctions qui proviennent de l'équation de Gauss*, C. R. Acad. Sci. Paris Sér. I Math. **92** (1881), 856–858 *Sur un système d'équations différentielles*, C. R. Acad. Sci. Paris Sér. I Math. **92** (1881), 1101–1103; *Sur certains systèmes d'équations différentielles*, C. R. Acad. Sci. Paris Sér. I Math. **92** (1881), 1404–1406.

[Hi] N. Hitchin, *Twistor spaces, Einstein metrics and isomonodromic deformations*, J. Differential Geom. **42** (1995), 30–112.

[HM] J. Harnad, and J. McKay, *Modular solutions to equations of generalized Halphen type*, Proc. Roy. Soc. London Ser. A (2000). (to appear)

[Oh] Y. Ohyama, *Systems of nonlinear differential equations related to second order linear equations*, Osaka J. Math. **33** (1996), 927–949; *Differential equations for modular forms with level three*, Osaka Univ., preprint, 1997.

[Ta] K. Takeuchi, *Arithmetic triangle groups*, J. Math. Soc. Japan **29** (1977), 91–106.

[To] K. P. Tod, *Self-dual Einstein metrics from the Painlevé* VI *equation*, Phys. Lett. A **190** (1994), no. 3-4, 221–224.

[WW] E. T. Whittaker, and G. N. Watson, *A course of modern analysis*, Chapter 21, 4th ed., Cambridge Univ. Press, London, N.Y., 1962.

TABLE 1. Triangular replicable functions.

Name	(a,b,c)	(λ,μ,ν)	ρ_0	ρ_1	F	$f(\tau)$
$1A \sim \Gamma$	$(\frac{1}{12},\frac{1}{12},\frac{2}{3})$	$(\frac{1}{3},\frac{1}{2},0)$	$\left[\begin{smallmatrix}0&-1\\1&-1\end{smallmatrix}\right]$	$\left[\begin{smallmatrix}0&-1\\1&0\end{smallmatrix}\right]$	$1728f-744$	$J=\dfrac{(\vartheta_2^8+\vartheta_3^8+\vartheta_4^8)^3}{54\vartheta_2^8\vartheta_3^8\vartheta_4^8}$
$2A$	$(\frac{1}{8},\frac{1}{8},\frac{3}{4})$	$(\frac{1}{4},\frac{1}{2},0)$	$\left[\begin{smallmatrix}0&-1\\2&-2\end{smallmatrix}\right]$	$\left[\begin{smallmatrix}0&-1\\2&0\end{smallmatrix}\right]$	$256f-104$	$\dfrac{(\vartheta_3^4+\vartheta_4^4)^4}{16\vartheta_2^8\vartheta_3^4\vartheta_4^4}$
$3A$	$(\frac{1}{6},\frac{1}{6},\frac{5}{6})$	$(\frac{1}{6},\frac{1}{2},0)$	$\left[\begin{smallmatrix}0&-1\\3&-3\end{smallmatrix}\right]$	$\left[\begin{smallmatrix}0&-1\\3&0\end{smallmatrix}\right]$	$108f-42$	$\dfrac{(\eta^{12}(\tau)+27\eta^{12}(3\tau))^2}{108\eta^{12}(\tau)\eta^{12}(3\tau)}$
$2B \sim \Gamma_0(2)$	$(\frac{1}{4},\frac{1}{4},\frac{1}{2})$	$(\frac{1}{2},0,0)$	$\left[\begin{smallmatrix}1&-1\\2&-1\end{smallmatrix}\right]$	$\left[\begin{smallmatrix}-1&0\\2&-1\end{smallmatrix}\right]$	$64f-40$	$1+\dfrac{1}{64}\left(\dfrac{\eta(\tau)}{\eta(2\tau)}\right)^{24}=\dfrac{(\vartheta_3^4(\tau)+\vartheta_4^4(\tau))^2}{\vartheta_2^8(\tau)}$
$3B \sim \Gamma_0(3)$	$(\frac{1}{3},\frac{1}{3},\frac{2}{3})$	$(\frac{1}{3},0,0)$	$\left[\begin{smallmatrix}1&-1\\3&-2\end{smallmatrix}\right]$	$\left[\begin{smallmatrix}-1&0\\3&-1\end{smallmatrix}\right]$	$27f-15$	$1+\dfrac{1}{27}\left(\dfrac{\eta(\tau)}{\eta(3\tau)}\right)^{12}$
$4C^* \sim \Gamma_0(4)$	$(\frac{1}{2},\frac{1}{2},1)$	$(0,0,0)$	$\left[\begin{smallmatrix}1&-1\\4&-3\end{smallmatrix}\right]$	$\left[\begin{smallmatrix}-1&0\\4&-1\end{smallmatrix}\right]$	$16f-8$	$\dfrac{1}{\lambda(2\tau)}=\dfrac{\vartheta_3^4(2\tau)}{\vartheta_2^4(2\tau)}=1+\dfrac{1}{16}\left(\dfrac{\eta(\tau)}{\eta(4\tau)}\right)^8$
$2a$	$(\frac{1}{6},\frac{1}{6},\frac{2}{3})$	$(\frac{1}{3},\frac{1}{3},0)$	$\left[\begin{smallmatrix}2&-3\\4&-4\end{smallmatrix}\right]$	$\left[\begin{smallmatrix}0&-1\\4&-2\end{smallmatrix}\right]$	$24\sqrt{3}i(2f-1)$	$\dfrac{\sqrt{3}i(e^{\pi i/3}\vartheta_3^4(2\tau)-\vartheta_2^4(2\tau))^3}{9\vartheta_2^4(2\tau)\vartheta_3^4(2\tau)\vartheta_4^4(2\tau)}$
$4a$	$(\frac{1}{4},\frac{1}{4},\frac{3}{4})$	$(\frac{1}{4},\frac{1}{4},0)$	$\left[\begin{smallmatrix}4&-5\\8&-8\end{smallmatrix}\right]$	$\left[\begin{smallmatrix}0&-1\\8&-4\end{smallmatrix}\right]$	$-16i(2f-1)$	$-\dfrac{i(\vartheta_3^2(2\tau)+i\vartheta_4^2(2\tau))^4}{8\vartheta_2^4(2\tau)\vartheta_3^2(2\tau)\vartheta_4^2(2\tau)}$
$6a$	$(\frac{1}{3},\frac{1}{3},\frac{5}{6})$	$(\frac{1}{6},\frac{1}{6},0)$	$\left[\begin{smallmatrix}6&-7\\12&-12\end{smallmatrix}\right]$	$\left[\begin{smallmatrix}0&-1\\12&-6\end{smallmatrix}\right]$	$6\sqrt{3}i(2f-1)$	$-\dfrac{\sqrt{3}i(\eta^6(2\tau)+3\sqrt{3}i\eta^6(6\tau))^2}{36\eta^6(2\tau)\eta^6(6\tau)}$

TABLE 2. Four vertex replicable functions (admitting a rational map of degree ≤ 4 to a triangular replicable function).

Name	(a_1, a_2, a_3)	ρ_1 ρ_2 ρ_3	$\sum_{i,j=1}^{3} r_{ij} v_i v_j$	F	$f(\tau)$
$6C$	$(-3, 0, 1)$	$\begin{pmatrix} 3 & -2 \\ 6 & -3 \end{pmatrix}$ $\begin{pmatrix} 3 & -1 \\ 12 & -3 \end{pmatrix}$ $\begin{pmatrix} -1 & 0 \\ 6 & -1 \end{pmatrix}$	$\frac{3}{4}v_1^2 + \frac{3}{4}v_2^2 + v_3^2 - \frac{1}{2}v_2v_3 - v_1v_3$	$4f+2$	$1 + \frac{1}{4}\frac{\eta^6(\tau)\eta^6(3\tau)}{\eta^6(2\tau)\eta^6(6\tau)}$
$6D$	$(\beta, \bar{\beta}, 1)$ $\beta := -\frac{3}{4} + \sqrt{2i}$	$\begin{pmatrix} 4 & -3 \\ 6 & -4 \end{pmatrix}$ $\begin{pmatrix} 2 & -1 \\ 6 & -2 \end{pmatrix}$ $\begin{pmatrix} -1 & 0 \\ 6 & -1 \end{pmatrix}$	$\frac{3}{4}v_1^2 + \frac{3}{4}v_2^2 + v_3^2 + \frac{131}{162}v_1v_2 - \frac{28 - 16\sqrt{2}i}{81}v_1v_3 - \frac{28 + 16\sqrt{2}i}{81}v_2v_3$	$4f$	$1 + \frac{1}{4}\frac{\eta^4(\tau)\eta^4(2\tau)}{\eta^4(3\tau)\eta^4(6\tau)}$
$6E \sim \Gamma_0(6)$	$(-\frac{1}{8}, 0, 1)$	$\begin{pmatrix} 5 & -3 \\ 12 & -7 \end{pmatrix}$ $\begin{pmatrix} 5 & -2 \\ 18 & -7 \end{pmatrix}$ $\begin{pmatrix} -1 & 0 \\ 6 & -1 \end{pmatrix}$	$v_1^2 + v_2^2 + v_3^2 - \frac{10}{9}v_2v_3 - \frac{8}{9}v_1v_3$	$8f-3$	$1 + \frac{1}{8}\frac{\eta^5(\tau)\eta(3\tau)}{\eta(2\tau)\eta^5(6\tau)}$
$6c$	$(-1, 1, 0)$	$\begin{pmatrix} 8 & -7 \\ 12 & -10 \end{pmatrix}$ $\begin{pmatrix} 2 & -1 \\ 12 & -4 \end{pmatrix}$ $\begin{pmatrix} -1 & 0 \\ 12 & -1 \end{pmatrix}$	$\frac{8}{9}v_1^2 + \frac{8}{9}v_2^2 + v_3^2 + \frac{16}{9}v_1v_2$	$i3\sqrt{3}f$	$-\frac{i}{3\sqrt{3}}\frac{\eta^6(2\tau)}{\eta^6(6\tau)}$
$8E \sim \Gamma_0(8)$	$(-1, 0, 1)$	$\begin{pmatrix} 3 & -2 \\ 8 & -5 \end{pmatrix}$ $\begin{pmatrix} 3 & -1 \\ 16 & -5 \end{pmatrix}$ $\begin{pmatrix} -1 & 0 \\ 8 & -1 \end{pmatrix}$	$v_1^2 + v_2^2 + v_3^2 - 2v_1v_3$	$4f$	$1 + \frac{1}{4}\frac{\eta^4(\tau)\eta^2(4\tau)}{\eta^2(2\tau)\eta^4(8\tau)} = \frac{\vartheta_3^2(2\tau) + \vartheta_4^2(2\tau)}{\vartheta_3^2(2\tau) - \vartheta_4^2(2\tau)}$
$9B \sim \Gamma_0(9)$	$(\omega, \bar{\omega}, 1)$ $\omega := e^{\frac{2\pi i}{3}}$	$\begin{pmatrix} 5 & -4 \\ 9 & -7 \end{pmatrix}$ $\begin{pmatrix} 2 & -1 \\ 9 & -4 \end{pmatrix}$ $\begin{pmatrix} -1 & 0 \\ 9 & -1 \end{pmatrix}$	$v_1^2 + v_2^2 + v_3^2 - v_1v_2 - (1-\omega)v_1v_3 - (1-\bar{\omega})v_2v_3$	$3f$	$1 + \frac{1}{3}\frac{\eta^3(\tau)}{\eta^3(9\tau)}$

Department of Mathematics and Statistics, Concordia University, 7141 Sherbrooke W., Montréal, Qué., H4B 1R6, Canada, and Centre de recherches mathématiques, Université de Montréal, C. P. 6128, succ. centre ville, Montréal, Qué., H3C 3J7, Canada.

E-mail address, J. Harnad: harnad@CRM.UMontreal.ca

E-mail address, J. McKay: mckay@cs.concordia.ca

Centre de Recherches Mathématiques
CRM Proceedings and Lecture Notes
Volume **25**, 2000

Symmetry Preserving Discretization of the Burgers Equation

R. Hernández Heredero, D. Levi, and P. Winternitz

ABSTRACT. We present a discrete Burgers equation, that is a second order nonlinear difference equation with two independent variables. It has the usual Burgers equation as its continuous limit. It is linearizable to the discrete heat equation via a discrete Cole-Hopf transformation. It has a five-dimensional symmetry algebra that in the continuous limit goes into the Lie algebra of the symmetry group of the continuous Burgers equation.

1. Introduction

At least two different points of view can be adopted in the study of symmetries of difference equations. One is to take the difference equation as given and then to develop and apply algebraic and analytical techniques that will make it possible to obtain a group of transformations, taking solutions into solutions. A sizable literature exists on this subject, in particular on symmetries that reduce to Lie point symmetries in the continuous limit [**4, 5, 8–10, 12–19, 21**].

Another approach is to start with a given differential equation and to discretize it in a manner preserving symmetries, in particular point symmetries. For linear equations this can be done in a reasonably straightforward way [**4, 14**]. For nonlinear equations this is more difficult. One method [**3**] involves the introduction of symmetry adapted lattices that "evolve" together with the solution of the equation.

In this contribution and a related article [**6**] we study discretizations that preserve symmetries, while making use of standard rectangular (constant) lattices. The method is restricted to the discretization of integrable, in particular linearizable equations and relies heavily on the existence of a related linear system.

The motivation for studying symmetries of difference equations is the same as for differential ones. Namely, to help solve the equations, or at least obtain particular solutions and properties of the solution space. While the theory is well understood for differential equations [**20, 22**], this is as yet far less true for difference ones.

2000 *Mathematics Subject Classification.* Primary: 35Q53; Secondary: 22E70, 39A12.

We thank Luc Vinet for helpful discussions.

Research grants from NSERC of Canada, FCAR du Quebec, MURST of Italy, CICYT PB95-0401 from Spain and NATO (CRG 960717) are gratefully acknowledged.

This is the final form of the paper.

Here we shall use the Burgers equation to show how an integrable discretization preserves all point symmetries of a differential equation, while modifying their point character.

The Burgers equation

$$u_t = u_{xx} + 2uu_x, \tag{1.1}$$

was originally introduced in the context of a statistical theory of turbulent motion in a fluid [**1**]. It is the simplest partial differential equation that describes physical phenomena that are simultaneously dissipative and nonlinear. Hence it has many applications in physics [**2**].

The Burgers equation is imminently suitable for our purposes for at least two reasons:

(1) It is directly linearizable by a transformation of the dependent variable, though the transformation is not a point one. Instead, it is the well-known Cole-Hopf transformation

$$u = \frac{w_x}{w}. \tag{1.2}$$

When u satisfies the Burgers equation, $w(x, t)$ satisfies the linear heat equation, and vice versa. So we have

$$w_t = w_{xx}, \qquad w_x = uw. \tag{1.3}$$

Another way of interpreting the situation is to view Eq. (1.3) as a Lax pair, with Eq. (1.1) as the compatibility condition $w_{tx} = w_{xt}$,

(2) The Burgers equation has a large symmetry group, i.e., group of Lie point transformations taking solutions into solutions. Indeed, the equation is invariant under space and time translations, Galilei transformations, dilations and projective transformations, generated by

$$\begin{gathered} P_1 = \partial_x, \quad P_2 = \partial_t, \quad B = 2t\partial_x - \partial_u, \\ D = 2t\partial_t + x\partial_x - u\partial_u, \quad K = t^2\partial_t + tx\partial_x - \left(tu + \frac{1}{2}x\right)\partial_u. \end{gathered} \tag{1.4}$$

Below we present a discretization of the Burgers equation that preserves the above properties. It is linearizable and it has a five-dimensional symmetry algebra. In the continuous limit its symmetries are the point ones, corresponding to Eq. (1.4).

2. The Discrete Burgers Equation

We use the following conventions and notations [**14**]. The dependent functions $\phi(x, t)$ and $u(x, t)$ will be continuous functions of continuous variables. The independent variables, x and t, while also continuous, will be "sampled" at discrete points. Thus, we shall consider a two-dimensional rectangular lattice with spacings σ_x and σ_t in the x and t directions, respectively. We shall use "discrete derivatives" Δ_x and Δ_t, and shift operators T_x and T_t. The corresponding definitions are

$$T_x f(x,t) = f(x + \sigma_x, t), T_t f(x,t) = f(x, t + \sigma_t), \tag{2.1}$$

$$\Delta_x = \frac{1}{\sigma_x}(T_x - 1), \qquad \Delta_t = \frac{1}{\sigma_t}(T_t - 1). \tag{2.2}$$

The three quantities Δ, T and σ are obviously not independent. We find it convenient to use all three of them in the same formula, but it must be kept in mind that formulas can be reshuffled using Eq. (2.2). Higher discrete derivatives are defined recursively.

Inspired by the continuous Burgers equation, we first write a discrete linear system, namely

$$\Delta_t \phi = \Delta_{xx} \phi, \tag{2.3}$$

$$\Delta_x \phi = u\phi, \tag{2.4}$$

consisting of the discrete heat equation and the discrete Cole-Hopf transformation, relating the functions $\phi(x,t)$ and $u(x,t)$. As in the continuous case, we can interpret eqs. (2.3) (2.4) as a Lax pair.

We use the second of the above equations to rewrite the first as

$$\Delta_t \phi = [\Delta_x u + uT_x u]\phi. \tag{2.5}$$

We then require the compatibility of (2.4) and (2.5), i.e., $\Delta_t \Delta_x \phi = \Delta_x \Delta_t \phi$ and obtain a condition on $u(x,t)$, which we shall call the "discrete Burgers equation":

$$\Delta_t u = \frac{1 + \sigma_x u}{1 + \sigma_t[\Delta_x u + uT_x u]} \Delta_x(\Delta_x u + uT_x u). \tag{2.6}$$

Equation (2.6) can be rewritten in many different forms, for instance we can use Eq. (2.2) to eliminate all discrete derivatives in terms of shift operators T and spacings σ, or alternatively, eliminate the spacings σ_x and σ_t.

The continuous limit of all the discrete equations is obtained by taking $\sigma_x \to 0$, $\sigma_t \to 0$. We have $\Delta_t \to \partial/\partial_t$, $T_t \to 1$, $\sigma_t \to 0$ and similarly for Δ_x, T_x and σ_x. In the continuous limit Eq. (2.6) goes into the usual Burgers equation in the form (1.1).

We mention that a related discrete Burgers equation was obtained earlier by Hirota [**7**]. A semidiscrete Burgers equation (with continuous time t) was also introduced earlier and its integrability was studied [**11**]. It can be reobtained here by taking the limit $\sigma_t \to 0$ for σ_x fixed, $\sigma_x \neq 0$. Using the Lax equation (2.3) we can derive a Bäcklund transformation for the discrete Burgers equation, in the same way as was done in the semidiscrete case [**11**]. The Bäcklund transformation relates a solution $u(x,t)$ of Eq. (2.5) to a new solution $\tilde{u}(x,t)$ of the same equation:

$$\tilde{u} = \frac{pu + (T_x u)(1 + \sigma_x u)}{p + 1 + \sigma_x u}, \tag{2.7}$$

where p is an arbitrary constant. In the continuous limit $\sigma_x \to 0$, $T_x \to 1$ the transformation becomes trivial; not so in the discrete case.

Let us now show that Eq. (2.6) has a five-dimensional symmetry algebra that has Eq. (1.4) as its continuous limit.

3. Symmetries of the Discrete Burgers Equation as Commuting Flows

To obtain the symmetries of the discrete Burgers equation we proceed with the same strategy as used for deriving the equation itself. We start from the symmetries of the discrete heat equation [**4, 14**], require their compatibility with the discrete Cole-Hopf transformation and express the result in terms of the function $u(x,t)$.

We make use of an evolutionary formalism for difference equations [**14**]. The symmetry algebra for a difference equation involving one dependent variable and two independent ones (x and t) will be realized by vector fields of the form

$$\widehat{X} = Q(x, t, T_x^a T_t^b u, T_x^c T_t^d \Delta_x u, T_x^e T_t^f \Delta_t u, \dots)\partial_u, \tag{3.1}$$

where $T_x^a T_t^b$ indicates that arbitrary shifts in x and t may be present. For linear difference equations a rather restrictive Ansatz on the form of the function Q provides symmetries that reproduce all point symmetries in the continuous limit, namely [**14**]

$$Q = \sum_{c,d} \xi(x,t) T_x^c T_t^d \Delta_x u + \sum_{e,f} \tau(x,t) T_x^e T_t^f \Delta_t u - \sum_{a,b} \psi(x,t) T_x^a T_t^b u. \tag{3.2}$$

Each element of the symmetry algebra provides a flow that is compatible with the studied equation, namely

$$u_\lambda = Q, \tag{3.3}$$

where λ is a group parameter that can also be viewed as a new evolution variable (new "time"). Both for differential and difference equations we can construct symmetries as commuting flows.

For the heat equation the elements of the symmetry algebra have the form (3.2). More specifically, an infinite-dimensional Lie algebra was obtained, involving arbitrary powers of shift operators [**14**]. A six-dimensional subalgebra was identified that is isomorphic to the symmetry algebra of the continuous heat equation. Both in the continuous and discrete case, we factor out an infinite-dimensional ideal, corresponding to the linear superposition principle. In terms of commuting flows we write a basis for this finite-dimensional Lie algebra as

$$\begin{aligned}
\phi_{\lambda_1} &= \Delta_t \phi, \\
\phi_{\lambda_2} &= \Delta_x \phi, \\
\phi_{\lambda_3} &= 2t T_t^{-1} \Delta_x \phi + x T_x^{-1} \phi + \frac{1}{2} \sigma_x T_x^{-1} \phi, \\
\phi_{\lambda_4} &= 2t T_t^{-1} \Delta_t \phi + x T_x^{-1} \Delta_x \phi + \left(1 - \frac{1}{2} T_x^{-1}\right) \phi, \\
\phi_{\lambda_5} &= t^2 T_t^{-2} \Delta_t \phi + t x T_t^{-1} T_x^{-1} \Delta_x \phi + \frac{1}{4} x^2 T_x^{-2} \phi \\
&\qquad + t\left(T_t^{-2} - \frac{1}{2} T_t^{-1} T_x^{-1}\right) \phi - \frac{1}{16} \sigma_x^2 T_x^{-2} \phi, \\
\phi_{\lambda_6} &= \phi.
\end{aligned} \tag{3.4}$$

In the continuous limit these flows yield the usual generators of time translations, space translations, Galilei transformations, dilations, projective transformations and the multiplication by a constant, respectively.

Each of these symmetries can be acted upon by arbitrary functions of the shift operators $f(T_x, T_t)$, obtaining further symmetries. In the continuous limit we have $f(T_x, T_t) = f(1,1) = \text{const.}$, so all these "higher" symmetries reduce to the six original ones for $\sigma_x \to 0$, $\sigma_t \to 0$. Among these symmetries two are particularly relevant, namely

$$\phi_\lambda = \phi_t, \quad \phi_\lambda = \phi_x. \tag{3.5}$$

These are the usual t and x translations, that can be obtained using the well known formula

$$\phi_z = \sum_{k=0}^{\infty} \frac{(-1)^k}{k+1} (T_z - 1)^k \Delta_z \phi, \tag{3.6}$$

(with $z = x$, or $z = t$).

All symmetries of the heat equation can be written symbolically as

$$\phi_\lambda = \mathcal{S}\phi, \tag{3.7}$$

where $\mathcal{S} = \mathcal{S}(x, t, \phi, T_x, T_t, \Delta_x, \Delta_t, \partial_x, \partial_t)$ is a linear operator that can in each case be read off from eqs. (3.4), (3.5).

We use the Cole-Hopf transformation to transform symmetries of the heat equation into those of the discrete Burgers equation. Let us first prove a general result.

Let Eq. (3.7) represent a symmetry of the discrete heat equation. Then the same operator $\mathcal{S}$ provides a symmetry of the discrete Burgers equation via the formula

$$u_\lambda = (1 + \sigma_x u)\Delta_x \left(\frac{\mathcal{S}\phi}{\phi}\right), \tag{3.8}$$

where $\mathcal{S}\phi/\phi$ is expressed entirely in terms of $u(x,t)$, its variations and their shifted values.

Indeed, let us require that the symmetry (3.7) and the Cole-Hopf transformation (2.4) be compatible. The compatibility condition

$$\frac{\partial}{\partial\lambda}(\Delta_x\phi) = \Delta_x\phi_\lambda,$$

implies

$$u_\lambda = \frac{\Delta_x(\mathcal{S}\phi) - u\mathcal{S}\phi}{\phi}. \tag{3.9}$$

On the other hand, a direct calculation yields

$$\begin{aligned}
\Delta_x\left(\frac{\mathcal{S}\phi}{\phi}\right) &= \frac{1}{\sigma_x}\left[\frac{T_x(\mathcal{S}\phi)}{T_x\phi} - \frac{\mathcal{S}\phi}{\phi}\right] \\
&= \frac{1}{\sigma_x(T_x\phi)\phi}\Big\{\phi\big[T_x(\mathcal{S}\phi) - \mathcal{S}\phi\big] - \big[(T_x\phi)(\mathcal{S}\phi) - \phi(\mathcal{S}\phi)\big]\Big\}, \\
\Delta_x\left(\frac{\mathcal{S}\phi}{\phi}\right) &= \frac{1}{T_x\phi}\big[\Delta_x(\mathcal{S}\phi) - u(\mathcal{S}\phi)\big].
\end{aligned}$$

Multiplying by $\sigma_x u$ and using the Cole-Hopf transformation again we obtain

$$\sigma_x u\Delta_x\left(\frac{\mathcal{S}\phi}{\phi}\right) = \frac{\Delta_x(\mathcal{S}\phi) - u(\mathcal{S}\phi)}{\phi} - \Delta_x\left(\frac{\mathcal{S}\phi}{\phi}\right). \tag{3.10}$$

Using (3.9), we replace the first term on the right hand side of (3.10) by u_λ, and obtain Eq. (3.8).

In order to show that the fraction $\mathcal{S}\phi/\phi$ can be expressed in terms of the function u, it is sufficient to write $\Delta_t\phi$, $\Delta_x\phi$, $T_x\phi$, $T_t\phi$, $T_x^{-1}\phi$, etc, as expressions depending on u, times ϕ (see Eq. (3.4)). The necessary formulas are obtained from Eq. (2.4) and (2.5), namely

$$\begin{gathered}
\Delta_x\phi = \phi u, \quad \Delta_t\phi = \phi v, \\
T_x\phi = \phi(1 + \sigma_x u), \quad T_t\phi = \phi(1 + \sigma_t v), \\
T_x^{-1}\phi = \phi T_x^{-1}\frac{1}{1 + \sigma_x u}, \quad T_t^{-1}\phi = \phi T_t^{-1}\frac{1}{1 + \sigma_t v},
\end{gathered} \tag{3.11}$$

where we have introduced the notation

$$v = \Delta_x u + uT_x u. \tag{3.12}$$

Applying the above result to the one-dimensional subalgebras $\phi_{\lambda_1}, \ldots, \phi_{\lambda_6}$ of Eq. (3.4), we obtain the corresponding symmetries of the discrete Burgers equation. A basis for this Lie algebra is given by the following flows:

$$u_{\lambda_1} = (1 + \sigma_t v)\Delta_t u, \tag{3.13}$$

$$u_{\lambda_2} = (1 + \sigma_x u)\Delta_x u, \tag{3.14}$$

$$u_{\lambda_3} = (1 + \sigma_x u)\Delta_x \left[2tT_t^{-1} \frac{u}{1 + \sigma_t v} + \left(x + \frac{1}{2}\sigma_x \right) T_x^{-1} \frac{1}{1 + \sigma_x u} \right], \tag{3.15}$$

$$u_{\lambda_4} = (1 + \sigma_x u)\Delta_x \left[2tT_t^{-1} \frac{v}{1 + \sigma_t v} + xT_x^{-1} \frac{u}{1 + \sigma_x u} - \frac{1}{2} T_x^{-1} \frac{1}{1 + \sigma_x u} \right], \tag{3.16}$$

$$\begin{aligned} u_{\lambda_5} = (1 + \sigma_x u)\Delta_x \Bigg[& t^2 T_t^{-1} \left(\frac{1}{1 + \sigma_t v} T_t^{-1} \frac{v}{1 + \sigma_t v} \right) \\ & + txT_x^{-1} \left(\frac{1}{1 + \sigma_x u} T_t^{-1} \frac{u}{1 + \sigma_t v} \right) \\ & + \frac{1}{4} \left(x^2 - \frac{\sigma_x^2}{4} \right) T_x^{-1} \left(\frac{1}{1 + \sigma_x u} T_x^{-1} \frac{1}{1 + \sigma_x u} \right) \\ & + tT_t^{-1} \left(\frac{1}{1 + \sigma_t v} T_t^{-1} \frac{1}{1 + \sigma_t v} \right) \\ & - \frac{1}{2} tT_x^{-1} \left(\frac{1}{1 + \sigma_x u} T_t^{-1} \frac{1}{1 + \sigma_t v} \right) \Bigg], \end{aligned} \tag{3.17}$$

$$u_{\lambda_6} = 0. \tag{3.18}$$

The quantity v is defined in Eq. (3.12).

Thus, the six-dimensional symmetry algebra of the discrete heat equation gives rise to a five-dimensional symmetry algebra of the discrete Burgers equation. The same is true in the continuous case.

The fact that the flows (3.13),...,(3.18) commute with the flow of the discrete Burgers equation (2.6) was also checked directly on a computer (using Mathematica).

In the continuous limit, Eq. (3.13),...,(3.18) go over correctly into the well known symmetries of the usual Burgers equation (1.1), namely time translations, space translations, Galilei boosts, dilations and projective transformations. The commutation relations in the discrete case are the same as in the continuous one.

We can show directly that the usual space and time translations are also symmetries of the discrete Burgers equation:

$$u_{\lambda_t} = u_t, \quad u_{\lambda_x} = u_x. \tag{3.19}$$

Indeed, it is easy to check that the corresponding λ-flows commute with the t-flow given by Eq. (2.6).

4. Reduction of the Discrete Burgers Equation to Single Variable Difference Equations

For a partial differential equation, symmetry reduction is a means for decreasing the number of independent variables in the equation. One way of viewing it is that one looks for fixed points of the λ-flow corresponding to a certain symmetry. Thus we have the surface condition

$$u_\lambda = Q(x, t, u, u_x, u_t, \dots) = 0. \tag{4.1}$$

This equation is to be solved in conjunction with the equation under consideration. Often one can solve Eq. (4.1) directly, in particular using the method of characteristics, if it is a first order linear or quasilinear equation. The result is then substituted into the original equation and this leads to a reduction. Alternatively, Eq. (4.1) can be used to eliminate all t-derivatives (or x-derivatives) from the equation, thus reducing it to an ordinary differential equation (with the other variable as a parameter).

Solutions obtained in this manner are called group invariant solutions [**20**]. They form a very important class of explicit solutions of nonlinear partial differential equations.

If Eq. (4.1) is implemented, together with the partial differential equation to be solved, the resulting system may have some further symmetries, inherited from the symmetry algebra $\mathcal{L}$ of the original equation. More specifically, let $X_e = Q\partial_u$ be the corresponding evolutionary vector field. The symmetries that will survive, once Eq. (4.1) is imposed, form a subalgebra $\mathcal{L}_0 \subset \mathcal{L}$, where $\mathcal{L}_0$ is the normalizer algebra of X_e:

$$\mathcal{L}_0 = \{Y_e \subset \mathcal{L} \mid [Y_e, X_e] = \lambda X_e\}, \quad \lambda \in \mathbb{R}. \tag{4.2}$$

Let us illustrate the situation using the continuous Burgers equation as an example. The symmetry algebra $\mathcal{L}$ in the usual vector field formalism has a basis given in Eq. (1.4).

From the commutation relations of these vector fields, we see that $\mathcal{L}$ is a semidirect sum of $\mathfrak{sl}(2, \mathbb{R})$ and an abelian Lie algebra:

$$\mathcal{L} \sim \{P_0, D, R\} \dotplus \{P_1, B\}. \tag{4.3}$$

As an example let us look at the reductions of the continuous Burgers equation by time translations P_0. Eq. (4.1) in this case is simply $u_\lambda = u_t = 0$. The Burgers equation reduces to the ordinary differential equation

$$u_{xx} + 2uu_x = 0. \tag{4.4}$$

We obtain three types of solutions:

$$u = \frac{1}{x}, \quad u = k \operatorname{arctanh}(kx), \quad u = k \arctan(kx), \tag{4.5}$$

depending on whether the first integral of the reduced equation is zero, positive or negative.

The normalizer of P_0 in the invariance algebra is $\operatorname{nor}\{P_0\} = \{D, P_1, P_0\}$, so we can use either D, or P_1, to perform a further reduction of Eq. (4.4). A reduction by P_1 leads to the trivial solution $u = u_0 = \text{const}$. Invariance under dilations provides the first of the three solutions in Eq. (4.5). The other two are invariant under time translations only.

Now let us go over to the discrete Burgers equation.

All of its symmetries can be written in the form (3.8) with $\mathcal{S}\phi/\phi$ expressed in terms of u. This allows us to write all the reduction formulas in the form $\Delta_x(\mathcal{S}\phi/\phi) = 0$. Hence, we can in all cases integrate once and write the analog of the surface condition (4.1) as

$$\frac{\mathcal{S}\phi}{\phi} = K(t), \tag{4.6}$$

and then rewrite Eq. (4.7) in terms of u. In general $\mathcal{S}$ is a linear combination of all the symmetry operators for the heat equation, i.e., the operators on the right-hand sides of Eq. (3.4). Here we shall just look at some individual basis elements of the Lie algebra.

A. Translations. For time translations we rewrite (3.13) as

$$u_{\lambda_1} = (1+\sigma_x u)\Delta_x\left(\frac{\Delta_t\phi}{\phi}\right). \tag{4.7}$$

Using (4.6) we have

$$\Delta_t\phi = K(t)\phi, \tag{4.8}$$

or in terms of u:

$$v = \Delta_x u + uT_x u = K(t). \tag{4.9}$$

The Burgers equation (2.6) can be written as

$$\Delta_t u = \frac{1+\sigma_x u}{1+\sigma_t v}\Delta_x v. \tag{4.10}$$

Eq. (4.9) then implies

$$\Delta_t u = 0, \quad K = K_0 = \text{const}. \tag{4.11}$$

Since ϕ satisfies the heat equation we rewrite Eq. (4.8) as

$$\Delta_{xx}\phi = K\phi. \tag{4.12}$$

This is a linear difference equation with constant coefficients, and we can easily solve it, putting $\phi = a^x$ and finding a.

For $K \neq 0$ the general solution is

$$\phi = c_1\big(1+\sqrt{K}\sigma_x\big)^{x/\sigma_x} + c_2\big(1-\sqrt{K}\sigma_x\big)^{x/\sigma_x}, \tag{4.13}$$

where c_1 and c_2 are arbitrary real constants for $K > 0$ and are complex, satisfying $c_2 = \bar{c_1}$ for $K < 0$.

For $K = 0$ the solution is

$$\phi = c_1 + c_2 x, \quad c_1, c_2 \in \mathbb{R}. \tag{4.14}$$

In all cases the corresponding invariant solution of the discrete Burgers equation is obtained via the Cole-Hopf transformation as $u = \Delta_x\phi/\phi$. In particular Eq. (4.14) yields a solutions invariant under dilations that can be written as

$$u = \frac{1}{x+\mu\sigma_x}, \quad \mu = \text{const}. \tag{4.15}$$

For space translations the result is trivial. From Eq. (3.14) we have directly $\Delta_x u = 0$, then the discrete Burgers equation implies $\Delta_t u = 0$ and hence $u = \text{const}$.

B. Galilei invariance. Substituting the appropriate $\mathcal{S}$ from Eq. (3.4) into (4.6) we obtain the linearized reduced equation

$$\begin{aligned}(4.16)\quad 2(t+\sigma_t)\Delta_x\phi + \left(x+\frac{1}{2}\sigma_x\right)T_x^{-1}\phi + \sigma_t\left(x+\frac{1}{2}\sigma_x\right)T_x^{-1}\Delta_{xx}\phi \\ = K(t)(\phi+\sigma_t\Delta_{xx}\phi).\end{aligned}$$

In terms of u the reduced equation is obtained using Eq. (3.11) and is

$$\begin{aligned}(4.17)\quad & 2tT_xu + x - K(t) + 2t\sigma_x uT_xu \\ & + \sigma_t\left(\frac{7}{2}T_xu + \frac{7}{2}\sigma_x uT_xu + xuT_xu + x\Delta_xu - \frac{3}{2}u\right) + \frac{3}{2}\sigma_x \\ & - K(t)\left[\sigma_xu + \sigma_t\left(T_x\Delta_xu + uT_x^2u - uT_xu\right) + T_xuT_x^2u + \sigma_x uT_xuT_x^2u\right] = 0.\end{aligned}$$

These are difference equations in one variable (namely x) only; t is a parameter. Equation (4.16) is a linear second order equation, but it has variable coefficients. Solving it is a nontrivial task which we shall not attempt here.

The situation is similar for the dilations and the projective transformation. We shall just present the reduced equations for ϕ and u, the first being a linearization (via the discrete Cole-Hopf transformation) of the second.

C. Dilational invariance. The linearized reduced equation is

$$\begin{aligned}(4.18)\quad & 2(t+\sigma_t)\Delta_{xx}\phi + xT_x^{-1}(\Delta_x\phi + \sigma_t\Delta_{xxx}\phi) \\ & + \left(1-\frac{1}{2}T_x^{-1}\right)(\phi+\sigma_t\Delta_{xx}\phi) = K(t)\left[\phi+\sigma_t\Delta_{xx}\phi\right].\end{aligned}$$

The reduced equation for u is:

$$\begin{aligned}(4.19)\quad & \frac{1}{2} + K(t) - u\left[x + \sigma_x(2-K(t)) - v_1\left(2t - \frac{3}{2}\sigma_t\right)\right] \\ & - u(T_xv_1)\left[2t + 4\sigma_x\sigma_t - x\sigma_t - K(t)\sigma_x\sigma_t\right] \\ & - (\Delta_xv_1)(2t - x\sigma_t) - \sigma_t\left(u - K(t)\right)T_xv_1 = 0,\end{aligned}$$

with

$$(4.20)\qquad v_{j+1} = \Delta_xv_j + uT_xv_j, \quad v_0 = u, \quad v_1 = v.$$

D. Projective invariance. The linearized reduced equation for ϕ and the reduced equation for u are respectively

$$\begin{aligned}(4.21)\quad & (t+2\sigma_t)^2\Delta_{xx}\phi + (t+2\sigma_t)xT_x^{-1}(\Delta_x\phi + \sigma_t\Delta_{xxx}\phi) \\ & + \left(\frac{1}{4}x^2 - \frac{1}{16}\sigma_x^2\right)T_x^{-2}(\phi + 2\sigma_t\Delta_{xx}\phi + \sigma_t^2\Delta_{xxxx}\phi) \\ & + (t+2\sigma_t)\left[\phi - \frac{1}{2}T_x^{-1}(\phi+\sigma_t\Delta_{xx}\phi)\right] \\ & = K(t)(\phi + 2\sigma_t\Delta_{xx}\phi + \sigma_t^2\Delta_{xxxx}\phi),\end{aligned}$$

and

$$
\begin{aligned}
(4.22)\quad & (t+2\sigma_t)^2(v_1+2\sigma_x v_2+\sigma_x^2 v_3)\\
&\quad + (t+2\sigma_t)(x+\sigma_x)(u+\sigma_x v_1+\sigma_t v_2+\sigma_x\sigma_t v_3)\\
&\quad + \frac{1}{4}\left[(x+2\sigma_x)^2-\frac{1}{4}\sigma_x^2\right](1+2\sigma_t v_1+\sigma_t^2 v_3)\\
&\quad + (t+2\sigma_t)\left(\frac{1}{2}+\frac{3}{2}\sigma_x u+\sigma_x^2 v_1-\frac{1}{2}\sigma_t v_1-\frac{1}{2}\sigma_x\sigma_t v_2\right)\\
&= K(t)(1+2\sigma_x u+\sigma_x^2 v_1+2\sigma_t v_1+4\sigma_x\sigma_t v_2+2\sigma_t\sigma_x^2 v_3\\
&\quad + \sigma_t^2 v_3+2\sigma_t^2\sigma_x v_4+\sigma_t^2\sigma_x^2 v_5).
\end{aligned}
$$

5. Conclusions

The main conclusion that we can draw from the example of the Burgers equation is that both the similarities and the differences between differential equations and difference ones must be taken into account.

The continuous and discrete equations are similar in that a large class of symmetries can be found using infinitesimal techniques. For differential equations this class includes all point symmetries. For difference equations the corresponding symmetries are not necessarily point ones. They only reduce to point symmetries in the continuous limit. In both cases one obtains a system of determining equations for the characteristic Q (see Eq. (3.1)) of the symmetry generator in its evolutionary form. The determining equations are linear, even if the studied equation is a nonlinear one.

Both for differential and difference equations the symmetry algebra can be viewed as an algebra of flows, commuting with that of the original equation. The corresponding evolutionary vector fields form a Lie algebra under commutation.

The applications are also similar. For instance, in this article we have used one-dimensional subalgebras of the symmetry algebra to reduce a difference equation in two variables, the discrete Burgers equation, to a difference equation in one variable only. Different mutually non conjugate subalgebras of the symmetry algebra provide different reductions, just as in the case of differential equations.

The differences between symmetries of discrete and continuous equations are mainly technical. The determining equations are themselves linear difference equations in the discrete case. In general, they may be difficult to solve. The problem of directly solving determining equations was avoided in this study by making use of the Cole-Hopf transformation and transforming the symmetries of the linear discrete heat equations into those of the Burgers equation. Already for the heat equation, most of the symmetries (3.7) involve shift operators and hence the function ϕ calculated at several points of the lattice.

The symmetries of the Burgers equation (3.15), (3.16) and (3.17) all involve the function $u(x,t)$ calculated at more than one point. Moreover, the right-hand sides of these equations are quite complicated. They of course satisfy the determining equations. However, we do not know of any reasonable Ansatz that would have yielded the solutions directly. In the continuous case it is usually easy to obtain the general solution of the determining equations. At this stage fewer tools are available for difference equations.

Before attacking the problem of solving overdetermined systems of linear difference equations (the determining equations) heads on, we hope to accumulate further insight by studying symmetries of integrable equations and making use of the corresponding linear Lax pairs.

References

1. J. M. Burgers, *Application of a model system to illustrate some points of the statistical theory of turbulence*, Nederl. Akad. Wetensch., Proc. **43** (1940), 2–12.
2. ———, *The nonlinear diffusion equation: Asymptotic solutions and statistical problems*, Reidel, Dordrecht, 1974.
3. V. Dorodnitsyn, *Transformation groups in Mesh spaces*, J. Soviet Math. **55** (1991), 1490–1517.
4. R. Floreanini and L. Vinet, *Lie symmetries of finite difference equations*, J. Math. Phys. **36** (1995), 7024–7042.
5. D. Gómez Ullate, S. Lafortune, and P. Winternitz, *Symmetries of discrete dynamical systems involving two species*, J. Math. Phys. **40** (1999), 2782–2804.
6. R. Hernández Heredero, D. Levi, and P. Winternitz, *Symmetries of the discrete Burgers equation*, J. Phys. A **32** (1999), 2685–2695.
7. R. Hirota, *Nonlinear partial difference equations. V. Nonlinear equations reducible to linear equations*, J. Phys. Soc. Japan **46** (1979), 312–319.
8. Z. Jiang, *Lie symmetries and their local determinancy for a class of differential-difference equations*, Phys. Lett. A **240** (1998), 137–143.
9. N. Joshi and P. J. Vassiliou, *The existence of Lie symmetries for first order analytic discrete dynamical systems*, J. Math. Anal. Appl. **195** (1995), 872–887.
10. S. Lafortune, L. Martina, and P. Winternitz, *Symmetries of Toda field theories and their reductions*, preprint CRM-2566, Montréal, 1998.
11. D. Levi, O. Ragnisco, and M. Bruschi, *Continuous and discrete matrix Burgers hierarchies*, Nuovo Cimento B **11** (1983), 33–51.
12. D. Levi and M. A. Rodríguez, *Symmetry groups of partial differential equations and of differential-difference equations: the Toda lattice versus the Korteweg-de Vries equation*, J. Phys. A **25** (1992), L875–L979.
13. D. Levi, L. Vinet, and P. Winternitz (eds.), *Symmetries and integrability of difference equations* (Estérel, 1994), CRM Proc. Lecture Notes, vol. 9, Amer. Math. Soc., Providence, RI, 1996.
14. ———, *Lie group formalism for difference equations*, J. Phys. A **30** (1997), 633–649.
15. D. Levi and P. Winternitz, *Continuous symmetries of discrete equations*, Phys. Lett. A **152** (1991), 335–338.
16. ———, *Symmetries and conditional symmetries of differential-difference equations*, J. Math. Phys. **34** (1993), 3713–3730.
17. ———, *Symmetries of discrete dynamical systems*, J. Math. Phys. **37** (1996), 5551–5576.
18. D. Levi and R. Yamilov, *Conditions for the existence of higher symmetries of evolutionary equations on a lattice*, J. Math. Phys. **38** (1997), 6648–6674.
19. S. Maeda, *The similarity method for difference equations*, IMA J. Appl. Math. **38** (1987), 129–134.
20. P. J. Olver, *Applications of Lie groups to differential equations*, Grad. Texts in Math., vol. 107, Springer, New York, 1986.
21. G. R. W. Quispel, H. W. Capel, and R. Sahadevan, *Continuous symmetries of differential-difference equations: The Kac–van Moerbeke equation and Painlevé reduction*, Phys. Lett. A **170** (1992), 379–383.
22. P. Winternitz, *Lie groups and solutions of nonlinear partial differential equations*, Integrable Systems, Quantum Groups and Quantum Field Theories (Salamanca, 1992) (A. Ibort and M. A. Rodriguez, eds.), NATO Adv. Sci. Inst. Ser. C Math. Phys. Sci., vol. 409 , Kluwer Acad. Publ., Dordrecht, 1993, pp. 429–495.

Departamento de Física Teórica II, Facultad de Ciencias Físicas, Universidad Complutense de Madrid, 28040 Madrid, Spain.
E-mail address: rafa@ciruelo.fis.ucm.es

Dipartimento di Fisica "Edoardo Amaldi", Universitá di Roma Tre and INFN, Sezione di Roma Tre, Via della Vasca Navale 84, 00146 Roma, Italy.
E-mail address: levi@amaldi.fis.uniroma3.it

Centre de recherches mathématiques, Université de Montréal, C.P. 6128, succ. Centre-ville, Montréal, Qué., H3C 3J7, Canada.
E-mail address: wintern@CRM.UMontreal.CA

Centre de Recherches Mathématiques
CRM Proceedings and Lecture Notes
Volume **25**, 2000

Singularity Confinement and Degree Growth

Jarmo Hietarinta and Claude Viallet

Abstract. We discuss the singularity confinement property using polynomial maps in projective variables. It is shown through examples that singularity confinement alone does not guarantee integrability, but in addition the degree growth of the map must be slow enough (polynomial rather than exponential). We also present a method by which one can easily find out when a nonautonomous map has its first cancellation.

1. Introduction

The Painlevé test has become an important tool for identifying differential equations whose solutions behave in a regular manner. It is quite natural to try to search for a similar test for difference equations or maps. In [**3**], a singularity confinement test was proposed as the analogue of the Painlevé test for discrete equations. This test turned out to be very useful in identifying well behaving maps, including discrete analogues of Painlevé equations [**6**].

However, the very concept of integrability is more problematic for discrete equations than for continuous ones, and as a consequence the singularity confinement test was met with some reserve. It was generally accepted that while singularity confinement is necessary it was not at all clear whether it is sufficient.

In [**4**] the present authors reported an example which passed the singularity confinement test but which nevertheless showed numerical chaos in a typical way. In this same paper attention was directed on the degree growth of the map and it was proposed that as an additional requirement the degree of an integrable map should not grow faster than polynomially (see also [**2**]). This means that when the map is viewed as a polynomial map in projective space there should be projective cancellations. These cancellations are intimately related to singularity confinement and for polynomial growth one sometimes just needs more of them.

2. The Singularity Confinement Test

We will choose our examples from the class of maps

$$x_{n+1} + x_{n-1} = a_n x_n + b_n + \frac{c_n}{x_n} + \frac{d_n}{x_n^2}, \tag{2.1}$$

2000 *Mathematics Subject Classification.* Primary: 37A99; Secondary: 39A99.
This is the final form of the paper.

where $a_n, \ldots, d_n$ are some functions of n. (We assume that c_n and/or d_n are nonzero, otherwise the map is linear.) It is clear that this equation can be iterated equally well in both directions and that potential problems arise only if for some k we obtain $x_k = 0$ (with x_{k-1} finite and nonzero). This would mean $x_{k+1} = \infty$ and then $x_{k+2} = \infty$ but in computing x_{k+3} we would encounter the undefined expression $\infty - \infty$.

The idea of the singularity confinement test is to check whether this potential singularity can be circumvented. Let us therefore assume that $x_1 = u$ which is finite and nonzero and that $x_0 = \varepsilon$, which is a small number and we will let it approach zero later. For this example let us further assume that in (2.1) $a_n = d_n = 1$, $b_n = c_n = 0$, $\forall n$, i.e.,

$$x_{n+1} + x_{n-1} = x_n + \frac{1}{x_n^2}.$$

We find

$$\begin{aligned} x_{-1} &= u, \\ x_0 &= \varepsilon, \\ x_1 &= \varepsilon^{-2} - u + \varepsilon, \\ x_2 &= \varepsilon^{-2} - u + \varepsilon^4 + \ldots, \\ x_3 &= -\varepsilon + 2\varepsilon^4 + \ldots, \\ x_4 &= u + 3\varepsilon + \ldots. \end{aligned}$$

Thus a closer analysis indicates that in this case "$\infty - \infty = 0$". Furthermore one can emerge from the singularity with the initial information u.

Let us now write the system as a polynomial map in $\mathbb{C}P^2$. First we write is as a first order system,

$$\begin{cases} x_{n+1} &= x_n - y_n + \dfrac{1}{x_n^2}, \\ y_{n+1} &= x_n, \end{cases}$$

and then we homogenize it by substituting

$$x_n = u_n/f_n, \quad y_n = v_n/f_n,$$

which yields

$$\begin{cases} \dfrac{u_{n+1}}{f_{n+1}} &= \dfrac{u_n - v_n}{f_n} + \dfrac{f_n^2}{u_n^2}, \\ \dfrac{v_{n+1}}{f_{n+1}} &= \dfrac{u_n}{f_n}. \end{cases}$$

Here we can eliminate denominators if we choose $f_{n+1} = f_n u_n^2$, this leads to

$$\begin{cases} u_{n+1} &= -u_n^2 v_n + u_n^3 + f_n^3, \\ v_{n+1} &= u_n^3, \\ f_{n+1} &= f_n u_n^2. \end{cases}$$

This map is in $\mathbb{C}P^2$ and therefore we have the identification $(u, v, f) \sim (\lambda u, \lambda v, \lambda f)$. The apparent degree growth of (2) is 3^n, where n is the iteration step. However, the identification rule means that common factors can be canceled and therefore the actual degree growth may be slower.

Let us now see how the singularity confinement looks from this point of view. In $\mathbb{C}P^2$ we have the following sequence

$$\begin{pmatrix}0\\u\\1\end{pmatrix},\to\begin{pmatrix}1\\0\\0\end{pmatrix},\to\begin{pmatrix}1\\-1\\0\end{pmatrix},\to\begin{pmatrix}0\\1\\0\end{pmatrix},\to\begin{pmatrix}0\\0\\0\end{pmatrix}.$$

In this case the first four points are regular points in the projective space but the last one is not in $\mathbb{C}P^2$ at all and therefore singular. To figure out how to proceed we must next study the behavior near it.

Starting from the point corresponding to $x_{-1} = u$, $x_0 = \varepsilon$ we obtain

$$\begin{pmatrix}\varepsilon\\u\\1\end{pmatrix}\to\begin{pmatrix}1-\varepsilon^2u+\varepsilon^3\\\varepsilon^3\\\varepsilon^2\end{pmatrix}\to\begin{pmatrix}1-3\varepsilon^2u+\dots\\1-3\varepsilon^2u+\dots\\\varepsilon^2-2\varepsilon^4u+\dots\end{pmatrix}$$
$$\to\begin{pmatrix}-\varepsilon^3+\dots\\1-9\varepsilon^2u+\dots\\\varepsilon^2+\dots\end{pmatrix}\to\begin{pmatrix}\varepsilon^8u+\dots\\-\varepsilon^9+\dots\\\varepsilon^8+\dots\end{pmatrix}=\begin{pmatrix}u+\dots\\-\varepsilon+\dots\\1+\dots\end{pmatrix}.$$

It is important here to note how we emerged from the singularity. The initial value was temporarily buried under ε^2, however, the cancellation of ε^8 from the vector (that was initially considered a singular point) made it finite and at the same time brought forward the initial value u. This is important: emergence from singularity is possible only if we can cancel enough factors of ε, but at the same time the cancellations also reduce the degree growth. The amount of cancellations required for singularity confinement is just not always enough to guarantee integrability, without any cancellations the degree of the n-th iterate would have been 3^n but for integrability one needs so many cancellations that the degree growth slows down from exponential to polynomial.

For the map we are discussing (2) we get the following sequence of degrees: 1, 3, 9, 27, 73, 195, 513, 1347, 3529, ... The first four degrees follow the 3^n rule but then we have the first cancellation: the fifth term would have been 81 but is only 73, corresponding to a cancellation of 8. This corresponds exactly to the factorization of ε^8 in the above calculation. (The later cancellations appearing in this sequence are not so easy to identify using ε expansion.)

From these first nine numbers of the degree sequence we can construct a generating function for the degrees, it has the form

$$\text{(2.2)}\quad g(x) = \frac{1+3\,x^3}{(1-x)\,(1+x)\,(x^2-3\,x+1)}$$
$$= 1+3\,x+9\,x^2+27\,x^3+73\,x^4+195\,x^5+513\,x^6$$
$$+\,1347\,x^7+3529\,x^8+9243\,x^9+24201\,x^{10}+\dots.$$

(There is an argument in [**1**] by which one can actually prove that this is the generating function.)

Let g_n be the degree of the n-th iterate, and λ the root of the denominator of g with smallest modulus. Thus λ determines the radius of convergence of g and therefore asymptotically $g_{n+1}/g_n \approx 1/\lambda$, and hence $g_n \approx \lambda^{-n}$. For (2.2) $1/\lambda = (3+\sqrt{5})/2 \approx 2.618$, and therefore we have exponential growth. Numerical studies confirm the prediction of nonintegrability and show typically chaotic behavior [**4**].

Let us compare the above with a truly integrable system, e.g., [6]

$$x_{n+1} + x_{n-1} = \frac{c}{x_n} + \frac{d}{x_n^2},$$

which is related to d-$P_{\rm I}$. This is also a third degree map in projective coordinates, but calculations show that cancellations are much stronger and produce the degrees 1, 3, 9, 19, 33, 51, 73, 99, 129, 163, 201, ... The generating function for this sequence is

$$g = \frac{1+3x^2}{(1-x)^3},$$

and in fact the degree growth follows the polynomial rule

$$g_n = 2n^2 + 1.$$

As another integrable example consider d-$P_{\rm I}$

$$x_{n+1} + x_{n-1} = b + \frac{c}{x_n},$$

with degree sequence 1, 2, 4, 7, 12, 25, 34, 44, 55, 68 the generating function

$$g = \frac{1+x^2+2x^4}{(1-x^3)(1-x)^2}.$$

Again the roots of the denominator have unit modulus.

3. Nonautonomous Systems

The above discussion was concerned with autonomous maps but we can use the same approach when trying to find those deautonomizations for which the system still passes the singularity confinement test.

Let us consider (2.1) but now with $a_n = -1$, $b_n = b$, $d_n = 0$, (d-$P_{\rm I}$)

$$x_{n+1} + x_n + x_{n-1} = b + \frac{c_n}{x_n}. \tag{3.1}$$

As a polynomial map in $\mathbb{C}P^2$ it reads

$$\begin{cases} u_{n+1} &= -u_n(u_n+v_n) + f_n(c_n f_n + b u_n), \\ v_{n+1} &= u_n^2, \\ f_{n+1} &= f_n u_n. \end{cases}$$

The singularity pattern of the original equation is

$$x_{-1} = x, \quad x_0 = 0, \quad x_1 = \infty, \quad x_2 = \ ?$$

and correspondingly in $\mathbb{C}P^2$ we have the sequence

$$\begin{pmatrix} 0 \\ x \\ 1 \end{pmatrix} \to \begin{pmatrix} 1 \\ 0 \\ 0 \end{pmatrix} \to \begin{pmatrix} 1 \\ -1 \\ 0 \end{pmatrix} \to \begin{pmatrix} 0 \\ 1 \\ 0 \end{pmatrix} \to \begin{pmatrix} 0 \\ 0 \\ 0 \end{pmatrix},$$

Expanding around the singularity gives the sequence

$$\begin{pmatrix}\varepsilon\\ x\\ 1\end{pmatrix}\to\begin{pmatrix}c_0+(b-x)\varepsilon+\dots\\ \varepsilon^2\\ \varepsilon\end{pmatrix}\to\begin{pmatrix}-c_0^2+\varepsilon c_0(2x-b)+\dots\\ c_0^2+2\varepsilon c_0(-x+b)+\dots\\ \varepsilon c_0+\varepsilon^2(-x+b)+\dots\end{pmatrix}$$
$$\to\begin{pmatrix}\varepsilon^2c_0^2(-c_0+c_1+c_2)+\dots\\ c_0^4+2\varepsilon c_0^3(-2x+b)\dots\\ -\varepsilon c_0^3+\varepsilon^2c_0^2(3x-2b)+\dots\end{pmatrix}\to\begin{pmatrix}\varepsilon^2c_0^6C_3+\varepsilon^3c_0^5K+\dots\\ \varepsilon^4c_0^4C_2^2+\dots\\ -\varepsilon^3c_0^5C_2+\dots\end{pmatrix},$$

where we have used the shorthand notation $C_2 := c_2+c_1-c_0$ and $C_3 := c_3-c_2-c_1+c_0$, $K=b(4C_3+c_0-c_2)-x(6C_3+c_0)$.

We now observe that we can in any case cancel ε^2. But from the singularity confinement point of view the crucial observation is that *if* $C_3=0$, $C_2\neq 0$ then we can cancel ε^3 and then take the $\varepsilon\to 0$ limit and get (after also factoring out $-c_0^5$)

$$\begin{pmatrix}c_0(x-b)+c_2b\\ 0\\ c_3\end{pmatrix}.$$

Thus we have emerged from the singularity and in particular recovered the initial data x. The condition $C_3=0$ reads

$$c_{n+3}-c_{n+2}-c_{n+1}+c_n=0.$$

This is a third order linear map with the general solution

$$c_n=\alpha+\beta n+\gamma(-1)^n,$$

which yields an integrable map.

Later emergence from the singularity is also possible. If $C_3\neq 0$ then at the seventh step we find

$$\begin{pmatrix}\varepsilon^2C_0^{14}C_2^2C_3^6C_6+\varepsilon^3C_0^{13}C_2C_3^5(bM+xN+C_6(\dots))+\dots\\ \varepsilon^4C_0^{12}C_2^4C_3^4C_5^2+\dots\\ \varepsilon^3C_0^{13}C_2^3C_3^5C_5+\dots\end{pmatrix},$$

where M and N are some polynomials of C_n. If now $C_6 := c_6-c_5-c_4+c_3-c_2-c_1+c_0=0$ we can divide out ε^3 and exit from the singularity with the initial datum x, as long as $N=C_3(4C_2C_3-2C_0C_3-C_0C_2)\neq 0$.

The condition $C_6=0$ has the solution $c_n=\sum_{\nu=1}^{6}\alpha_\nu X_\nu^n$ where X_ν is a root of

$$X^6-X^5-X^4+X^3-X^2-X+1=0.$$

Note that 1 is not a root of this equation, furthermore, there is a root of modulus <1 and since this expression appears in the denominator of the generating function [**5**] we have exponential growth.

The conclusion here is that there are many ways to deautonomize an integrable autonomous system while preserving the singularity confinement property. However, a deautonomized map with a later confinement is probably chaotic. It seems that integrability is only associated with confinement at the first possibility, in any case no counterexamples are known for this conjecture.

4. Condition for the First Cancellation

We have shown above that the number of steps needed for singularity confinement can vary. In any case one needs some cancellations, and we will next derive the condition for the first cancellation to take place after N iteration steps.

We shall consider the map

$$x_{n+1} + x_{n-1} = a_n x_n + b_n + \frac{c_n}{x_n},$$

where x_n is the dynamical variable and a, b, c are some fixed functions of n, of which we assume that $c_n \neq 0$, $\forall n$. In projective coordinates the map reads

$$\begin{cases} u_{n+1} &= -v_n u_n + a_n u_n^2 + b_n u_n f_n + c_n f_n^2, \\ v_{n+1} &= u_n^2, \\ f_{n+1} &= u_n f_n. \end{cases} \tag{4.1}$$

When this map it iterated the degrees grow like 2^n, unless some cancellations take place. The iterations are supposed to start with $u_0 = x, v_0 = y, f_0 = z$ which are the indeterminates. The assumption that the first cancellation takes place at the N'th iteration means

$$\gcd(u_N, v_N, f_N) \neq 1, \text{ but } \gcd(u_k, v_k, f_k) = 1, \quad \forall k < N.$$

Let us first derive some lemmas that will be useful in the analysis.

LEMMA 4.1. $\gcd(u_k, f_k) = 1 \iff \gcd(u_{k+1}, v_{k+1}, f_{k+1}) = 1.$

PROOF. From (4.1) we find $\gcd(u_k, f_k)$ divides $\gcd(u_{k+1}, v_{k+1}, f_{k+1})$, from which follows "$\Leftarrow$". On the other hand, if $\gcd(u_k, f_k) = 1$ then $\gcd(v_{k+1}, f_{k+1}) = u_k$, but if $c_k \neq 0$ then $u_k \nmid u_{k+1}$ and therefore $\gcd(u_{k+1}, v_{k+1}, f_{k+1}) = 1$. □

Using this lemma we find that the first cancellation takes place at N if

$$\gcd(u_k, f_k) = 1, \quad \forall k < N-1, \qquad \gcd(u_{N-1}, f_{N-1}) \neq 1.$$

LEMMA 4.2.

$$\gcd(u_k, f_k) = 1 \Rightarrow \gcd(u_{k+1}, f_{k+1}) = \gcd(-v_k + a_k u_k, f_k).$$

PROOF. Recalling that if $\gcd(u, f) = 1$ and $\gcd(u, b) = 1$ then $\gcd(ux + bf, fu) = \gcd(x, f)$, it then follows that

$$\begin{aligned} \gcd(u_{k+1}, f_{k+1}) &= \gcd(u_k(-v_k + a_k u_k + b_k f_k) + a_k f_k^2, u_k f_k), \\ &= \gcd(-v_k + a_k u_k + b_k f_k, f_k) = \gcd(-v_k + a_k u_k, f_k). \quad \square \end{aligned}$$

This lemma again takes us one step forward so that we now find the condition for first cancellation at step N to be

$$\gcd(-v_k + a_k u_k, f_k) = 1, \quad \forall k < N-2, \qquad \gcd(-v_{N-2} + a_{N-2} u_{N-2}, f_{N-2}) \neq 1.$$

LEMMA 4.3. *If* $\gcd(u_k, f_k) = 1$ *then*

$$\gcd(\alpha_{k+1} v_{k+1} + \beta_{k+1} u_{k+1}, f_{k+1}) = \begin{cases} u_k, & \text{if } \beta_{k+1} = 0, \\ \gcd(\alpha_k v_k + \beta_k u_k, f_k) & \text{otherwise,} \end{cases}$$

where

$$\begin{cases} \alpha_{n-1} &= -\beta_n, \\ \beta_{n-1} &= \alpha_n + a_{n-1} \beta_n. \end{cases} \tag{4.2}$$

PROOF. If $\beta_{k+1} = 0$ then $\gcd(\alpha_{k+1}v_{k+1}, f_{k+1}) = \gcd(\alpha_{k+1}u_k^2, u_k f_k) = u_k$. If $\beta_{k+1} \neq 0$ then

$$\begin{aligned}
&\gcd(\alpha_{k+1}v_{k+1} + \beta_{k+1}u_{k+1}, f_{k+1}) \\
&\quad = \gcd(u_k(\alpha_{k+1}u_k + \beta_{k+1}(-v_k + a_k u_k + b_k f_k)) + \beta_{k+1}a_k f_k^2, f_k u_k), \\
&\quad = \gcd(-\beta_{k+1}v_k + (\alpha_{k+1} + \beta_{k+1}a_k)u_k, f_k). \quad \square
\end{aligned}$$

Collecting the results of the 3 lemmas we get

THEOREM 4.4. *First cancellation takes place after N steps iff in the map* (4.2) *the initial values $\beta_{N-2} = a_{N-2}$, $\alpha_{N-2} = -1$ imply $\beta_1 = 0$ and $\beta_k \neq 0$, $\forall 1 < k \leq N-2$.*

As the first result we find that we can have the first cancellation at $N = 3$ provided that $\beta_1 = a_1 = 0$. If this is to hold with any starting point n we must take $a_n \equiv 0$, this is what happens with the map (2).

Later cancellation is possible for other values of a_n. We find from (4.2) that $\beta_{N-3} = -1 + a_{N-3}a_{N-2}$ and thus first cancellation is at $N = 4$ if $a_{n+1}a_n = 1$. This happens for (3.1), which has $a_n = -1$. The condition for the first cancellation at the $N = 5$ step is $a_{n+1}a_n a_{n-1} = a_{n+1} + a_{n-1}$, one solution of which is $c = \sqrt{2}$.

If we only consider constant a_n then $a_n = 2$, $\forall n$ is special in that then $\alpha_n = -1$, $\beta_n = 1$, $\forall n$ and therefore there will be *no cancellations ever*. This may therefore be considered as a maximally nonintegrable map.

It should be noted that a cancellation of factors is necessary but not sufficient for singularity confinement. Thus this result can be used just to narrow down the set of maps that are candidates for singularity confinement.

5. Conclusions

The above discussions show that the singularity confinement question can be best studied by viewing the rational map as a polynomial map in projective space. It is then clear that:

- Some cancellations are possible even without singularity confinement.
- For singularity confinement one needs so many cancellations that the initial data can be recovered.
- For integrability one needs so many cancellations that the degree of the map grows only polynomially.

It would be interesting to find more examples of maps satisfying only some of the listed properties.

References

[1] M. P. Bellon and C.-M. Viallet, *Algebraic entropy*, Comm. Math. Phys. **204** (1999), 425–437, (chao-dyn/9805006).

[2] G. Falqui and C.-M. Viallet, *Singularity, complexity, and quasi-integrability of rational mappings*, Comm. Math. Phys. **154** (1993), 111–125.

[3] B. Grammaticos, A. Ramani, and V. Papageorgiou, *Do integrable mappings have the Painlevé property*?, Phys. Rev. Lett. **67** (1991), 1825–1828.

[4] J. Hietarinta and C. Viallet, *Singularity confinement and chaos in discrete systems*, Phys. Rev. Lett. **81** (1998), 325–328.

[5] ———, *Discrete Painlevé* I *and singularity confinement in projective space*, Chaos Solitons and Fractals **11** (2000), 29–32.

[6] A. Ramani, B. Grammaticos, and J. Hietarinta, *Discrete versions of the Painlevé equation*, Phys. Rev. Lett. **67** (1991), 1829–1831; B. Grammaticos, F. W. Nijhoff, and A. Ramani, *Discrete Painlevé equations*, The Painlevé Property, One Century Later, CRM Series in Mathematical Physics, Springer, Berlin, 1999, pp. 413–516.

DEPARTMENT OF PHYSICS, UNIVERSITY OF TURKU, FIN–20014 TURKU, FINLAND.
E-mail address: hietarin@utu.fi

UNIVERSITÉ PARIS VI, BOÎTE 126, 4 PLACE JUSSIEU, F–75252 PARIS CEDEX 05, FRANCE.
E-mail address: viallet@lpthe.jussieu.fr

Centre de Recherches Mathématiques
CRM Proceedings and Lecture Notes
Volume **25**, 2000

Time-Discretization of Soliton Equations

Ryogo Hirota and Masataka Iwao

ABSTRACT. A method of time-discretizing soliton equations is presented. The method is based on the bilinear formalism. The soliton equations are transformed into the bilinear forms through the dependent variable transformations. Time-discretization of the bilinear forms are easily performed using the gauge invariance of the bilinear forms but difficulties arise from transforming the bilinear equations into the nonlinear difference equations of ordinary form.

1. Introduction

In a series of papers [**2**]–[**7**] one of the authors (Hirota) has proposed a method of discretizing soliton equations. The method uses the bilinear formalism and follows three steps. First a given equation is transformed into the bilinear form by a dependent variable transformation. Secondly the bilinear equation is discretized with the help of the gauge-invariance of the equation. Thirdly the bilinear difference equation is transformed back into the nonlinear difference equation by the associated dependent variable transformation.

Method of Discretizing Soliton Equations.

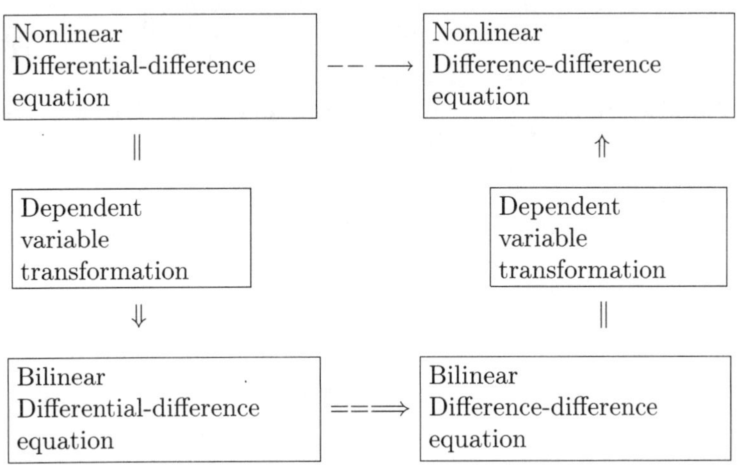

2000 *Mathematics Subject Classification.* Primary: 35Q51; Secondary: 35Q53.
This is the final form of the paper.

We shall discretize the following soliton equations in order to explain the method.

(1) Lotka-Volterra equation [**10**]

$$\frac{du_n}{dt} = u_n(u_{n-1} - u_{n+1}).$$

(2) Discrete-KdV equation (discrete in space

$$\frac{d\widehat{u}_n}{dt} = -(\widehat{u}_n)^2(\widehat{u}_{n-1} - \widehat{u}_{n+1}).$$

(3) Self-dual nonlinear network equation [**8**]

$$\frac{dv_n}{dt} = [1 + (v_n)^2](v_{n-1} - v_{n+1}).$$

(4) Hybrid equation

$$\frac{dw_n}{dt} = \left[1 + bw_n + c(w_n)^2\right](w_{n-1} - w_{n+1}).$$

2. Lotka-Volterra and KdV Equations

In this section we consider time-discretization of the Lotka-Volterra and the KdV equations of the following forms.

(1) Lotka-Volterra equation

$$\frac{du_n}{dt} = u_n(u_{n-1} - u_{n+1}). \tag{2.1}$$

(2) Discrete KdV equation

$$\frac{d\widehat{u}_n}{dt} = -(\widehat{u}_n)^2(\widehat{u}_{n-1} - \widehat{u}_{n+1}). \tag{2.2}$$

2.1. Transformation into the bilinear form. The Lotka-Volterra equation (2.1) and the discrete KdV equation (2.2) are transformed into the same bilinear form

$$D_t f_n \cdot f_{n+1} - f_{n-1}f_{n+2} + f_n f_{n+1} = 0, \tag{2.3}$$

through the dependent variable transformations.

(1) Lotka-Volterra equation

$$u_n = \frac{f_{n-1}f_{n+2}}{f_n f_{n+1}}. \tag{2.4}$$

(2) Discrete KdV equation

$$\widehat{u}_n = \frac{f_{n-1}f_{n+1}}{f_n^2}. \tag{2.5}$$

2.2. Miura transformation. We have the Miura transformation which maps solutions of the discrete KdV equation to solutions of the Lotka-Volterra equation.

The dependent variable transformations (2.4) and (2.5) give the following relation

$$
\begin{aligned}
u_n &= \frac{f_{n-1}f_{n+2}}{f_nf_{n+1}}, \\
&= \frac{f_{n-1}f_{n+1}}{f_nf_n}\frac{f_nf_{n+2}}{f_{n+1}f_{n+1}}, \\
&= \widehat{u}_n\widehat{u}_{n+1}.
\end{aligned} \tag{2.6}
$$

Hence we have

$$u_n = \widehat{u}_n\widehat{u}_{n+1}, \tag{2.7}$$

which is the Miura transformation.

2.3. Discretization of the bilinear equation. We shall discretize the bilinear equation (2.3) in this subsection.

Time-discretization of the bilinear equation follows three steps:

(1) Replace the bilinear differential operator

$$D_tf\cdot g \equiv \frac{\mathrm{d}f}{\mathrm{d}t}g - f\frac{\mathrm{d}g}{\mathrm{d}t}, \tag{2.8}$$

by the difference operator:

$$\Delta_tf\cdot g \equiv [\Delta_tf]g - f[\Delta_tg], \tag{2.9}$$

where Δ_t is the forward difference operator defined by

$$\Delta_tf = \big[f(t+\delta) - f(t)\big]/\delta, \tag{2.10}$$

where δ is the time-interval.

(2) Check the gauge invariance of the bilinear equation. It is known that the bilinear equation

$$F(\Delta_t, \Delta_n, \cdots)f\cdot f = 0, \tag{2.11}$$

is invariant under the gauge transformation

$$f \to f\exp(\omega t + pn), \tag{2.12}$$

where ω and p are constants.

(3) Check the soliton solution. We know by experience that the soliton solution of the discrete-time bilinear equation is the same as that of the continuous-time bilinear equation except the dispersion relation.

We discretize the bilinear form of the Lotka-Volterra equation (2.3) following the three steps.

STEP 1. We replace the bilinear differential operator $D_tf_n\cdot f_{n+1}$ by the difference operator $\Delta_tf_n\cdot f_{n+1}$.

$$
\begin{aligned}
&\Delta_tf_n\cdot f_{n+1} \\
&\quad= \delta^{-1}\big\{\big[f_n(t+\delta) - f_n(t)\big]f_{n+1}(t) - f_n(t)\big[f_{n+1}(t+\delta) - f_{n+1}(t)\big]\big\}, \\
&\quad= \delta^{-1}\big[f_n(t+\delta)f_{n+1}(t) - f_n(t)f_{n+1}(t+\delta)\big], \\
&\quad= \delta^{-1}[f_n^{m+1}f_{n+1}^m - f_n^mf_{n+1}^{m+1}],
\end{aligned} \tag{2.13}
$$

where $t = m\delta$, m being integers.

We obtain a time-discretized bilinear form

$$f_n^{m+1} f_{n+1}^m - f_n^m f_{n+1}^{m+1} - \delta[f_{n-1}^m f_{n+2}^m - f_n^m f_{n+1}^m] = 0. \tag{2.14}$$

STEP 2. Check the gauge invariance. Let us transform the dependent variable f_n^m as follow

$$f_n^m \to f_n^m \exp(pn + qm), \tag{2.15}$$

where p and q are constants. Then the bilinear form (2.14) becomes

$$\begin{aligned}[f_n^{m+1} f_{n+1}^m - f_n^m f_{n+1}^{m+1}] &\exp[2(pn + qm) + p + q] \\ &= -\delta[f_{n-1}^m f_{n+2}^m - f_n^m f_{n+1}^m] \exp[2(pn + qm) + p] = 0,\end{aligned} \tag{2.16}$$

which is different from Eq. (2.14). Accordingly Eq. (2.14) is not gauge-invariant. Gauge-invariant bilinear forms are given by

$$f_n^{m+1} f_{n+1}^m - f_n^m f_{n+1}^{m+1} - \delta[f_{n-1}^m f_{n+2}^{m+1} - f_n^m f_{n+1}^{m+1}] = 0, \tag{2.17}$$

or by

$$f_n^{m+1} f_{n+1}^m - f_n^m f_{n+1}^{m+1} - \delta[f_{n-1}^{m+1} f_{n+2}^m - f_n^{m+1} f_{n+1}^m] = 0. \tag{2.18}$$

The second bilinear equation can be transformed into the first by the coordinates transformation:

$$m \to -m, \tag{2.19}$$

$$\delta \to -\delta. \tag{2.20}$$

Hereafter we use the first bilinear equation (2.17) only.

STEP 3. Check the soliton solution. We know two-soliton solution to the bilinear form

$$D_t f_n \cdot f_{n+1} - f_{n-1} f_{n+2} + f_n f_{n+1} = 0, \tag{2.21}$$

is given by

$$f_n = 1 + \exp(\eta_1) + \exp(\eta_2) + a_{12} \exp(\eta_1 + \eta_2), \tag{2.22}$$

where

$$\exp(\eta_j) = c_j P_j^n \exp(q_j t), \tag{2.23}$$

$$q_j = P_j^{-1} - P_j, \quad \text{for} \quad j = 1, 2, \tag{2.24}$$

$$a_{12} = \frac{(P_1 - P_2)^2}{(P_1 P_2 - 1)^2}, \tag{2.25}$$

where c_j and P_j for $j = 1, 2$ are parameters related to the positions and amplitudes of solitons.

Two-soliton solution to the discrete bilinear equation (2.17)

$$f_n^{m+1} f_{n+1}^m - f_n^m f_{n+1}^{m+1} - \delta[f_{n-1}^m f_{n+2}^{m+1} - f_n^m f_{n+1}^{m+1}] = 0,$$

is obtained by a perturbational method. We find

$$f_n^m = 1 + \exp(\eta_1) + \exp(\eta_2) + a_{12} \exp(\eta_1 + \eta_2), \tag{2.26}$$

where

$$\exp(\eta_j) = c_j P_j^n Q_j^m, \tag{2.27}$$

$$Q_j = \frac{1+\delta P_j^{-1}}{1+\delta P_j}, \quad \text{for} \quad j = 1, 2, \tag{2.28}$$

$$a_{12} = \frac{(P_1 - P_2)^2}{(P_1 P_2 - 1)^2}. \tag{2.29}$$

Note that the 2-soliton solutions of the discrete-time equation is the same as that of the continuous-time equation except the dispersion relation.

The continuous-time dispersion relation is

$$q_j = P_j^{-1} - P_j,$$

while the discrete-time dispersion relation is

$$Q_j = \frac{1+\delta P_j^{-1}}{1+\delta P_j}.$$

2.4. Transformation into the ordinary form. We shall transform the discrete-time bilinear equation into the ordinary form using the dependent variable transformation.

We have obtained the discrete-time bilinear equation

$$f_n^{m+1} f_{n+1}^m - f_n^m f_{n+1}^{m+1} - \delta[f_{n-1}^m f_{n+2}^{m+1} - f_n^m f_{n+1}^{m+1}] = 0, \tag{2.30}$$

First we transform it back into the discrete-time Lotka-Volterra equation.

We use the dependent variable transformation

$$u_n^m = \frac{f_{n-1}^m f_{n+2}^{m+1}}{f_n^m f_{n+1}^{m+1}}, \tag{2.31}$$

which is a discrete-time analogue of the dependent variable transformation (2.4)

$$u_n = \frac{f_{n-1} f_{n+2}}{f_n f_{n+1}}.$$

Then dividing Eq. (2.30) by $f_n^m f_{n+1}^{m+1}$ we obtain

$$\frac{f_n^{m+1} f_{n+1}^m}{f_n^m f_{n+1}^{m+1}} - 1 + \delta - \delta = \frac{f_{n-1}^m f_{n+2}^{m+1}}{f_n^m f_{n+1}^{m+1}} = 0, \tag{2.32}$$

which is transformed by using Eq. (2.31) into

$$1 - \delta + \delta = u_n^m = \frac{f_n^{m+1} f_{n+1}^m}{f_n^m f_{n+1}^{m+1}}. \tag{2.33}$$

On the other hand, Eq. (2.31) gives

$$\frac{u_n^{m+1}}{u_n^m} = \frac{f_{n-1}^{m+1} f_{n+2}^{m+2}}{f_n^{m+1} f_{n+1}^{m+2}} \cdot \frac{f_n^m f_{n+1}^{m+1}}{f_{n-1}^m f_{n+2}^{m+1}}, \tag{2.34}$$

$$= \frac{f_{n-1}^{m+1} f_n^m}{f_{n-1}^m f_n^{m+1}} \cdot \frac{f_{n+1}^{m+1} f_{n+2}^{m+2}}{f_{n+1}^{m+2} f_{n+2}^{m+1}},$$

which is expressed by using Eq. (2.33) as

$$= \frac{1-\delta+\delta u_{n-1}^m}{1-\delta+\delta u_{n+1}^{m+1}},$$

which is rewritten as

$$u_n^{m+1} - u_n^m = \hat{\delta}(u_n^m u_{n-1}^m - u_n^{m+1} u_{n+1}^{m+1}), \tag{2.35}$$

where

$$\hat{\delta} = \frac{\delta}{1-\delta},$$

which is the discrete-time Lotka-Volterra equation.

Next we tranform the discrete-time bilinear equation into the discrete-time KdV equation.

Using the dependent variable transformation

$$\widehat{u}_n^m = \frac{f_{n-1}^m f_{n+1}^{m+1}}{f_n^m f_n^{m+1}}, \tag{2.36}$$

which is a discrete-time analogue of the dependent variable transformation (2.5)

$$\widehat{u}_n = \frac{f_{n-1} f_{n+1}}{f_n^2},$$

we have

$$\begin{aligned} 1 - \delta \widehat{u}_n^m \widehat{u}_{n+1}^m &= 1 - \delta \frac{f_{n-1}^m f_{n+1}^{m+1}}{f_n^m f_n^{m+1}} \cdot \frac{f_n^m f_{n+2}^{m+1}}{f_{n+1}^m f_{n+1}^{m+1}}, \\ &= 1 - \delta \frac{f_{n-1}^m f_{n+2}^{m+1}}{f_n^{m+1} f_{n+1}^m}, \end{aligned} \tag{2.37}$$

which is transformed by using the discrete-time bilinear equation (2.30)

$$f_n^{m+1} f_{n+1}^m - f_n^m f_{n+1}^{m+1} - \delta[f_{n-1}^m f_{n+2}^{m+1} - f_n^m f_{n+1}^{m+1}] = 0,$$

into

$$1 - \delta \widehat{u}_n^m \widehat{u}_{n+1}^m = \frac{(1-\delta) f_n^m f_{n+1}^{m+1}}{f_n^{m+1} f_{n+1}^m}. \tag{2.38}$$

On the other hand we have

$$\begin{aligned} \frac{\widehat{u}_n^{m+1}}{\widehat{u}_n^m} &= \frac{f_{n-1}^{m+1} f_{n+1}^{m+2}}{f_n^{m+1} f_n^{m+2}} \cdot \frac{f_n^m f_n^{m+1}}{f_{n-1}^m f_{n+1}^{m+1}}, \\ &= \frac{f_{n-1}^{m+1} f_n^m}{f_{n-1}^m f_n^{m+1}} \cdot \frac{f_n^{m+1} f_{n+1}^{m+2}}{f_n^{m+2} f_{n+1}^{m+1}}. \end{aligned} \tag{2.39}$$

By using Eq. (2.38) we obtain

$$\frac{\widehat{u}_n^{m+1}}{\widehat{u}_n^m} = \frac{1 - \delta \widehat{u}_n^{m+1} \widehat{u}_{n+1}^{m+1}}{1 - \delta \widehat{u}_{n-1}^m \widehat{u}_n^m}, \tag{2.40}$$

which is rewritten as

$$\frac{1 - \delta \widehat{u}_n^{m+1} \widehat{u}_{n+1}^{m+1}}{\widehat{u}_n^{m+1}} = \frac{1 - \delta \widehat{u}_{n-1}^m \widehat{u}_n^m}{\widehat{u}_n^m}, \tag{2.41}$$

or

$$\frac{1}{\widehat{u}_n^{m+1}} - \frac{1}{\widehat{u}_n^m} = \delta[\widehat{u}_{n+1}^{m+1} - \widehat{u}_{n-1}^m], \tag{2.42}$$

which is the discrete-time KdV equation.

2.5. Miura transformation. We have the Miura transformation

$$u_n = \widehat{u}_n \widehat{u}_{n+1},$$

which maps solutions of the discrete KdV equation to solutions of the Lotka-Volterra equation.

Solutions to the discrete-time KdV equation and to the discrete Lotka-Volterra equation are expressed by f_n^m

$$\widehat{u}_n^m = \frac{f_{n-1}^m f_{n+1}^{m+1}}{f_n^m f_n^{m+1}}, \tag{2.43}$$

and

$$u_n^m = \frac{f_{n-1}^m f_{n+2}^{m+1}}{f_n^m f_{n+1}^{m+1}}.$$

We rewrite the expression of u_n^m as follows

$$u_n^m = \frac{(f_{n-1}^m f_{n+1}^{m+1})/(f_n^m f_n^{m+1})}{(f_n^m f_{n+2}^{m+1})/(f_{n+1}^m f_{n+1}^{m+1})} = \frac{f_{n+1}^{m+1} f_n^m}{f_n^{m+1} f_{n+1}^m}, \tag{2.44}$$

which is expressed, by using Eqs. (2.38) and (2.43), as

$$= (1-\delta)\frac{\widehat{u}_n^m \widehat{u}_{n+1}^m}{1 - \delta \widehat{u}_n^m \widehat{u}_{n+1}^m}. \tag{2.45}$$

Hence the Miura transformation is given by

$$u_n^m = (1-\delta)\frac{\widehat{u}_n^m \widehat{u}_{n+1}^m}{1 - \delta \widehat{u}_n^m \widehat{u}_{n+1}^m}. \tag{2.46}$$

3. The Self-Dual Nonlinear Network Equation

Next we consider the self-dual nonlinear network equation

$$\frac{dv_n}{dt} = \left[1 + (v_n)^2\right](v_{n-1} - v_{n+1}). \tag{3.1}$$

Let

$$v_n = \frac{d\phi_n}{dt}. \tag{3.2}$$

Then Eq. (3.1) is expressed as

$$\begin{aligned} \frac{d\tan^{-1} v_n}{dt} &= v_{n-1} - v_{n+1}, \\ &= \frac{d}{dt}[\phi_{n-1} - \phi_{n+1}], \end{aligned} \tag{3.3}$$

which is integrated with respect to time to give

$$\tan^{-1} v_n = \phi_{n-1} - \phi_{n+1}, \tag{3.4}$$

which is equal to

$$\frac{d\phi_n}{dt} = \tan(\phi_{n-1} - \phi_{n+1}), \tag{3.5}$$

which we call a potential form of the self-dual nonlinear network equation.

3.1. Transformation into the bilinear form. We use the following dependent variable transformation

$$\tan\phi_n = \frac{g_n}{f_n}, \tag{3.6}$$

in order to transform the potential form of the self-dual nonlinear network equation into the bilinear form. We have

$$\frac{\mathrm{d}\phi_n}{\mathrm{d}t} = \frac{D_t g_n \cdot f_n}{f_n^2 + g_n^2}, \tag{3.7}$$

$$\tan(\phi_{n-1} - \phi_{n+1}) = \frac{\tan\phi_{n-1} - \tan\phi_{n+1}}{1 + \tan\phi_{n-1}\tan}, \tag{3.8}$$

$$\phi_{n+1} = \frac{g_{n-1}f_{n+1} - g_{n+1}f_{n-1}}{f_{n+1}f_{n-1} + g_{n+1}g_{n-1}}. \tag{3.9}$$

Hence we have

$$\frac{D_t g_n \cdot f_n}{f_n^2 + = g_n^2} = \frac{g_{n-1}f_{n+1} - g_{n+1}f_{n-1}}{f_{n+1}f_{n-1} + g_{n+1}g_{n-1}}, \tag{3.10}$$

which is decoupled to a set of bilinear equations

$$\begin{cases} D_t g_n \cdot f_n = g_{n-1}f_{n+1} - g_{n+1}f_{n-1}, \\ f_{n+1}f_{n-1} + g_{n+1}g_{n-1} = f_n^2 + g_n^2. \end{cases} \tag{3.11}$$

3.2. Discretization of the bilinear equation. Following the same steps 1 and 2 described in the previous section we discretize the first equation:

$$D_t g_n \cdot f_n = g_{n-1}f_{n+1} - g_{n+1}f_{n-1},$$

as

$$\delta^{-1}[g_n^{m+1}f_n^m - g_n^m f_n^{m+1}] = g_{n-1}^m f_{n+1}^{m+1} - g_{n+1}^{m+1} f_{n-1}^m, \tag{3.12}$$

or

$$\delta^{-1}[g_n^{m+1}f_n^m - g_n^m f_n^{m+1}] = g_{n-1}^{m+1} f_{n+1}^m - g_{n+1}^m f_{n-1}^{m+1}, \tag{3.13}$$

which are transformed into each other by the coordinates transformation. We choose the first equation hereafter.

There are several discrete forms of the second equation. We choose the following forms out of them

$$f_{n+1}^m f_{n-1}^m + g_{n+1}^m g_{n-1}^m = f_n^m f_n^m + g_n^m g_n^m, \tag{3.14}$$

or

$$f_{n+1}^{m+1} f_{n-1}^m + g_{n+1}^{m+1} g_{n-1}^m = f_n^{m+1} f_n^m + g_n^{m+1} g_n^m. \tag{3.15}$$

The third step is to check the soliton solution. We have a set of bilinear equations

$$\begin{cases} D_t g_n \cdot f_n = g_{n-1}f_{n+1} - g_{n+1}f_{n-1}, \\ f_{n+1}f_{n-1} + g_{n+1}g_{n-1} = f_n^2 + g_n^2, \end{cases} \tag{3.16}$$

whose two-soliton solution is given by

$$f_n = 1 + a_{12}\exp(\eta_1 + \eta_2), \tag{3.17}$$

$$g_n = \exp(\eta_1) + \exp(\eta_2), \tag{3.18}$$

where

$$\exp(\eta_j) = c_j P_j^n \exp(q_j t), \tag{3.19}$$

$$q_j = P_j^{-1} - P_j, \quad \text{for} \quad j = 1, 2, \tag{3.20}$$

$$a_{12} = -\frac{(P_1 - P_2)^2}{(P_1 P_2 - 1)^2}, \tag{3.21}$$

where c_j and P_j for $j = 1, 2$ are parameters related to the positions and amplitudes of solitons.

The discrete bilinear equations

$$\begin{cases} \delta^{-1}[g_n^{m+1} f_n^m - g_n^m f_n^{m+1}] = g_{n-1}^m f_{n+1}^{m+1} - g_{n+1}^{m+1} f_{n-1}^m, \\ f_{n+1}^m f_{n-1}^m + g_{n+1}^m g_{n-1}^m = f_n^m f_n^m + g_n^m g_n^m, \end{cases} \tag{3.22}$$

and

$$\begin{cases} \delta^{-1}[g_n^{m+1} f_n^m - g_n^m f_n^{m+1}] = g_{n-1}^m f_{n+1}^{m+1} - g_{n+1}^{m+1} f_{n-1}^m, \\ f_{n+1}^{m+1} f_{n-1}^m + g_{n+1}^{m+1} g_{n-1}^m = f_n^{m+1} f_n^m + g_n^{m+1} g_n^m, \end{cases} \tag{3.23}$$

exhibit the same two-soliton solution which is expressed as follows

$$f_n^m = 1 + a_{12} \exp(\eta_1 + \eta_2), \tag{3.24}$$

$$g_n^m = \exp(\eta_1) + \exp(\eta_2), \tag{3.25}$$

where

$$\exp(\eta_j) = c_j P_j^n Q_j^m, \tag{3.26}$$

$$Q_j = \frac{1 + \delta P_j^{-1}}{1 + \delta P_j}, \quad \text{for} \quad j = 1, 2, \tag{3.27}$$

$$a_{12} = -\frac{(P_1 - P_2)^2}{(P_1 P_2 - 1)^2}. \tag{3.28}$$

3.3. Transformation into the ordinary form. We have two sets of bilinear equations:

$$\begin{cases} \delta^{-1}[g_n^{m+1} f_n^m - g_n^m f_n^{m+1}] = g_{n-1}^m f_{n+1}^{m+1} - g_{n+1}^{m+1} f_{n-1}^m, \\ f_{n+1}^m f_{n-1}^m + g_{n+1}^m g_{n-1}^m = f_n^m f_n^m + g_n^m g_n^m \end{cases}$$

and

$$\begin{cases} \delta^{-1}[g_n^{m+1} f_n^m - g_n^m f_n^{m+1}] = g_{n-1}^m f_{n+1}^{m+1} - g_{n+1}^{m+1} f_{n-1}^m, \\ f_{n+1}^{m+1} f_{n-1}^m + g_{n+1}^{m+1} g_{n-1}^m = f_n^{m+1} f_n^m + g_n^{m+1} g_n^m. \end{cases}$$

We consider the second set of bilinear equations first.

A ratio of the second set of bilinear equations becomes

$$\delta^{-1} \frac{g_n^{m+1} f_n^m - g_n^m = f_n^{m+1}}{f_n^{m+1} f_n^m + g_n^{m+1} g_n^m} = \frac{g_{n-1}^m f_{n+1}^{m+1} - g_{n+1}^{m+1} f_{n-1}^m}{f_{n+1}^{m+1} f_{n-1}^m + g_{n+1}^{m+1} g_{n-1}^m},$$

which is equal to

$$\delta^{-1} \frac{(g_n^{m+1})/(f_n^{m+1}) - (g_n^m)/(f_n^m)}{1 + (g_n^{m+1})/(f_n^{m+1})(g_n^m)/(f_n^m)} \frac{(g_{n-1}^m)/(f_{n-1}^m) - (g_{n+1}^{m+1})/(f_{n+1}^{m+1})}{1 + (g_{n+1}^{m+1})/(f_{n+1}^{m+1})(g_{n-1}^m)/(f_{n-1}^m)}. \tag{3.29}$$

Using the dependent variable transformation

$$\tan \phi_n^m = \frac{g_n^m}{f_n^m}, \tag{3.30}$$

Eq. (3.29) is transformed back into a discrete equation of ϕ_n^m

$$\tan(\phi_n^{m+1} - \phi_n^m) = \delta \tan(\phi_{n-1}^m - \phi_{n+1}^{m+1}), \tag{3.31}$$

which is a time-discretization of the potential form of the self-dual network equation[9].

It is very difficult to transform the first set of bilinear equations

$$\begin{cases} \delta^{-1}[g_n^{m+1} f_n^m - g_n^m f_n^{m+1}] = g_{n-1}^m f_{n+1}^{m+1} - g_{n+1}^{m+1} f_{n-1}^m, \\ f_{n+1}^m f_{n-1}^m + g_{n+1}^m g_{n-1}^m = f_n^m f_n^m + g_n^m g_n^m, \end{cases} \tag{3.32}$$

into the ordinary form.

We have not succeeded it yet to transform it into the ordinary form.

4. Hybrid Equation

We have discretized the following three equations:

(1) Lotka-Volterra equation

$$\frac{du_n}{dt} = u_n(u_{n-1} - u_{n+1}).$$

(2) Discrete-KdV equation

$$\frac{d\widehat{u}_n}{dt} = -(\widehat{u}_n)^2 = (\widehat{u}_{n-1} - \widehat{u}_{n+1}).$$

(3) Self-dual nonlinear network equation

$$\frac{dv_n}{dt} = \left[1 + (v_n)^2\right](v_{n-1} - v_{n+1}).$$

We consider a hybrid equation [1]

$$\frac{dw_n}{dt} = \left[1 + bw_n + c(w_n)^2\right](w_{n-1} - w_{n+1}), \tag{4.1}$$

which includes these three equations as special cases:

(1) Lotka-Volterra equation, $u_n = 1 + w_n$, $b = 1$, $c = 0$.
(2) Discrete KdV equation, $\widehat{u}_n = 1 + w_n$, $b = 2$, $c = 1$ and $t \to -t$.
(3) Self-dual nonlinear network equation, $v_n = w_n$, $b = 0$, $c = 1$.

4.1. Transformation into the bilinear form. We use the dependent variable transformation

$$w_n = \frac{G_n}{F_n}. \tag{4.2}$$

Then Eq. (4.1) becomes

$$\frac{D_t G_n \cdot = F_n}{F_n^2} = \left(1 + b\frac{G_n}{F_n} + c\frac{G_n^2}{F_n^2}\right)\left(\frac{G_{n-1}}{F_{n-1}} - \frac{G_{n+1}}{F_{n+1}}\right), \tag{4.3}$$

or

$$\frac{D_t G_n \cdot F_n}{F_n^2} = \frac{(F_n^2 + bG_nF_n + cG_n^2 = F_n^2)}{F_n^2} \frac{(G_{n-1}F_{n+1} - G_{n+1}F_{n-1})}{F_{n+1}F_{n-1}}, \tag{4.4}$$

which is decoupled to a set of bilinear equations

$$\begin{cases} D_t G_n \cdot F_n = G_{n-1}F_{n+1} - G_{n+1}F_{n-1}, \\ F_{n-1}F_{n+1} = F_n^2 + bG_nF_n + cG_n^2. \end{cases} \tag{4.5}$$

2-soliton solution is expressed as

$$\begin{aligned}F_n &= 1 + a_{10}\exp(\eta_1) + a_{01}\exp(\eta_2),\\ &= +a_{20}\exp(2\eta_1) + a_{11}\exp(\eta_1+\eta_2) + a_{02}\exp(2\eta_2)\\ &\quad + a_{21}\exp(2\eta_1+\eta_2) + a_{12}\exp(\eta_1+2\eta_2) + a_{22}\exp(2\eta_1+2\eta_2),\\ G_n &= \exp(\eta_1) + \exp(\eta_2) + b_{11}\exp(\eta_1+\eta_2)\\ &\quad + b_{21}\exp(2\eta_1+\eta_2) + b_{12}\exp(\eta_1+2\eta_2),\end{aligned} \tag{4.6}$$

where

$$\exp(\eta_j) = c_j P_j^n \exp(q_j t),$$

for $j = 1, 2$, and the coefficients $a_{j,k}$ and $b_{j,k}$ for $j, k = 0, 1, 2$ are constants.

4.2. Discretization of the bilinear equation. We obtain the following two sets of discrete-time bilinear equations following the the same steps 1 and 2 as before

$$\begin{cases} G_n^{m+1}F_n^m - G_n^m F_n^{m+1} = \delta(G_{n-1}^m F_{n+1}^{m+1} - G_{n+1}^{m+1}F_{n-1}^m),\\ F_{n-1}^m F_{n+1}^m = F_n^m F_n^m + bG_n^m F_n^m + cG_n^m G_n^m,\end{cases} \tag{4.7}$$

and

$$\begin{cases} G_n^{m+1}F_n^m - G_n^m F_n^{m+1} = \delta(G_{n-1}^m F_{n+1}^{m+1} - G_{n+1}^{m+1}F_{n-1}^m),\\ F_{n-1}^{m+1} F_{n+1}^m = F_n^m F_n^{m+1} + bG_n^{m+1} F_n^m + cG_n^m = G_n^{m+1}.\end{cases} \tag{4.8}$$

The third step is to check the soliton solution. We find that the second set of discrete-time bilinear equations

$$\begin{cases} G_n^{m+1}F_n^m - G_n^m F_n^{m+1} = \delta(G_{n-1}^m F_{n+1}^{m+1} - G_{n+1}^{m+1}F_{n-1}^m),\\ F_{n-1}^{m+1} F_{n+1}^m = F_n^m F_n^{m+1} + bG_n^{m+1} F_n^m + cG_n^m G_n^{m+1},\end{cases} \tag{4.9}$$

does not exhibits two-soliton solution. Accordingly we choose the first set.

The first set of bilinear equations

$$\begin{cases} G_n^{m+1}F_n^m - G_n^m F_n^{m+1} = \delta(G_{n-1}^m F_{n+1}^{m+1} - G_{n+1}^{m+1}F_{n-1}^m),\\ F_{n-1}^m F_{n+1}^m = F_n^m F_n^m + bG_n^m F_n^m + cG_n^m G_n^m,\end{cases} \tag{4.10}$$

is shown to have two-soliton solution which has the same form as that of the continuous-time bilinear equation except the dispersion relation.

4.3. Transformation into the ordinary form. We transform the first set into the ordinary form by introducing an auxiliary variable Γ_n^m.

Let

$$w_n^m = \frac{G^m}{F^m}. \tag{4.11}$$

Then the first bilinear equation

$$G_n^{m+1}F_n^m - G_n^m F_n^{m+1} = \delta(G_{n-1}^m F_{n+1}^{m+1} - G_{n+1}^{m+1}F_{n-1}^m), \tag{4.12}$$

gives

$$\begin{aligned}\frac{G_n^{m+1}}{F_n^{m+1}} - \frac{G_n^m}{F_n^m} &= \delta\frac{F_{n-1}^m F_{n+1}^{m+1}}{F_n^m F_n^{m+1}} = \left(\frac{G_{n-1}^m}{F_{n-1}^m} - \frac{G_{n+1}^{m+1}}{F_{n+1}^{m+1}}\right),\\ &= \delta\Gamma_n^m = \left(\frac{G_{n-1}^m}{F_{n-1}^m} - \frac{G_{n+1}^{m+1}}{F_{n+1}^{m+1}}\right),\end{aligned} \tag{4.13}$$

where

$$\Gamma_n^m \equiv= \frac{F_{n-1}^m F_{n+1}^{m+1}}{F_n^m F_n^{m+1}}. \tag{4.14}$$

The second bilinear equation

$$F_{n-1}^m F_{n+1}^m = F_n^m F_n^m + bG_n^m F_n^m + cG_n^m G_n^m, \tag{4.15}$$

gives

$$\frac{F_{n-1}^m F_{n+1}^m}{F_n^m F_n^m} = 1 + bw_n^m + c(w_n^m)^2. \tag{4.16}$$

Hence we have

$$\begin{aligned} \frac{\Gamma_n^m}{\Gamma_{n+1}^m} &= \frac{F_{n-1}^m F_{n+1}^{m+1}}{F_n^m F_n^{m+1}} \frac{F_{n+1}^m F_{n+1}^{m+1}}{F_n^m F_{n+2}^{m+1}}, \\ &= \frac{F_{n-1}^m F_{n+1}^m}{F_n^m F_n^m} \frac{F_{n+1}^{m+1} F_{n+1}^{m+1}}{F_n^{m+1} F_{n+2}^{m+1}}, \\ &= \frac{1 + bw_n^m + c(w_n^m)^2}{1 + bw_{n+1}^{m+1} + c(w_{n+1}^{m+1})^2}. \end{aligned} \tag{4.17}$$

Accordingly the first set of bilinear equations is transformed into the ordinary form

$$\begin{cases} w_n^{m+1} - w_n^m = \delta\Gamma_n^m (w_{n-1}^m - w_{n+1}^{m+1}), \\ \Gamma_n^m = \frac{1+bw_n^m+c(w_n^m)^2}{1+bw_{n+1}^{m+1}+c(w_{n+1}^{m+1})^2} \Gamma_{n+1}^m. \end{cases} \tag{4.18}$$

Finally we remark that there are some bilinear equations which can be transformed into discrete forms without destroying the integrability but it is very difficult to transform them back into the ordinary forms.

Ohta have found that a double sine-Gordon equation

$$\frac{\partial^2 u}{\partial x \partial y} = \sin u + a^2 \left(\int^x \frac{\partial}{\partial y} \sin u dx \right) \cos u, \tag{4.19}$$

is transformed into the following set of bilinear equations

$$\begin{cases} D_x D_y f_n \cdot f_n = 2(f_n f_n - f_{n+1} f_{n-1}), \\ D_x D_y g_n \cdot g_n = 2(g_n g_n - g_{n+1} g_{n-1}), \\ (aD_y + 1) f_n \cdot g_n - f_{n+1} g_{n-1} = 0, \\ (aD_y + 1) g_n \cdot f_n - g_{n+1} f_{n-1} = 0, \end{cases} \tag{4.20}$$

through the dependent variable transformation

$$u = \frac{\mathrm{i}}{2} \log \frac{g_n f_{n+1}}{f_n g_{n+1}}, \tag{4.21}$$

which exhibits N-soliton solution expressed by Wronskian determinants (Y. Ohta, Annual Meeting of Physical Society of Japan, March 30, 1992).

It is easy to discretize the bilinear equations. However we have not succeeded it yet to transform them into the ordinary form(s).

5. Summary

A method of time-discretizing soliton equations based on the bilinear formalism is presented. It is easy to discretize the bilinear forms. Difficulties arise from transforming the bilinear forms into the ordinary forms.

References

1. A. S. Cârstea, *On the dynamics of rational solutions for the 1D Volterra system*, Phys. Lett. A **233** (1997), 378–382.
2. Ryogo Hirota, *Nonlinear partial difference equations.* I. *A difference analogue of the Korteweg-de Vries Equations*, J. Phys. Soc. Japan **43** (1977), 1424–1433.
3. ______, *Nonlinear partial difference equations.* II. *Discrete-time Toda equation*, J. Phys. Soc. Japan **43** (1977), 2074–2078.
4. ______, *Nonlinear partial difference equations.* III. *Discrete sine-Gordon equation*, J. Phys. Soc. Japan **43** (1977), 2079–2086.
5. *'Nonlinear partial difference equations.* IV. *Bäcklund transformation for the discrete-time Toda equation*, J. Phys. Soc. Japan **45** (1978), 321–332.
6. ______, *Nonlinear partial difference equations.* V. *Nonlinear equations reducible to linear equations*, J. Phys. Soc. Japan **46** (1979), 312–319.
7. ______, *Discrete analogue of a generalized Toda equation*, J. Phys. Soc. Japan **50** (1981), 3785–3791.
8. ______, *Exact N-soliton solution of nonlinear lumped self-dual network equations*, J. Phys. Soc. Japan **35** (1973), 289–294.
9. ______, *Discretization of the potential modified KdV equation*, J. Phys. Soc. Japan **67** (1998), 2234–2236.
10. Ryogo Hirota and Junkichi Satsuma, *N-soliton solutions of nonlinear network equations describing a Volterra system*, J. Phys. Soc. Japan **40** (1976), 891–900.

Department of Information and Computer Science, School of Science and Engineering, Waseda University,Tokyo 169, Japan.

E-mail address, Ryogo Hirota: roy@hirota.info.waseda.ac.jp

E-mail address, Masataka Iwao: iwao@hirota.info.waseda.ac.jp

Centre de Recherches Mathématiques
CRM Proceedings and Lecture Notes
Volume **25**, 2000

Bäcklund Transformations for the Hénon–Heiles and Garnier Systems

A. N. W. Hone, V. B. Kuznetsov, and O. Ragnisco

ABSTRACT. The Hénon–Heiles (case (ii)) and Garnier systems may both be considered as reductions of the KdV hierarchy. Using this connection we construct Bäcklund transformations for these systems as suitable restrictions of the standard Darboux-Crum transformation for Schrödinger operators.

1. Introduction

Bäcklund transformations (BTs) are an important aspect of the theory of integrable systems which have traditionally been studied in the context of evolution equations. However, more recently there has been much interest in discrete systems or integrable mappings [**1, 13**]. Within the modern approach to separation of variables (reviewed by Sklyanin in [**12**]) this has led to the study of BTs for finite-dimensional Hamiltonian systems [**9**]. The latter are canonical transformations including a Bäcklund parameter λ, and apart from being interesting integrable mappings in their own right they also lead to separation of variables when n such mappings are applied to an integrable system with n degrees of freedom. The sequence of Bäcklund parameters λ_j together with a set of conjugate variables μ_j constitute the separation variables, and satisfy a new property called *spectrality* introduced in [**9**]. We proceed to illustrate these ideas with two new examples of BTs for particle systems, namely the case (ii) integrable Hénon–Heiles system and the Garnier system. The former has two degrees of freedom, while the latter is an n-body system, but both arise as suitable reductions of the KdV hierarchy. Thus the BTs we derive are obtained as a reduction of the standard BT for KdV, which arises from the Darboux-Crum transformation [**2**] for Schrödinger operators.

As is well known, the Crum transformation [**2**] consists of mapping the operator $\partial_t^2+Q-\lambda$ to another operator $\partial_t^2+\widetilde{Q}-\lambda$ by factorizing the former and then reversing the order of factorization. Given an eigenfunction ϕ satisfying

$$(\partial_t^2 + Q - \lambda)\phi = 0$$

2000 *Mathematics Subject Classification.* Primary: 70-xx; Secondary: 39-xx.

ANWH thanks the Leverhulme Trust for providing a Study Abroad Studentship in Rome, and is grateful to J. Harnad and Y. Suris for useful conversations. VBK's visit to Rome was supported by the Istituto Nazionale di Fisica Nucleare.

This is the final form of the paper.

we may set $y = \big(\log[\phi]\big)_t$ and then

$$Q = -y_t - y^2 + \lambda, \quad \widetilde{Q} = y_t - y^2 + \lambda; \tag{1.1}$$

for $\lambda = 0$ this is just the Miura map for KdV. Also given another eigenfunction ψ of the Schrödinger operator with potential Q for a different spectral parameter u we have

$$(\partial_t^2 + Q - u)\psi = 0, \quad (\partial_t^2 + \widetilde{Q} - u)\widetilde{\psi} = 0,$$

where the transformation to the new eigenfunction $\widetilde{\psi}$ and its derivative may be given in matrix form as

$$\begin{pmatrix} \widetilde{\psi} \\ \widetilde{\psi_t} \end{pmatrix} = k \begin{pmatrix} -y & 1 \\ y^2 + u - \lambda & -y \end{pmatrix} \begin{pmatrix} \psi \\ \psi_t \end{pmatrix}, \tag{1.2}$$

for any constant k.

We are now ready to discuss in general how this transformation restricts to the finite-dimensional Hamiltonian systems expressed in terms of canonically conjugate variables (q_j, p_j). The Hénon–Heiles system arises as the stationary flow of fifth-order KdV [**4**], while the Garnier system is a restricted flow of the KdV hierarchy [**15**]. In both cases the potential of the Schrödinger operator (the KdV dependent variable) becomes a function of the coordinates $Q(q_j)$ and from this alone it is possible to deduce a canonical transformation (integrable mapping) to new variables $(\widetilde{q_j}, \widetilde{p_j})$ so that $\widetilde{Q} = Q(\widetilde{q_j})$ and the Hamiltonians are preserved. Both systems have a Lax representation $L_t = [N, L]$ where

$$N = \begin{pmatrix} 0 & 1 \\ u - Q(q_j) & 0 \end{pmatrix}, \quad L = \begin{pmatrix} -\Pi_t & 2\Pi \\ -\Pi_{tt} + 2(u - Q)\Pi & \Pi_t \end{pmatrix}; \tag{1.3}$$

$\Pi = \Pi(q_j, p_j; u)$ is a suitable rational function of u which depends on which particular system we are considering. We shall present the appropriate Π in each case below. Of course the derivatives of Π can be rewritten using the equations of motion.

There are other common properties of these two systems (these properties should be shared by any restricted or higher stationary flow of KdV). Note that by adding the two equations (1.1) we may express y purely as a function of the coordinates related by the BT, viz.

$$y(q_j, \widetilde{q_j}) = \pm\sqrt{\lambda - \frac{1}{2}(Q + \widetilde{Q})}. \tag{1.4}$$

This turns out to be extremely convenient when we come to write the generating function $F(q_j, \widetilde{q_j}) = \sum_j p_j dq_j - \widetilde{p_j}\widetilde{q_j}$ of the BT, for in both cases y appears in a natural way; we present the appropriate generating functions below. Also, denoting by M the matrix that appears in (1.2), the transformation of the Lax matrix $L = L(q_j, p_j; u)$ under the BT is given by

$$ML = \tilde{L}M,$$

where $\tilde{L} = L(\widetilde{q_j}, \widetilde{p_j}; u)$. Given this identity the invariance of the Hamiltonians under the BT follows from the preservation of the spectral curve $\det\big(v - L(u)\big) = 0$, since the Hamiltonians are just the coefficients of the latter equation.

Finally if we write the (hyperelliptic) spectral curve as

$$v^2 = R(u)$$

then the spectrality property is satisfied for these systems, in the sense that if we define the conjugate variable to λ by

$$\mu = \frac{\partial F}{\partial \lambda},$$

then we find that (up to a constant of proportionality) μ is an eigenvalue of $L(\lambda)$ with eigenvector $(1, y)^T$, or in other words $\mu = \sqrt{R(\lambda)}$ so that (λ, μ) is a point on the spectral curve. We can also write y explicitly in terms of both the old and the new variables related by the BT, thus:

$$(1.5) \qquad y(q_j, p_j) = \frac{\Pi_t(\lambda) + \mu}{2\Pi(\lambda)}, \quad y(\widetilde{q}_j, \widetilde{p}_j) = -\frac{(\widetilde{\Pi}_t(\lambda) - \mu)}{2\widetilde{\Pi}(\lambda)};$$

clearly we denote $\widetilde{\Pi}(\lambda) = \Pi(\widetilde{q}_j, \widetilde{p}_j; \lambda)$.

2. The Hénon–Heiles System

The integrable case (ii) of the Hénon–Heiles system has Hamiltonian

$$h = \frac{1}{2}(p_1^2 + p_2^2) + \frac{1}{2}q_1 q_2^2 + q_1^3 - \frac{1}{2}\ell^2 q_2^{-2} - \frac{1}{2}c q_1;$$

actually in the above we have included two Casimirs ℓ^2 and c, so that the above system in fact corresponds to the reduction of fifth-order KdV to travelling waves with speed c (the stationary flow has $c = 0$). This is almost the same as one of the nonautonomous Hamiltonians considered in [**5, 6**]; in the nonautonomous case the Casimir ℓ (there denoted λ) corresponds to the discrete time shifted under the Bäcklund transformation, while c is replaced by the continuous time (denoted z). The link between fifth-order KdV and the type (ii) system and the reduction of the Lax pair from KdV was noted by Fordy in [**4**]; actually this connection was already anticipated in work of Weiss [**14**], who used Painlevé analysis to derive a BT and associated linear problem (a similar result also appears in [**10**]). However, none of these authors wrote a BT explicitly as a canonical transformation with parameter.

For this system the connection with the KdV dependent variable (potential of Schrödinger operator) is given by the formula

$$Q \equiv q_1,$$

while for the Lax pair we take

$$\Pi(u) = 4u + 2q_1 - \frac{1}{4}q_2^2 u^{-1}.$$

This differs by a factor of u^{-1} from the Lax matrix derived from KdV as in [**4**], but this factor is essential to ensure spectrality. After some calculation the generating function for the BT can be expressed explicitly as

$$F(q_j, \widetilde{q}_j) = \ell \tan^{-1}(Z/\ell) - Z + \frac{16}{5}y^5 + \left(4\lambda(q_1 + \widetilde{q}_1) - 2q_1\widetilde{q}_1 + \frac{1}{2}(q_2^2 + \widetilde{q}_2^{\,2}) - c\right)y.$$

In the above y is to be interpreted as a function of the coordinates given by the formula (1.4), i.e.

$$y = \pm\sqrt{\lambda - \frac{1}{2}(q_1 + \widetilde{q}_1)},$$

and

$$Z^2 = \ell^2 + \lambda q_2^2 \widetilde{q}_2^2.$$

For the special case $\lambda = 0$ this formula degenerates and it is necessary to replace the Z terms by the expression $\ell \log[q_2\widetilde{q_2}]$; in this case the generating function of the BT for the nonautonomous system was calculated in [**5**]. When taking the continuum limit of this discrete transformation with a small parameter $\varepsilon \to 0$, it is necessary to take λ with leading order behaviour $1/\varepsilon^2$, and take the negative branch of the square root in (1.4)). A detailed discussion of this limit (in a more general setting) will be found in [**7**].

3. The Garnier System

The Garnier system (see e.g. [**11**]) has the Hamiltonian

$$h = \frac{1}{2}\sum_j p_j^2 + \frac{1}{2}\Big(\sum_j q_j^2\Big)^2 - \frac{1}{2}\sum_j \lambda_j q_j^2.$$

Writing the Newton equations for the q_j as

$$q_{j,tt} + 2(\sum_k q_k^2)q_j = \lambda_j q_j,$$

it is apparent that each q_j is an eigenfunction of a Schrödinger operator with potential

$$Q = 2\sum_j q_j^2,$$

for eigenvalue λ_j. The above corresponds to a restricted flow of KdV [**15**]. Applying the Darboux-Crum transformation we wish to preserve this restriction so that

$$\widetilde{Q} = 2\sum_j \widetilde{q_j}^2.$$

But clearly from (1.2) the eigenfunctions q_j and their derivatives must transform as

$$\begin{pmatrix} \widetilde{q_j} \\ \widetilde{p_j} \end{pmatrix} = k_j \begin{pmatrix} -y & 1 \\ y^2 + \lambda_j - \lambda & -y \end{pmatrix} \begin{pmatrix} q_j \\ p_j \end{pmatrix},$$

for suitable constants k_j. Actually in order for this to define a canonical transformation it turns out that

$$k_j^2 = (\lambda - \lambda_j)^{-1},$$

which means that the above matrices are elements of SL(2).

For the Lax matrix of the Garnier system we take

$$\Pi(u) = \sum_j \frac{q_j^2}{u - \lambda_j} + 1,$$

and then spectrality is satisfied with the generating function for the BT given by

$$F(q_j, \widetilde{q_j}) = \sum_j k_j^{-1} q_j \widetilde{q_j} - \frac{1}{3}y^3,$$

where from (1.4) we have $y = \pm\sqrt{\lambda - \sum_j (q_j^2 + \widetilde{q_j}^2)}$.

4. Conclusions

We have found canonical transformations (BTs with parameter) for the Hénon–Heiles and Garnier systems which may equivalently by viewed as exact discretizations. The extension to the quantum case, together with the link to separation of variables, will be treated in [**8**]. It would also be interesting to look at BTs with parameter in the nonautonomous case [**6**], where deformation with respect to the Bäcklund parameter would probably have to be introduced (corresponding to the associated isomonodromy problem).

References

1. M. Bruschi, O. Ragnisco, P. M. Santini, and T. Gui-Zhang, *Integrable symplectic maps*, Physica D **49** (1991), 273–294.
2. M. M. Crum, *Note on functions of exponential type in a half-plane*, Quart. J. Math. Oxford Ser. (2) **6** (1955), 283–287.
3. J. C. Eilbeck, V. Z. Enolskii, V. B. Kuznetsov, and D. V. Leykin, *Linear r-matrix algebra for systems separable in parabolic coordinates*, Phys. Lett. A **180** (1993), 208–214.
4. A. P. Fordy, *The Hénon–Heiles systems revisited*, Physica D **52** (1991), 204–210.
5. A. N. W. Hone, *Integrable systems and their finite-dimensional reductions*, Ph.D thesis, Univ. Edinburgh, 1996.
6. ———, *Nonautonomous Hénon–Heiles systems*, Physica D **118** (1998), 1–16.
7. ———, *Exact discretization of the Ermakov-Pinney equation*, Phys. Lett. A **263** (1999), 347–354.
8. A. N. W. Hone, V. B. Kusnetsov, and O. Ragnisco, *Bäcklund transformations for many-body systems related to* KdV, J. Phys. A **32** (1999), L299–L306.
9. V. B. Kuznetsov and E. K. Sklyanin, *On Bäcklund transformations for many-body systems*, J. Phys. A **31** (1998), 224-2251.
10. A. C. Newell, M. Tabor, and Y. B. Zeng, *A unified approach to Painlevé expansions*, Physica D **29** (1987) 1–68.
11. S. Rauch-Wojciechowski, *Hamiltonian structures and complete integrability in analytical mechanics*, Soliton Theory: A Survey of Results (A. P. Fordy, ed.), Nonlinear Sci. Theory Appl., Manchester Univ. Press, 1990, pp. 235-272.
12. E. K. Sklyanin, *Separation of variables. New trends*, Progr. Theoret. Phys. Suppl. **118** (1995), 35–60.
13. A. P. Veselov, *Integrable maps*, Russian Math. Surveys **46** (1991), 1–51.
14. J. Weiss, *Bäcklund transformation and the Hénon–Heiles system*, Phys. Lett. A **105** (1984), 387-389.
15. Y. Zeng and Y. Li, *The constraints of potentials and the finite-dimensional integrable systems*, J. Math. Phys. **30** (1989), 1679–1689.

Dipartimento di Fisica "E. Amaldi", Universita' di Roma Tre, Via della Vasca Navale 84, 00146 Roma, Italy.

E-mail address: `ahone@maths.adelaide.edu.au`

Department of Mathematical and Computational Physics, Institute of Physics, St. Petersburg University, St. Petersburg 198904, Russia.

E-mail address: `vadim@amsta.leeds.ac.uk`

Dipartimento di Fisica "E. Amaldi", Universita' di Roma Tre, Via della Vasca Navale 84, 00146 Roma, Italy.

E-mail address: `ragnisco@amaldi.fis.uniroma3.it`

Centre de Recherches Mathématiques
CRM Proceedings and Lecture Notes
Volume **25**, 2000

Irregular Singular Behaviour in the First Discrete Painlevé Equation

N. Joshi

ABSTRACT. We consider divergent asymptotic series solutions of the difference equation $w_z(w_{z+\alpha} + w_z + w_{z-\alpha}) = z + \gamma w_z$ (where α, γ are constants) in the limit as z approaches infinity. We show how to prove that these diverge factorially and exhibit their hidden exponentially small perturbations. We also announce a new method of summation that gives a *unique* sum of such divergent series.

1. Introduction

In this paper, we consider divergent asymptotic series solutions of the difference equation

$$w(\overline{w} + w + \underline{w}) = z + \gamma w, \tag{1.1}$$

in the limit $z \to \infty$ where $w = w(z)$, $\overline{w} := w(z + \alpha)$, $\underline{w} := w(z - \alpha)$, α and γ are given constants, and $z \in \mathbb{C}$.

This is the well known first discrete Painlevé equation (dP1). Its continuum limit is the classical second-order ODE known as the first Painlevé equation. Both are integrable in the sense of being solvable through an associated linear system of equations.

dP1 appears as a model in two-dimensional quantum gravity [**3, 4**] where the asymptotic behaviour as $z \to \infty$ is of physical interest. This limit of dP1 has been studied by several authors [**5, 6, 8**]. It is known that the generic asymptotic behaviour in this limit is given by a discrete sampling of elliptic functions. Joshi [**6**] pointed out that dP1 also possesses nongeneric behaviours given by algebraic series expansions in z.

In §2, we show that such a series is a formal solution of dP1 and derive the recursion relation for its coefficients a_n. In §3, we prove an asymptotic result for a_n as $n \to \infty$. This shows that the series diverges for all $|z| \gg 1$. In §4, we deduce the formal exponentially small terms hidden in such a series and calculate the region of asymptotic validity of the series. Moreover, by extending an approach developed

---

2000 *Mathematics Subject Classification.* Primary 39A10; Secondary 30E15, 41A60.

The author was supported by ARC Grant #A69803721 and Fellowship #F69700172.

This is the final form of the paper.

for nonlinear ODEs, we announce a method of *unique* summation of such series along a "Stokes ray" in the z-plane.

2. Existence of Formal Series

Here we show that formal power-series solutions of Eq. (1.1) exist. It is straight forward to extend this to any equation of the type

$$w(\overline{w} + w + \underline{w}) = P(z) + \gamma w,$$

for rational $P(z)$.

LEMMA 2.1. *Eq.* (1.1) *admits a formal series solution of the form*

$$w(z) = \sum_{m=0}^{\infty} a_m z^{(1-m)/2}, \quad a_0 = \frac{1}{\sqrt{3}}, \tag{2.1}$$

where for $0 \leq k \leq 5$, *the coefficients* a_k *are given by*

$$\begin{aligned}
&3{a_0}^2 - 1 = 0,\\
&(6a_1 - \gamma)a_0 = 0,\\
&6a_0a_2 + 3a_1^2 - \gamma a_1 = 0,\\
&6a_0a_3 + (6a_1 - \gamma)a_2 = 0,\\
&6a_0a_4 + (6a_1 - \gamma)a_3 + 3{a_2}^2 - \alpha^2{a_0}^2/2 = 0,\\
&6a_0a_5 + (6a_1 - \gamma)a_4 + 6a_2a_3 - \alpha^2a_0a_1/2 = 0,
\end{aligned}$$

and for $m \geq 6$, *recursion relation governing* a_m *is*

$$\begin{aligned}
6a_0a_m &= -3\sum_{k=1}^{m-1} a_ka_{m-k} + \gamma a_{m-1}\\
&\quad - 2\sum_{p=1}^{[m/4]}\sum_{l=0}^{m-4p} \frac{\alpha^{2p}}{(2p)!}a_la_{m-4p-l}\frac{\Gamma(2p+(l-1)/2)}{\Gamma((l-1)/2)}.
\end{aligned} \tag{2.2}$$

Note here that the square brackets in the limits of summation denote integer parts. The proof is by direct substitution.

3. Divergence of the Formal Series

The lemma below shows that the coefficients a_m in the formal series (2.1) grow factorially with m and, therefore, that this series is divergent. For simplicity, we consider only the case $\gamma = 0$. However, the argument can be extended to $\gamma \neq 0$ without difficulty.

For $\gamma = 0$, and integer k, the recursion relation Eq. (2.2) and the results for $a_0, \ldots, a_5$ give

$$a_{4k+\mu} = 0, \quad \text{for } \mu = 1, 2, 3.$$

Let $m = 4n$ and $a_{4n} =: b_n$. Then Eq. (2.2) becomes

$$
\begin{aligned}
(3.1)\qquad 6b_0 b_n =& -4\alpha^2 b_0 (n-1)^2 b_{n-1} - 3\sum_{k=2}^{n-2} b_k b_{(n-k)} \\
& -\alpha^2 \sum_{k=1}^{n-2} b_k b_{n-1-k}(2k+1/2)(2k-1/2) \\
& -2\sum_{p=2}^{n}\sum_{k=0}^{n-p} \frac{\alpha^{2p}}{(2p)!}\frac{\Gamma(2(k+p)-1/2)}{\Gamma(2k-1/2)} b_k b_{n-p-k}.
\end{aligned}
$$

First we sketch a heuristic deduction of the asymptotic behaviour of b_n. The first term on the right of Eq. (3.1) indicates that to leading order

$$
b_n \approx -\frac{2}{3}\alpha^2 (n-1)^2 b_{n-1}.
$$

Summation shows that

$$
\begin{aligned}
b_n &\approx \left(-\frac{2}{3}\alpha^2\right)^{n-1} \big((n-1)!\big)^2 b_1 \\
&\approx \left(-\frac{\alpha^2}{6}\right)^{n-1} \Gamma(2n-1/2) d_1, \quad n \to \infty,
\end{aligned}
$$

for some constant d_1. Every other term in the right side of Eq. (3.1) is smaller than this for $n \gg 1$. We prove this below.

LEMMA 3.1. *For $m \gg 1$, and $\gamma = 0$, the coefficients a_m, $m = 4n$, of the series in Eq. (2.1) have the asymptotic behaviour given by*

$$
a_{4n} \approx \left(-, \frac{\alpha^2}{6}\right)^{n-1} \Gamma(2n-1/2) d_1,
$$

for some nonzero constant d_1.

PROOF. Let $E_0 = b_0$ and for $n \geq 1$ define

$$
E_n := \frac{b_n}{\Gamma(2n-1/2)} \left(-\frac{\alpha^2}{6}\right)^{-(n-1)}.
$$

Then Eq. (3.1) becomes

$$
\begin{aligned}
E_n =& \left(1 - \frac{1}{16(n-1)^2}\right)^{-1} E_{n-1} \\
&+ \frac{3}{\alpha^2 b_0} \sum_{k=2}^{n-2} \frac{\Gamma(2k-1/2)\Gamma(2(n-k)-1/2)}{\Gamma(2n-1/2)} E_k E_{n-k} \\
&- \frac{6}{\alpha^2 b_0} \sum_{k=1}^{n-2} \frac{\Gamma(2k-1/2)\Gamma(2(n-k)-5/2)}{\Gamma(2n-1/2)} (4k^2 - 1/4) E_k E_{n-1-k} \\
&+ \frac{2}{\alpha^2 b_0} \sum_{p=2}^{n}\sum_{k=0}^{n-p} (-6)^p \frac{\Gamma(2(n-p-k)-1/2)\Gamma(2(k+p)-1/2)}{\Gamma(2p+1)\Gamma(2n-1/2)} \\
&\times E_k E_{n-p-k}.
\end{aligned}
$$

To bound the coefficients in the sums on the right, we use the Beta function

$$B(x, y) = \int_0^1 t^{x-1}(1-t)^{y-1}\, dt,$$

(see, e.g., Ref. [**9**]) which satisfies

$$B(x, y) = \frac{\Gamma(x)\Gamma(y)}{\Gamma(x+y)}, \quad \Re(x) > 0,\ \Re(y) > 0. \tag{3.2}$$

Elementary arguments show that for $x \geq 1$, $y \geq 1$, $B(x, y) \leq 1$. Moreover, the derivatives of the integrand with respect to t, x, and y, show that, for $x < N$, $B(x, N-x)$ (which is the reciprocal of the binomial coefficient in the case when x and N are integers) has a unique minimum at $x = N/2$ with maxima at $x = 1, N-1$.

Define

$$G(k, p) := 6^p \frac{\Gamma(2(n-p-k) - 1/2)\Gamma(2(k+p) - 1/2)}{\Gamma(2p+1)\Gamma(2n - 1/2)}.$$

By Eq. (3.2), we get

$$\begin{aligned}
G(k, p) &= 6^p B\big(2(k+p) - 1/2, 2(n-p-k) - 1/2\big)\frac{\Gamma(2n-1)}{\Gamma(2p+1)\Gamma(2n-1/2)} \\
&\leq \frac{\Gamma(2(n-p) - 1/2)\Gamma(2p - 1/2)}{\Gamma(2n - 1/2)} \\
&= B\big(2p - 1/2, 2(n-p) - 1/2\big)\frac{\Gamma(2n-1)}{\Gamma(2n - 1/2)} \\
&\leq \frac{\Gamma(7/2)\Gamma(2n - 9/2)}{\Gamma(2n - 1/2)} \\
&= \frac{\Gamma(7/2)}{(2n-3/2)(2n-5/2)(2n-7/2)(2n-9/2)} \\
&= \mathcal{O}\left(\frac{1}{n^4}\right),
\end{aligned}$$

where, in the second step, we also used the inequality: $6^p/\Gamma(2p+1) < 1\ \forall p \geq 2$. Similar arguments show that

$$\frac{\Gamma(2k - 1/2)\Gamma(2(n-k) - 5/2)}{\Gamma(2n - 1/2)}(4k^2 - 1/4) = \mathcal{O}\left(\frac{1}{n^4}\right),$$

and

$$\frac{\Gamma(2k - 1/2)\Gamma(2(n-k) - 1/2)}{\Gamma(2n - 1/2)} = \mathcal{O}\left(\frac{1}{n^4}\right).$$

Direct summation gives

$$\left|\frac{3}{\alpha^2 b_0}\sum_{k=2}^{n-2}\frac{\Gamma(2k-1/2)\Gamma(2(n-k)-1/2)}{\Gamma(2n-1/2)}\right| \le \frac{3\sqrt{3}}{16|\alpha|^2}\frac{\Gamma(7/2)}{(n-4)^3},$$

$$\left|\frac{6}{\alpha^2 b_0}\sum_{k=1}^{n-2}\frac{\Gamma(2k-1/2)\Gamma(2(n-k)-5/2)}{\Gamma(2n-1/2)}\right| \le \frac{6\sqrt{3}}{16|\alpha|^2}\frac{\Gamma(7/2)}{(n-3)^3},$$

$$\left|\frac{2}{\alpha^2 b_0}\sum_{p=2}^{n}\sum_{k=0}^{n-p}(-6)^p\frac{\Gamma(2(n-p-k)-1/2)}{\Gamma(2p+1)}\right.$$
$$\left.\times\frac{\Gamma(2(k+p)-1/2)}{\Gamma(2n-1/2)}\right| \le \frac{3\sqrt{3}}{32|\alpha|^2}\frac{\Gamma(7/2)}{(n-9/4)^2}.$$

In the following, we lower bound each of $(n-3)^2$, $(n-4)^3$, $2(n-9/4)^2$ occurring in the right sides here by $2(n-5)^2$.

For simplicity, assume $3\sqrt{3}\Gamma(7/2)/\big(16|\alpha|^2\big)=4/3$. Let ξ be an upperbound for $|E_1|$. Assume that this is a maximum also of every $|E_p|$ for $2\le p\le n-1$. We have

$$\sum_{k=6}^{n}\frac{1}{(k-5)^2}\le\frac{1}{5}-\frac{1}{n-5},$$

$$\big(1-1/\big(16(n-1)^2\big)\big)^{-1}<+1\,/\,\big(15(n-1)^2\big),$$

$$\prod_{n=1}^{\infty}\left(1+\frac{1}{15(n-1)^2}\right)=\frac{\sqrt{15}}{\pi}\sinh\left(\frac{\pi}{\sqrt{15}}\right)=1.11333\ldots.$$

(See Ref. [**9**]) for identities such as the last one above; it can also be confirmed by using symbolic programs such as `Mathematica`.) Therefore, earlier bounds on coefficients of Eq. (3.2) give

$$|E_n|\le\delta+\frac{\xi}{75}+\xi^2:=f(\xi),$$

where

$$\delta:=\frac{\sqrt{15}}{\pi}\sinh\left(\frac{\pi}{\sqrt{15}}\right)|E_1|<(37/75)^2.$$

This last inequality ensures that the quadratic $f(\xi)-\xi$ has a negative mininum value at $\xi=37/75$ $\big(>|E_1|\big)$. Hence we get $|E_n|\le 37/75$ for all n. Similar arguments show that $|E_n|$ is lowerbounded by a nonzero number.

To prove that $|E_n|$ has a limit as $n\to\infty$, we consider the differences E_n-E_{n-1}. From the bounds obtained above, it follows that

$$E_n-E_{n-1}=\mathcal{O}\left(\frac{1}{n^2}\right).$$

Hence, the summation of the sequence $\{E_n-E_{n-1}\}$, converges (since it is majorized by $\sum 1/n^2$). Therefore, E_n has a limit. □

4. Sector of Validity

At first glance, the asymptotic series Eq. (2.1) appears valid for any large enough $|z|$. (Note that in this section, we consider generic $\gamma\ne 0$.) However, a perturbation argument shows that this is not the case. Take any solution W that is asymptotic to such a series and write a neighbouring solution w as

$$w = W + v.$$

Substitution into dP1 shows that

$$\overline{v} + 4v + \underline{v} = \frac{1}{a_0 z^{1/2}}\left(1 - \frac{a_1}{a_0 z^{1/2}} + O(1/z)\right)\left(\gamma v - v\left(\overline{v} + v + \underline{v}\right)\right). \tag{4.1}$$

The linear homogeneous equation

$$\overline{v} + 4v + \underline{v} = 0$$

has solutions λ^z where $\lambda = -2 \pm \sqrt{3}$. For Eq. (2.1) to be asymptotically valid, we need the perturbation v be small. If z has positive real part, we must choose $\lambda = -2 + \sqrt{3}$ which satisfies $|\lambda| < 1$. We assume this below.

Write

$$v =: \lambda^{z/\alpha} e^{\nu(z)}.$$

Then we get

$$\lambda \exp\bigl(2(\overline{\nu} - \nu)\bigr) + \left(4 - \frac{\gamma}{a_0 z^{1/2}} + O(1/z)\right)\exp(\overline{\nu} - \nu) = -\lambda^{-1} + O(\lambda^{z/\alpha} e^{\nu(z)}). \tag{4.2}$$

Here we have used the assumption that $\exp(\underline{\nu} - \nu) \approx \exp\bigl(-(\overline{\nu} - \nu)\bigr)$ which is equivalent to assuming that ν has algebraic behaviour in z. (This is standard in asymptotic theory, see e.g. Chapter 5 of [**1**].)

Taking logarithms of both sides in Eq. (4.2) gives

$$\overline{\nu} - \nu = \frac{\gamma}{2z^{1/2}} - \frac{\gamma^2(\sqrt{3}+3)}{24z} + O(1/z^{3/2}),$$

where we have used the values $a_0 = 1/\sqrt{3}$ and $a_1 = \gamma/6$ found in §2.

Summation gives

$$\nu = \frac{\gamma}{\alpha z^{1/2}} - \frac{\gamma^2(\sqrt{3}+3)}{24\alpha}\log(z) + O(1/z^{1/2}).$$

Therefore, the solution w admits exponentially small perturbations of the form

$$C_0 \lambda^{z/\alpha} \exp\bigl((\gamma/\alpha) z^{1/2} - \{\gamma^2(\sqrt{3}+3)/(24\alpha)\}\log(z)\bigr)\{1 + O(1/z^{1/2})\}$$

for arbitrary constant C_0.

Such terms remain small in regions in the complex z-plane, near infinity, in which $z = r\exp(i\theta)$ satisfies

$$\Re\bigl((\gamma/\alpha) z^{1/2} - \{\gamma^2(\sqrt{3}+3)/(24\alpha)\}\log(z)\bigr) < 0.$$

Assume $\gamma > 0$, and recall that $\log(\lambda) < 0$. Letting $\xi = \cos(\theta/2)$, this condition becomes

$$\xi^2 + \frac{\gamma r^{-1/2}}{2\log(\lambda)}\xi > 1/2.$$

The two solutions of this quadratic equation give to leading order

$$\frac{1}{\sqrt{2}} < \xi \le 1, \quad \text{or} \quad -1 \le \xi < -\frac{1}{\sqrt{2}}.$$

In other words, we have asymptotic validity in sectors that are bounded to leading-order by

$$-\frac{\pi}{4} < \theta < \frac{\pi}{4}, \quad \text{or} \quad \frac{3\pi}{4} < \theta < \frac{5\pi}{4}.$$

The asymptotic corrections to these bounding rays show that in fact the sectors are slightly wider in each case. The exponentially small $\exp(\nu)$ is minimal along the rays $\theta = 0, \pi$. These are called the *Stokes' rays.*

Despite the divergence of the series (2.1), it still provides accurate approximations of the solution for sufficiently large z in the sectors described above. Multisummability theorems for difference equations (see [**2**]) show that such a series may be summed in its region of asymptotic validity. However, such sums are not necessarily unique.

In other words, dP1 has solutions that cannot be uniquely identified from their divergent asymptotic power series. We call this the *identification problem.* We have solved this problem, in each of the sectors described above, by using optimal asymptotics.

That is, define

$$W^{[M]}(z) = \sum_{m=0}^{M-1} a_m z^{(1-m)/2},$$

and a corresponding sequence of numbers z_M defined by the requirement that the last term of $W^{[M]}$ be minimal at z_M. Such a sum is called an optimal truncation of (2.1). We use these sums as the initial value at z_M (and at z_{M+1}) to define a sequence of solutions $w_M(z)$. Our main result is that this sequence has a uniform limit as $M \to \infty$. The result is a unique solution of dP1 and is asymptotic to (2.1). In particular, its asymptotic behaviour is purely this series along a Stokes' ray where $C_0 = 0$.

The method is similar to the one used to solve an identification problem for the first Painlevé equation in Ref. [**7**]. The adaptation to difference equations, in particular to dP1, will be described in another paper.

References

1. C. Bender and S. Orszag, *Advanced mathematical methods for scientists and engineers*, McGraw-Hill, New York, 1978.
2. B. L. J. Braaksma and B. F. Faber, *Multisummability for some classes of difference equations*, Ann. Inst. Fourier (Grenoble) **46** (1996), 183–217.
3. E. Brezin and V. Kazakov, *Exactly solvable field theories of closed strings*, Phys. Lett. B **236** (1990, 144–150.
4. A. S. Fokas, A. R. Its, and A. V. Kitaev, *Discrete Painlevé equations and their appearance in quantum gravity*, Comm. Math. Phys. **142** (1991), 313–344.
5. ———, *The isomonodromy approach to matrix models in 2D quantum gravity*, Comm. Math. Phys. **147** (1992), 395–430.
6. N. Joshi, *Local asymptotics of the first discrete Painlevé equation as the discrete independent variable approaches infinity*, Math. Analysis and Appl. **4** (1997), 124–133.
7. Nalini Joshi and Martin D. Kruskal, *A natural sum of divergent asymptotic series solutions of the first Painlevé equation*, 1998. `http://www.adelaide.edu.au/pure/iga/iga1998-16.ps`.
8. V. L. Vereschagin, *Asymptotics of solutions of the discrete string equation*, Physica D **95** (1996), 268–282.
9. E. T. Whittaker and G. N. Watson, *A course of modern analysis*, Cambridge Univ. Press, Cambridge, 1962.

DEPARTMENT OF PURE MATHEMATICS, UNIVERSITY OF ADELAIDE, ADELAIDE SA 5005, AUSTRALIA.

E-mail address: `Nalini.Joshi@adelaide.edu.au`

Centre de Recherches Mathématiques
CRM Proceedings and Lecture Notes
Volume **25**, 2000

Rational and Special Solutions of the P_{II} Hierarchy

Nicolai A. Kudryashov and Andrew Pickering

ABSTRACT. We give an approach to finding rational and special solutions of the P_{II} hierarchy using the relationship between this hierarchy and the Schwarzian P_{II} hierarchy obtained via the singular manifold method.

As is well known, at the turn of the century Painlevé and his school discovered six ordinary differential equations (ODEs) that defined new functions. This was achieved by classifying second order ODEs of a certain form having what is today referred to as the Painlevé property (the general solution should be free of movable critical points). These six functions are now called the Painlevé transcendents.

Recently [**8**] one of the present authors made the attempt to find new functions other than the Painlevé transcendents, again determined by nonlinear ODEs. With this aim, the P_{I} and P_{II} hierarchies of ODEs were introduced using the modified Korteveg-de Vries (mKdV) and Schwarzian KdV (SKdV) hierarchies.

Here we will be interested in the connections between Miura maps and truncated Painlevé expansions and we will see how such connections can be useful in deriving rational and special solutions of the P_{II} hierarchy [**1, 5**]. We also give for the first time a Schwarzian version of this hierarchy. We then show how our approach can be used to generate sequences of rational solutions and special integrals for each member of this hierarchy. Rational solutions have not previously been constructed for the higher order members of this hierarchy.

Given two differential equations, say

$$D[u] = 0, \tag{1}$$

and

$$E[z] = 0, \tag{2}$$

we say that we have a map of equation (2) into (1) if we have a differential substitution of the form

$$u = F[z, z_x, \dots], \tag{3}$$

such that for some differential operator $\widehat{C}$,

$$D[u] = \widehat{C}E[z]. \tag{4}$$

2000 *Mathematics Subject Classification.* Primary: 34M55; Secondary: 37K35.
This is the final form of the paper.

Differential substitutions of the form (3) map solutions of (2) into solutions of (1); examples of such maps include Miura maps, linearising transformations such as the Cole–Hope transformation, and truncated Painlevé expansions.

Let us take the KdV hierarchy in the form

$$\omega_t + \mathcal{R}^n \omega_x = \omega_t + K_{2n+1}[\omega] = \omega_t + \partial_x b^{n+1}[\omega] + 0, \quad n = 0, 1, \ldots, \tag{5}$$

where $\partial_x = \partial/\partial x$ and the recursion operator R is given (4) by

$$\mathcal{R} = \partial_x^2 + 4\omega + 2\omega_x \partial_x^{-1}, \tag{6}$$

and let us consider the two successive modifications

$$\omega = \Omega[u] = u_x - u^2, \tag{7}$$

and

$$u = U[\varphi] = \frac{1}{2}\left(\frac{\varphi_{xx}}{\varphi_x}\right). \tag{8}$$

Then we have of course the well-known relations

$$\begin{aligned} \mathcal{R}\Omega'[u] &= \Omega'[u]\overline{\mathcal{R}}, \\ \omega &= \Omega[u], \end{aligned} \tag{9}$$

where

$$\overline{\mathcal{R}} = \partial_x^2 - 4u^2 - 4u_x \partial_x^{-1} u, \tag{10}$$

and

$$\begin{aligned} \overline{\mathcal{R}}U'[\varphi] &= U'[\varphi]\widetilde{\mathcal{R}}, \\ u &= U[\varphi], \end{aligned} \tag{11}$$

where

$$\widetilde{\mathcal{R}} = \varphi_x \partial_x^{-1} \varphi_x \partial_x \varphi_x^{-1} \partial_x \varphi_x^{-1} \partial_x, \tag{12}$$

and $\Omega'[u]$ and $U'[\varphi]$ denote the Fréchet derivatives of $\Omega[u]$ and $U[\varphi]$ respectively;

$$\Omega'[u] = \partial_x - 2u, \quad U'[\varphi] = \frac{1}{2}\partial_x \varphi_x^{-1} \partial_x. \tag{13}$$

We then obtain the mKdV and SKdV hierarchies as

$$u_t + \overline{\mathcal{R}}^n u_x = u_t + \overline{K}_{2n+1}[u] = u_t + \partial_x(\partial_x + 2u) b^n[u_x - u^2] = 0, \quad n = 0, 1, \ldots, \tag{14}$$

and

$$\varphi_t + \widetilde{\mathcal{R}}^n \varphi_x = \varphi_t + \widetilde{K}_{2n+1}[\varphi] = \varphi_t + 2\varphi_x b^n \left[\frac{1}{2}\{\varphi; x\}\right] = 0, \quad n = 0, 1, \ldots, \tag{15}$$

respectively, where in (15) $\{\varphi; x\}$ denotes the Schwarzian derivative of φ,

$$\{\varphi; x\} = \left(\frac{\varphi_{xx}}{\varphi_x}\right)_x - \frac{1}{2}\left(\frac{\varphi_{xx}}{\varphi_x}\right)^2, \tag{16}$$

introduced through the composition of (7) and (8),

$$\omega = u_x - u^2 = \frac{1}{2}\{\varphi; x\}. \tag{17}$$

Corresponding to (4) we have of course the relations

$$\omega_t + K_{2n+1}[\omega] = \Omega'[u]\left(u_t + \overline{K}_{2n+1}[u]\right), \tag{18}$$

and

$$u_t + \overline{K}_{2n+1}[u] = U'[\varphi]\big(\varphi_t + \widetilde{K}_{2n+1}[\varphi]\big). \tag{19}$$

It is the SKdV hierarchy that we used in [**10**] to generate rational solutions of the mKdV and KdV hierarchies. The mKdV and KdV hierarchies have the same sequence of singular manifold equations (see [**13**] for definitions) and for zero value of the spectral parameter this sequence becomes precisely the SKdV hierarchy (15). The connection between Miura maps and truncated Painlevé expansions can be seen using the invariance of the SKdV hierarchy under the Möbius group; sending $\varphi \to -1/\varphi$ gives that

$$u = -\frac{\varphi_x}{\varphi} + \frac{1}{2}\left(\frac{\varphi_{xx}}{\varphi_x}\right), \tag{20}$$

is a solution of the mKdV hierarchy provided that (15) holds. Given that (8) is also a solution of the mKdV hierarchy we see that (20) defines an auto-Bäcklund transformation between two solutions u and u_1 of the mKdV hierarchy as

$$u = -\frac{\varphi_x}{\varphi} + u_1, \quad u_1 = \frac{1}{2}\left(\frac{\varphi_{xx}}{\varphi_x}\right). \tag{21}$$

The invariance of the mKdV hierarchy under $u \to -u$ then gives a second auto-Bäcklund transformation between two solutions v and v_1 as

$$v = \frac{z_x}{z} + v_1, \quad v_1 = -\frac{1}{2}\left(\frac{z_{xx}}{z_x}\right), \tag{22}$$

for any solution z of

$$S_{2n+1}[z] \equiv z_t + 2z_x b^n\left[\frac{1}{2}\{z;x\}\right] = 0. \tag{23}$$

Expressions (21) and (22) are of course the Painlevé expansions for the principal families of the mKdV hierarchy truncated at constant level. From (19) we find that corresponding to (20) we have the relation

$$u_t + \overline{K}_{2n+1}[u] = \frac{1}{2}\partial_x\left(\frac{1}{\varphi_x}\partial_x - \frac{2}{\varphi}\right)S_{2n+1}[\varphi], \tag{24}$$

which gives precisely the result of substituting the truncated Painlevé expansion (21) into the mKdV hierarchy. Similarly for the truncation (22). (Details in [**10**].)

Following the approach in [**9**] we make a double iteration of this constant level truncation and identify $u_1 = v$ to obtain

$$\varphi_x = \frac{z^2}{z_x}, \tag{25}$$

which is the formula used by Weiss to obtain rational solutions of the KdV equation. This formula allows the iterative construction of solutions of (23) via

$$z_{k+1,x} = \frac{z_k^2}{z_{k,x}}. \tag{26}$$

This double iteration of the Weiss auto-Bäcklund transformation makes use of the two principal families of the mKdV hierarchy and greatly simplifies the derivation of the iterative formula (26). Being a double iteration of the constant level truncation, this approach is different from those based on the use of two singular manifolds [**4, 12**]. (The mapping (26) of course corresponds to just one iteration.)

Now we consider the application of our approach to hierarchies of ODEs, namely the P_{II} hierarchy [**1, 5, 8**] and a Schwarzian $\to P_{\mathrm{II}}$ hierarchy introduced here. We begin with the SKdV hierarchy (15), and we make the similarity reduction

$$\varphi(x,t) = \left[(2n+1)t\right]^{m_n/(2n+1)}\psi(X), \quad X = x\left[(2n+1)t\right]^{-1/(2n+1)}, \tag{27}$$

where m_n is a constant; this then yields (here $' = d/dX$)

$$m_n\psi - X\psi' + 2\psi' b^n\left[\frac{1}{2}\{\psi; X\}\right] = 0. \tag{28}$$

Under the action of the operator $(\psi')^{-1}d/dX$, the sequence of ODEs (28) becomes

$$\left(\frac{d}{dX} + 2V\right)b^n[V' - V^2] - XV - \alpha_n = 0, \tag{29}$$

where we have made the substitution

$$V = \frac{1}{2}\left(\frac{\psi''}{\psi'}\right), \tag{30}$$

and we have set

$$\alpha_n = \frac{1}{2}(1 - m_n). \tag{31}$$

The sequence of ODEs (29) is the hierarchy of higher order P_{II} equations [**1, 5, 8**]. We now consider the iteration which corresponds to our iteration $\varphi_k \to \varphi_{k+1}$ of the SKdV hierarchy. We take

$$\varphi_k = \left[(2n+1)t\right]^{m_n/(2n+1)}\psi_k(X), \tag{32}$$

and

$$\varphi_{k+1} = \left[(2n+1)t_{2n+1}\right]^{M_n/(2n+1)}\psi_{k+1}(X). \tag{33}$$

The iterative formula

$$\varphi_{k+1,x} = \frac{\varphi_k^2}{\varphi_{k,x}}, \tag{34}$$

then yields

$$M_n = m_n + 2, \tag{35}$$

and

$$\psi'_{k+1} = \frac{\psi_k^2}{\psi'_k}. \tag{36}$$

Note that $M_n = m_n + 2$ corresponds to $\alpha_n^{\mathrm{new}} = \alpha_n^{\mathrm{old}} - 1$. This shift on α_n can be understood by recalling that the iteration (33) was derived by sending $\varphi \to -1/\varphi$ in (8) to obtain (21), and then sending $u \to -u$ to obtain (22); mirroring this sequence of transformations for ψ and V gives the above shift on α_n. Thus we have derived an alternative form for the Bäcklund transformation for the P_{II} hierarchy [**1**]. We will use this to generate sequences of rational solutions and special integrals for the P_{II} hierarchy (and corresponding solutions of the Schwarzian hierarchy). These sequences of special integrals can be found in [**3**].

This shift on α_n and the mapping (36) were in fact obtained for equation (28) in the special case $n = 1$ in [**14**]. However this approach involved the direct consideration of additional symmetries of equation (28) (with $n = 1$).

Bäcklund transformations for the $P_{\rm II}$ hierarchy are given in [**1**]. A derivation of these Bäcklund transformations, using a new extension of the Painlevé truncation method, can be found in [**3**]. Bäcklund transformations for members of the $P_{\rm II}$ hierarchy have also been discussed in [**6**] and [**11**]. Further applications and extensions of the approach in [**3**] can be found in [**7**].

In order to generate sequences of rational solutions, we begin with the solution

$$\psi_0 = X, \tag{37}$$

of (28) for $m_n = 1$; corresponding to this we have the solution

$$V_0 = 0, \tag{38}$$

of (29) for $\alpha_n = 0$. We then obtain successive solutions ψ_k of (28) for $m_n = 2k+1$,

$$(2k+1)\psi_k - X\psi_k' + 2\psi_k' b^n\left[\frac{1}{2}\{\psi_k; X\}\right] = 0, \tag{39}$$

and successive solutions V_k of the $P_{\rm II}$ hierarchy for $\alpha_n = -k$,

$$\left(\frac{d}{dX} + 2V_k\right)b^n[V_k' - V_k^2] - XV_k + k = 0. \tag{40}$$

We now consider, separately for each value of n, the iterative construction of rational solutions to the ODEs (39) and (40), beginning with the solutions (37) and (38) for $k = 0$. We note that the $P_{\rm II}$ hierarchy has the symmetry $(V, \alpha_n) \to (-V, -\alpha_n)$ and thus, begining with $V = 0$ and $\alpha_n = 0$, we are able to construct rational solutions for any integer value of α_n. Here, without loss of generality, we construct such solutions only for negative integer values of α_n. At each step we renormalise so that ψ_k is the quotient of two monic polynonials. We also at each step introduce a constant of integration which is fixed by substitution back into (39).

In addition, we can give the following Lemma:

LEMMA 1. *The function ρ_k given by*

$$\rho_k = X^{2k+1} + C_k, \quad k = 0, 1, 2, \ldots, \tag{41}$$

is a solution of the equation (39) *for any $n > k$ provided that*

$$C_k = 0, \tag{42}$$

and is a solution for $n = k$ provided that

$$C_k = (-1)^{k+1} 2^k \prod_{l=0}^{k-1} \frac{(2l+1)[k(k+1) - l(l+1)]}{(l+1)}. \tag{43}$$

There are no solutions of (39) *of the form* (41) *for $n < k$.*

PROOF. We can see that

$$2b^n\left[\frac{1}{2}\{\rho_k; x\}\right] = (-1)^n \frac{2^n}{X^{2n}} \prod_{l=0}^{n-1} \frac{(2l+1)[k(k+1) - l(l+1)]}{(l+1)}, \tag{44}$$

and that this vanishes for $n > k$. We also see that

$$(2k+1)\rho_k - X\rho_k' = (2k+1)C_k, \tag{45}$$

and so it is clear that substitution of (41) into (39) gives (42) for $n > k$ and (43) for $n = k$. For $n < k$ we obtain that $2\rho'_k b^n\left[\frac{1}{2}\{\rho_k; X\}\right]$ does not vanish and is proportional to $X^{2(k-n)}$, and so we cannot have a solution of the form (41). $\square$

This Lemma is useful for determining the additive constant in the first non-monomial solution of (38) obtained via our scheme; further solutions are then obtained using the iterative formula (36).

For $n = 1$ we find for $k = 1, 2, 3, 4$ the sequence of solutions of (39)

$$\tag{46} \begin{gathered} \psi_1 = X^3 + 4, \quad \psi_2 = \frac{X^6 + 20X^3 - 80}{X}, \quad \psi_3 = \frac{X^{10} + 60X^7 + 11200X}{X^3 + 4}, \\ \psi_4 = \frac{X^{15} + 140X^{12} + 2800X^9 + 78400X^6 - 3136000X^3 - 6272000}{X^6 + 20X^3 - 80}, \end{gathered}$$

corresponding to which we have the following solutions of (40)

$$\tag{47} \begin{gathered} V_1 = \frac{1}{X}, \quad V_2 = \frac{2(X^3 - 2)}{X(X^3 + 4)}, \quad V_3 = \frac{3(X^8 + 8X^5 + 160X^2)}{(X^3 + 4)(X^6 + 20X^3 - 80)}, \\ V_4 = \frac{4(X^{15} + 50X^{12} + 1000X^9 - 22400X^6 - 112000X^3 - 224000)}{X(X^6 + 20X^3 - 80)(X^9 + 60X^6 + 11200)}. \end{gathered}$$

For $n = 2$ we find solutions of (39) given by

$$\tag{48} \begin{gathered} \psi_1 = X^3, \quad \psi_2 = X^5 - 144, \quad \psi_3 = \frac{X^{10} - 1008X^5 - 48384}{X^3}, \\ \psi_4 = \frac{X^{15} - 4032X^{10} - 3048192X^5 + 146313216}{X(X^5 - 144)}, \end{gathered}$$

corresponding to which we have the following solutions of (40)

$$\tag{49} \begin{gathered} V_1 = \frac{1}{X}, \quad V_2 = \frac{2}{X}, \quad V_3 = \frac{3(X^5 + 96)}{X(X^5 - 144)}, \\ V_4 = \frac{4(X^{15} - 72X^{10} + 217728X^5 - 1741824)}{X(X^5 - 144)(X^{10} - 1008X^5 - 48384)}. \end{gathered}$$

For $n = 3$ we obtain the solutions

$$\tag{50} \begin{gathered} \psi_1 = X^3, \quad \psi_2 = X^5, \quad \psi_3 = X^7 + 14400, \\ \psi_4 = \frac{X^{14} + 129600X^7 - 373248000}{X^5}, \end{gathered}$$

corresponding to which we have the solutions of (40)

$$\tag{51} V_1 = \frac{1}{X}, \quad V_2 = \frac{2}{X}, \quad V_3 = \frac{3}{X}, \quad V_4 = \frac{4(X^7 - 10800)}{X(X^7 + 14400)}.$$

For $n = 4$ we obtain the solutions

$$\tag{52} \psi_1 = X^3, \quad \psi_2 = X^5, \quad \psi_3 = X^7, \quad \psi_4 = X^9 - 2822400,$$

corresponding to which we have the solutions of (40)

$$\tag{53} V_1 = \frac{1}{X}, \quad V_2 = \frac{2}{X}, \quad V_3 = \frac{3}{X}, \quad V_4 = \frac{4}{X}.$$

We note that the iteration (36) sends $X^{2p+1} \to X^{2p+3} + C_{p+1}$. Our Lemma then tells us that beginning with $\psi_0 = X$ the first n solutions of (39) found using our iteration are $\psi_1 = X^3$, $\psi_2 = X^5$, ..., $\psi_n = X^{2n+1} + C_n$ where C_n is given by (43). Corresponding solutions of the hierarchy (40) are then given by (30) as $V_p = p/X$, $p = 1, 2, \ldots, n$. These last correspond of course to the leading order behaviours of the $n^{\rm th}$ member of the $P_{\rm II}$ hierarchy.

The solutions (49) have been given in [**11**]. However, rational solutions for higher members of the $P_{\rm II}$ hierarchy ($n > 2$) have not previously been constructed (except for trivial examples where the rational solution corresponds to a leading order behaviour in the Painlevé analysis of the ODE [**5**]).

We now see how our approach can also be used to obtain sequences of special solutions, and corresponding special integrals, of the $P_{\rm II}$ hierarchy (29). We begin by noting that in the case $\alpha_n = \frac{1}{2}$ ($m_n = 0$), equation (28) becomes

$$b^n[V' - V^2] = \frac{X}{2}, \tag{54}$$

which then gives a special integral of the $P_{\rm II}$ hierarchy defined, as observed in [**8**] and [**3**], in terms of a lower order member of the P_I hierarchy,

$$b^n[F] = \frac{X}{2}, \tag{55}$$

(54) and (55) being obtained from (28) via

$$\frac{1}{2}\{\psi; X\} = V' - V^2 = F. \tag{56}$$

(Note that in the case $\alpha_n = \frac{1}{2}$ the constant of integration in the general integral of (29), with $V = \frac{1}{2}\psi''/\psi'$, cannot be removed by a shift on ψ.) Solutions of the $P_{\rm II}$ hierarchy for $\alpha_n = \frac{1}{2}$ can then be obtained from (56), which can be linearized by the well known transformation $V = -\omega'/\omega$ onto

$$\omega'' + F\omega = 0. \tag{57}$$

The iteration (36) breaks down when (54) holds. Thus in order to iterate to find sequences of special solutions we need an iteration that does not break down on solutions of (54). That is, we need the second independent Bäcklund transformation of the $P_{\rm II}$ hierarchy which gives $\alpha_n^{\rm new} = \alpha_n^{\rm old} + 1$ (here without loss of generality we discuss special solutions and special integrals only for $\alpha_n > 0$). This last is easily derived using a variation on the above approach incorporating the symmetry $(V, \alpha_n) \to (-V, -\alpha_n)$ of the $P_{\rm II}$ hierarchy. Here however, rather than give this derivation, we show how the iteration (36) can be used to obtain the sequence of special integrals which correspond to these special solutions. We do this by using (36) to express the special integral (54) of the $P_{\rm II}$ hierarchy for $\alpha_n = \frac{1}{2}$ in terms of solutions of the Schwarzian $P_{\rm II}$ hierarchy for α_n any positive half-integer. Rewriting these expressions in terms of solutions V of the $P_{\rm II}$ hierarchy itself then recovers the sequence of special integrals obtained in [**3**] using the standard form of the Bäcklund transformation. Here we give as examples derivations of the special integrals for $\alpha_n = 3/2$ for $n = 1$ and $n = 2$.

Let us begin with $n = 1$. Equation (28) with $\alpha_1 = \frac{1}{2}$ ($m_1 = 0$) gives (54) for $n = 1$ in terms of ψ. Substituting $\psi' = \eta^2/\eta'$ then allows us to express this special integral in terms of a solution η of (28) for $\alpha_1 = 3/2$. Thus we obtain

$$\left(\frac{\eta}{\eta'}\right)^3 + \left[2\left(\frac{\eta''}{2\eta'}\right)^2 + X\right]\left(\frac{\eta}{\eta'}\right)^2 - 4\left(\frac{\eta''}{2\eta'}\right)\left(\frac{\eta}{\eta'}\right) + 2 = 0. \tag{58}$$

We can rewrite this in terms of a solution V of P_{II} for $\alpha_1 = 3/2$ using the identities

$$\frac{\eta''}{2\eta'} = V \quad \text{and} \quad \frac{\eta}{\eta'} = V' - V^2 - \frac{1}{2}X, \tag{59}$$

(see (28)). The resulting special integral of P_{II} is as given in [**3**].

Similarly for $n = 2$. Substituting $\psi' = \eta^2/\eta'$ into (28) for $\alpha_2 = \frac{1}{2}$ ($m_2 = 0$) allows us to express this special integral of fourth order P_{II} in terms of a solution of the Schwarzian version of fourth order P_{II} for $\alpha_2 = 3/2$. This then gives

$$\left(\frac{\eta}{\eta'}\right)^3 + [4VV'' - 2V'^2 - 6V^4 + X]\left(\frac{\eta}{\eta'}\right)^2 + 4[2V^3 - V'']\left(\frac{\eta}{\eta'}\right) + 4[V' - V^2] = 0, \tag{60}$$

where for ease of writing we have already set $\eta''/2\eta' = V$. The corresponding special integral of fourth order P_{II} is then recovered by substituting (see (28))

$$\frac{\eta}{\eta'} = V''' - 2VV'' + V'^2 - 6V^2V' + 3V^4 - \frac{1}{2}X. \tag{61}$$

Thus (60) is a compact form of the special integral for $\alpha_2 = 3/2$ given in [**3**].

Let us now summarise the results presented here. We have derived an alternative form of the Bäcklund transformation for the P_{I} hierarchy, phrased in terms of a corresponding Schwarzian hierarchy. We have then shown how this form of the Bäcklund transformation can be used to derive rational solutions and special integrals of both the Schwarzian hierarchy and of course the P_{II} hierarchy itself.

References

1. H. Airault, *Rational solutions of Painlevé equations*, Stud. Appl. Math. **61** (1979), 31–53.
2. F. Calogero, *A method to generate solvable nonlinear evolution equations*, Lett. Nuovo Cimento (2) **14** (1975), 443–447; P. J. Olver, *Evolution equations possessing infinitely many symmetries*, J. Math. Phys. **18** (1977), 1212–1215.
3. P. A. Clarkson, N. Joshi and A. Pickering, *Bäcklund transformations for the second Painlevé hierarchy: A modified truncation approach*, Inverse Problems **15** (1999), 175–187.
4. P. G. Estévez, P. R. Gordoa, L. Martínez Alonso, and E. M. Reus, *Modified singular manifold expansion: Application to the Boussinesq and Mikhailov–Shabat systems*, J. Phys. A **26** (1993), 1915–1925.
5. H. Flaschka and A. C. Newell, *Monodromy and spectrum-preserving deformations*. I, Comm. Math. Phys. **76** (1980), 65–116.
6. V. I. Gromak, *Bäcklund transformations of Painlevé equations and their applications*, The Painlevé Property One Century Later (R. Conte, ed.), CRM Series in Mathematical Physics, Springer, New York, 1999.
7. P. R. Gordoa, N. Joshi and A. Pickering, *Mappings preserving locations of movable poles: A new extension of the truncation method to ordinary differential equations*, Nonlinearity **12** (1999), 955–968.
8. N. A. Kudryashov, *The first and second Painlevé equations of higher order and some relations between them*, Phys. Lett. A **224** (1997), 353–360.
9. ———, *Method for deriving rational solutions of some nonlinear evolution equations*, J. Phys. A **30** (1997), 5445–5453.
10. N. A. Kudryashov and A. Pickering, *Rational solutions for Schwarzian integrable hierarchies*, J. Phys. A **31** (1998), 9505–9518.

11. N. A. Kudryashov and M. B. Soukharev, *Uniformization and transcendence of solutions for the first and second Painlevé hierarchies*, Phys. Lett. A **137** (1998), 206–216.
12. M. Musette and R. Conte, *The two-singular-manifold method.* I. *Modified Korteveg-de Vries and sine-Gordon equations*, J. Phys. A **27** (1994), 3895–3913.
13. A. Pickering, *The singular manifold method revisited*, J. Math. Phys. **37** (1996), 1894–1927.
14. J. Weiss, *Bäcklund transformation and the Henon–Heiles system*, Phys. Lett. A **105** (1984), 387–389.

Department of Applied Mathematics, Moscow Engineering Physics Institute, 31 Kashirskoe Shosse, Moscow 115409, Russia.

E-mail address: nkud@dpt31.mephi.msk.su

Area de Fisica Teórica, Facultad de Ciencias, Edificio de Fisica, Universidad de Salamanca, 37008 Salamanca, Spain.

E-mail address: andrew@sonia.usal.es

Centre de Recherches Mathématiques
CRM Proceedings and Lecture Notes
Volume **25**, 2000

Linearisable Systems and the Gambier Approach

S. Lafortune, B. Grammaticos, and A. Ramani

ABSTRACT. A systematic study of the discrete second order projective system is presented, complemented by the integrability analysis of the associated multilinear mapping. Moreover, we show how we can obtain third order integrable equations as the coupling of a Riccati equation with second order Painlevé equations. This is done in both continuous and discrete cases.

1. Introduction

Integrability is a far too general term. In order to fix the ideas we can just present three most common types of integrability, which suffice in order to explain the properties of the majority of integrable systems [**9**]. These three types are:

- Reduction to quadrature through the existence of the adequate number of integrals of motion.
- Reduction to linear differential systems through a set of local transformations.
- Integration through IST techniques. This last case is mediated by the existence of a Lax pair (a linear system the compatibility of which is the nonlinear equation under integration) which allows the reduction of the nonlinear equation to a linear integrodifferential one. The above notions can be extended *mutadis mutandis* to the domain of discrete systems.

This paper will focus on the second type of integrability, usually referred to as linearisability. The prototype of the linearisable equations is the Riccati equation. In differential form this equation writes:

$$w' = \alpha w^2 + \beta w + \gamma, \tag{1.1}$$

which is linearisable through a discrete version of the Cole–Hopf transformation. Similarly, the discrete Riccati equation, which assumes the form of a homographic mapping:

$$\overline{x} = \frac{\alpha x + \beta}{\gamma x + \delta}, \tag{1.2}$$

2000 *Mathematics Subject Classification.* Primary: 39A10; Secondary: 34A34.

S. Lafortune acknowledges two scholarships: one from NSERC (National Science and Engineering Research Council of Canada) for his Ph.D. and one from "Programme de Soutien de Cotutelle de Thèse de doctorat du Gouvernement du Québec" for his stay in Paris.

This is the final form of the paper.

where x stands for x_n, $\overline{x}$ for x_{n+1} (and, of course, $\underline{x}$ for x_{n-1}), can be also linearized through a discrete version of the Cole–Hopf transformation.

The extension of the Riccati to higher orders can be and has been obtained [**1**]. The simplest linearisable system at N dimensions is the projective Riccati which assumes the form:

$$w'_\mu = a_\mu + \sum_\nu b_{\mu\nu} w_\nu + w_\mu \sum_\nu c_\nu w_\nu \quad \text{with } \mu = 1, \ldots, N. \tag{1.3}$$

In two dimensions the projective Riccati system can be cast into the second order equation:

$$w'' = -3ww' - w^3 + q(t)(w' + w^2). \tag{1.4}$$

The discrete analog of the projective Riccati does exist and is studied in detail in [**7**]. The corresponding form is:

$$\overline{x}_\mu = \frac{a_\mu + x_\mu + \sum_\nu b_{\mu\nu} x_\nu}{1 - \sum_\nu c_\nu x_\nu}. \tag{1.5}$$

Again in two dimensions the discrete projective Riccati can be written as a second-order mapping of the form

$$\alpha \overline{x} x \underline{x} + \beta \overline{x} x + \gamma x \underline{x} + \delta \overline{x} \underline{x} + \varepsilon x + \zeta \overline{x} + \eta \underline{x} + \theta = 0 \tag{1.6}$$

which was first introduced in [**11**]. The coefficients α, $\beta, \ldots, \theta$ are not totally free. Although the linearisability constraints have been obtained in [**11**], the study of mappings of the form (1.6) was not complete. In the present work we intend the study of (1.6) in its general form.

Another point must be brought to attention here. In the continuous case the study of second order equations has revealed the relation of the linearisable equations (1.4) to the Gambier equation [**2**]. The latter is obtained as a system of two coupled Riccati in cascade

$$\begin{aligned} y' &= -y^2 + qy + c, \\ w' &= aw^2 + nyw + \sigma, \end{aligned} \tag{1.7}$$

where n is an integer. It contains as a special case the linearisable equation (1.4) which is obtained from (1.7) for $n = 1$ and $a = -1$, $c = 0$ and $\sigma = 0$. The discrete analog of the Gambier mapping was introduced in [**3**] and in full generality in [**4**].

In the present work we shall also address the question of the construction of integrable third order systems in the spirit of Gambier. Namely we shall start with a second order integrable equation and couple it with a Riccati (or a linear) first order (also integrable) equation. This enterprise may easily assume staggering proportions. In order to limit the scope of our investigation we shall consider coupled systems where the dependent variable enters only in a polynomial way. This leads naturally to the coupling of a Painlevé ($\mathbb{P}$) I or II to a Riccati.

In the next section we shall analyse (1.6) and show how one can isolate the integrable cases through the use of the singularity confinement criterion. In Section 3 we will present how a third order integrable equation can be constructed from the coupling of a Riccati and a Painlevé equation (in the discrete and the continuous case).

2. Linearisable Mappings as Discrete Projective Systems

In [**11**] we have introduced projective system as a way to linearize a second-order mapping. (The general theory of discrete projective systems has been recently presented in [**7**]). In this older work of ours we have focused on a three-point mapping that can be obtained from a 3×3 projective system. The main idea was to consider the system:

$$\begin{pmatrix} \overline{u} \\ \overline{v} \\ \overline{w} \end{pmatrix} = \begin{pmatrix} p_{11} & p_{12} & p_{13} \\ p_{21} & p_{22} & p_{23} \\ p_{31} & p_{32} & p_{33} \end{pmatrix} \begin{pmatrix} u \\ v \\ w \end{pmatrix}, \tag{2.1}$$

and conversely

$$\begin{pmatrix} \underline{u} \\ \underline{v} \\ \underline{w} \end{pmatrix} = \begin{pmatrix} m_{11} & m_{12} & m_{13} \\ m_{21} & m_{22} & m_{23} \\ m_{31} & m_{32} & m_{33} \end{pmatrix} \begin{pmatrix} u \\ v \\ w \end{pmatrix}, \tag{2.2}$$

where the matrix M is obviously related to the matrix P through $\overline{M} = P^{-1}$. Introducing the variable $x = u/v$ and the auxiliary $y = w/v$ we can rewrite (2.1) and (2.2) as

$$\overline{x} = \frac{p_{11}x + p_{12} + p_{13}y}{rp_{21}x + p_{22} + p_{23}y}, \tag{2.3}$$

$$\underline{x} = \frac{m_{11}x + m_{12} + m_{13}y}{m_{21}x + m_{22} + m_{23}y}. \tag{2.4}$$

(Since the m_{3i}, p_{3i} do not appear in (2.3), (2.4) and we can simplify M, P by taking $m_{33} = p_{33} = 1$ and $m_{31} = p_{31} = m_{32} = p_{32} = 0$). Finally eliminating y between (2.3) and (2.4) we obtain the mapping:

$$\alpha\overline{x}x\underline{x} + \beta\overline{x}x + \gamma x\underline{x} + \delta\overline{x}\underline{x} + \varepsilon x + \zeta\overline{x} + \eta\underline{x} + \theta = 0, \tag{2.5}$$

where the α, $\beta, \ldots, \theta$ are related to the m, p's.

Equation (2.5) will be the starting point of the present study. Our question will be when is an equation of this form integrable. (Clearly the relation to the projective system works only for a particular choice of the parameters). In order to investigate the integrability of (2.5) we shall use the singularity confinement approach that was introduced in [**5**]. What are the singularities of (2.5)? Given the form of (2.5) it is clear that diverging x does not lead to any difficulty. However, another (subtler) difficulty arises whenever x_{n+1} is defined independently of x_{n-1}. In this case the mapping "loses one degree of freedom". Thus the singularity condition is

$$\frac{\partial x_{n+1}}{\partial x_{n-1}} = 0,$$

which leads to :

$$(\alpha x + \delta)(\varepsilon x + \theta) = (\beta x + \zeta)(\gamma x + \eta). \tag{2.6}$$

Equation (2.6) is the condition for the appearance of a singularity. Given the invariance of (2.5) under homographic transformations it is clear that one can use them in order to simplify (2.6). Several choices exist but the one we shall make here is to choose the roots of (2.6) so as to be equal to 0 and ∞, unless of course

(2.6) has two equal roots, in which case we bring them both to 0. Let us examine the distinct root case. For the roots of (2.6) to be 0 and ∞ we must have:

$$\begin{aligned} \alpha\varepsilon &= \beta\gamma, \\ \delta\theta &= \zeta\eta. \end{aligned} \tag{2.7}$$

The generic mapping of the form (2.5) has $\alpha\theta \neq 0$ and we can take $\alpha = \theta = 1$ (by the appropriate scaling of x and a division). We have thus,

$$\overline{x}x\underline{x} + \beta\overline{x}x + \gamma x\underline{x} + \zeta\eta\overline{x}\underline{x} + \beta\gamma x + \zeta\overline{x} + \eta\underline{x} + 1 = 0. \tag{2.8}$$

Nongeneric cases do exist as well and have been examined in detail in [**6**].

In order to investigate the integrability of the mapping (2.8) we shall apply the singularity confinement criterion. Here the singularities are by construction 0 and ∞. Following the results of [**11**] we require confinement in just one step. We require to have an indeterminate form 0/0 at the step following the singularity. This leads to the condition $\beta = \zeta = 0$. We thus obtain the mapping:

$$\overline{x}x\underline{x} + \gamma x\underline{x} + \eta\underline{x} + 1 = 0, \tag{2.9}$$

or, solving for $\overline{x}$:

$$\overline{x} = -\gamma + \frac{\eta\underline{x} + 1}{x\underline{x}}, \tag{2.10}$$

where γ and η are free. This is indeed integrable: we can show that (2.10) can be obtained from the projective system (2.1), (2.2) provided we take

$$P = \begin{pmatrix} -(\gamma q + 1) & q\overline{q} & 1 \\ q & 0 & 0 \\ 0 & 0 & 1 \end{pmatrix}, \tag{2.11}$$

and $M = \underline{P}^{-1}$ provided q is some solution of the equation $\overline{q}q\underline{q} + q\underline{q}\eta + \underline{q}\gamma + 1 = 0$. Integrable but nonlinearisable cases of (2.8) do also exist: they have been identified and presented in detail in [**6**].

Before closing this section let us present the continuous limits of the linearisable equation we have identified above. For (2.9) we put $x = -1 + \nu w$, $\gamma = 3 + \nu^2 p$, $\eta = \gamma + \nu^3 q$ and we obtain at the limit $\nu \to 0$ the equation:

$$w'' = 3ww' - w^3 + pw + q. \tag{2.12}$$

This is equation #6 in the Painlevé/Gambier classification [**8**] (in noncanonical form) and precisely the one that can be obtained from a $N = 2$ projective Riccati system.

3. Constructing Integrable Third Order Systems: The Gambier Approach

The key idea of Gambier was to construct an integrable second order equation by suitably coupling two integrable first order ones. The latter were well-known: at first order the only integrable (in the sense of having the Painlevé property) ordinary differential equations are either linear or of Riccati type. The Gambier equation is precisely the coupling of two Riccati in cascade.

In [**10**], we extend this idea for third order systems. We couple Painlevé second order equations with the Riccati equation both in the continuous and the discrete cases.

The coupling of a Painlevé equation with a Riccati was first considered by Chazy. He examined the coupling of $\mathrm{P_I}$:

$$w'' = 6w^2 + z, \tag{3.1}$$

with a Riccati:

$$y' = \alpha y^2 + \lambda w + \gamma, \tag{3.2}$$

where α, β, λ, γ are functions of z. This coupling is additive: it is indeed the only coupling that is compatible with integrability. Chazy found that (3.2) must have the form:

$$y' = \frac{1-k^2}{4} y^2 + w + \gamma, \tag{3.3}$$

where $k = 6m + n$. Chazy found the following necessary integrability constraints:

$$\begin{aligned} n &= 2 \qquad \gamma = 0, \\ n &= 3 \qquad \gamma' = 0, \\ n &= 4 \qquad \gamma'' = \mu\gamma^2 + \nu z, \\ n &= 5 \qquad \gamma''' = \mu\gamma\gamma' + \nu, \end{aligned} \tag{3.4}$$

where μ and ν are specific numerical constants. It turns out that for $k = n$ they are also sufficient. For $k = 6m+1$ the first condition appears at $k = 7$. In this case the constraint reads:

$$\gamma^{(5)} = 48\gamma\gamma''' + 120\gamma'\gamma'' - \frac{2304}{5}\gamma'\gamma^2 - 24z\gamma' - 48\gamma. \tag{3.5}$$

This equation has the Painlevé property and is thus expected to be integrable.

In [**10**] we have presented the coupling of a Riccati to other integrable second order differential equations.

To construct integrable discrete systems in the same spirit as Gambier we need a detailed knowledge of the forms of the equations to be coupled and an integrability detector. The second order mappings which play the role of the Painlevé equations in the discrete domain have been the object of numerous detailed studies and we are now in possession of discrete forms of all the equations of the Painlevé/Gambier classification. The discrete integrability detector is based on the singularity confinement.

We consider the coupling of a discrete Riccati for the variable y:

$$\overline{y} = \frac{(\alpha x + \beta)y + (\eta x + \theta)}{(\varepsilon x + \zeta)y + (\gamma x + \delta)}, \tag{3.6}$$

(where $\alpha, \beta, \ldots, \theta$ depend in general on n) the coefficients of which depend linearly on x, the solution of the discrete $\mathrm{P_{II}}$. The mapping (3.6) can be brought under canonical form through the application of homographic transformations on y to either:

$$\overline{y} = \frac{(\alpha x + \beta)y + 1}{y + (\gamma x + \delta)}, \tag{3.7}$$

or:

$$\overline{y}(\gamma x + \delta) - y(\alpha x + \beta) - 1 = 0. \tag{3.8}$$

In [**10**] we examined in detail the coupling of (3.7) and (3.8) with either d-P_I or d-P_{II} (under various forms). Here, we will focus on a particular example of a coupling to d-P_{II}.

How does one apply the singularity confinement criterion to a mapping such as (3.7) when x is given by some discrete equation like d-P_I or d-P_{II}? The singularity manifests itself by the fact that $\overline{y}$ is independent of y i.e. when

$$(\gamma x + \delta)(\alpha x + \beta) = 1. \tag{3.9}$$

This quadratic equation has two roots X_1, X_2. The confinement condition is for y to recover the lost degree of freedom. This can be done if y assumes an indeterminate form $0/0$. This means that x at this stage must again satisfy (3.9) and moreover be such that the denominator (or, equivalently, the numerator) vanishes.

Let us assume now that for some n we have $x_n = X_1$. The confinement requirement is that k steps later $x_{n+k} = X_2$. Starting from $x_n = X_1$ and some initial datum x_{n-1}, we can iterate the mapping for x and obtain x_{n+k} as a complicated function of x_{n-1} and X_1. Since x_{n+k} depends on the free parameter x_{n-1} there is no hope for x_{n+k} to be equal to X_2 if X_1 is a generic point for the mapping of x. The only possibility is that both X_1 and X_2 be special values. What are the special values of this equation depends on its details, but clearly in the case of the discrete Painlevé's we shall examine here, these values can only be the ones related to the singularities. To be more specific, let us examine d-P_{II}:

$$\overline{x} + \underline{x} = \frac{zx + a}{1 - x^2}. \tag{3.10}$$

The only special values of x are the ones related to the singularity $x_n = \pm 1$, $x_{n+1} = \infty$, $x_{n+2} = \mp 1$ while $\ldots, x_{n-2}, x_{n-1}$ and $x_{n+3}, x_{n+4}, \ldots$ are finite. This means that the two roots of (3.9) must be two of $\{+1, \infty, -1\}$ and moreover that confinement must occur in two steps. The precise implementation of singularity confinement requires that the denominator of (3.7) at $n+2$ vanishes (and because of (3.9) this ensures that the numerator vanishes as well). Moreover, we must make sure that the lost degree of freedom (i.e. the dependence on y) is indeed recovered through the indeterminate form.

The singularity patterns of (3.10) are

$$\{\pm 1, \infty, \mp 1\}. \tag{3.11}$$

This means that the singularity condition (3.9) must have ± 1 as roots. As a result we have:

$$\delta = -\beta/(\alpha^2 - \beta^2), \tag{3.12}$$

$$\gamma = \alpha/(\alpha^2 - \beta^2). \tag{3.13}$$

The two different patterns lead to a first confinement conditions given by:

$$\beta = k\alpha, \tag{3.14}$$

where k is a constant with binary freedom which we will ignore from now. The second condition:

$$\underline{\alpha}\alpha^2\overline{\alpha} = \frac{1}{(1 - k^2)^2}. \tag{3.15}$$

This equation can be solved by linearisation just by taking the logarithm of both sides. More examples of couplings of discrete equations can be found in [**10**].

4. Conclusion

In the previous sections we have first investigated 3-point mappings that are integrable through linearisation. Our analysis was guided by the analogy with the continuous situation and results of ours on $N = 2$ projective systems. We have presented an analysis of the linearisable mapping and identified one of its integrable form. Moreover we have presented an approach for the construction of integrable third order systems through the coupling of a second order equation to a Riccati or a linear first order equation. Thus we have extended the Gambier approach (first used in his derivation of the second order ODE that bears his name) to higher order systems. We have applied this coupling method to both continuous and discrete systems.

References

1. R. L. Anderson, J. Harnad, and P. Winternitz, *Systems of ordinary differential equations with superposition principles*, Physica D **4** (1982), 164–182.
2. B. Gambier, *Sur les équations différentielles du second ordre et du premier degré dont l'intégrale générale est à points critiques fixes*, Acta Math. **33** (1910), 1–55.
3. B. Grammaticos and A. Ramani, *The Gambier mapping*, Physica A **223** (1996), 125–136.
4. B. Grammaticos, A. Ramani, and S. Lafortune, *The Gambier mapping, revisited*, Physica A **253** (1998), 260–270.
5. B. Grammaticos, A. Ramani, and V. Papageorgiou, *Do integrable mappings have the Painlevé property?*, Phys. Rev. Lett. **67** (1991), 1825–1828.
6. B. Grammaticos, A. Ramani, K. M.Tamizhmani, and S. Lafortune, *Again, linearisable mappings*, Physica A **252** (1998), 138–150.
7. B. Grammaticos, A. Ramani, and P. Winternitz, *Discretizing families of linearisable equations*, Phys. Lett. A **245** (1998), 382–388.
8. E. L. Ince, *Ordinary differential equations*, Dover, New York, 1956.
9. M. D. Kruskal, A. Ramani, and B. Grammaticos, *Singularity analysis and its relation to complete, partial and nonintegrability*, NATO Adv. Sci. Inst. Ser. C Math. Phys. Sci. **310**, Kluwer 1989, 321–372.
10. S. Lafortune, B. Grammaticos, and A. Ramani, *Constructing third order integrable systems: The Gambier approach*, Inverse Problems **14** (1998), 287–298.
11. A. Ramani, B. Grammaticos, and G. Karra, *Linearisable mappings*, Physica A **180** (1992), 115–127.

Current address: Centre de recherches mathématiques, Université de Montréal, Montréal, Qué., H3C 3J7, Canada.

LPTM ET GMPIB, UNIVERSITÉ PARIS VII, TOUR 24-14, 5E ÉTAGE, 75251 PARIS, FRANCE.
E-mail address: `lafortus@crm.umontreal.ca`

LPN, UNIVERSITÉ PARIS VII, TOUR 24-14, 5^{e}ÉTAGE, 75251 PARIS, FRANCE.
E-mail address: `grammati@paris7.jussieu.fr`

CPT, ÉCOLE POLYTECHNIQUE, CNRS, UPR 14, 91128 PALAISEAU, FRANCE.
E-mail address: `ramani@cpht.polytechnique.fr`

Centre de Recherches Mathématiques
CRM Proceedings and Lecture Notes
Volume **25**, 2000

Difference Equations of Poincaré–Perron Type and Central Two-Point Connection Problems

Wolfgang Lay

ABSTRACT. Difference equations of Poincaré–Perron type are intimately related to central two-point connection problems of ordinary second-order linear differential equations with polynomial coefficients. We study the case when the solutions of the differential equation are to connect an ordinary point with an irregular singularity. In this case the difference equation also becomes of irregular type.

Our main result is that there is a certain location of the two relevant points and there are certain expansions (called Jaffé expansions) that result in this sort of difference equations with the help of which it is possible to formulate the exact eigenvalue condition. The main mathematical tool is the Birkhoff set of the difference equation.

1. Introduction

Consider the linear second-order ordinary differential equation

$$\frac{d^2 f(x)}{dx^2} + Q(x;E)f(x) = 0, \quad x \in \mathbb{C}, \tag{1.1}$$

where $Q(x;.)$ is a rational function in x. We suppose that $Q(x;.)$ has a pole of higher order than two and thus is an irregular singularity of (1.1) the s-rank (see [**10**] and [**11**]) of which is $R > 1$, $R \in \mathbb{N}$. Moreover, we suppose that (1.1) has a regular singularity at infinity and that there is no singularity within the unit circle $|x| = 1$. However, there may be further singularities of (1.1). The coefficient $Q(x;E)$ depends on a parameter E that serves as an eigenvalue parameter in the following. For this aim it is sufficient to suppose that Q depends on E as

$$Q = Q_1 + EQ_2, \quad Q_2 \neq 0. \tag{1.2}$$

We are going to study an eigenvalue problem of (1.1), (1.2) on the real interval $[0,1]$. To formulate it we have to study the asymptotic behaviour of solutions at the relevant singularity $x = 1$.

2000 *Mathematics Subject Classification.* Primary: 34M35; Secondary: 30B30.

I am indebted to Dipl. Phys. Karlheinz Bay for the collaboration in elaborating the idea of the proof as well as for carrying out some of its calculations. Moreover, I am pleased to acknowledge some fruitful discussions with Dr. Michael Zaiser and with Professor Dr. Serguei Yu. Slavyanov on the topic presented above.

This is the final form of the paper.

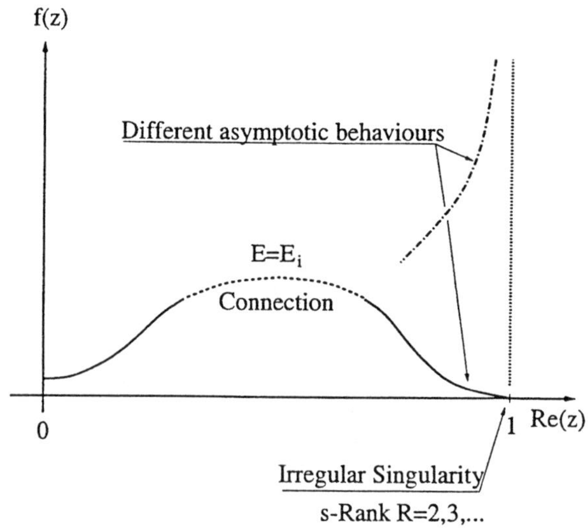

FIGURE 1. The central two-point connection problem.

The general solution of (1.1) near $x = 1$ may be written asymptotically as

$$\lim_{x\to 1} f(x) \sim A(E)\exp\left(+\frac{\alpha}{(1-x)^{R-1}} + \sum_{1}^{R-2}\frac{\alpha_j}{(1-x)^j}\right)\cdot(1-x)^{\alpha_0} \tag{1.3}$$
$$+ B(E)\exp\left(-\frac{\alpha}{(1-x)^{R-1}} + \sum_{1}^{R-2}\frac{\alpha_j}{(1-x)^j}\right)\cdot(1-x)^{\alpha_0},$$

where A and B are constants in x and α is real positive. The fact that there is a fundamental system of equation (1.1) the particular solutions of which are identifiable by their asymptotic behaviours for $x \to 1$ along the real axis gives rise to a certain boundary-value problem which we are going to study in the following.

The origin $x = 0$ is an ordinary point of (1.1). Therefore, all particular solutions are holomorphic there and thus may be expanded in a power series

$$f(x) = \sum_{n=0}^{\infty} a_n x^n. \tag{1.4}$$

For fixing the particular solution of (1.1) by (1.4), one has to choose the values of a_0 and of a_1.

Now, we go to the other side of the relevant interval, i.e. to $x = 1$. Suppose, the limiting process in (1.3) is taken from within the unit circle approaching the singularity at $x = 1$ radially. We may then look for those values of $E = E_i$, $i = 1, 2, \ldots$ for which the coefficient $A(E)$ in (1.3) is zero when $f(x)$ is represented by the series (1.4)

$$A(E) = 0 \quad \text{for} \quad E = E_i.$$

This means that the series in (1.4) represents a function that must decrease exponentially for $x \to 1$.

This is a particular boundary-value problem that is characterized by the fact that at one end of the relevant interval (here $[0, 1]$) is located a singularity of the differential equation (1.1). This sort of boundary problems is called central two-point connection problem (CTCP) and may be generalized in several respects. First, we have to state that the singularity in our problem is of irregular type while the second endpoint of the relevant interval $[0,1]$ is an ordinary point of the underlying differential equation. However, our approach may be generalized to the cases when this point also is a singularity either regular or irregular. Besides this, both singularities may be regular ones. The various situations are treated and may be seen from [**4–8**].

The reader may find sketched the above-discussed situation in Fig. 1.

To solve the CTCP we make an ansatz that exhibits the required characteristic features discussed above:

$$f(x) = \exp\left(-\frac{\alpha}{(1-x)^{R-1}} + \sum_{1}^{R-2} \frac{\alpha_j}{(1-x)^j}\right) \cdot (1-x)^{\alpha_0} \cdot g(x), \tag{1.5}$$

with

$$g(x) = \sum_{n=0}^{\infty} b_n x^n. \tag{1.6}$$

The radius of convergence of the power series in this ansatz is unity. By determining b_0, b_1 in the power series (1.6) of the ansatz (1.5) the boundary condition at $x = 0$ can be fulfilled, immediately. The problem now is to find the eigenvalue condition for which the boundary condition at the other end of the relevant interval at $x = 1$ is met. This boundary condition means that the solution $f(x)$ must behave as the pre-factor in (1.5) for $x \to 1$. Whether this is the case or not is governed by the asymptotic behaviour of b_n in (1.6) for $n \to \infty$ which we are going to study, now.

2. Difference Equations of Poincaré–Perron Type

By inserting (1.5), (1.6) into (1.1) the b_n become solutions of an *irregular* difference equation of *Poincaré–Perron* type the order of which is R:

$$\sum_{l=0}^{R} \Omega_l(n) b_{n+l} = 0, \quad \lim_{n\to\infty} \Omega_l(n) = \Omega_l, \tag{2.1}$$

where

$$\Omega_l(n) = \Omega_l + \frac{\alpha_l}{n} + \frac{\beta_l}{n^2}, \quad (\Omega_l, \alpha_l, \beta_l \text{ are constants in } n),$$

and

$$\sum_{l=0}^{R} \Omega_l = 0, \quad \Omega_R \neq 0, \quad \Omega_0 \neq 0.$$

In (2.1), we have neglected $R-2$ initial equations such that prescribing b_0 and b_1 means that all the following b_n, $n = 2, 3, \ldots\ R-1,\ R,\ R+1 \ldots$ may be calculated recursively.

We are going to study the particular solutions of the difference equation (2.1). Defining $t(n) := (b_{n+1})/(b_n)$ all the particular solutions of this difference equation may be distinguished by characteristic equations

$$\sum_{l=0}^{R} \left(\Omega_l + \frac{\alpha_l}{n}\right) t^l = 0, \tag{2.2}$$

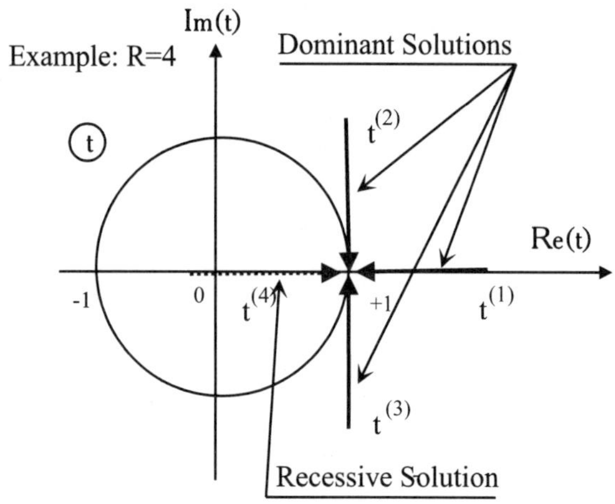

FIGURE 2. Particular solutions of the difference equation presented in the complex t-plane divided into dominant ones (solid arrows) and recessive ones (broken arrow).

the index-asymptotic solutions of which are given by

$$t^{(k)}(n) = 1 + \frac{\gamma_k}{n^{\frac{1}{R}}} + O\left(\frac{1}{n^{\frac{2}{R}}}\right), \quad \text{for} \quad n \to \infty, \quad k = 1, 2, \ldots, R, \tag{2.3}$$

where the distinct values of γ_k are solutions of the Rth-order algebraic equation

$$\gamma^R = -\sum_{l=0}^{R} \alpha_l. \tag{2.4}$$

(2.3) means that the asymptotic behaviour of all particular solutions of (2.1) is exponential for $n \to \infty$. There is an asymptotic representation of a fundamental system of the difference equation (2.1) for $n \to \infty$ that is called Birkhoff set (see [**13**, p. 274]). It has the form

$$s^{(k)}(n) = \exp\left[\sum_{j=1}^{R-1} \gamma_{kj} n^{\frac{j}{R}}\right] n^r \left[1 + \frac{C_{1k}}{n^{\frac{1}{R}}} + \frac{C_{2k}}{n^{\frac{2}{R}}} + \cdots + \frac{C_{ik}}{n^{\frac{i}{R}}} + \ldots\right], \tag{2.5}$$
$$k = 1, 2, \ldots, R,$$

where r and the C_{ik} are constants to be determined by inserting (2.5) into (2.1).

The totality of particular solutions $b_n^{(k)}$ of the difference equation (2.1) we divide into two sets for the following namely the recessive solutions $b_n^{(k',\mathrm{rec})}$ and the dominant solutions $b_n^{(k",\mathrm{dom})}$. The recessive ones are characterized by the fact that their related $t^{(k')}(n)$ (cf. (2.3) and Fig. 2) tend to unity from *inside* the unit cirlce in the complex t-plane while for the dominant ones the related $t^{(k")}(n)$ tend to unity from *outside* the unit circle. The general solution of (2.1) thus may be written as

$$b_n = \sum_{k'} L_{k'}^{\mathrm{rec}}(E) b_n^{(k',\mathrm{rec})} + \sum_{k"} L_{k"}^{\mathrm{dom}}(E) b_n^{(k",\mathrm{dom})}. \tag{2.6}$$

The summution runs over the whole sets of recessive and dominant particular solutions, respectively.

3. Eigenvalue Condition

For the eigenvalue condition of the CTCP it is crucial that the role of the function $g(x)$ in the ansatz (1.5) changes when the parameter E becomes an eigenvalue E_i. This role-changing is deeply related to the two sets of particular solutions of (2.1) defined above. For this to see we consider the asymptotic behaviour of $g(x)$ for $x \to 1$. If E is no eigenvalue we have

$$g(x) \sim \exp\left(+2\frac{\alpha}{(1-x)^{R-1}}\right) \text{ as } x \to 1, \tag{3.1}$$

as a leading order asymptotics. This may be seen from (1.3) and (1.5). As soon as E becomes an eigenvalue this behaviour changes and becomes

$$g(x) \sim O(1) \text{ as } x \to 1, \tag{3.2}$$

because all the coefficients $L_{k"}^{\mathrm{dom}}(E)$ for the dominant solutions of (2.1) have to vanish in order to get the exponentially decreasing behaviour of the differential equation represented by the exponential factor in (1.5). This means, that the power series (1.6) at $x = 1$ becomes convergent, thus

$$\sum_{n=0}^{\infty} b_n, \tag{3.3}$$

is finite.

As a result, the exact eigenvalue for the CTCP may be written in the form

$$L_{k"}^{\mathrm{dom}}(E) = 0 \text{ for } E = E_i, \quad \text{and for all } k". \tag{3.4}$$

The crucial mathematical item in our theory is the relation between the solutions (1.5) of the differential equation (1.1) and the Birkhoff set (2.5) that we have exhibited in a formal manner above. In the following we will write it more rigorously in the form of the following

PROPOSITION. *If a (complex-valued) function behaves like the asymptotic factor in* (1.5) *for* $x \to 1$ *(this is the right-hand side of* (1.5) *withouth* $g(x)$*), then the coefficients of its expansion in a power series about* $x = 0$ *(that is supposed to be convergent within the unit circle) behave like the Birkhoff set* (2.5) *for* $n \to \infty$.

4. Proof

The following lines are devoted to the proof of the above-written proposition. In order to prove the proposition we need a lemma:

LEMMA. *Consider the function*

$$h(x) := e^{\alpha/((1-x)^s)}(1-x)^{-\beta}, \quad s = 1, 2, 3 \ldots, x \in \mathbb{C}, \tag{4.1}$$

with α, β *being complex constants. The exponential function in* (4.1) *has an essential singularity at* $x = 1$. *The function* $h(x)$ *may be expanded in a converging power series*

$$h(x) = \sum_{n=0}^{\infty} c_n x^n, \tag{4.2}$$

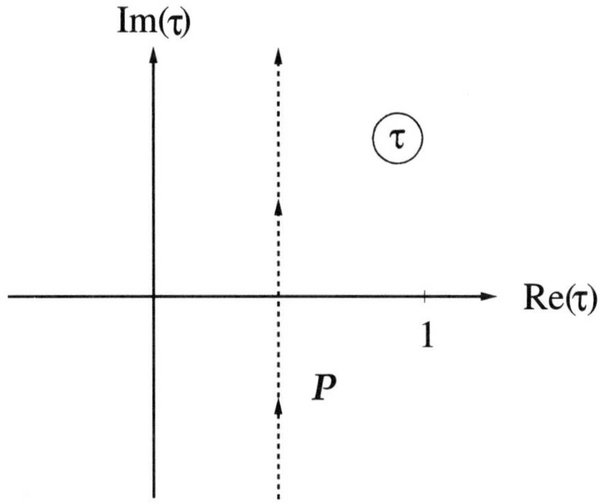

FIGURE 3. Integration path $\mathcal{P}$.

the radius of convergence is unity. Then, there is the following integral representation of the coefficients of the series in (4.2) (*cf.* [**2**, Eq. (12)]):

$$c_n = \frac{1}{2\pi\imath}\int_{\mathcal{P}} \tau^{-\beta}(1-\tau)^{-(n+1)} e^{\frac{\alpha}{\tau^s}}\,\mathrm{d}\tau. \tag{4.3}$$

The path of integration $\mathcal{P}$ in the complex τ-plane is a line parallel to the imaginary axis starting at minus infinity crossing the real axis at positive values (*cf.* Fig. 3).

In the following we give a proof of this lemma.

PROOF OF THE LEMMA. Following Perron [**9**] we expand (4.1) in powers of α according to

$$e^{\alpha/((1-x)^s)}(1-x)^{-\beta} = \sum_{\nu=0}^{\infty} \frac{\alpha^\nu}{\nu!}(1-x)^{-(\beta+s\nu)}. \tag{4.4}$$

The series in (4.4) converges within the unit circle $|x| < 1$. Expanding the term

$$(1-x)^{-(\beta+s\nu)}$$

for $\nu = 0, 1, 2, 3\ldots$ to powers in x, the right hand side of (4.4) yields the scheme

$$\begin{array}{ccccccccc}
 & \frac{\alpha^0}{0!}1 & + & \frac{\alpha^0}{0!}(\beta+0s)\frac{x}{1!} & + & \frac{\alpha^0}{0!}(\beta+0s)(\beta+0s+1)\frac{x^2}{2!} & + \ldots \\
+ & \frac{\alpha^1}{1!}1 & + & \frac{\alpha^1}{1!}(\beta+1s)\frac{x}{1!} & + & \frac{\alpha^1}{1!}(\beta+1s)(\beta+1s+1)\frac{x^2}{2!} & + \ldots \\
+ & \frac{\alpha^2}{2!}1 & + & \frac{\alpha^3}{2!}(\beta+2s)\frac{x}{1!} & + & \frac{\alpha^2}{2!}(\beta+2s)(\beta+2s+1)\frac{x^2}{2!} & + \ldots \\
+ & \frac{\alpha^3}{3!}1 & + & \frac{\alpha^3}{3!}(\beta+3s)\frac{x}{1!} & + & \frac{\alpha^3}{3!}(\beta+3s)(\beta+3s+1)\frac{x^2}{2!} & + \ldots \\
+ & \vdots & + & \vdots & + & \vdots & + \ddots
\end{array} \tag{4.5}$$

where ν numbers the lines of the scheme. Denoting the columns in (4.5) by n on closer examination of (4.5) and in connection with Euler's Beta function defined by (cf. [**1**, p. 258])

$$B(x,n) := \frac{(n-1)!}{x(x+1)\ldots(x+n-1)} = \frac{\Gamma(x)\Gamma(n)}{\Gamma(x+n)}, \quad n = 1, 2, \ldots,$$

we find that the n-th column of the scheme (4.5) is given by

$$\sum_{\nu=0}^{\infty} \frac{\alpha^\nu}{\nu!} \frac{(n-1)!}{B(\beta+s\nu,n)} \frac{x^n}{n!}.$$

With respect to Weierstrass' "Doppelreihensatz" [**12**] we rewrite (4.4) as

$$\sum_{n=0}^{\infty} x^n \frac{1}{n} \sum_{\nu=0}^{\infty} \frac{\alpha^\nu}{\nu!} \frac{1}{B(\beta+s\nu,n)} =: \sum_{n=0}^{\infty} c_n x^n, \quad c_0 = e^\alpha, \tag{4.6}$$

where the coefficients of the series on the right-hand side of (4.6)—given by the expansion of (4.1) in (4.2)—yield

$$c_n = \frac{1}{n} \sum_{\nu=0}^{\infty} \frac{\alpha^\nu}{\nu!} \frac{1}{B(\beta+s\nu,n)} = \frac{1}{\Gamma(n+1)} \sum_{\nu=0}^{\infty} \frac{\alpha^\nu}{\nu!} \frac{\Gamma(\beta+s\nu+n)}{\Gamma(\beta+s\nu)}. \tag{4.7}$$

To get an integral representation of (4.7) we consider the well-known integral representation of the Beta function (cf. [**1**, p. 258])

$$B(\widetilde{\beta}, \widetilde{\gamma}-\widetilde{\beta}) = \int_0^1 v^{\widetilde{\beta}-1}(1-v)^{\widetilde{\gamma}-\widetilde{\beta}-1} \mathrm{d}v = \frac{\Gamma(\widetilde{\beta})\Gamma(\widetilde{\gamma}-\widetilde{\beta})}{\Gamma(\widetilde{\gamma})}, \tag{4.8}$$

where $\Re(\widetilde{\beta}) > 0$ and $\Re(\widetilde{\gamma}-\widetilde{\beta}) > 0$.

If $\mathcal{L}$ is a loop in the v-plane starting from zero below the positive real axis encircling the point $v = 1$ anti-clockwise thus crossing the real axis from negative imaginary parts to positive ones for $\Re(v) > 1$ and returning to the origin we get (cf. [**9**] and Fig. 4)

$$\begin{aligned} \int_{\mathcal{L}} v^{\widetilde{\beta}-1}(v-1)^{\widetilde{\gamma}-\widetilde{\beta}-1} \mathrm{d}v \\ = \left(\mathrm{e}^{-\pi\imath(\widetilde{\gamma}-\widetilde{\beta}-1)} - \mathrm{e}^{\pi\imath(\widetilde{\gamma}-\widetilde{\beta}-1)}\right) \cdot \int_0^1 v^{\widetilde{\beta}-1}(1-v)^{\widetilde{\gamma}-\widetilde{\beta}-1} \mathrm{d}v. \end{aligned} \tag{4.9}$$

Applying Euler's "Ergänzungssatz" for the gamma function (cf. [**1**, p. 256])

$$\Gamma(\widetilde{\gamma}-\widetilde{\beta})\Gamma\big(1-(\widetilde{\gamma}-\widetilde{\beta})\big) = \frac{\pi}{\sin \pi(\widetilde{\gamma}-\widetilde{\beta})}$$

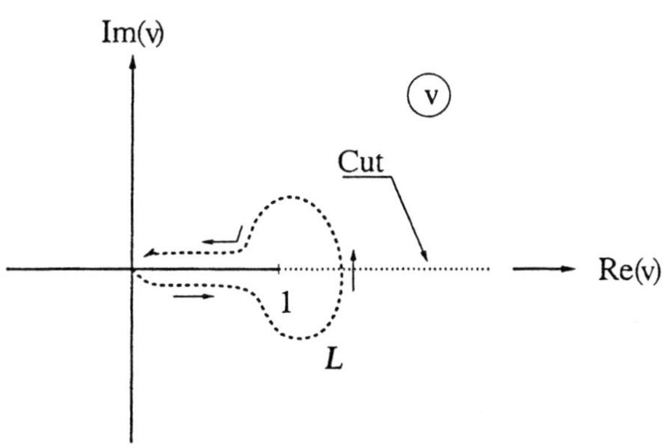

FIGURE 4. Integration path $\mathcal{L}$.

to the right-hand side of (4.8) and inserting this into (4.9) thus replacing the integral on the right-hand side we get

$$\int_{\mathcal{L}} v^{\widetilde{\beta}-1}(v-1)^{\widetilde{\gamma}-\widetilde{\beta}-1}\mathrm{d}v = 2\pi\imath \frac{\Gamma(\widetilde{\beta})}{\Gamma(\widetilde{\gamma})\Gamma(1-\widetilde{\gamma}+\widetilde{\beta})}. \tag{4.10}$$

In this formula on both sides are written entire transcendental functions in $\widetilde{\gamma}$. Therefore, we may drop the condition $\Re(\widetilde{\gamma}-\widetilde{\beta}) > 0$ while $\Re(\widetilde{\beta}) > 0$ still must hold.

Adding the exponential factor $\exp(\alpha v^s)$ to the integrand of the integral on the left-hand side of (4.10), expanding this term in powers of α and integrating by parts with respect to ν, we eventually obtain

$$\begin{aligned}\int_{\mathcal{L}} v^{\widetilde{\beta}-1}(v-1)^{\widetilde{\gamma}-\widetilde{\beta}-1}e^{\alpha v^s}\mathrm{d}v &= \sum_{\nu=0}^{\infty}\frac{\alpha^\nu}{\nu!}\int_{\mathcal{L}} v^{\widetilde{\beta}+s\nu-1}(v-1)^{\widetilde{\gamma}-\widetilde{\beta}-1}\mathrm{d}v,\\ &= 2\pi\imath\sum_{\nu=0}^{\infty}\frac{\alpha^\nu}{\nu!}\frac{\Gamma(\widetilde{\beta}+s\nu)}{\Gamma(\widetilde{\gamma}+s\nu)\Gamma(1-\widetilde{\gamma}+\widetilde{\beta})}.\end{aligned} \tag{4.11}$$

A comparison of the coefficients in (4.6) with (4.11) shows that $\widetilde{\beta} = \beta + n$, $\widetilde{\gamma} = \beta$ and it follows

$$\frac{1}{\Gamma(n+1)}\sum_{\nu=0}^{\infty}\frac{\alpha^\nu}{\nu!}\frac{\Gamma(\beta+s\nu+n)}{\Gamma(\beta+s\nu)} = \frac{1}{2\pi\imath}\int_{\mathcal{L}} v^{\beta+n-1}(v-1)^{-(n+1)}e^{\alpha v^s}\mathrm{d}v,$$

whereby the condition $\Re(\widetilde{\beta}) > 0$ can be always fulfilled for sufficiently large n. The transformation $\tau := 1/v$ yields the required formula. □

PROOF OF THE PROPOSITION. According to asymptotics we evaluate the integral (4.3) by the saddle point method (see e.g. [**3**, p. 77]). This means that if we have an integral of the form

$$\int_{\mathcal{L}} \exp(\psi(\tau))\mathrm{d}\tau,$$

and if τ_0 is a saddle point thus is a solution of

$$\frac{\mathrm{d}\psi(\tau)}{\mathrm{d}\tau} = 0, \tag{4.12}$$

then the contribution of this saddle point to the integral can be evaluated by

$$\sqrt{\pi}\left|\frac{1}{2}\frac{\mathrm{d}^2\psi(\tau)}{\mathrm{d}\tau^2}\right|_{\tau=\tau_0}^{-1/2} e^{\psi(\tau_0)}. \tag{4.13}$$

From (4.1) and (4.12) we see that the locations of the saddle points τ_0 are given by the solutions of

$$\frac{\mathrm{d}\psi(\tau)}{\mathrm{d}\tau} = -\frac{s\alpha}{\tau^{s+1}} + \frac{n}{1-\tau} = 0,$$

or in leading order asymptotics:

$$\tau(n)^{s+1} \sim \frac{s\alpha}{n} \quad \text{for} \quad n\to\infty. \tag{4.14}$$

Eq. (4.14) shows that if α is real positive there is always a saddle point on the positive real axis. If α is real negative there is never a saddle point on the positive real axis while there are always two complex maximum solutions being complex conjugate to one another. The fact that there is a nonzero imaginary part generates an exponentially decreasing function (4.1).

Elaborating only the saddle point of (4.3) which has the largest real part gives the leading order behaviour of c_n thus one term of the sort (4.13). Elaborating all the saddle points yields the Birkhoff set (2.5) (cf. [**4, 5, 8**]) thus the complete asymptotic factors of the large n-behaviour of c_n (see examples below).

After having evaluated the integral (4.3) according to (4.13) the ratio of two consecutive terms $(b_{n+1})/(b_n)$ in leading order results in $t(n)$ according to (2.3) □

EXAMPLES 1. As an example, lets have a look at the integral (4.3) with $s = 1$ or $R = 2$. There are two saddle points the locations of which are given in the asymptotic limit for large n by

$$\tau(n)_{1,2} \approx \pm\left(\frac{\alpha}{n}\right)^{1/2}.$$

Taking into account both saddle points and expanding the exponential terms in powers of $n^{1/2}$ we obtain the asymptotic behaviour of the coefficients according to

$$c_n \approx \frac{1}{2\sqrt{\pi}}\alpha^{1/4-\beta/2}n^{\beta/2-3/4}e^{+2(\alpha n)^{1/2}}\sum_{k=0}^{\infty}C_{k,1}n^{-k/2} + \frac{1}{2\sqrt{\pi}}\alpha^{1/4-\beta/2}n^{\beta/2-3/4}e^{-2(\alpha n)^1/2}\sum_{k=0}^{\infty}C_{k,2}n^{-k/2},$$

where $C_{k,1}$, $C_{k,2}$ are constants. This corresponds to the results of Perron [**9**].

As a second example we consider the case $R = 4$. The saddle points now are located at

$$\tau(n)_{1,2} \approx \pm\left(\frac{3\alpha}{n}\right)^{1/4} - \frac{1}{4}\left(\frac{3\alpha}{n}\right)^{1/2}, \quad \tau(n)_{3,4} \approx \tau(n)_{1,2}\imath.$$

Here, we have to take in account the order $n^{1/4}$ for the locations of the approximate saddle points. The calculation of the coefficients then yields

$$\begin{aligned} c_n \approx{}& \frac{1}{2\sqrt{\pi}}\alpha^{1/8-\beta/4}\exp\left(\frac{4}{3}(3\alpha)^{1/4}n^{3/4} + \frac{1}{2}(3\alpha)^{2/4}n^{2/4}\right) \\ &\times \exp\left(\frac{5}{24}(3\alpha)^{3/4}n^{1/4}n^{\beta/4-5/8}\right)\sum_{k=0}^{\infty}C_{k,1}n^{-k/4} \\ &+ \frac{1}{2\sqrt{\pi}}\alpha^{1/8-\beta/4}\exp\left(-\frac{4}{3}(3\alpha)^{1/4}n^{3/4} + \frac{1}{2}(3\alpha)^{2/4}n^{2/4}\right) \\ &\times \exp\left(-\frac{5}{24}(3\alpha)^{3/4}n^{1/4}n^{\beta/4-5/8}\right)\sum_{k=0}^{\infty}C_{k,2}n^{-k/4} \\ &+ \frac{1}{2\sqrt{\pi}}\alpha^{1/8-\beta/4}\exp\left(+\frac{4}{3}(3\alpha)^{1/4}n^{3/4}\imath - \frac{1}{2}(3\alpha)^{2/4}n^{2/4}\right) \\ &\times \exp\left(-\frac{5}{24}(3\alpha)^{3/4}n^{1/4}\imath n^{\beta/4-5/8}\right)\sum_{k=0}^{\infty}C_{k,3}n^{-k/4} \\ &+ \frac{1}{2\sqrt{\pi}}\alpha^{1/8-\beta/4}\exp\left(-\frac{4}{3}(3\alpha)^{1/4}n^{3/4}\imath - \frac{1}{2}(3\alpha)^{2/4}n^{2/4}\right) \\ &\times \exp\left(\frac{5}{24}(3\alpha)^{3/4}n^{1/4}\imath n^{\beta/4-5/8}\right)\sum_{k=0}^{\infty}C_{k,4}n^{-k/4}, \end{aligned}$$

where $C_{k,i}$, $i = 1, 2, 3, 4$ are constants.

From these examples we see that the locations τ_0 of the saddle points are directly related to the particular solutions of the difference equations and presented in the t-plane above.

References

1. M. Abramowitz and I. A. Stegun, *Handbook of mathematical functions*, 9th ed., Dover Publications, Inc., New York, 1972.
2. M. A. Barkatou and F. Jung, *Formal solutions of linear differential and difference equations*, Programmirovanie 1997, no. 1, 24–42 (Russian); English transl., Program. Comput. Software **23** (1997), no. 1, 17–30
3. N. G. de Bruijn, *Asymptotic methods in analysis* 2nd ed., North-Holland Publ. Co., Amsterdam, 1961.
4. W. Lay, *The central two-point connection problem of Heun's class of differential equations*, Theoret. and Math. Phys. **101** (1994), no. 3, 1413–1418.
5. ______, *Das zentrale Zweipunkt-Verbindungsproblem für die Klasse der Heunschen Differentialgleichungen*, Habilitation thesis, Stuttgart, 1995.
6. ______, *The central two-point connection problem in quantum mechanics*, Z. Agnew. Math. Mech. **78** (1998), no. 3, pp. S991–S992.
7. W. Lay, A. Ya. Kazakov, and S. Yu. Slavyanov, *Equivalent eigenvalue problems for the differential equations of Heun's class*, Algebra i Analiz **8** (1996), no. 1, pp. 129–141 (Russian).
8. W. Lay and S. Yu. Slavyanov, *The central two-point connection problem for the Heun class of* ODE, J. Phys. A **31** (1998) pp. 4249–4261.
9. O. Perron, *Über das Verhalten einer ausgearteten hypergeometrischen Reihe bei unbegrenztem Wachstum eines Parameters*, J. für die reine und angewandte Math. **151** (1921), 63–78.
10. S. Yu. Slavyanov, W. Lay, and A. Seeger, *Heun's differential equations* (A. Ronveaux, ed.), Oxford Sci. Publ.,The Clarendon Press, Oxford Univ. Press, New York, 1995, pp. 289–323.
11. ______, *A generalization of the Riemann scheme for equations of the hypergeometric and Heun classes*, Internat. J. Math. Ed. Sci. Techn. **28** (1997), no. 5, 641–660.
12. K. Weierstrass, *Zur Theorie der Potenzreihen*, Mathematische Werke. I. Abhandlungen 1 Mayer & Muller, Berlin, 1894, pp. 67–74.
13. J. Wimp, *Computations with recurrence relations*, Pitman (Advanced Publishing Program), Boston, Mass.-London, 1984.

Institut fuer Theoretische und Angewandte Physik, Universitaet Stuttgart, Pfaffenwaldring 57, 70550 Stuttgart, Germany.

E-mail address: `lay@itap.physik.uni-stuttgart.de`

Centre de Recherches Mathématiques
CRM Proceedings and Lecture Notes
Volume **25**, 2000

Self-Similarity in Spectral Problems and q-Special Functions

Igor Loutsenko and Vyacheslav Spiridonov

ABSTRACT. Some similarity symmetries of the factorization chains for one-dimensional differential and finite-difference Schrödinger equations are discussed. Properties of the potentials defined by self-similar reductions of these chains are reviewed. In particular, their algebraic structure, relations to q-special functions, infinite soliton systems, supersymmetry, coherent states, orthogonal polynomials, one-dimensional Ising chains and random matrices are outlined.

1. Introduction

Spectral problems of various types play an important role in classical and quantum mechanics and in the theory of integrable systems. Depending on the situation they may be based upon differential, finite-difference or integral equations, etc. Roughly speaking all such problems can be approximated by a search of eigenvalues of some matrices [**53**]. This leads to a number of common features in the methods of treatment of different spectral problems. We choose as a basic object of discussion the ordinary one-dimensional Schrödinger equation

$$L\psi(x) = -\psi_{xx}(x) + u(x)\psi(x) = \lambda\psi(x). \tag{1.1}$$

The eigenvalue problem for the operator L is completely defined by imposing some boundary conditions upon $\psi(x)$. However, it is convenient to consider (1.1) without such additional constraints and assume that $x, \lambda \in C$. When a nonformal meaning of the eigenvalues and eigenfunctions will be necessary, we assume that $\psi(x) \in L^2(R)$.

There are several connections of (1.1) with finite-difference equations and discrete systems—the main subject of the present workshop. First, the scheme of

---

2000 *Mathematics Subject Classification.* 33D80, 81R30, 82B20.

The authors are indebted to many people for interest to the results reviewed in this paper, stimulating discussions and a collaboration, in particular, to R. Askey, Yu. Berest, R. Hirota, A. Iserles, K. Kajiwara, A. Orlov, T. Shiota, S. Skorik, S. Suslov, C. Tracy, L. Vinet and A. Zhedanov. V.S. is greatful to D. Levi and O. Ragnisco for inviting to the SIDE III workshop and a kind hospitality. The first author is supported by a fellowship from NSERC (Canada). The second author is partially supported by the RFBR (Russia) grant 97-01-00281 and the INTAS grant 96-0700.

This is the final form of the paper.

treatment of (1.1) to be described below is universal, e.g., it can be easily reformulated for the second order difference equation. Second, finite-difference equations appear already in the consideration of (1.1) for different potentials. In the space of all equations of this form finite-differences enter through the symmetry transformations mapping Schrödinger equations onto each other:

$$u(x) \to \widetilde{u}(x), \quad \psi(x) \to \widetilde{\psi}(x), \quad \lambda \to \widetilde{\lambda}.$$

Third, in some cases the discrete spectrum eigenfunctions satisfy simple discrete equations. E.g., for $u(x) \propto x^2$ one gets from the condition of square integrability of eigenfunctions $\lambda_n \propto n$, $\psi_n(x) \propto H_n(x)e^{-x^2/2}$, where $H_n(x)$ are Hermite polynomials satisfying three-term recurrence relation. Fourth, there is a direct relation of (1.1) to some lattice models of statistical mechanics.

The paper is organized as follows. In the next several sections we review briefly some old results on self-similar potentials, their coherent states and the relation to q-special functions. Then we outline briefly a generalization of the construction to orthogonal polynomials. In the last two sections we present our recent results on the appearance of one-dimensional Ising chains and random matrices in the context of the Schrödinger equation (1.1). The list of references given in the end does not pretend to be complete, it contains mostly the papers which influenced somehow the work of authors.

2. Self-Similar Spectra

Under the similarity transformations in d-dimensional Euclidean space one counts usually the rotations, dilations and shifts of the coordinates, $x_i \to q_{ij}x_j + a_i$, $i, j = 1, \ldots, d$, i.e. the affine transformations. In the $d = 1$ case this is reduced to $x \to qx + a$, which will be used below under the assumption that the parameters q and a are *fixed* (for $q \neq 1$ one may set $a = 0$ without loss of generality).

Let us consider separately the effects of translations and scalings upon the Schrödinger operator L with potential $u(x)$, eigenfunctions $\{\psi_j(x)\}$ and the spectrum $\{\lambda_j\}$ (the use of the subscript j for notation of the spectrum is symbolic). After application of some symmetry transformation we get the operator $\widetilde{L} = -d^2/dx^2 + \widetilde{u}(x)$ with the potential $\widetilde{u}(x)$, unnormalized eigenfunctions $\{\widetilde{\psi}_j(x)\}$ and the eigenvalues $\{\widetilde{\lambda}_j\}$. So, the shift $x \to x + a$ brings in

$$\widetilde{u}(x) = u(x+a), \quad \{\widetilde{\psi}_j(x)\} = \{\psi_j(x+a)\}, \quad \{\widetilde{\lambda}_j\} = \{\lambda_j\}.$$

The spectrum does not change under this transformation due to the absence of distinguished points in the boundary conditions. In the case of scaling transformations one replaces in (1.1) the variable x by qx, then multiplies the resulting equation by q^2 in order to bring it to the canonical form and gets the operator $\widetilde{L}$ with

$$\widetilde{u}(x) = q^2 u(qx), \quad \{\widetilde{\psi}_j(x)\} = \{\psi(qx)\}, \quad \{\widetilde{\lambda}_j\} = \{q^2\lambda_j\}.$$

Now one gets a scaled form of the spectrum $\lambda \to q^2\lambda$. For $|q| > 1$ the spectral points are stretched with respect to the $\lambda = 0$ point, and, vice versa, for $|q| < 1$ we get a squeezed spectrum. The complete affine transformation of the spectral parameter requires joining of a shift $\lambda \to \lambda + h$ which is easily reached by the shift of the potential

$$\widetilde{u}(x) = u(x) + h, \quad \{\widetilde{\psi}_j(x)\} = \{\psi_j(x)\}, \quad \{\widetilde{\lambda}_j\} = \{\lambda_j + h\}.$$

The combination of these three transformations gives

$$\widetilde{u}(x)=q^2u(qx+a)+h,\quad \{\widetilde{\psi}_j(x)\}=\{\psi_j(qx+a)\},\quad \{\widetilde{\lambda}_j\}=\{q^2\lambda_j+h\},$$

i.e. 'trivial' similarity transformations of the spectral data. The key constructive idea, bringing in a nontrivial content into the above considerations, consists in a sticking to the systems with *self-similar* spectra for which $\{\widetilde{\lambda}_j\}=\{\lambda_j\}$. Note that such a constraint is applicable to any spectral problem, not just to the Schrödinger equation.

Analyze first the $q=1$ case. The condition $\{\lambda_j+h\}=\{\lambda_j\}$ shows that the discrete spectrum should appear in the form of a number of independent bilateral arithmetic progressions with the increment equal to h. In the simplest case the number of such progression is finite ($=N$) and one can parametrize the eigenvalues as solutions of the equation $\lambda_j+h=\lambda_{j+N}$ or $\lambda_{pN+k}=hp+\lambda_k$, $k=1,\dots N$, $p\in Z$. The continuous spectrum may appear in the form of arithmetic progressions of permitted bands. For instance, if one requires that the addition of h to potential is equivalent to an isospectral transformation $x\to x+a$, i.e. $u(x)+h=u(x+a)$, then $u(x)$ is a sum of a periodic potential $u_p(x+a)=u_p(x)$ and the Airy part $\propto x$ known to lead to continuous spectrum filling the whole line, $u(x)=u_p(x)+hx/a$.

Consider now the $\{q^2\lambda_i\}=\{\lambda_i\}$ case. Formally, the discrete spectrum appears in the form of bilateral geometric progressions accumulating near the $\lambda=0$ point from below or above. Enumeration of pure point spectra in case of the finite number of such progressions may be based upon the relation $q^2\lambda_j=\lambda_{j+N}$ giving $\lambda_{pN+k}=\lambda_k q^{2p}$, $k=1,\dots,N$, $p\in Z$. The continuous spectrum may arise in the form of infinite number of permitted bands concentrating near the zero.

For instance, the described situation may take place if $q^2u(qx)=u(x)$. A solution of this equation is given by $u(x)=h(x)/x^2$, where $h(x)$ is a function periodic in the logarithmic scale $h(qx)=h(x)$. For $h=\text{const.}$ the wave functions $\psi(x)$ are given in terms of the Bessel functions. For $h<-1/4$ there is the "fall onto the center" phenomenon and under certain conditions discrete spectrum consists of one bilateral geometric progression of negative eigenvalues [**12**]. The bad point of this and of the previous example is that the potentials $u(x)$ are not bounded from below. Let us impose such a demand and assume that $u(x)$ is bounded from below and nonsingular. Evidently this condition does not permit the infinite negative spectrum, i.e. it is necessary to delete the lower tails of our geometric (or arithmetic) progressions. However, the formula $q^2\lambda_j=\lambda_{j+N}$ still works and gives $\lambda_{pN+k}=\lambda_k q^{2p}$, $p=0,1,\dots$. We have thus a partially self-similar spectrum:

$$\{q^2\lambda_j\}=\{\lambda_j\}/(\lambda_1,\dots,\lambda_N),$$

i.e. the scaling $x\to qx$ leads to the deletion of N lowest eigenvalues from the spectrum. Below we outline a procedure for building a class of potentials with this property.

Note that for the $q^2\neq 1$ case the geometric progressions of positive eigenvalues of the abstract operator L are bounded from below and no truncation is necessary. However, one has to go beyond the Schrödinger equation for building systems with such completely self-similar spectrum.

3. Infinite-Soliton Systems

The inverse scattering method and the general theory of solitons (see, e.g., [**1, 11, 17, 54**]) provide us a guide for construction of potentials with prefixed spectral properties. The compatibility condition of two linear equations

$$L\psi(x,t) = \lambda\psi(x,t), \quad \psi_t(x,t) = A\psi(x,t),$$

has the operator form $L_t = [A, L]$. When $L = -\partial_x^2 + u(x,t)$ and $A = -\partial_x^3 + 6u\partial_x + 3u_x$ one arrives at the KdV equation $u_t + u_{xxx} - 6uu_x = 0$.

If the potential $u(x) \to 0$ for $x \to \infty$ sufficiently fast then according to the inverse scattering method the potential $u(x)$ can be uniquely reconstructed from the reflection coefficient $R(k)$, discrete spectrum eigenvalues $\{\lambda_j\}$ and the normalization constants $\{\theta_j^{(0)}\}$ for corresponding eigenfunctions $\{\psi_j(x)\}$. For $R(k) = 0$ and finite number of eigenvalues one arrives at the N-soliton potentials. The corresponding solution of the KdV equation can be represented in the form

$$u(x,t) = -2\partial_x^2 \ln \tau_N(x,t), \tag{3.1}$$

where τ_N is the determinant of a $N \times N$ matrix C,

$$\begin{gathered} \tau_N = \det C, \quad C_{ij} = \delta_{ij} + \frac{2\sqrt{k_i k_j}}{k_i + k_j} e^{(\theta_i + \theta_j)/2}, \\ \theta_i = k_i x - k_i^3 t + \theta_i^{(0)}, \quad i, j = 1, 2, \ldots, N. \end{gathered} \tag{3.2}$$

Parameters k_i describe amplitudes of solitons related to eigenvalues of L as $\lambda_i = -k_i^2/4$, $\theta_i^{(0)}/k_i$ are zero time phases of solitons and k_i^2 are their velocities.

Let us extend this solution to the pN-soliton one by imposing the constraint $k_{j+N} = qk_j$, $q < 1$, and take the $p \to \infty$ limit. This leads to the self-similar spectrum $k_{pN+m} = q^p k_m$, $m = 1, \ldots, N$, i.e. to the infinite-soliton wavetrains whose amplitudes form N geometric progressions. The question of convergency of such a limit requires a special consideration.

Applying scaling transformations to derived potentials one has to scale the time t too, which gives $\tilde{u}(x,t) = q^2 u(qx, q^3 t)$. The phases of this potential are $\theta_j(qx, q^3 t) = k_{j+N} x - k_{j+N}^3 t + \theta_j^{(0)}$. There are two "drawbacks" of the resulting expression. First, there are infinitely many arbitrary parameters $\theta_j^{(0)}$, i.e. there is a functional freedom in the definition of $u(x)$. Second, there is a mixup between soliton amplitudes and phases—the solitons with the amplitudes k_{j+N} have now the zero time phases $\theta_j^{(0)}$ instead of $\theta_{j+N}^{(0)}$. Both these shortcomings are removed if we impose the constraint $\theta_{j+N}^{(0)} = \theta_j^{(0)}$, i.e. $\theta_j(qx, q^3 t) = \theta_{j+N}(x,t)$, which leads to truly self-similar infinite-soliton systems—the dilation $x \to qx$, $t \to q^3 t$ just deletes N solitons, corresponding to the lowest eigenvalues of L. Such potential contains a finite number of parameters $k_1, \ldots, k_N$, q, $\theta_1^{(0)}, \ldots, \theta_N^{(0)}$ and should be characterized by some "finite" equations. There are two equivalent approaches to the construction of these equations. The algebraic one, based upon the factorization method, and the functional-analytic one, relying upon the Darboux transformations.

4. The Factorization Method

This method is well known in quantum mechanics [**26**], it was discussed already by Schrödinger. Within this approach one takes a chain of Schrödinger operators, $L_j = -d^2/dx^2 + u_j(x)$, and factorizes them as products of the first-order differential operators up to some constants λ_j:

$$L_j = A_j^+ A_j^- + \lambda_j, \quad A_j^{\pm} = \mp d/dx + f_j(x). \tag{4.1}$$

Then the neighboring L_j are tied to each other through the abstract factorization chain

$$L_{j+1} = A_{j+1}^+ A_{j+1}^- + \lambda_{j+1} = A_j^- A_j^+ + \lambda_j,$$

i.e. one passes from L_j to L_{j+1} just by the permutation of operator factors. This gives the intertwining relations $A_j^- L_j = L_{j+1} A_j^-$, $L_j A_j^+ = A_j^+ L_{j+1}$. As a result, if one has $L_j \psi^{(j)} = \lambda \psi^{(j)}$, then $\psi^{(j+1)} \propto A_j^- \psi^{(j)}$. Indeed, $L_{j+1}(A_j^- \psi^{(j)}) = A_j^- L_j \psi^{(j)} = \lambda(A_j^- \psi^{(j)})$. Under particular conditions the action of the operator A_j^- leads to removing or addition of an eigenvalue in the spectrum of the operator L_{j+1} with respect to the spectrum of L_j. There may take place an isospectral situation as well. If zero modes of all A_j^- are normalizable, then λ_j, $\lambda_{j+1}, \ldots$ form the discrete spectrum of L_j.

Let us define the operators

$$M_j^- = A_{j+N-1}^- \cdots A_{j+1}^- A_j^-, \quad M_j^+ = A_j^+ A_{j+1}^+ \cdots A_{j+N-1}^+.$$

They relate eigenfunctions of the operators L_j and L_{j+N} due to the intertwining relations

$$L_{j+N} M_j^- = M_j^- L_j, \quad M_j^+ L_{j+N} = L_j M_j^+.$$

It is easy to see that the products of operators $M_j^\pm$ should commute either with L_j or with L_{j+N}. Indeed, one has the equalities

$$M_j^+ M_j^- = \prod_{k=0}^{N-1} (L_j - \lambda_{j+k}), \quad M_j^- M_j^+ = \prod_{k=0}^{N-1} (L_{j+N} - \lambda_{j+k}), \tag{4.2}$$

which, together with the previous relations, look almost as an algebra of symmetries.

The main advantage of this method is that it is not tied to any particular type of spectral problem. One may realize the abstract factorization chain with the help of differential, difference, integral, etc. operators.

5. Darboux Transformations

One starts from the semi-discrete LA-pair

$$L_j \psi^{(j)} = \lambda \psi^{(j)}, \qquad L = -d^2/dx^2 + u_j(x),$$
$$\psi^{(j+1)} = A_j^- \psi^{(j)}, \quad A_j^- = d/dx + f_j(x).$$

The compatibility condition yields the intertwining relation $A_j^- L_j = L_{j+1} A_j^-$, the resolution of which gives $u_j = f_j^2 - f_{jx} + \lambda_j, u_{j+1} = u_j + 2f_{jx} = f_j^2 + f_{jx} + \lambda_j$. The substitution $f_j = -\phi_x^{(j)}/\phi^{(j)}$ converts the relation between u_j and f_j into the equation $-\phi_{xx}^{(j)} + u_j \phi^{(j)} = \lambda_j \phi^{(j)}$. As a result one gets the original Darboux transformation: $\psi^{(j+1)} = \psi_x^{(j)} - (\phi_x^{(j)}/\phi^{(j)})\psi^{(j)}$.

The compatibility condition (or the factorization chain) is equivalent to the following infinite chain of nonlinear differential-difference equations

$$\big(f_j(x) + f_{j+1}(x)\big)_x + f_j^2(x) - f_{j+1}^2(x) = \mu_j \equiv \lambda_{j+1} - \lambda_j. \tag{5.1}$$

One may search solutions of this equation in the form of power series in j. The finite term expansion occurs only if $f_j(x) = a(x)j + b(x) + c(x)/j$, where a, b, c are some elementary functions of x [**26**]. The resulting Schrödinger equation is solved in terms of the ${}_2F_1$ hypergeometric function.

6. Polynomial Supersymmetry

Upon the two neighboring operators L_j, L_{j+1} of the factorization chain a simple superalgebra describing the boson-fermion symmetry is realized. Let us introduce the 2×2 matrix Hamiltonian $H = -d^2/dx^2 + f_j^2(x) - f_{jx}(x)\sigma_3$ and define the supercharges $Q^\pm = A_j^\mp(\sigma_1 \pm i\sigma_2)/2$ (σ_k are the Pauli matrices). Then one has $\{Q^+, Q^-\} = H$, $[H, Q^\pm] = (Q^\pm)^2 = 0$. This construction was generalized in [**39**] to a symmetry between particles with parastatistics (para-supersymmetry). In the simplest case one takes the Hamiltonian given by 3×3 diagonal matrix with the entries L_{j-1}, L_j, L_{j+1}. A symmetry between the boson and a parafermion of the second order is described now by some polynomial (cubic) relations between H and the corresponding generalized supercharges.

As a next step in the "nonlinearization" of supersymmetry the following polynomial supersymmetry algebra has been derived in [**3**]

$$\{Q^+, Q^-\} = P(H), \quad [H, Q^\pm] = (Q^\pm)^2 = 0,$$

where $Q^\pm = M_1^\mp(\sigma_1 \pm i\sigma_2)/2$ and $P(H)$ is a polynomial of H:

$$P(H) = \prod_{k=1}^{N}(H - \lambda_k), \quad H = \begin{pmatrix} L_1 & 0 \\ 0 & L_{N+1} \end{pmatrix}.$$

Since the supercharges have now more than one zero mode, the Witten index does not characterize the supersymmetry breaking. One may deform this algebra via the affine transformations and impose various natural constraints upon H. This gives an alternative way of derivation (which we are not describing here) of self-similar potentials considered in the next section.

7. Self-Similar Potentials

The infinite soliton potentials with self-similar spectra considered above appear as self-similar solutions of the chain (5.1). The simplest $N = 1$ case has been considered in [**40, 42**]. The general class of these potentials is defined by the following q-periodic reduction [**43**]:

(7.1) $$f_{j+N}(x) = qf_j(qx), \quad \mu_{j+N} = q^2\mu_j.$$

Note that the solitonic interpretation is valid only if $0 < q^2 < 1$ and there are no singularities on the whole line $-\infty < x < \infty$, which is not true for arbitrary q, x, $\lambda_j \in C$. What are the properties of these potentials ?

Algebraically, these are the systems whose symmetries are described by the quantum algebras. Indeed, the equations (4.2) form a closed algebra in the cases when the operators L_j and L_{j+N} are related to each other in some way, e.g., if they are similar to each other $L_j \sim L_{j+N}$. The q-periodic closure corresponds to the following constraint $L_{j+N} = q^2 T L_j T^{-1}$, where T is the scaling operator $T\psi(x) = \sqrt{|q|}\psi(qx)$. This gives $u_{j+N}(x) = q^2 u_j(qx)$.

For real $q \neq 0$, T is the unitary operator $T^\dagger = T^{-1}$. Substituting the operator constraint into (4.2) one arrives at the following polynomial algebra

(7.2) $$LB^\pm = q^{\pm 2}B^\pm L, \quad B^+B^- = \prod_{k=1}^{N}(L - \lambda_k), \quad B^-B^+ = \prod_{k=1}^{N}(q^2L - \lambda_k),$$

where we have denoted $L \equiv L_1$, $B^- \equiv T^{-1}M_1$, $B^+ \equiv M_1^+T$. For $N=1$ this is a q-analog of the harmonic oscillator algebra [**33**], for $N=2$ one gets the $su_q(1,1)$ algebra, etc.

Analytically, we deal with a class of functions appearing from solutions of the following system of nonlinear differential-q-difference equations:

$$\big(f_1(x)+f_2(x)\big)_x + f_1^2(x) - f_2^2(x) = \mu_1, \dots,$$
$$\big(f_N(x)+qf_1(qx)\big)_x + f_N^2(x) - q^2 f_1^2(qx) = \mu_N.$$

A flavor of the structure of the general solution is obtained from consideration of various limiting cases, when $f_j(x)$ are expressed through known functions. The simplest situation is obtained when $f_j(x) = c_j =$ const. and arbitrary q. This leads to $L = -d^2/dx^2$, i.e. to the free nonrelativistic particle, which is interpreted in this way as a q-algebraic system. When $f_1(x)$ is not singular at $x=0$, the crystal base limit $q \to 0$ results in the general N-soliton potential. One can take the limit $q \to 1$ in such a way that the geometric progressions in the spectrum are converted into the arithmetic ones. Then for $N = 1, 2$ one gets the potentials $u(x) \propto x^2, ax^2 + b/x^2$. For $N = 3, 4$ the functions $f_j(x)$ are expressed through the Painlevé IV and V transcendents [**2, 51**]. For $q = -1$ one gets a similar situation but now the functions are constrained to obey certain structure under the parity transformation. It is possible to consider the limit from Schrödinger equation to classical mechanics, in which case the self-similar potentials are determined by the much more simple functions [**15, 47**].

For q a primitive root of unity, $q^n = 1, q \neq \pm 1$, one gets the finite gap potentials (there are some peculiarities for even n) with additional (quasi)crystallographic symmetries [**41**]. The appearance of Painlevé functions and of the finite-gap potentials in a similar setting was discussed also in [**16, 52**]. It is natural to refer to the whole class of functions emerging for general initial conditions and $0 < |q| < 1$ as the continuous q-Painlevé functions. Because of the emergence of quantum algebras, the derived class of self-similar functions represents a new set of q-special functions defined upon the differential equations. They differ in structure from the basic hypergeometric series [**18**], but there is a connection with them via the coherent states.

Let us remark that the Darboux transformations allow one to remove or add only a finite number of levels from/to the spectrum of a given Schrödinger operator. If one finds an intertwining operator deleting or adding permitted bands from/to the given spectrum, then the systems with a self-similar infinite gap spectrum can be constructed from the requirement that the action of this intertwining operator is equivalent to a simple dilation of the coordinate x.

8. Coherent States

There are many definitions of coherent states in physics. The purely group-theoretical approach uses the orbits of groups lying behind the symmetries of a taken physical system. For simple spectrum generating algebras, as in the harmonic oscillator or singular oscillator cases, these states may be defined as eigenfunctions of the lowering operator. In this approach coherent states play the role of generating functions of irreducible representations of the underlying algebra. Below we outline briefly the results of application of the latter definition to the nonlinear algebra of symmetries (7.2). A more detailed discussion of the structure of these coherent states can be found in [**44**].

The action of operators $B^{\pm}$ upon abstract eigenstates of the operator L, $L|\lambda\rangle = \lambda|\lambda\rangle$, has the form:

$$B^-|\lambda\rangle = \prod_{k=1}^{N} \sqrt{\lambda - \lambda_k}|\lambda q^{-2}\rangle, \quad B^+|\lambda\rangle = \prod_{k=1}^{N} \sqrt{\lambda q^2 - \lambda_k}|\lambda q^2\rangle.$$

Let $0 < q < 1$ and $\lambda_k < 0$. Then the $\lambda < 0$ eigenstates of L are formed from up to N lowest weight unitary irreducible representations of the algebra (7.2). Indeed, B^- is the lowering operator for $\lambda < 0$ states and $\sqrt{\lambda - \lambda_k}$ becomes complex for $\lambda < \min\{\lambda_k\}$. Since $B^-|\lambda_k\rangle = 0$, this problem does not arise for a special choice of λ. Namely, if the point $\lambda < 0$ belongs to the spectrum of L, it must be of the form $\lambda_{pN+k} \equiv \lambda_k q^{2p}$ and the corresponding state is $|\lambda_{pN+k}\rangle \propto (B^+)^p|\lambda_k\rangle$ (we assume that $|\lambda_k\rangle$ are normalizable).

Coherent states for these series are defined as eigenstates of the lowering operator B^-:

$$B^-|\alpha\rangle_-^{(k)} = \alpha|\alpha\rangle_-^{(k)}, \quad k = 1, \dots, N. \tag{8.1}$$

There are N such states since they are defined for each lowest weight series separately. Representing $|\alpha\rangle_-^{(k)}$ as a superposition of the states $|\lambda_{pN+k}\rangle$ one finds

$$|\alpha\rangle_-^{(k)} \propto \sum_{p=0}^{\infty} C_p^{(k)} \alpha^p |\lambda_{pN+k}\rangle \propto {}_N\varphi_{N-1}\left(\begin{matrix} 0, \dots, 0 \\ b_1^k, \dots, b_{N-1}^k \end{matrix}; q^2, z\right)|\lambda_k\rangle,$$

where ${}_N\varphi_{N-1}$ is a basic hypergeometric series with the operator argument $z = (-1)^N \alpha B^+/\lambda_1 \dots \lambda_N$ and the parameters $\{q^2, b_j^k\} = \{q^2\lambda_k/\lambda_j\}$. The general definition of the series of this type is [**18**]

$${}_r\varphi_s\left(\begin{matrix} a_1, a_2, \dots, a_r \\ b_1, b_2, \dots, b_s \end{matrix}; q, z\right) = \sum_{n=0}^{\infty} \frac{(a_1, a_2, \dots, a_r; q)_n}{(q, b_1, \dots, b_s; q)_n} \left[(-1)^n q^{n(n-1)/2}\right]^{1+s-r} z^n,$$

where r and s are arbitrary positive integers, and $a_1, \dots, a_r$, $b_1, \dots, b_s$ are free parameters. We use also the following compact notations $(a; q)_n = (1-a)\dots(1-q^{n-1})$ and $(a_1, a_2, \dots, a_n; q)_n = (a_1; q)_n (a_2; q)_n \cdots (a_n; q)_n$. These coherent states are normalizable if the complex variable α lies inside of the circle $|\alpha|^2 < |\lambda_1 \dots \lambda_N|$. General ${}_r\varphi_s$ series are related to coherent states of the rational generalization of the algebra (7.2) [**38**].

The $\lambda = 0$ eigenstates of L are simultaneously eigenstates of the $B^{\pm}$ operators, i.e. they have a direct interpretation as coherent states. This is a special degenerate representation of the algebra (7.2).

Unusual coherent states appear from the nonhighest weight representations of (7.2) corresponding to the $\lambda > 0$ eigenstates of L. Let $E_0 > 0$ be a discrete spectrum point. Then the action of $B^{\pm}$ generates discrete spectrum of L in the form of one bilateral geometric progression $E_0 q^{2n}, n \in Z$ (such situation cannot take place for the Schrödinger equation). In principle there may be an *arbitrary* number of such progressions because $\sqrt{\lambda - \lambda_k}$ is real for arbitrary $\lambda > 0$ and there are no truncation conditions as in the $\lambda < 0$ case.

Since for $\lambda > 0$ the lowering operator is B^+, the coherent states should be defined as eigenstates of this operator (instead of B^-):

$$B^+|\alpha\rangle_+ = \alpha|\alpha\rangle_+. \tag{8.2}$$

These coherent states can be expanded over the series of eigenstates of L with positive eigenvalues $|E_0 q^{2n}\rangle$:

$$|\alpha\rangle_+ \propto \sum_{n=-\infty}^{\infty} C_n \alpha^n |E_0 q^{2n}\rangle \propto {}_0\psi_N \left({0, \ldots, 0 \atop \lambda_1/E_0, \ldots, \lambda_N/E_0}; q^2, z \right) |E_0\rangle,$$

where ${}_0\psi_N$ is a bilateral q-hypergeometric series with the operator argument $z = \alpha B^-/(-E_0)^N$. The general representative of these series is defined as follows [**18**]:

$${}_r\psi_s \left({a_1, \ldots, a_r \atop b_1, \ldots, b_s}; q, z \right) = \sum_{n=-\infty}^{\infty} \frac{(a_1, \ldots, a_r; q)_n}{(b_1, \ldots, b_s; q)_n} \left((-1)^n q^{n(n-1)/2}\right)^{s-r} z^n.$$

Coherent states $|\alpha\rangle_+$ are normalizable if the parameter α lies outside of the previously indicated circle $|\alpha|^2 > |\lambda_1 \ldots \lambda_N|$.

When the $\lambda > 0$ region is occupied by continuous spectrum, coherent states are defined as eigenfunctions of B^+ again, but in this case it is necessary to use integrals in the expansion of $|\alpha\rangle_+$ over the states $|\lambda\rangle$. Since the continuous spectrum may be considered as a continuous direct sum of irreducible representations with bilateral geometric progressions formed by spectral points, there are now infinitely many coherent states. Let $N = 1$ and the states $|\lambda\rangle$ are not degenerate. Then, normalizing $\lambda_1 = 1/(q^2 - 1)$, we have

$$|\alpha\rangle_+^{(s)} = C(\alpha) \int_0^\infty \frac{\lambda^{\gamma_s} |\lambda\rangle\, d\lambda}{\sqrt{(-\lambda q^2(1-q^2); q^2)_\infty}},$$

where

$$\gamma_s = \frac{2\pi i s - \ln(\alpha q^2 \sqrt{1-q^2})}{\ln q^2}, \qquad s \in \mathbf{Z}.$$

These states are normalizable for $|\alpha|^2 > 1/(1-q^2)$, and have the unit norm for

$$|C|^{-2} = \int_0^\infty \frac{\lambda^{-\tau}\, d\lambda}{(-\lambda q^2(1-q^2); q^2)_\infty} = \frac{\pi}{\sin \pi\tau} \frac{(q^{2\tau}; q^2)_\infty (q^2(1-q^2))^{\tau-1}}{(q^2; q^2)_\infty},$$

where $\tau = \ln|\alpha q \sqrt{1-q^2}|/\ln q$. The last integral is calculated exactly being a particular subcase of a Ramanujan q-beta integral [**18**].

Schrödinger operators with self-similar potentials have the continuous spectrum for $\lambda > 0$. It is doubly degenerate, which leads to duplication of the number of coherent states. An instructive example is provided by the free nonrelativistic particle model for which $L = -d^2/dx^2$. The corresponding q-harmonic oscillator algebra generators have the form

$$B^- = T^{-1}(d/dx + 1/\sqrt{1-q^2}), \quad B^+ = (-d/dx + 1/\sqrt{1-q^2})T,$$
$$B^- B^+ - q^2 B^+ B^- = 1, \qquad L = B^+ B^- - 1/(1-q^2).$$

The eigenvalue problem $B^- \psi_\alpha^-(x) = \alpha \psi_\alpha^-(x)$ leads to the pantograph equation, which was analyzed in detail in [**28**]. The corresponding results show that in accordance with the purely algebraic consideration there are no normalizable coherent states of this type. However, there are infinitely many normalizable eigenstates of the operator B^+ determined as appropriate solutions of the advanced pantograph equation [**28**]

$$\frac{d}{dx}\psi_\alpha^+(x) = -\alpha q^{-3/2}\psi_\alpha^+(q^{-1}x) + \frac{q^{-1}}{\sqrt{1-q^2}}\psi_\alpha^+(x).$$

An important property of these states is that they are determined by C^∞ but not analytical functions at the $x = 0$ point. Consideration of the $N > 1$ symmetry algebras in this realization leads to the generalized pantograph equations [**27**].

As a general conclusion to this section we would like to stress that all three types of objects associated with q-special functions in their standard meaning [**18**]: the ordinary and bilateral basic hypergeometric series, and Ramanujan type integrals, show up in the context of self-similar spectral problems.

9. Second Order Difference Equation

Let us describe briefly a realization of the algebra of symmetries (7.2) upon the discrete Schrödinger equation or three term recurrence relation for orthogonal polynomials [**45, 46, 48**]. One considers an infinite chain of Jacobi matrices L_j and the corresponding eigenvalue problems:

$$L_j\psi_n^j \equiv \psi_{n+1}^j + u_n^j\psi_{n-1}^j + b_n^j\psi_n^j = \lambda\psi_n^j, \quad n, j \in Z. \tag{9.1}$$

In the case of orthogonal polynomials, one considers (9.1) only for $n > 0$ and imposes the boundary conditions $\psi_0^j(\lambda) = 0$, $\psi_1^j(\lambda) = \lambda - b_1^j$. Similar to the continuous case the forward discrete time step is defined by the Christoffel's transformation to kernel polynomials [**50**]:

$$\psi_n^{j+1} = \frac{\psi_{n+1}^j + C_n^{j+1}\psi_n^j}{\lambda - \lambda_{j+1}} \equiv \frac{S_{j+1}\psi_n^j}{\lambda - \lambda_{j+1}}. \tag{9.2}$$

The backward transformation was analyzed by Geronimus [**21**] and it has the form:

$$\psi_n^{j-1} = \psi_n^j + A_n^j\psi_{n-1}^j \equiv R_j\psi_n^j. \tag{9.3}$$

Here A_n^j and C_n^j are discrete analogs of the superpotentials $f_j(x)$ and S_j, R_j are the first order difference operators. The compatibility conditions of the $j \to j \pm 1$ moves yield the factorizations $L_j = S_jR_j + \lambda_j$, $L_{j-1} = R_jS_j + \lambda_j$, or $u_n^j = A_n^jC_n^j$, $b_n^j = A_{n+1}^j + C_n^j + \lambda_j$. The abstract factorization chain in this case is equivalent to the following set of nonlinear finite-difference equations:

$$A_n^jC_{n-1}^j = A_n^{j-1}C_n^{j-1}, \quad A_n^j + C_n^j + \lambda_j = A_{n+1}^{j-1} + C_n^{j-1} + \lambda_{j-1}, \tag{9.4}$$

known to define a discrete-time Toda lattice. Note that this is not an isospectral flow since the constants λ_j determine the character of the change of the spectrum of the taken Jacobi matrix after the Christoffel or Geronimus transformations.

Discrete Schrödinger equation analogs of the self-similar reductions (7.1) were described in [**46**]:

$$A_n^{j+N} = qA_{n+k}^j, \quad C_n^{j+N} = qC_{n+k}^j, \quad \lambda_{j+N} = q\lambda_j, \tag{9.5}$$

where k is an integer. When n is considered as continuous, k may be taken as a continuous variable as well. This closure is associated with classical, semi-classical and, so-called, Laguerre-Hahn class of orthogonal polynomials on linear and q-linear grids [**34**]. The corresponding recurrence coefficients are related to ordinary and q-analogs of some discrete Painlevé transcendents [**22, 46**]. The algebra of symmetries of these systems is derived along the same lines as in the continuous case, however, in the present case the compact versions of algebras, like $su_q(2)$, are allowed as well. The coherent states are defined in the same way as in the continuous case. For some explicit examples see, e.g., [**5**].

The following discrete-time Volterra lattice has been derived in [**48**]:

$$D_n^j(D_{n-1}^j - \beta_j) = D_n^{j-1}(D_{n+1}^{j-1} - \beta_{j-1}). \tag{9.6}$$

It can be mapped upon the discrete-time Toda lattice (9.4) via the following quadratic relation:

$$A_n^j = D_{2n}^j D_{2n+1}^j, \quad C_n^j = (D_{2n+1}^j - \beta_j)(D_{2n+2}^j - \beta_j), \quad \lambda_j = \text{const.} - \beta_j^2, \tag{9.7}$$

which generalizes well-known relation between the ordinary Toda and Volterra lattices. There is also the second similar mapping

$$A_n^j = D_{2n-1}^j D_{2n}^j, \quad C_n^j = (D_{2n}^j - \beta_j)(D_{2n+1}^j - \beta_j), \tag{9.8}$$

with the same connection between λ_j and β_j as given before. As shown in [**48**], this discrete-time Volterra lattice is related to the g-algorithm proposed by Bauer in numerical analysis [**8**]. There exists an ansatz of semi-separation of discrete variables in (9.6) which leads [**48**] to recurrence coefficients of the Askey-Wilson polynomials—the most general set of classical orthogonal polynomials [**6**] and, simultaneously, of the polynomials considered by Askey and Ismail [**4**].

An interesting discrete symmetry for the chain (5.1) has been described in [**2**]. It is associated with a freedom in the intermediate steps of two-step discrete-time shifts $j \to j+2$ and related to the statement on permutability of a sequence of Bäcklund-Darboux transformations [**11**] for fixed set of corresponding parameters. In [**45**] an analog of this symmetry for the discrete time Toda and Volterra lattices has been derived.

Let us describe briefly this refactorization symmetry for the equation (9.6). Let $\beta_j \neq 0$ for any j, then:

$$\widetilde{D}_n^j = \frac{1}{\beta_{j-1}}\left(\beta_j D_n^j + \frac{(\beta_j^2 - \beta_{j-1}^2)(D_n^j D_{n+1}^j + D_n^{j-1} D_{n+1}^{j-1})}{\beta_j(\beta_j - D_{n-1}^j - D_{n+1}^j) + \beta_{j-1}(\beta_{j-1} - D_n^{j-1} - D_{n+2}^{j-1})}\right), \tag{9.9}$$

$$\widetilde{D}_n^{j-1} = \frac{1}{\beta_j}\left(\beta_{j-1} D_n^{j-1} - \frac{(\beta_j^2 - \beta_{j-1}^2)(D_{n-1}^j D_n^j + D_{n-1}^{j-1} D_n^{j-1})}{\beta_j(\beta_j - D_{n-2}^j - D_n^j) + \beta_{j-1}(\beta_{j-1} - D_{n-1}^{j-1} - D_{n+1}^{j-1})}\right). \tag{9.10}$$

The change of spectral parameters β_j looks as follows:

$$\widetilde{\beta}_j = \beta_{j-1}, \quad \widetilde{\beta}_{j-1} = \beta_j. \tag{9.11}$$

If one substitutes into (9.6) instead of D_n^j, D_n^{j-1} and β_j, β_{j-1} the tilded variables and keeps all other D_n^k, β_k, $k \neq j, j-1$, fixed, then the resulting equation will be satisfied automatically, i.e. we have a discrete symmetry. The transformation laws for superpotentials A_n^j, C_n^j, describing a similar symmetry for the discrete-time Toda lattice, follow from the maps (9.7), (9.8). If $\beta_j = 0$ or $\beta_{j-1} = 0$ (j is fixed), then there appears some additional freedom, we refer for details to [**45**].

We have sketched only the simplest types of discrete-time integrable systems. For an analysis of ordinary Toda lattice and its self-similar reductions see, e.g., [**30, 31**]. For more complicated examples and their applications see [**23, 29, 36, 37**]. Recently self-similar reductions of spectral transformation chains have been considered for biorthogonal rational functions in [**49**].

10. One-Dimensional Ising Chains

Tau-function of the N-soliton solution of the KdV equation (3.2) can be represented in the following Hirota form, which was widely discussed in the literature (see, e.g., [**1, 24**] and references therein):

$$\tau_N = \sum_{\mu_i=0,1} \exp\Big(\sum_{1\le i<j\le N} A_{ij}\mu_i\mu_j + \sum_{i=1}^{N} \theta_i\mu_i \Big). \tag{10.1}$$

Here the phase shifts A_{ij} are expressed via the spectral variables k_i in a simple way

$$e^{A_{ij}} = \frac{(k_i - k_j)^2}{(k_i + k_j)^2}. \tag{10.2}$$

In [**32**] it was noticed that this expression for τ_N defines the grand partition function of the lattice gas model if $\theta_i = \theta^{(0)} = \text{const}$. A comparison with [**9**] shows that μ_i play the role of the filling factors of the lattice sites by molecules and $\theta^{(0)}$ is a chemical potential. The constants A_{ij} are proportional to the interaction energies of molecules.

Simultaneously, τ_N (10.1) describes the partition function of a particular one-dimensional Ising chain:

$$Z_N = \sum_{\sigma_i=\pm 1} e^{-\beta E}, \quad \beta = \frac{1}{kT}, \tag{10.3}$$

$$E = \sum_{1\le i<j\le N} J_{ij}\sigma_i\sigma_j - \sum_{i=1}^{N} H_i\sigma_i,$$

where N is the number of spins $\sigma_i = \pm 1$, J_{ij} are the exchange constants, H_i is an external magnetic field, T is the temperature and k is the Boltzmann constant. Indeed, after the substitution into (10.1) of the spin variables $\mu_i = (\sigma_i + 1)/2$, some simple calculations yield

$$\tau_N = e^{\Phi} Z_N, \quad \Phi = \frac{1}{4}\sum_{i<j} A_{ij} + \frac{1}{2}\sum_{j=1}^{N} \theta_j, \tag{10.4}$$

with the following relations between soliton parameters and Ising chain characteristics

$$\beta J_{ij} = -\frac{1}{4}A_{ij}, \quad \beta H_i = \frac{1}{2}\theta_i + \frac{1}{4}\sum_{j=1, i\neq j}^{N} A_{ij}. \tag{10.5}$$

A similar relation with Ising chains is valid for the whole KP hierarchy and some other differential and difference nonlinear integrable equations. The corresponding tau-functions have the form (10.1) for different choice of the phase shifts A_{ij} and the phases θ_i, a list of such equations can be found, e.g., in [**1, 24**].

A crucial observation of [**32**] is that the condition of self-similarity of the spectrum is related to translational invariance of the spin exchange constants J_{ij} of the infinite Ising chain induced by the KdV equation. In the simplest case one demands translational invariance of the system with respect to the shift by one site $j \to j+1$ which means that $J_{i+1,j+1} = J_{ij}$. As a result, the exchange J_{ij} or phase shifts A_{ij} depend only on the distance between the sites $|i-j|$. This natural physical requirement forces k_i to form one geometric progression

$$(10.6) \qquad k_i = k_1 q^{i-1}, \quad q = e^{-2\alpha}, \quad A_{ij} = 2\ln\big|\tanh\alpha(i-j)\big|,$$

where k_1 and $q < 1$ are free parameters. One may demand also the invariance with respect to shifts by a multiple of the lattice site, when $J_{i+M,j+M} = J_{ij}$, and this leads to the general self-similar spectra $k_{j+M} = qk_j$.

In fact the translational invariance in not exact for finite chain $1 \le j \le N$. The limit $N \to \infty$ corresponds to the thermodynamic limit. It gives an infinite soliton potential with self-similar discrete spectra. The coordinate x and time t (and higher "times" of the corresponding hierarchies) are interpreted as parameters of the magnetic field H_i. The x, t dependent part of H_i decays exponentially fast for $i \to \infty$, since $q < 1$. Therefore in the thermodynamic limit only the values of constants $\theta_i^{(0)}$ are relevant for the partition function. The formalism allows us to treat the M-periodic magnetic fields, $H_{i+M} = H_i$, which for $M = 1$ is just the homogeneity condition.

For the KdV equation case one has $0 < \big|\tanh\alpha(i-j)\big| < 1$ and $J_{ij} = -A_{ij}/4\beta > 0$, i.e. an antiferromagnetic Ising chain. A similar situation takes place for general M-periodic case. Such interaction has a long distance character but the intensity falls off exponentially fast. The absence of phase transitions in such systems for nonzero temperatures is well known.

For $\alpha \to 0$ or $q \to 1$ the phase shifts $A_{ij} \propto J_{ij}/kT$ are diverging. If one renormalizes the exchange constants $J_{ij}^{\rm ren} = J_{ij}(q^{-1} - q)$ and the temperature $kT_{\rm ren} = kT(q^{-1} - q)$, then the interaction energy of a single spin with all other ones will be finite for $q \to 1$. As a result, in this limit one actually gets a nonlocal interaction model at a low effective temperature. Note that for imitation of the change of the temperature it is necessary to change simultaneously the magnetic field $H = H^{\rm ren}/(q^{-1} - q)$.

The limit $q \to 0$ gives $J_{ij}^{\rm ren} \propto \delta_{i+1,j}$, or the high temperature nearest neighbor interaction Ising chain (if H is renormalized). If H is kept finite then this limit corresponds to the noninteracting spins. So, the formalism provides only a partial description of the partition function in a two-dimensional subspace of (T, H, q). For fixed q the temperature T has a prescribed value and one may normalize the "KdV temperature" to $\beta = 1$.

Using the Wronskian formula for the representation of N-soliton potential [**13**], we were able to calculate the partition function for the translationally invariant Ising chains in a homogeneous magnetic field. Omitting technical details, which can be found in [**32**], for $M = 1$ we get $Z_N \to \exp(-N\beta f_I)$ for $N \to \infty$, where the free energy per site f_I has the form

$$(10.7) \qquad -\beta f_I(q, H) = \ln \frac{2(q^4; q^4)_\infty \cosh\beta H}{(q^2; q^2)_\infty^{1/2}} + \frac{1}{4\pi}\int_0^{2\pi} d\nu \ln\big(\big|\rho(\nu)\big|^2 - q\tanh^2\beta H\big),$$

$$\left|\rho(\nu)\right|^2 = \frac{(q^2e^{i\nu};q^4)^2_\infty (q^2e^{-i\nu};q^4)^2_\infty}{(q^4e^{i\nu};q^4)^2_\infty (q^4e^{-i\nu};q^4)^2_\infty}\frac{1}{4\sin^2(\nu/2)} = q\frac{\theta_4^2(\nu/2,q^2)}{\theta_1^2(\nu/2,q^2)}.$$

Here $\theta_{1,4}(y,q)$ are the standard Jacobi theta-functions of the argument y and base q (our base is q^2). The original expression for the density function $\rho(\nu)$ is

$$\rho(\nu) = \frac{(q^2;q^4)^2_\infty}{(q^4;q^4)^2_\infty}\sum_{k=-\infty}^{\infty}\frac{e^{i\nu k-\varepsilon k}}{1-q^{4k+2}} = \frac{(q^2e^{i\nu-\varepsilon};q^4)_\infty (q^2e^{-i\nu+\varepsilon};q^4)_\infty}{(e^{i\nu-\varepsilon};q^4)_\infty (q^4e^{-i\nu+\varepsilon};q^4)_\infty}, \tag{10.8}$$

where a small regularization parameter $\varepsilon \to 0$ is introduced for absolute convergency of the infinite sum. The last equality in (10.8) was obtained with the help of the ${}_1\psi_1$ Ramanujan's bilateral basic hypergeometric series sum [**18**]:

$$\sum_{n=-\infty}^{\infty}\frac{(a;q)_n}{(b;q)_n}z^n = \frac{(q,b/a,az,q/az;q)_\infty}{(b,q/a,z,b/az;q)_\infty}.$$

This fact demonstrates again the general statement given earlier that the q-special functions are always behind the self-similar spectral problems.

The derivative of $f_I(H)$ with respect to H yields the total magnetization of the lattice:

$$\begin{aligned} m(H) &= -\partial_H f_I = \lim_{N\to\infty}\frac{1}{N}\sum_{i=1}^{N}\langle\sigma_i\rangle, \\ &= \tanh\beta H\left(1-\frac{1}{\pi}\int_0^\pi \frac{\theta_1^2(\nu,q^2)\,d\nu}{\theta_4^2(\nu,q^2)\cosh^2\beta H - \theta_1^2(\nu,q^2)\sinh^2\beta H}\right). \end{aligned} \tag{10.9}$$

The function $m(H)$ grows monotonically with H, its qualitative properties are predicted by the general theorems on the behavior of 1D systems with the fast decaying interactions.

The main drawback of the described scheme is that the exact calculation is valid only for a fixed temperature in the partition function (q is fixed). One may try to replace (10.6) by $A_{ij} = 2n\ln\left|(k_i-k_j)/(k_i+k_j)\right|$, where n is a sequence of integers, to change the temperature and look for integrable models with such phase shifts. This is not a simple task since there are many constraints involved into the resolution of this problem. We have found only one more example [**32**] for $n=2$ corresponding to a subcase of the N-soliton solution of the KP-equation of B-type (BKP). The BKP equation has the form (see, e.g., [**14, 25**])

$$\frac{\partial}{\partial x_1}\left(9\frac{\partial u}{\partial x_5} - 5\frac{\partial^3 u}{\partial x_3\partial x_1^2} + \frac{\partial^5 u}{\partial x_1^5} - 30\frac{\partial u}{\partial x_3}\frac{\partial u}{\partial x_1} + 30\frac{\partial u}{\partial x_1}\frac{\partial^3 u}{\partial x_1^3} + 60\left(\frac{\partial u}{\partial x_1}\right)^3\right) - 5\frac{\partial^2 u}{\partial x_3^2} = 0. \tag{10.10}$$

τ-function of this integrable equation generates partition functions of quite general Ising chains. The exchange constants have now the form

$$\beta J_{ij} = -\frac{1}{4}A_{ij}, \quad e^{A_{ij}} = \frac{(a_i-a_j)(b_i-b_j)(a_i-b_j)(b_i-a_j)}{(a_i+a_j)(b_i+b_j)(a_i+b_j)(b_i+a_j)}, \tag{10.11}$$

with the energy and partition function given by (10.3). For $a_i = b_i = k_i/2$ one gets the KdV-inspired model for $\beta = n = 2$.

Translational invariance of the spin lattice, $J_{ij} = J(i-j)$, results in the following spectral self-similarity

$$a_i = q^{i-1}, \quad b_i = bq^{i-1}, \quad q = e^{-2\alpha}, \tag{10.12}$$

where we set $a_1 = 1$ and assume that $\alpha > 0$. This gives the exchange

$$\beta J_{ij} = -\frac{1}{4} \ln \frac{\tanh^2 \alpha(i-j) - (b-1)^2/(b+1)^2}{\coth^2 \alpha(i-j) - (b-1)^2/(b+1)^2}.$$

Since this expression is invariant with respect to the inversion $b \to b^{-1}$, the parameter b is restricted to the unit disk $|b| \leq 1$. It is not difficult to see that for $-1 < b < -q$ one has now the ferromagnetic, i.e. $J_{ij} < 0$. For two other physical regions $q < b \leq 1$ and $b = e^{i\phi} \neq -1$, one has the antiferromagnetic, i.e. $J_{ij} > 0$.

In the thermodynamic limit $N \to \infty$ the partition function can be represented as the determinant of a Toeplitz matrix, which is diagonalized by the discrete Fourier transformation. Using again the ${}_1\psi_1$ sum, we have found the free energy per site for the homogeneous magnetic field $H_i = H$:

$$-\beta f_I(H) = \frac{1}{4} \ln \frac{(q, q, bq, q/b; q)_\infty}{(-q, -q, -bq, -q/b; q)_\infty} + \frac{1}{4\pi} \int_0^{2\pi} d\nu \ln |2\rho(\nu)|,$$

where

$$\rho(\nu) = \cosh 2\beta H + \frac{(-q;q)_\infty^2}{(-e^{i\nu}, -qe^{-i\nu}; q)_\infty} \left(\frac{(b^{-1}e^{i\nu}, qbe^{-i\nu}; q)_\infty}{(b^{-1}, qb; q)_\infty} + \frac{(be^{i\nu}, qb^{-1}e^{-i\nu}; q)_\infty}{(b, qb^{-1}; q)_\infty} \right). \tag{10.13}$$

The magnetization is obtained after taking the derivative with respect to H

$$m(H) = \tanh 2\beta H \left(1 - \frac{1}{\pi} \int_0^\pi \frac{d\nu}{1 + d(\nu) \cosh 2\beta H} \right), \tag{10.14}$$

where

$$d(\nu) = \frac{(qb, q/b; q)_\infty (b^{-1/2} - b^{1/2}) \theta_2(\nu, q^{1/2})}{(-q;q)_\infty^2 2 \operatorname{Im} \theta_1(\nu - (i/2) \ln b, q^{1/2})},$$

for $q < b \leq 1$,

$$d(\nu) = \frac{(qb, q/b; q)_\infty (|b|^{-1/2} + |b|^{1/2}) \theta_2(\nu, q^{1/2})}{(-q;q)_\infty^2 2 \operatorname{Re} \theta_2(\nu - (i/2) \ln |b|, q^{1/2})},$$

for $-1 < b < -q$, and

$$d(\nu) = \frac{(qe^{i\phi}, qe^{-i\phi}; q)_\infty 2 \sin(\phi/2) \theta_2(\nu, q^{1/2})}{(-q;q)_\infty^2 [\theta_1(\nu + \phi/2, q^{1/2}) - \theta_1(\nu - \phi/2, q^{1/2})]},$$

for $b = e^{i\phi}$. Here $\theta_2(\nu, q^{1/2})$ is another Jacobi θ-function. Taking the limit $b \to 1$ one may find the magnetization for the second value of discrete temperature of the KdV equation inspired spin chain, $n = 2$. The details of calculations and graphical representation of $m(H)$ for some values of the parameters can be found in [**32**]. In a somewhat similar way one may consider the $M > 1$ periodic systems as well.

11. Random Matrices

Random matrices were employed by Wigner and Dyson for studying spectra of complex systems with many degrees of freedom. When the density of levels is high enough, excitation energies can be described statistically. The detailed structure of hamiltonians is not known and statistical characteristics of the system are described by averaging over ensembles of random matrices. A fundamental constraint is that the probability distributions should be invariant under basic symmetries such as parity, rotation and time-reversal transformations [**35**].

Gaussian ensembles use a real symmetric, Hermitian or self-dual Hermitian random matrix H with the statistically independent elements H_{ik}. The probability density $P(H)$ for H_{ik} to lie in a unit volume is proportional to $\exp(-a\mathrm{tr}H^2+b\mathrm{tr}H)$, with the measure invariant under orthogonal, unitary or symplectic transformations. Abandoning the condition of statistical independence one may pass to ensembles with more complicated $P(H)$.

Dyson has introduced circular ensembles of unitary random $n\times n$ matrices S with eigenvalues $\varepsilon_j = e^{i\phi_j}$, $j=1,\dots,n$, such that the behavior of the phases ϕ_i is equivalent to the distribution of eigenvalues of a system. The condition of invariance of the measure under all unitary automorphisms $S\to USW$, where U,W are arbitrary unitary matrices, determines the probability distribution. Any S can be represented in the form $S=U^{-1}EU$, E a diagonal matrix with the eigenvalues ε_j and U some unitary matrix. Then, using the invariance of probability measure, one can gauge away the U-dependent part of S and integrate out the "angular variables" in the probability distribution. As a result, the eigenvalue distribution becomes

$$P\,d\phi_1\dots d\phi_n \propto \prod_{i<j}|\varepsilon_i-\varepsilon_j|^2\,d\phi_1\dots d\phi_n.$$

One can weaken the condition of invariance of the measure under arbitrary left or right translations and leave only pure unitary transformations. Then the invariant measure is not uniquely defined and the eigenvalue distribution may be written as

$$P\,d\phi_1\dots d\phi_n = f(\phi_1,\dots,\phi_n)\prod_{i<j}|\varepsilon_i-\varepsilon_j|^2\,d\phi_1\dots d\phi_n,$$

where the function f is symmetric under permutation of its arguments. One of such ensembles has been considered by Gaudin long time ago [**20**]. In this model one has the following probability law for eigenvalues

$$P\,d\phi_1\dots d\phi_n \propto \prod_{i<j}\left|\frac{\varepsilon_i-\varepsilon_j}{\varepsilon_i-z\varepsilon_j}\right|^2 d\phi_1\dots d\phi_n, \qquad (11.1)$$

interpolating between the Dyson unitary ensemble ($z=0$) and the uniform distribution ($z=1$).

The same model can be interpreted also as a Coulomb gas on a circle with the partition function

$$Z_n \propto \int_0^{2\pi}\cdots\int_0^{2\pi} d\phi_1\dots d\phi_n \exp\Big(-\beta\sum_{i<j}V(\phi_i-\phi_j)\Big),$$

where $\beta = 1/kT$ is fixed and the potential energy is

$$\beta V(\phi_i - \phi_j) = \ln\left(1 + \frac{\sinh^2\gamma}{\sin^2((\phi_i - \phi_j)/2)}\right), \quad z = e^{-2\gamma}. \tag{11.2}$$

The grand partition function of this model corresponds to the τ-function of the KP hierarchy in the specific infinite soliton limit [**32**]. The finite soliton solutions provide thus a discretization of the system, namely, a lattice gas on the circle. For example, one may take unitary $n \times n$ matrices S with the eigenvalues equal to N-th roots of unity, i.e. $\phi_j = 2\pi m_j/N, m_j = 1, \ldots, N$. The measure is taken to be continuous in the "angular" variables and discrete in eigenvalues, i.e. one takes the sums over ϕ_j instead of the integrals. The completely continuous model is recovered for $N \to \infty$, ϕ_j fixed:

$$\left(\frac{2\pi}{N}\right)^n \sum_{m_1=1}^{N} \cdots \sum_{m_n=1}^{N} \underset{N\to\infty}{\longrightarrow} \int_0^{2\pi} d\phi_1 \ldots \int_0^{2\pi} d\phi_n. \tag{11.3}$$

For $n \times n$ matrices the lattice partition function is

$$Z_n(N,z) = \left(\frac{2\pi}{N}\right)^n \sum_{m_1=1}^{N} \cdots \sum_{m_n=1}^{N} \prod_{1\le i<j\le n} \left|\frac{\varepsilon_i - \varepsilon_j}{\varepsilon_i/\sqrt{z} - \sqrt{z}\varepsilon_j}\right|^2,$$

$$\varepsilon_j = \exp\frac{2i\pi m_j}{N}, \quad z = e^{-2\gamma}.$$

The grand canonical ensemble partition function is

$$\begin{aligned} Z(z,\theta) &= \sum_{n=0}^{N} \frac{Z_n(N,z)e^{\theta n}}{n!} \\ &= \sum_{\mu_m=0,1} \exp\left(\sum_{1\le m<k\le N} A_{mk}\mu_m\mu_k + (\theta+\eta)\sum_{m=1}^{N}\mu_m\right), \end{aligned} \tag{11.4}$$

where $\eta = \ln(2\pi/N)$ enters as an addition to the chemical potential θ and

$$A_{mk} = \ln\frac{\sin^2(\pi(m-k)/N)}{\sin^2(\pi(m-k)/N) + \sinh^2\gamma} = \ln\frac{(a_m - a_k)(b_m - b_k)}{(a_m + b_k)(b_m + a_k)},$$

is the KP phase shift with the restricted choice of parameters

$$a_m = e^{2i\pi m/N}, \quad b_m = -za_m, \ m = 1, 2, \ldots, N. \tag{11.5}$$

So, the grand partition function of this discrete circular unitary matrix model coincides with the particular KP N-soliton τ-function at zero hierarchy "times". In the thermodynamical limit $N \to \infty$ the relation (11.3) takes place and we get the matrix model (11.1). The root of unity discretization has been already considered by Gaudin himself [**19**], where the connection with Ising chains was noticed as well but the relation with integrable equations was not established.

The BKP equation suggests a further generalization of this matrix model with the probability law

$$P\, d\phi_1 \ldots d\phi_n \propto \prod_{i<j} \left|\frac{\varepsilon_i - \varepsilon_j}{\varepsilon_i + \varepsilon_j}\right|^2 \left|\frac{\varepsilon_i + z\varepsilon_j}{\varepsilon_i - z\varepsilon_j}\right|^2 d\phi_1 \ldots d\phi_n. \tag{11.6}$$

The grand partition function is also defined by (11.4) where A_{mk} is given by (10.11). In order to escape singularities appearing from zeros $\varepsilon_i + \varepsilon_j = 0$ for $\phi_j = \phi_i + \pi$, the parameters a_m, b_m have to be restricted. The choice (11.5) works for odd N only, which makes the $N \to \infty$ limit problematic. Another option is to replace m/N in (11.5) by $m/2N$, which corresponds to the half-circle system. Comparing with the connection of BKP equation to Ising chains described in the previous section we see, that now we have spins on the circle, q is a root of unity and the exchange $J_{mk} = J(m-k)$ is

$$\beta J(m) = -\frac{1}{4} \ln \frac{1 + \sinh^2 \gamma / \cos^2(\pi m/N)}{1 + \sinh^2 \gamma / \sin^2(\pi m/N)}.$$

An important difference between the $|q| < 1$ and $|q| = 1$ Ising chains is that in the former case the thermodynamic limit $N \to \infty$ does not lead necessarily to the continuous space models as in the circular case. Also the relation of noncompact Ising chains to hermitian random matrices is not so straightforward. In particular, in order to get nontrivial weight functions one has to take inhomogeneous magnetic field corresponding to nonzero hierarchy "times". Calculation of the partition function is more involved in this situation.

12. Conclusions

Self-similarity in spectral problems serves as a key to special functions, in particular, to q-special functions. We hope that the brief review given in this paper gives a sufficient number of arguments for convincing in this vague statement. As to the notion of exact solvability of a given equation, it requires a more rigorous group-theoretical definition in terms of the operator (differential, difference, etc.) Galois theory. In this approach one fixes in advance the basic fields of functions entering spectral problems and asks about the transcendence of solutions of the resulting equations over these fields. In the scheme we have presented the situation is slightly different—the basic fields of functions are defined as solutions of complicated nonlinear equations appearing after self-similar reductions of some infinite systems of equations. Usually this gives simultaneously a part of solutions of the underlying linear spectral equations in quadratures over the basic field of functions, but determination of the transcendence of the general solution remains open. Therefore these two approaches should be used in conjunction. One of the problems appearing on this route is to give a complete characterization of the changes in the structure of corresponding (differential, ...) Galois groups resulting from the Darboux-type spectral transformations [**44**].

Self-similar spectra considered here are based upon linear map of the spectral parameter $\lambda \to q\lambda + h$. This leads naturally to spectra containing a finite number of geometric progressions and "Painlevé-type" transcendental functions on the background. A natural generalization involves polynomial maps $\lambda \to P(\lambda)$ leading in general to chaotic structure of the spectrum. The situation now is much more complicated and only the simplest possibilities were considered in the literature. Note that, similar to the linear case, the spectra partially invariant with respect to the map $\lambda \to P(\lambda)$ and its inverses are permitted too, but strong difficulties arise in the characterization of functions involved in these systems already in the completely self-similar setting (see, e.g., [**7, 10**]). The scalings of discrete grids, or decimations entering at this stage hint upon the relevance of structures hidden in wavelets and quasicrystals. Perhaps all these subjects are naturally unified within an extended approach to the concept of self-similarity.

References

1. M. J. Ablowitz and H. Segur, *Solitons and the inverse scattering transform*, SIAM Studies in Applied Mathematics, vol. 4, SIAM, Philadelphia, Pa., 1981.
2. V. E. Adler, *Cuttings of polygons*, Funktsional. Anal. i Prilozhen. **27** (1993), no. 2, 79–82 (Russian); English transl., Functional Anal. Appl. **27** (1993), no. 2, 141–143.
3. A. A. Andrianov, M. V. Ioffe, and V. P. Spiridonov, *Higher-derivative supersymmetry and the Witten index*, Phys. Lett. A **174** (1993), 273–279.
4. R. Askey and M. E. H. Ismail, *Recurrence relations, continued fractions and orthogonal polynomials*, Mem. Amer. Math. Soc. **49** (1984), no. 300.
5. R. Askey and S. K. Suslov, *The q-harmonic oscillator and an analogue of the Charlier polynomials*, J. Phys. A **26** (1993), L693–L698.
6. R. Askey and J. Wilson, *Some basic hypergeometric orthogonal polynomials that generalize Jacobi polynomials*, Mem. Amer. Math. Soc. **319** (1985).
7. M. F. Barnsley, J. S. Geronimo, and A. N. Harrington, *On the invariant sets of a family of quadratic maps*, Comm. Math. Phys. **88** (1982), 479–501.
8. F. L. Bauer, *The g-algorithm*, J. Soc. Indust. Appl. Math. **8** (1960), 1–17.
9. R. J. Baxter, *Exactly solved models in statistical Mechanics*, Academic Press, London, 1982.
10. D. Bessis and P. Moussa, *Orthogonality properties of iterated polynomial mappings*, Comm. Math. Phys. **88** (1983), 503–529.
11. F. Calogero and A. Degasperis, *Spectral transform and solitons*, North-Holland, Amsterdam, 1982.
12. K. M. Case, *Singular potentials*, Phys. Rev. **80** (1950), 797–806.
13. M. M. Crum, *Associated Sturm-Liouville systems*, Quart. J. Math. Oxford Ser. (2) **6** (1955), 121–127.
14. E. Date, M. Jimbo, M. Kashiwara, and T. Miwa, *Transformation groups for soliton equations* IV. *A new hierarchy of soliton equations of* KP-*type*, Physica D **4** (1982), 343–365.
15. V. M. Eleonsky and V. G. Korolev, *On the nonlinear generalization of the Fock method*, J. Phys. A **28** (1995), 4973–4985.
16. H. Flaschka, *A commutator representation of Painlevé equations*, J. Math. Phys. **21** (1980), 1016–1018.
17. C. S. Gardner, J. H. Greene, M. D. Kruskal, and R. M. Miura, *Method for solving the Korteweg-de Vries equation*, Phys. Rev. Lett. **19** (1967), 1095–1097.
18. G. Gasper and M. Rahman, *Basic hypergeometric series*, Cambridge Univ. Press, Cambridge, 1990.
19. M. Gaudin, *Gaz coulombien discret à une dimension*, J. Phys. (France) **34** (1973), 511–522.
20. ———, *Une famille à une paramètre d'ensembles unitaires*, Nuclear Phys. B **85** (1966), 545–575.
21. Ya. L. Geronimus, *On the polynomials orthogonal with respect to a given number sequence and a theorem by W. Hahn*, Izvestiya Akad. Nauk SSSR. Ser. Mat. **4** (1940), 215–228.
22. B. Grammaticos, V. Papageorgiou, and A. Ramani, *Discrete dressing transformations and Painlevé equations*, Phys. Lett. A **235** (1997), 475–479; V. Papageorgiou, talk at this conference.
23. R. Hirota, *Discrete two-dimensional Toda molecule equation*, J. Phys. Soc. Japan **56** (1987), 4285–4288, and the review talk at this conference.
24. ———, *Direct methods in soliton theory*, Solitons (R. K. Bullough and P. J. Caudrey, eds.), Springer, Berlin-New York, 1980.
25. ———, *Soliton solutions to the* BKP *equations* I. *The Pfaffian technique*, J. Phys. Soc. Japan **58** (1989), 2285–2296.
26. L. Infeld and T. E. Hull, *The factorization method*, Rev. Modern Phys. **23** (1951), 21–68.
27. A. Iserles, *On the generalized pantograph functional-differential equation*, European J. Appl. Math. **4** (1993), 1–38.
28. T. Kato and J. B. McLeod, *The functional-differential equation* $y'(x) = ay(\lambda x) + by(x)$, Bull. Amer. Math. Soc. (N.S.) **77** (1971), 891–937.
29. I. Krichever, O. Lipan, P. Wiegmann, and A. Zabrodin, *Quantum integrable models and discrete classical Hirota equations*, Comm. Math. Phys. **188** (1997), 267–304.
30. D. Levi, *Nonlinear differential-difference equations as Bäcklund transformations*, J. Phys. A **14** (1981), 1083–1098.

31. D. Levi and P. Winternitz, *Continuous symmetries of discrete equations*, Phys. Lett. A **152** (1991), 335–338; *Symmetries and conditional symmetries of differential-difference equations*, J. Math. Phys. **34** (1993), 3713–3730.
32. I. Loutsenko and V. Spiridonov, *Self-similar potentials and Ising models*, JETP Lett. **66** (1997), 789–795; *Spectral self-similarity, one-dimensional Ising chains and random matrices*, Nuclear Phys. B **538** (1999), 731–758.
33. A. J. Macfarlane, *On q-analogues of the quantum harmonic oscillator and quantum group* $SU(2)_q$, J. Phys. A **22** (1989), 4581–4588.
34. A. P. Magnus, *Special nonuniform lattice (snul) orthogonal polynomials on discrete dense sets of points*, J. Comput. Appl. Math. **65** (1995), 253–265.
35. M. L. Mehta, *Random matrices*, 2nd ed., Academic Press, Boston, 1991.
36. F. Nijhoff, O. Ragnisco, and V. Kuznetsov, *Integrable time-discretization of the Ruisenaars-Schneider model*, Comm. Math. Phys. **176** (1996), 681–700.
37. S. P. Novikov and I. A. Dynnikov, *Discrete spectral symmetries of low-dimensional differential operators and difference operators on regular lattices and two-dimensional manifolds*, Uspekhi Mat. Nauk **52** (1997), 175–234.
38. A. Odzijewicz, *Quantum algebras and q-special functions related to coherent states maps of the disk*, Comm. Math. Phys. **192** (1998), 183–215.
39. V. A. Rubakov and V. P. Spiridonov, *Parasupersymmetric quantum mechanics*, Modern Phys. Lett. A **3** (1988), 1337–1347.
40. A. Shabat, *The infinite dimensional dressing dynamical system*, Inverse Problems **8** (1992), 303–308.
41. S. Skorik and V. Spiridonov, *Self-similar potentials and the q-oscillator algebra at roots of unity*, Lett. Math. Phys. **28** (1993), 59–74; *On the spectra of hyperelliptic potentials*, Phys. Lett. A **190** (1994), 90–95.
42. V. Spiridonov, *Exactly solvable potentials and quantum algebras*, Phys. Rev. Lett. **69** (1992), 398–401.
43. ———, *Deformation of supersymmetric and conformal quantum mechanics through affine transformations*, Proc. of the International Workshop on Harmonic Oscillators (College Park, 1992) (D. Han, Y. S. Kim, and W. W. Zachary, eds.), NASA Conf. Publ. 3197, 1993, pp. 93–108; *Symmetries of the self-similar potentials*, Comm. Theoret. Phys. **2** (1993), 149–163.
44. ———, *Universal superpositions of coherent states*, CRM–1913, Univ. de Montréal, Canada, 1993; *Coherent states of the q-Weyl algebra*, Lett. Math. Phys. **35** (1995), 179–185; *Universal superpositions of coherent states and self-similar potentials*, Phys. Rev. A **52** (1995), 1909–1935.
45. ———, *Symmetries of factorization chains for the discrete Schrödinger equation*, J. Phys. A **30** (1997), L15–L21.
46. V. Spiridonov, L. Vinet, and A. Zhedanov, *Difference Schrödinger operators with linear and exponential discrete spectra*, Lett. Math. Phys. **29** (1993), 63–73; *Spectral transformations, self-similar reductions and orthogonal polynomials*, J. Phys. A **30** (1997), 7621–7637.
47. V. Spiridonov and A. Zhedanov, *Symmetry preserving quantization and self-similar potentials*, J. Phys. A **28** (1995), L589–L595.
48. ———, *Discrete Darboux transformations, discrete time Toda lattice, and the Askey-Wilson polynomials*, Methods Appl. Anal. **2** (1995), 369–398; *Discrete time Volterra chain and classical orthogonal polynomials*, J. Phys. A **30** (1997), 8727–8737.
49. ———, *Spectral transformation chains and some new biorthogonal rational functions*, Comm. Math. Phys. (to be published).
50. G. Szegö, *Orthogonal polynomials*, 2nd ed., Amer. Math. Soc., Providence, RI, 1959.
51. A. P. Veselov and A. B. Shabat, *A dressing chain and the spectral theory of Schrödinger operator*, Funktsional. Anal. i Prilozhen. **27** (1993), no. 2, 1–21 (Russian); English transl., Functional Anal. Appl. **27** (1993), no. 2, 81–96.
52. J. Weiss, *Periodic fixed points of Bäcklund transformations*, J. Math. Phys. **31** (1987), 2025–2039.
53. J. H. Wilkinson, *The algebraic eigenvalue problem*, Clarendon Press, Oxford, 1965.
54. V. E. Zakharov, S. V. Manakov, S. P. Novikov, and L. P. Pitaevski, *Theory of solitons*, Moscow, Nauka, 1980.

Department of Physics, Joseph Henry Laboratories, Jadwin Hall, Princeton University, P.O. Box 708, Princeton, NJ 08544-0708, USA.
E-mail address: loutseni@feynman.princeton.edu

Bogoliubov Laboratory of Theoretical Physics, Joint Institute for Nuclear Research, Dubna, Moscow region 141980, Russia.
E-mail address: svp@thsun1.jinr.ru

Centre de Recherches Mathématiques
CRM Proceedings and Lecture Notes
Volume **25**, 2000

Lie Point Symmetries of Discrete and Continual SU(∞) Toda Field Theory

L. Martina

ABSTRACT. The relation between the symmetry groups of the discrete and the continual SU(∞) Toda field theory is studied. Several symmetry reductions for the two cases are compared and solved.

This is a short note containing some recent results concerning the Lie-point symmetries of the SU(∞) Toda field theory, which is represented either as a discrete lattice or as its "continual" realization, i.e. the so-called heavenly equation

$$\partial_x \partial_y u = -\partial_z^2 e^u, \tag{1}$$

solved by Saveliev [**17, 18**] and Ward [**21**]. Equation (1) has applications in the theory of gravitational instantons [**21**] and in the theory of the area-preserving diffeomorphisms [**2**]. On the other hand, the lattice of Toda fields we want to study is [**11, 19, 20**]

$$\partial_x \partial_y u_n = -\exp(u_{n-1}) + 2\exp(u_n) - \exp(u_{n+1}), \quad -\infty < n < +\infty. \tag{2}$$

The coupling constants form a tridiagonal infinite dimensional matrix, which can be interpreted as the extension for $N \to \infty$ of the Cartan matrix of the simple Lie algebra A_N, whose generators of the Chevalley–Weyl basis $\{H_i, E_i^{\pm}\}$ $i = 1, \dots, N$ have the commutation relations

$$[H_i, H_j] = 0, \quad [E_i^+, E_j^+] = \delta_{ij} H_i, \quad [H_i, E_j^{\pm}] = \pm K_{ij} E_j^{\pm}. \tag{3}$$

For each finite N one could introduce a convenient real parameter Δn in order to rescale the generators as

$$\widetilde{H}_i = \frac{H_i}{\Delta n^3}, \quad \widetilde{E}_i^{\pm} = \frac{E_i^{\pm}}{\Delta n^2}. \tag{4}$$

2000 *Mathematics Subject Classification.* Primary: 22E70; Secondary: 22E65.

This work was partially supported by the NATO CRG 960717, by the italian INFN and by the Murst in the project SINTESI. The author would like to acknowledge the Centre des Recherches Mathèmatiques for the warm hospitality and P. Winternitz and D. Levi for many helpful discussions. Finally, the author thanks G. Soliani for many helpful discussions.

This is the final form of the paper.

Then the structure constants rescale in such a way that in the distribution sense for $N \to \infty$ one has

$$\frac{\delta_{ij}}{\Delta n} \to \delta(z - z'), \quad \frac{K_{ij}}{\Delta n^3} \to -\delta''(z - z'). \tag{5}$$

In this sense the Toda field lattice (2) is related to (1). Thus, it is quite natural to investigate the common properties of the Lie point transformations for both systems.

The point-symmetry analysis of the heavenly equation has been performed in [**1**]. The resulting symmetry group is the direct product of the conformal transformation group in 2 dimensions times a finite group, containing only translations and scaling in the z variable. The infinitesimal generators are

$$\begin{cases} \widehat{X}(\xi) = \xi(x)\partial_x - \partial_x\xi\partial_u, \\ \widehat{Y}(\eta) = \eta(y)\partial_y - \partial_y\eta\partial_u, \\ \widehat{Z} = \partial_z, \quad \widehat{T} = z\partial_z + 2\partial_u, \end{cases} \tag{6}$$

where ξ and η are arbitrary, regular functions of their argument. The symmetry group of the heavenly equation is the direct sum

$$\mathrm{Vir}^x \oplus \mathrm{Vir}^y \oplus \{\widehat{Z}, \widehat{T}\}, \tag{7}$$

of two copies of the centerless Virasoro algebra, which is a simple infinite Lie algebra [**5, 6**] locally isomorphic to the real diffeomorphisms algebra on the real axis [**3**], and of the 2-dimensional solvable Lie algebra generated by translations and scaling in the z variable.

The general results concerning the point—transformations group analysis of lattice systems of the Toda type are reported elsewhere, but they are based on the paper by Levi and Winternitz [**9**], with the aim to develop computational methods of the symmetry group of the difference differential equations. Here we will use the standard method for dealing with systems of differential equations [**13,14**], adapted to the presence of infinitely many dependent variables.

This peculiar structure of the system (2) implies that the group of the point transformations is the Lie group is $\mathrm{Vir}^x \oplus \mathrm{Vir}^y$ generated by

$$\begin{cases} \widehat{X}(\xi) = \xi(x)\partial_x - \partial_x\xi \sum_n \partial_{u_n}, \\ \widehat{Y}(\eta) = \eta(y)\partial_y - \partial_y\eta \sum_n \partial_{u_n}. \end{cases} \tag{8}$$

No further continuous symmetries are allowed.

Systems of the type (1) and (2) have solutions which depends on two arbitrary functions, since they admit general conformal transformations. They can be defined of Liouvillian type in the terminology of [**23**]. The commutation relations satisfied by the vector fields (8) are given by

$$\left[\widehat{X}(\xi_1), \widehat{X}(\xi_2)\right] = \widehat{X}(\xi_1\xi_{2x} - \xi_{1x}\xi_2), \tag{9}$$

and analogous relations among the $\widehat{Y}(\eta)$'s. The one-parameter subgroup generated by $\widehat{X}(\xi)$ for (2) is

$$\widetilde{x}_\lambda(x) = F^{-1}\left[\lambda + F(x)\right], \quad F(s) = \int \frac{ds}{\xi(s)}, \quad \widetilde{x}_0 = x, \tag{10}$$

$$\widetilde{u}_i(\widetilde{x}_\lambda, y) = u_i\big(x(\widetilde{x}_\lambda), y\big) - \ln\left(\frac{\xi(\widetilde{x}_\lambda)}{\xi(x(\widetilde{x}_\lambda))}\right). \tag{11}$$

Taking account of (1), one has to replace $u_i(x,y) \to u(x,y,z)$. The group transformations generated by the subalgebra $\{\widehat{Z}, \widehat{T}\}$ is simply given by

$$\begin{aligned} &\exp(\lambda\widehat{Z}): \quad z \to z+\lambda, \quad u \to u, \\ (12) \qquad &\exp(\lambda\widehat{T}): \quad z \to ze^{\lambda}, \quad u \to u+2\lambda. \end{aligned}$$

Finally, we mention also that the systems (1) and (2) admit a discrete group of invariance, which cannot be obtained by any infinitesimal method. The common group is generated by the parity transformation $x \to -x$, $y \to -y$, the permutation $x \leftrightarrow y$. Equation (1) admits also reflections in z, while (2) admits the discrete reflection $n \to -n$ and the integer translation $\widehat{Z}_n$ of n

$$(13) \qquad n' = n+N, \;\; N \in \mathbb{Z}.$$

It is natural to perform the identification $\widehat{Z}_n = \exp(n\widehat{Z})$, and look at the translational invariance on the lattice as a special case of the local continuous invariance in z for (1). To prove if further discrete symmetries exist for (2), one should use the method developed in [**8**], in order to generalize the Lie group formalism to difference equation.

In conclusion, the general structure of the symmetry group is

$$(14) \qquad G = G_d \dot{\times} G_c,$$

with G_c the continuous invariant subgroup of G and G_d the discrete one. Subgroups of the continuous group G_c can be used, for instance, to reduce the number of continuous variables in the lattices. In particular, it can be reduced to a pure difference equation. Subgroups of G_d can be used to impose the periodicity in n, i.e. $u_{n+N} = u_n$ for some fixed N.

The central step in finding the possible symmetry reductions concerns the classification of the subalgebras of the symmetry algebra. Going through the symmetry reductions procedure, one has to find the invariants corresponding to each subalgebra listed. Assuming some of them as the arguments of the remaining invariants, after solving with respect to u (respectively, u_n), one can look for the reduced symmetry equations. The subalgebras have to be chosen in such a way that the dependent variables u can be expressed in terms of subgroup invariants. This is equivalent to look for subgroups which have orbits of codimension 1, 2, 3 for the heavenly equation, since it has three independent variables. On the other hand, for the Toda lattice (2) one has two independent variables and, then, only 1- and 2-dimensional subalgebras are of interest for us. However, in this case, further reductions can be obtained by combining the continuous and the discrete, or even only the discrete one, constraining the u_n among them, but leaving the same number of independent variables [**9**]. In particular we will explore the relation among the symmetry reductions allowed by (1) and the combined continuous-discrete reductions of the system (2).

As an intermediate but important step in this analysis, one has to classify all the possible subalgebras into conjugacy classes, under the action of the symmetry group by using the methods developed for finite-dimensional [**15, 16**] and for infinite-dimensional Lie algebras (see [**22**] for a review). Here first we report the result of the classification for the conformal symmetry algebra common to both the systems.

The conjugacy classes of all finite dimensional subalgebras of Vir^x (the same can be done for Vir^y) under the action of $Ad(\mathrm{Vir}^x)$ are only three, precisely

$$\{\widehat{X}(1)\}, \quad \{\widehat{X}(1), \widehat{X}(x)\}, \tag{15}$$

$$\{\widehat{X}(1), \widehat{X}(x), \widehat{X}(x^2)\} \equiv \mathrm{sl}(2). \tag{16}$$

Since Vir^x is simple, there is no any subalgebra of splitting type, in the terminology of [**22**]. Then, we consider the low dimensional subalgebras containing elements both from Vir^x and from Vir^y, following the Goursat's twisting method described in [**22**]. We will distinguish the so-called twisted subalgebras, recalling that we are factorig out with respect to the discrete permutation subgroup.

There is only one 1-dimensional subalgebra of twisted type

$$\{\widehat{X}(1) \pm \widehat{Y}(1)\}. \tag{17}$$

The 2-dimensional untwisted subalgebras are given only by

$$\{\widehat{X}(1), \widehat{Y}(1)\}. \tag{18}$$

There are no 2-dimensional abelian twisted subalgebras, and the 2-dimensional solvable subalgebras are in the following list

$$\{\widehat{X}(1), \widehat{X}(x) \pm \widehat{Y}(1)\}, \quad \{\widehat{X}(1) \pm \widehat{Y}(1), \widehat{X}(x) + \widehat{Y}(y)\}. \tag{19}$$

In the case of the algebra (6) we have to consider also the subalgebras related to the z variable. We have the following list of subalgebras of $\{\widehat{Z}, \widehat{T}\}$

$$\{\widehat{Z}\}, \quad \{\widehat{T}\}. \tag{20}$$

The 1-dimensional twisted subalgebras are

$$\begin{gathered} \{\widehat{X}(1) \pm \widehat{Z}\}, \quad \{\widehat{X}(1) \pm \widehat{T}\}, \quad \{\widehat{X}(1) \pm \widehat{T}\}, \\ \{\widehat{X}(1) + \varepsilon'\widehat{Y}(1) + \varepsilon''\widehat{Z}\}, \quad \{\widehat{X}(1) + \varepsilon'\widehat{Y}(1) + \varepsilon''\widehat{T}\}, \\ \{\widehat{X}(1) + \varepsilon'\widehat{Y}(1) + \varepsilon''\widehat{T}\}. \end{gathered} \tag{21}$$

The 2-dimensional non twisted subalgebras are

$$\begin{gathered} \{\widehat{X}(1), \widehat{Z}\}, \quad \{\widehat{X}(1) \pm \widehat{T}\}, \\ \{\widehat{X}(1) \pm \widehat{Y}(1), \widehat{Z}\}, \quad \{\widehat{X}(1) \pm \widehat{Y}(1), \widehat{T}\}. \end{gathered} \tag{22}$$

Finally, the 2-dimensional twisted subalgebras are

$$\begin{gathered} \{\widehat{X}(1) + \varepsilon'\widehat{Z}, \widehat{Y}(1) + \varepsilon''\widehat{Z}\}, \quad \{\widehat{X}(1) + \varepsilon'\widehat{T}, \widehat{Y}(1) + \varepsilon''\widehat{T}\}, \\ \{\widehat{X}(1), \widehat{X}(x) + \varepsilon\widehat{Z}\}, \quad \{\widehat{X}(1), \widehat{X}(x) + a\widehat{T}\}, \\ \{\widehat{X}(1) + a\widehat{Z}, \widehat{X}(x) + \widehat{T}\}, \quad \{\widehat{X}(1), \widehat{X}(x) \pm \widehat{Y}(1) + \varepsilon\widehat{Z}\}, \\ \{\widehat{X}(1), \widehat{X}(x) \pm \widehat{Y}(1) + a\widehat{T}\}, \quad \{\widehat{X}(1) \pm \widehat{Z}, \widehat{X}(x) \pm \widehat{Y}(1) + \widehat{T}\}, \\ \{\widehat{X}(1) \pm \widehat{Y}(1), \widehat{X}(x) + \widehat{Y}(y) + \varepsilon\widehat{Z}\}, \quad \{\widehat{X}(1) \pm \widehat{Y}(1), \widehat{X}(x) + \widehat{Y}(y) + a\widehat{T}\}, \\ \{\widehat{X}(1) \pm \widehat{Y}(1) + a\widehat{Z}, \widehat{X}(x) + \widehat{Y}(y) + \widehat{T}\}; \quad \varepsilon', \varepsilon'' = \pm 1, 0, \neq (0,0), \; a \neq 0, \; \varepsilon = \pm 1. \end{gathered} \tag{23}$$

We do not need to discuss higher dimensional symmetry subalgebras.

Since the conformal subgroups of the Eqs. (1) and (2) are isomorphic, they lead to the same type of reductions. Meaning that in all cases the variables n and z can be interchanged. For instance, the invariance under the subgroup of translations generated by $\{\widehat{X}(1)\}$ implies both $u_n = u_n(y)$ and $u = u(y, z)$, respectively. Thus, (2) reduces to the a pure difference equation, the solutions of which are

$$u_n = \ln\big(n\alpha(y) + \beta(y)\big), \tag{24}$$

for arbitrary functions α and β. The solution (24) holds also for the equation (1), after the replacement $n \to z$.

Analogously, the invariance under the subgroup of "Galilei" boost $\{\widehat{X}(1) \pm \widehat{Y}(1)\}$ implies

$$u_n(x, y) = u_n(\zeta), \quad \zeta = x \mp y. \tag{25}$$

The Toda lattice reduces to the ordinary system

$$\mp\partial_\zeta^2 u_n = -\exp(u_{n-1}) + 2\exp(u_n) - \exp(u_{n+1}), \tag{26}$$

which admits several classes of solutions [**7,10,19,20**]. One of them takes the form

$$u_n = \ln\frac{\mp n^2 + an + b}{\zeta^2}, \tag{27}$$

where a and b are arbitrary real constants. Also in this case one can easily check that (27) is a solution of (1) after the substitution $n \to z$.

On the other hand, the generator of the translations $\widehat{Z}$ commutes with the conformal algebra. Then, the corresponding subgroup elements commute. A special case of them is provided by $\widehat{Z}_n$ (notice that this is not the case for the "usual" Toda, as can be seen in [**9**]). By consequence, all the subgroups containing conformal and $\exp\lambda\widehat{Z}$ transformations provide the same reductions, in particular for $\lambda = n$. The invariant solutions in the continual and discrete case are related by the exchange $z \leftrightarrow n$. For instance, the first of the subalgebras (23) has the invariants

$$\xi = y + \varepsilon'\varepsilon''x - \varepsilon''z, \ u. \tag{28}$$

Then one can easily find the corresponding invariant solutions

$$u(x, y, z) = f^{-1}(a\xi + b), \tag{29}$$

$$f(u) = \varepsilon'\varepsilon''u + e^u, \ a, b \in \mathbf{R}. \tag{30}$$

The corresponding reduction on the discrete system (2) provides the difference equation

$$u_{n+1} - 2u_n + u_{n-1} = -a^2\frac{e^{u_n}}{(\varepsilon'\varepsilon'' + e^{u_n})}. \tag{31}$$

The situation is sightly different when we are dealing with the transformation generated by $\widehat{T}$. Indeed, a corresponding discrete group does not seem to exist. The reason is that on the lattice there is no any natural length scale. However, by analogy with the continual case we could study the transformation on the lattice

$$\widetilde{n} = nN, \quad \widetilde{u}_{\widetilde{n}} = 2\log N + u_n. \tag{32}$$

Its action on (2) is

$$\partial_x\partial_y\widetilde{u}_{\widetilde{n}} = \frac{1}{N^2}\big(-e^{\widetilde{u}_{\widetilde{n}-N}} + 2e^{\widetilde{u}_{\widetilde{n}}} - e^{\widetilde{u}_{\widetilde{n}+N}}\big). \tag{33}$$

Then, this transformation rescales the strength of the interaction. Moreover, if we look at its invariants, actually invariant of the group $\exp N\widehat{T}$, one finds

$$u_n - 2\log n = U(x, y). \tag{34}$$

Substitution into (2) leads to the Liouville equation

$$\partial_x \partial_y U = -e^U, \tag{35}$$

the general solution of which is well known [**12**] and, for instance, takes the form

$$U = \log\left(-2\frac{\partial_x v \partial_y v}{v^2}\right), \quad \text{where} \quad \partial_x \partial_y v = 0. \tag{36}$$

In the context of the heavenly Eq. (1), particular symmetry invariant solutions with respect to $\widehat{T}$ are interpreted as Eguchi–Hanson gravitational instantons [**4**]. The corresponding discrete analogous is given by

$$u_n = \log\frac{n^2}{(1+xy)^2}. \tag{37}$$

There are many other interesting reductions, but we will give a complete account on them elsewhere.

References

1. E. Alfinito, G. Soliani, and L. Solombrino, *The symmetry structure of the heavenly equation*, Lett. Math. Phys. **41** (1997), 379–389.
2. I. Bakas, *Area-preserving diffeomorphisms and higher spin fields in two dimensions*, Supermembranes and Physics in 2 + 1-dimensions (M. Duff, C. Pope and E. Sezgin, eds.) (Trieste, 1989), World Scientific, River Edge, NJ, 1990, pp. 352–362.
3. E. Cartan, *Œuvres complétes.* III, vol. 2, CNRS, Paris, 1984.
4. T. Eguchi, P. B. Gilkey, and A. J. Hanson, *Gravitation, gauge theory and differential geometry*, Phys. Rep. **66** (1980), 213–393.
5. P. Goddard and D. Olive, *Kac–Moody and Virasoro algebras in relation to quantum physics*, Internat. J. Modern Phys. A **1** (1986), 303–414.
6. V. G. Kac, *Infinite dimensional Lie algebras. An introduction*, Progr. Math., vol. 44, Birkhuser Boston, Inc., Boston, Mass., 1983.
7. D. Levi and A. Rodriguez, *Symmetry group of partial differential equations and of differential—difference equations: The Toda lattice versus the Korteweg-de Vries equation*, J. Phys. A **25** (1992), L975–L979.
8. D. Levi, L. Vinet, and P. Winternitz, *Lie group formalism for difference equations*, J. Phys. A **30** (1997), 633–649.
9. D. Levi and P. Winternitz, *Symmetries and conditional symmetries of differential-difference equations*, J. Math. Phys. **34** (1993), 3713–3730.
10. ———, *Continuous symmetries of discrete equations*, Phys. Lett. A **152** (1991), 335–338.
11. A. N. Leznov and M. S. Saveliev, *Representation of zero curvature for the system of nonlinear partial differential equations $x_{\alpha,z\bar{z}} = \exp(kx)_\alpha$ and its integrability*, Lett. Math. Phys. **3** (1979),489–494.
12. J. Liouville, *Sur léquation aux différences partielles $(d^2 \log\lambda)/(du\,dv) \pm (\lambda)/(2a^2) = 0$*, J. Math. Pures Appl. (9) **18** (1853), 71–72.
13. P. J. Olver, *Applications of Lie groups to differential equations*, Grad. Texts in Math., vol. 107, Springer-Verlag, New York-Berlin, 1986.
14. L. V. Ovsiannikov, *Group analysis of differential equations*, Academic Press Inc., New York-London, 1982. (Russian)
15. J. Patera, R. T. Sharp, P. Winternitz, and H. Zassenhaus, *Invariants of real low dimensional Lie algebras*, J. Math. Phys. **17** (1976), 986–993.
16. J. Patera, P. Winternitz, and H. Zassenhaus, *Continuous subgroups of the fundamental groups of physics. I. General method and the Poincaré group*, J. Math. Phys. **16** (1975), 1597–1614; *Continuous subgroups of the fundamental groups of physics* II. *The similitude group*, J. Math. Phys. **16** (1975), 1615–1624; *Quantum numbers for particles in de Sitter space*, J. Math. Phys. **17** (1976), 717–728.
17. M. V. Saveliev, *Integro-differential nonlinear equations and continual Lie algebras*, Comm. Math. Phys. **121** (1989), 283–290.

18. ———, *On the integrability problem of a continuous Toda system*, Teoret. Mat. Fiz. **92** (1992), 1024–1031.
19. M. Toda, *Vibration of a chain with nonlinear interaction*, J. Phys. Soc. Japan **22** (1967), 431–436.
20. ———, *Nonlinear waves and solitons*; Chinese transl., Math. Appl. (Chinese Ser.), vol. 5, Kluwer Academic Publishers Group, Dordrecht; KTK Scientific Publishers, Tokyo, 1989.
21. R. S. Ward, *Einstein–Weyl spaces and* $\mathrm{SU}(\infty)$ *Toda fields*, Classical Quantum Gravity **7** (1990), L95–L98.
22. P. Winternitz, *Group theory and exact solutions of partially integrable differential systems*, Partially Integrable Evolution Equations in Physics (R. Conte and N. Boccara, eds.) (Les Houches, 1989), NATO Adv. Sci. Inst. Ser. C, vol. 310, Kluwer Acad. Publ., Dordrecht, 1990, pp. 515–567.
23. A. V. Ziber, N. H. Ibragimov, and A. B. Shabat, *Equation of Liouville type*, Soviet Math. Dokl. **20** (1979), 1183–1187.

DIPARTIMENTO DI FISICA DELL'UNIVERSITÀ AND INFN SEZIONE DI LECCE, ITALY.
E-mail address: martina@le.infn.it

Centre de Recherches Mathématiques
CRM Proceedings and Lecture Notes
Volume **25**, 2000

A Note on Integrable Systems Related to Discrete Time Toda Lattice

Ken-ichi Maruno, Kenji Kajiwara, and Masayuki Oikawa

ABSTRACT. Bilinear equations for some discrete integrable systems related to Toda lattice are presented by using the singularity patterns obtained through singularity confinement test. We also present two extension of relativistic Toda lattice equation, the random time and ultradiscrete analogues.

1. Introduction

It is well-known that Hirota's direct method provides us with a powerful tool for constructing a wide class of exact solutions of continuous or semi-discrete integrable systems. The dependent variable in the direct method is called the τ function, and is known as one of the most fundamental objects in the theory of nonlinear integrable systems [**6, 12**].

The crucial step of the direct method is to find a suitable dependent variable transformation. Although sometimes this step requires "art of discovery", usually $u \propto (\log \tau)_{xx}$ or $u = F/G$ may work for continuous case.

Now let us consider the discrete case, taking the discrete KdV equation (dK dV) [**5**],

$$x_{j-1}^{i+2} - x_j^i = \frac{\varepsilon}{\delta}\left(\frac{1}{x_{j-1}^{i+1}} - \frac{1}{x_j^{i+1}}\right), \tag{1.1}$$

as an example. It is known that Eq. (1.1) is bilinearized as

$$\delta\tau_j^{i+1}\tau_{j+1}^{i-2} - \varepsilon\tau_j^{i-1}\tau_{j+1}^{i-2} + (\varepsilon - \delta)\tau_{j+1}^{i-1}\tau_j^i = 0, \tag{1.2}$$

through the dependent variable transformation,

$$x_j^i = \frac{\tau_j^i\tau_{j+1}^{i-2}}{\tau_j^{i-1}\tau_{j+1}^{i-1}}. \tag{1.3}$$

In fact, Hirota first discretized the bilinear form of KdV equation to get Eq. (1.2), then dK dV equation (1.1) was derived from the bilinear equation (1.2).

Now suppose that we are given only the nonlinear form of the dK dV equation (1.1), and we want to find the soliton solutions. Then, how can we guess the

2000 *Mathematics Subject Classification.* Primary: 37K10; Secondary: 37J35.
This is the final form of the paper.

dependent variable transformation (1.3) from Eq. (1.1)? Even if we know the result of the corresponding continuous case, it is far from obvious. This is one aspect of difficulties in studying the discrete integrable systems.

The singularity confinement(SC) test which was proposed by Grammaticos *et al.* [**2**] solves this difficulty. This was first pointed out by Ramani *et al.*, and they apply this technique for the discrete Painlevé equations [**13**].

In this contribution, we apply this technique to some discrete soliton equations and show that the bilinearization and the construction of solutions for the discrete soliton equations are done through elementary and systematic calculations. We also present two extension of relativistic Toda lattice (RTL) equation, random time RTL and ultradiscrete RTL equation.

2. Bilinearization of Discrete Soliton Equations

2.1. Bilinearization through SC test. Let us consider the dK dV equation (1.1) again, SC test is performed in the following manner. Suppose that x_j^i happens to be 0. This gives rise to singularities for some other points on the lattice. More concretely, let us put

$$x_j^i = 0 + \xi, \tag{2.1}$$

and start the iteration, assuming that other necessary values are finite. Then we immediately find that x_j^{i+1} and x_{j-1}^{i+1} become $O(\xi^{-1})$, i.e. infinite. Furthermore, we find that x_{j-1}^{i+2} becomes 0. Now what will happen to x_{j-1}^{i+3}? We need x_j^{i+2}, which is found to be finite, and x_{j-1}^{i+1}, which is 0 for the denominator, x_j^{i+1}, which is infinite, for the numerator. Thus we have $\infty - 1/0$ for x_{j-1}^{i+3}. However, it happens that those singularities are cancelled out, which yields that x_{j-1}^{i+3} becomes finite. Moreover, we also find that values on other lattice points become finite. Thus, we conclude that the singularities are confined, i.e. the dK dV equation (1.1) passes the SC test. The important information for the bilinearization is obtained from the singularity pattern which appeared in the above SC test. Let us write it symbolically as

$$\{x_j^i, x_j^{i+1}, x_{j-1}^{i+1}, x_{j-1}^{i+2}\} \to \{0, \infty, \infty, 0\}, \tag{2.2}$$

and assume that this pattern is caused by the vanishing of the τ function $\tau_j^i = 0$. Then we easily see that if x_j^i is expressed as

$$x_j^i = \frac{\tau_j^i \tau_{j+1}^{i-2}}{\tau_j^{i-1} \tau_{j+1}^{i-1}}, \tag{2.3}$$

the singularity pattern (2.2) is recovered. Equation (2.3) is the same as Eq. (1.3). Thus, substituting Eq. (2.3) into dK dV equation (1.1), we get the bilinear equation (1.2). Then, various solutions, such as soliton or algebraic solutions, are obtained from Eq. (1.2) by the standard technique [**7**].

The merit of this method is that we can obtain the bilinear equations by routine work. Using algebraic manipulation software, such as REDUCE, Maple, or Mathematica is quite effective.

2.2. Suris lattice related to Toda lattice. Suris presented a new simple lattice related to Toda lattice [**15**],

$$\ddot{q}_n = (\nu + \dot{q}_n)\big(\exp(q_{n-1} - q_n) - \exp(q_n - q_{n+1})\big). \tag{2.4}$$

We call this Suris lattice. He also presented discrete analogues of Suris lattice

$$\frac{\exp(q_n^{t+1}-q_n^t)-1}{\exp(q_n^t-q_n^{t-1})-1}=\frac{1+h\exp(q_{n-1}^{t-1}-q_n^t)}{1+h\exp(q_n^t-q_{n+1}^{t+1})}, \tag{2.5}$$

and

$$\frac{\exp(q_n^{t+1}-q_n^t)-1}{\exp(q_n^t-q_n^{t-1})-1}=\frac{1+h\exp(q_{n-1}^{t}-q_n^t)}{1+h\exp(q_n^t-q_{n+1}^{t})}. \tag{2.6}$$

Let us consider bilinearization of above lattices by using SC. For convenience, we transform the discrete Suris lattices into

$$\frac{1-I_n^{t+1}}{1-I_n^t}=\frac{I_n^{t+1}}{I_n^t}\frac{1+hV_n^t}{1+hV_{n+1}^{t+1}}, \tag{2.7a}$$

$$I_n^tV_{n+1}^{t+1}=I_{n+1}^{t+1}V_{n+1}^t, \tag{2.7b}$$

and

$$\frac{1-I_n^{t+1}}{1-I_n^t}=\frac{I_n^{t+1}+hV_n^{t+1}}{I_n^t+hV_{n+1}^t}, \tag{2.8a}$$

$$I_n^tV_{n+1}^{t+1}=I_{n+1}^{t+1}V_{n+1}^t, \tag{2.8b}$$

through the variable transformations,

$$I_n^t=\exp(q_n^{t-1}-q_n^t),\quad V_n^t=\exp(q_{n-1}^{t-1}-q_n^t). \tag{2.9}$$

Applying the SC test to Eqs. (2.7b), we have two singularity patterns according which happens to vanish V_n^t or I_n^t-1. They are the following singularity patterns:

$$\{V_n^t,V_{n+1}^t,V_{n+1}^{t+1}\}\to\{0,-1/h,\infty\},\qquad \{I_n^t,I_n^{t+1},I_{n+1}^{t+1}\}\to\{0,\infty,1\}, \tag{2.10a}$$

$$\{V_{n+1}^t,V_{n+1}^{t+1},V_{n+2}^{t+1}\}\to\{\infty,-1/h,0\},\quad \{I_n^t,I_{n+1}^t,I_{n+1}^{t+1}\}\to\{1,\infty,0\}. \tag{2.10b}$$

We assume that these singularity patterns are caused by the vanishing of the discrete dependent variables F_n^t and G_n^t at the lattice point (n,t), respectively. Then we easily see that if I_n^t and V_n^t are expressed as

$$I_n^t=\alpha\frac{F_n^tG_{n-1}^{t-1}}{F_n^{t-1}G_{n-1}^t}=1+\beta\frac{F_{n-1}^{t-1}G_n^t}{F_n^{t-1}G_{n-1}^t}, \tag{2.11a}$$

$$V_n^t=\gamma\frac{F_n^tG_{n-2}^{t-1}}{F_{n-1}^{t-1}G_{n-1}^t}=-1/h+\delta\frac{F_{n-1}^tG_{n-1}^{t-1}}{F_{n-1}^{t-1}G_{n-1}^t}, \tag{2.11b}$$

(we note $q_n^t=\log(G_{n-1}^t)/(F_n^t))$ the singularity patterns are recovered. Thus we get the bilinear equations,

$$F_n^{t-1}G_{n-1}^t+\beta F_{n-1}^{t-1}G_n^t=\alpha F_n^tG_{n-1}^{t-1}, \tag{2.12a}$$

$$-(1/h)F_{n-1}^{t-1}G_{n-1}^t+\delta F_{n-1}^tG_{n-1}^{t-1}=\gamma F_n^tG_{n-2}^{t-1}. \tag{2.12b}$$

Similarly, for Eqs. (2.8b), we have the following two singularity patterns according which happen to vanish V_n^t or I_n^t-1.

$$\{V_n^t,V_{n+1}^t,V_{n+1}^{t+1}\}\to\{0,-1/h,\infty\},\qquad \{I_n^t,I_n^{t+1},I_{n+1}^{t}\}\to\{0,\infty,1\}, \tag{2.13a}$$

$$\{V_{n+1}^t,V_{n+1}^{t-1},V_{n+2}^{t}\}\to\{-1/h,\infty,0\},\quad \{I_n^t,I_{n+1}^t,I_{n+1}^{t-1}\}\to\{1,0,\infty\}. \tag{2.13b}$$

We assume again that these singularity patterns are caused by the vanishing of the discrete dependent variables F_n^t and G_n^t at the lattice point (n, t), respectively. Then we see that if I and V are expressed as

$$I_n^t = \alpha \frac{F_n^t G_{n-1}^t}{F_n^{t-1} G_{n-1}^{t+1}} = 1 + \beta \frac{F_{n-1}^t G_n^t}{F_n^{t-1} G_{n-1}^{t+1}}, \tag{2.14a}$$

$$V_n^t = \gamma \frac{F_n^t G_{n-2}^t}{F_{n-1}^{t-1} G_{n-1}^{t+1}} = -1/h + \delta \frac{F_{n-1}^t G_{n-1}^t}{F_{n-1}^{t-1} G_{n-1}^{t+1}}, \tag{2.14b}$$

(we note $q_n^t = \log(G_{n-1}^{t+1})/(F_n^t))$ the singularity patterns are recovered. Thus we get the bilinear equations,

$$F_n^{t-1} G_{n-1}^{t+1} + \beta F_{n-1}^t G_n^t = \alpha F_n^t G_{n-1}^t, \tag{2.15a}$$

$$-(1/h) F_{n-1}^{t-1} G_{n-1}^{t+1} + \delta F_{n-1}^t G_{n-1}^t = \gamma F_n^t G_{n-2}^t. \tag{2.15b}$$

We can easily construct the N-soliton solutions using perturbational or Casoratian techniques. We note that these bilinear equations are the Bäcklund transformations of the Toda equation.

In continuous limit of the above dependent variable transformation, we can get the dependent variable transformation

$$q_n = \log \frac{G_{n-1}}{F_n}. \tag{2.16}$$

Substituting this transformation into Eq. (2.4), we have the bilinear equations of continuous system (2.4),

$$\ddot{F}_n F_n - (\dot{F}_n)^2 - \xi_1 F_{n+1} F_{n-1} = \xi_2 (F_n)^2, \tag{2.17a}$$

$$\ddot{G}_n G_n - (\dot{G}_n)^2 - \xi_1 G_{n+1} G_{n-1} = \xi_2 (G_n)^2, \tag{2.17b}$$

$$F_n \dot{G}_{n-1} - \dot{F}_n G_{n-1} + \nu F_n G_{n-1} = \xi_1 F_{n-1} G_n. \tag{2.17c}$$

Their bilinear equations are those for the Toda lattice and the Bäcklund transformation of Toda lattice.

2.3. Discrete time relativistic Toda lattice equation. The relativistic Toda lattice equation is very important in the field of mathematical physics [**14**]. The discrete time relativistic Toda lattice equation was presented by Suris [**16**]:

$$\frac{\delta \exp(q_n^{t+1} - q_n^t) - 1}{\delta \exp(q_n^t - q_n^{t-1}) - 1} = \frac{1 + g^2 \exp(q_{n-1}^t - q_n^t)}{1 + g^2 \exp(q_n^t - q_{n+1}^t)} \times \frac{1 + (g^2)/\delta \exp(q_n^t - q_{n+1}^{t+1})}{1 + (g^2)/\delta \exp(q_{n-1}^{t-1} - q_n^t)}, \tag{2.18}$$

where $\delta = \exp(\mathrm{ch})$, c denotes the light speed, h is the parameter specifying the interval, and g is the coupling constant. This is rewritten as

$$I_n^t = \frac{I_{n-1}^{t-1} V_n^t}{V_n^{t-1}},$$

$$V_n^t = -\frac{\delta}{g^2} + \frac{\delta + g^2 V_{n-1}^{t-1}}{g^2} \frac{\delta - I_{n-1}^t}{\delta - I_{n-1}^{t-1}} \frac{I_{n-1}^{t-1} + g^2 V_n^{t-1}}{I_{n-1}^t + g^2 V_{n-1}^t},$$

through the transformation

$$I_n^t = \exp(q_n^{t-1} - q_n^t), \quad V_n^t = \exp(q_{n-1}^{t-1} - q_n^t). \tag{2.19}$$

By applying the SC test, we obtain the following results.

Singularity Patterns:

(2.20a)
$$\{I_n^t, I_n^{t+1}, I_{n+1}^t, I_{n+1}^{t+1}\} \to \{0, \infty, \infty, 0\},$$
$$\{V_n^t, V_{n+1}^t, V_{n+1}^{t+1}, V_{n+2}^{t+1}\} \to \{0, \infty, \infty, 0\},$$

(2.20b)
$$\{I_n^t\} \to \{\delta\}, \qquad \{V_{n+1}^t\} \to \left\{-\frac{\delta}{g^2}\right\},$$

(2.20c)
$$\{I_{n+1}^{t+1}\} \to \{\delta\}, \quad \{V_{n+1}^{t+1}\} \to \left\{-\frac{\delta}{g^2}\right\}.$$

Dependent Variable Transformation:

(2.21a)
$$I_n^t = \delta + \xi_1 \frac{G_n^t \overline{G}_{n-1}^{t-1}}{F_n^{t-1} F_{n-1}^t} = \xi_2 \frac{F_n^t F_{n-1}^{t-1}}{F_n^{t-1} F_{n-1}^t},$$

(2.21b)
$$V_n^t = -\frac{\delta}{g^2} + \xi_3 \frac{G_{n-1}^t \overline{G}_{n-1}^{t-1}}{F_{n-1}^t F_{n-1}^{t-1}} = \xi_4 \frac{F_n^t F_{n-2}^{t-1}}{F_{n-1}^t F_{n-1}^{t-1}}.$$

Bilinear Equations:

(2.22a)
$$\delta F_{n-1}^{t+1} F_n^t - \xi_2 F_n^{t+1} F_{n-1}^t + \xi_1 \overline{G}_{n-1}^t G_n^{t+1} = 0,$$

(2.22b)
$$(\delta/g^2) F_n^t F_n^{t+1} + \xi_4 F_{n-1}^t F_{n+1}^{t+1} - \xi_3 \overline{G}_n^t G_n^{t+1} = 0,$$

(2.22c)
$$\xi_2 F_n^t F_n^t + g^2 \xi_4 F_{n-1}^t F_{n+1}^t - \xi_5 G_n^t \overline{G}_n^t = 0.$$

It is shown that these bilinear equations are nothing but reductions of the Bäcklund transformations of the discrete time Toda lattice equation. Based on this fact, the N-soliton solutions are constructed in the determinant form [**8**].

3. Extension of Discrete Time Relativistic Toda Lattice Equation

3.1. Random time relativistic Toda lattice equation. Hirota presented a random time Toda equation where the time interval of the discrete time Toda equation is replaced by random variables, and gave the Lax pair of the random-time Toda equation, and showed it has higher-order conserved quantities in spite of the randomness introduced into the equation [**3**].

We present the random time relativistic Toda lattice equation (rRTL) and its Casorati determinant solution.

The rRTL equation is given by the form

(3.1)
$$\frac{\delta^{\alpha+1}\exp(q_n^{t_{\alpha+1}} - q_n^{t_\alpha}) - 1}{\delta^\alpha \exp(q_n^{t_\alpha} - q_n^{t_{\alpha-1}}) - 1} = \frac{1 + g^2 \exp(q_{n-1}^{t_\alpha} - q_n^{t_\alpha})}{1 + g^2 \exp(q_n^{t_\alpha} - q_{n+1}^{t_\alpha})} \times \frac{1 + (g^2)/(\delta^{\alpha+1}) \exp(q_n^{t_\alpha} - q_{n+1}^{t_{\alpha+1}})}{1 + (g^2)/(\delta^\alpha) \exp(q_{n-1}^{t_{\alpha-1}} - q_n^{t_\alpha})},$$

where $\delta^\alpha = \exp(ch^\alpha)$. The bilinear equations for this are

(3.2a)
$$\delta^{\alpha+1} F_{n-1}^{t_{\alpha+1}} F_n^{t_\alpha} - \xi_2{}^{\alpha+1} F_n^{t_{\alpha+1}} F_{n-1}^{t_\alpha} + \xi_1{}^{\alpha+1} \overline{G}_{n-1}^{t_\alpha} G_n^{t_{\alpha+1}} = 0,$$

(3.2b)
$$F_n^{t_\alpha} F_n^{t_{\alpha+1}} + (g^2/\delta^{\alpha+1}) \xi_4{}^{\alpha+1} F_{n-1}^{t_\alpha} F_{n+1}^{t_{\alpha+1}} - \xi_3{}^{\alpha+1} \overline{G}_n^{t_\alpha} G_n^{t_{\alpha+1}} = 0,$$

(3.2c)
$$\xi_2{}^\alpha F_n^{t_\alpha} F_n^{t_\alpha} + g^2 \xi_4{}^\alpha F_{n-1}^{t_\alpha} F_{n+1}^{t_\alpha} - \xi_5{}^\alpha G_n^{t_\alpha} \overline{G}_n^{t_\alpha} = 0.$$

Here, the parameters ξ have to satisfy the conditions

$$\xi_2{}^{\alpha} = \xi_4{}^{\alpha} = 1, \qquad \frac{\xi_1{}^{\alpha+1}}{\xi_1{}^{\alpha}} = \frac{\xi_3{}^{\alpha+1}}{\xi_3{}^{\alpha}}. \tag{3.3}$$

The N-soliton solution for the rRTL (3.1) is expressed by the Casorati determinant as

$$q_n^{t_\alpha} = \log \frac{F_{n-1}^{t_\alpha}}{F_n^{t_\alpha}}, \tag{3.4}$$

$$F_n^{t_\alpha} = \det\left|\phi_i^{(n+j-1)}(t_\alpha)\right|_{1\le i,j\le N} \tag{3.5}$$

$$= \begin{vmatrix} \phi_1^{(n)}(t_\alpha) & \phi_1^{(n+1)}(t_\alpha) & \cdots & \phi_1^{(n+N-1)}(t_\alpha) \\ \phi_2^{(n)}(t_\alpha) & \phi_2^{(n+1)}(t_\alpha) & \cdots & \phi_2^{(n+N-1)}(t_\alpha) \\ \vdots & \vdots & \cdots & \vdots \\ \phi_N^{(n)}(t_\alpha) & \phi_N^{(n+1)}(t_\alpha) & \cdots & \phi_N^{(n+N-1)}(t_\alpha) \end{vmatrix},$$

where

$$\phi_i^{(n+j-1)}(t_\alpha) = p_i^{j-1}\prod_{\zeta}^{\alpha}(1-a_\zeta p_i)^{-1}\left(\frac{p_i}{1-bp_i}\right)^n + \left(\frac{1}{p_i}\right)^{j-1}\prod_{\zeta}^{\alpha}\left(1-\frac{a_\zeta}{p_i}\right)^{-1}\left(\frac{1/p_i}{1-b/p_i}\right)^n, \tag{3.6}$$

is the entry at the intersection of the ith row and the jth column, p_i are arbitrary constants and

$$g^2 = \frac{b^2}{1-b^2}, \qquad \delta^{\alpha+1} = -\frac{b}{a_{\alpha+1}-b}. \tag{3.7}$$

The products represent, for example,

$$\begin{aligned} \prod_{\zeta}^{\alpha}(1-a_\zeta p_i)^{-1} &= \prod_{\zeta=1}^{\alpha}(1-a_\zeta p_i)^{-1}, \quad \alpha\ge 1, \\ &= 1, \quad \alpha = 0, \\ &= \prod_{\zeta=\alpha+1}^{0}(1-a_\zeta p_i), \quad \alpha\le -1. \end{aligned} \tag{3.8}$$

The auxiliary variables G_n^t and $\overline{G}_n^t$ are given by

$$G_n^{t_\alpha} = \begin{vmatrix} \psi_1^{(n)}(t_\alpha) & \psi_1^{(n+1)}(t_\alpha) & \cdots & \psi_1^{(n+N-1)}(t_\alpha) \\ \psi_2^{(n)}(t_\alpha) & \psi_2^{(n+1)}(t_\alpha) & \cdots & \psi_2^{(n+N-1)}(t_\alpha) \\ \vdots & \vdots & \cdots & \vdots \\ \psi_N^{(n)}(t_\alpha) & \psi_N^{(n+1)}(t_\alpha) & \cdots & \psi_N^{(n+N-1)}(t_\alpha) \end{vmatrix}, \tag{3.9a}$$

$$\overline{G}_n^{t_\alpha} = \begin{vmatrix} \overline{\psi}_1^{(n)}(t_\alpha) & \overline{\psi}_1^{(n+1)}(t_\alpha) & \cdots & \overline{\psi}_1^{(n+N-1)}(t_\alpha) \\ \overline{\psi}_2^{(n)}(t_\alpha) & \overline{\psi}_2^{(n+1)}(t_\alpha) & \cdots & \overline{\psi}_2^{(n+N-1)}(t_\alpha) \\ \vdots & \vdots & \cdots & \vdots \\ \overline{\psi}_N^{(n)}(t_\alpha) & \overline{\psi}_N^{(n+1)}(t_\alpha) & \cdots & \overline{\psi}_N^{(n+N-1)}(t_\alpha) \end{vmatrix}, \tag{3.9b}$$

where

$$\begin{aligned}(3.10a)\qquad &\psi_i^{(n+j-1)}(t_\alpha)\\ &= p_i^j \prod_\zeta^\alpha (1-a_\zeta p_i)^{-1}\left(\frac{p_i}{1-bp_i}\right)^{n-1}\\ &\qquad + \left(\frac{1}{p_i}\right)^j \prod_\zeta^\alpha \left(1-a_\zeta/p_i\right)^{-1}\left(\frac{1/p_i}{1-b/p_i}\right)^{n-1},\end{aligned}$$

$$\begin{aligned}(3.10b)\qquad &\overline{\psi}_i^{(n+j-1)}(t_\alpha)\\ &= p_i^{j-2} \prod_\zeta^\alpha (1-a_\zeta p_i)^{-1}\left(\frac{p_i}{1-bp_i}\right)^{n+1}\\ &\qquad + \left(\frac{1}{p_i}\right)^{j-2} \prod_\zeta^\alpha \left(1-a_\zeta/p_i\right)^{-1}\left(\frac{1/p_i}{1-b/p_i}\right)^{n+1}.\end{aligned}$$

The above result is derived by considering a reduction from the more general τ function,

$$(3.11)\quad \tau_n(t_\alpha, s_\beta, l_\gamma) = \begin{vmatrix} \phi_1^{(n)}(t_\alpha, s_\beta, l_\gamma) & \phi_1^{(n+1)}(t_\alpha, s_\beta, l_\gamma) & \dots & \phi_1^{(n+N-1)}(t_\alpha, s_\beta, l_\gamma) \\ \phi_2^{(n)}(t_\alpha, s_\beta, l_\gamma) & \phi_2^{(n+1)}(t_\alpha, s_\beta, l_\gamma) & \dots & \phi_2^{(n+N-1)}(t_\alpha, s_\beta, l_\gamma) \\ \vdots & \vdots & \dots & \vdots \\ \phi_N^{(n)}(t_\alpha, s_\beta, l_\gamma) & \phi_N^{(n+1)}(t_\alpha, s_\beta, l_\gamma) & \dots & \phi_N^{(n+N-1)}(t_\alpha, s_\beta, l_\gamma) \end{vmatrix},$$

where $\phi_i^{(n)}$'s are arbitrary functions satisfying the dispersion relations,

$$(3.12a)\qquad \Delta_{t_\alpha}\phi_j^{(n)}(t_\alpha) \equiv \frac{\phi_j^{(n)}(t_\alpha)-\phi_j^{(n)}(t_{\alpha-1})}{a_\alpha} = \phi_j^{(n+1)}(t_\alpha),$$

$$(3.12b)\qquad \Delta_{s_\beta}\phi_j^{(n)}(s_\beta) \equiv \frac{\phi_j^{(n)}(s_\beta)-\phi_j^{(n)}(s_{\beta-1})}{b_\beta} = \phi_j^{(n+1)}(s_\beta),$$

$$(3.12c)\qquad \Delta_{l_\gamma}\phi_j^{(n)}(l_\gamma) \equiv \frac{\phi_j^{(n)}(l_\gamma)-\phi_j^{(n)}(l_{\gamma-1})}{d_\gamma} = \phi_j^{(n-1)}(l_\gamma),$$

$$(3.13)\qquad a_\alpha = t_\alpha - t_{\alpha-1},\quad b_\beta = s_\beta - s_{\beta-1},\quad d_\gamma = l_\gamma - l_{\gamma-1},\quad \alpha,\beta,\gamma \in \mathbf{Z}.$$

For example, we can choose $\phi_i^{(n)}$ as

$$\begin{aligned}(3.14)\quad \phi_i^{(n)}(t_\alpha, s_\beta, l_\gamma) &= p_i^n \prod_\zeta^\alpha (1-a_\zeta p_i)^{-1} \prod_\zeta^\beta (1-b_\zeta p_i)^{-1} \prod_\zeta^\gamma \left(1-\frac{d_\zeta}{p_i}\right)^{-1}\\ &\quad + q_i^n \prod_\zeta^\alpha (1-a_\zeta q_i)^{-1} \prod_\zeta^\beta (1-b_\zeta q_i)^{-1} \prod_\zeta^\gamma \left(1-\frac{d_\zeta}{q_i}\right)^{-1}.\end{aligned}$$

The τ function (3.11) satisfies the bilinear equation for the discrete two-dimensional Toda lattice in t, l and n [**4, 9**],

$$
\begin{aligned}
(3.15)\quad & a_{\alpha+1}d_{\gamma+1}\tau_{n-1}(t_\alpha, s_\beta, l_{\gamma+1})\tau_{n+1}(t_{\alpha+1}, s_\beta, l_\gamma) \\
& \qquad + (1 - a_{\alpha+1}d_{\gamma+1})\tau_n(t_\alpha, s_\beta, l_\gamma)\tau_n(t_{\alpha+1}, s_\beta, l_{\gamma+1}) \\
& \qquad\qquad - \tau_n(t_{\alpha+1}, s_\beta, l_\gamma)\tau_n(t_\alpha, s_\beta, l_{\gamma+1}) = 0.
\end{aligned}
$$

After Freeman and Nimmo [**1**], we use the following notation for simplicity:

$$
(3.16)\quad |n_{1_{t_1,s_1,l_1}}, n_{2_{t_2,s_2,l_2}}, \ldots, n_{N_{t_N,s_N,l_N}}|
= \begin{vmatrix}
\phi_1^{(n_1)}(t_1,s_1,l_1) & \phi_1^{(n_2)}(t_2,s_2,l_2) & \cdots & \phi_1^{(n_N)}(t_N,s_N,l_N) \\
\phi_2^{(n_1)}(t_1,s_1,l_1) & \phi_2^{(n_2)}(t_2,s_2,l_2) & \cdots & \phi_2^{(n_N)}(t_N,s_N,l_N) \\
\vdots & \vdots & \cdots & \vdots \\
\phi_N^{(n_1)}(t_1,s_1,l_1) & \phi_N^{(n_2)}(t_2,s_2,l_2) & \cdots & \phi_N^{(n_N)}(t_N,s_N,l_N)
\end{vmatrix}.
$$

By using this notation, Eq. (3.11) is rewritten as

$$
(3.17)\qquad \tau_n(t_\alpha, s_\beta, l_\gamma) = |n_{t_\alpha,s_\beta,l_\gamma}, n+1_{t_\alpha,s_\beta,l_\gamma}, \ldots, n+N-1_{t_\alpha,s_\beta,l_\gamma}|.
$$

In the case that there is no possibility of confusion, we also suppress the suffix,

$$
(3.18)\qquad \tau_n(t_\alpha, s_\beta, l_\gamma) = |n, n+1, \ldots, n+N-1|.
$$

By using the Casoratian technique developed in Refs. [**11**] and [**10**], it is possible to derive the following difference formulas for the τ function,

$$
\begin{aligned}
&(3.19a)\quad \tau_n(t_{\alpha+1}, s_\beta, l_\gamma) = |n, \ldots, n+N-2, n+N-1_{t_{\alpha+1}}|, \\
&(3.19b)\quad \tau_n(t_\alpha, s_{\beta+1}, l_\gamma) = |n, \ldots, n+N-2, n+N-1_{s_{\beta+1}}|, \\
&(3.19c)\quad \tau_n(t_\alpha, s_\beta, l_{\gamma+1}) = |n_{l_{\gamma+1}}, n+1, \ldots, n+N-2, n+N-1|, \\
&(3.19d)\quad a_{\alpha+1}\tau_n(t_{\alpha+1}, s_\beta, l_\gamma) = |n, \ldots, n+N-2, n+N-2_{t_{\alpha+1}}|, \\
&(3.19e)\quad b_{\beta+1}\tau_n(t_\alpha, s_{\beta+1}, l_\gamma) = |n, \ldots, n+N-2, n+N-2_{s_{\beta+1}}|, \\
&(3.19f)\quad d_{\gamma+1}\tau_n(t_\alpha, s_\beta, l_{\gamma+1}) = |n+1_{l_{\gamma+1}}, n+1, \ldots, n+N-2, n+N-1|, \\
&(3.19g)\quad (a_{\alpha+1} - b_{\beta+1})\tau_n(t_{\alpha+1}, s_{\beta+1}, l_\gamma) \\
&\qquad\qquad = |n, \ldots, n+N-3, n+N-2_{s_{\beta+1}}, n+N-2_{t_{\alpha+1}}|, \\
&(3.19h)\quad (1 - a_{\alpha+1}d_{\gamma+1})\tau_n(t_{\alpha+1}, s_\beta, l_{\gamma+1}) \\
&\qquad\qquad = |n_{l_{\gamma+1}}, n+1, \ldots, n+N-2, n+N-1_{t_{\alpha+1}}|, \\
&(3.19i)\quad (1 - b_{\beta+1}d_{\gamma+1})\tau_n(t_\alpha, s_{\beta+1}, l_{\gamma+1}) \\
&\qquad\qquad = |n_{l_{\gamma+1}}, n+1, \ldots, n+N-2, n+N-1_{s_{\beta+1}}|.
\end{aligned}
$$

Now we consider the following identity for $2N \times 2N$ determinant,

$$
(3.20)\qquad \left|\begin{array}{c|c|c}
n \ldots n+N-2 & \emptyset & n+N-1 \;\; n+N-1_{s_{beta+1}} \;\; n+N-1_{t_{\alpha+1}} \\
\hline
\emptyset & n+1 \ldots n+N-2 & n+N-1 \;\; n+N-1_{s_{beta+1}} \;\; n+N-1_{t_{\alpha+1}}
\end{array}\right| = 0,
$$

where $\emptyset$ means an empty block. Applying the Laplace expansion to the left-hand side, we obtain

$$
\begin{aligned}
(3.21)\quad & |n,\dots,n+N-2,n+N-1| \\
&\quad \times |n+1,\dots,n+N-2,n+N-1_{s_{\beta+1}},n+N-1_{t_{\alpha+1}}| \\
&\qquad - |n,\dots,n+N-2,n+N-1_{s_{\beta+1}}| \\
&\quad \times |n+1,\dots,n+N-2,n+N-1,n+N-1_{t_{\alpha+1}}| \\
&\qquad + |n,\dots,n+N-2,n+N-1_{t_{\alpha+1}}| \\
&\quad \times |n+1,\dots,n+N-2,n+N-1,n+N-1_{s_{\beta+1}}| = 0,
\end{aligned}
$$

from which we get a bilinear difference equation,

$$
\begin{aligned}
(3.22)\quad & (a_{\alpha+1}-b_{\beta+1})\tau_n(t_\alpha,s_\beta,l_\gamma)\tau_{n+1}(t_{\alpha+1},s_{\beta+1},l_\gamma) \\
&\qquad - a_{\alpha+1}\tau_n(t_\alpha,s_{\beta+1},l_\gamma)\tau_{n+1}(t_{\alpha+1},s_\beta,l_\gamma) \\
&\qquad\qquad + b_{\beta+1}\tau_n(t_{\alpha+1},s_\beta,l_\gamma)\tau_{n+1}(t_\alpha,s_{\beta+1},l_\gamma) = 0,
\end{aligned}
$$

by using the difference formulas (3.18) and (3.19a). Similarly, we obtain

$$
\begin{aligned}
(3.23)\quad & (a_{\alpha+1}-b_{\beta+1})d_{\gamma+1}\tau_{n-1}(t_\alpha,s_\beta,l_{\gamma+1})\tau_{n+1}(t_{\alpha+1},s_{\beta+1},l_\gamma) \\
&\qquad + (1-a_{\alpha+1}d_{\gamma+1})\tau_n(t_\alpha,s_{\beta+1},l_\gamma)\tau_n(t_{\alpha+1},s_\beta,l_{\gamma+1}) \\
&\qquad\qquad - (1-b_{\beta+1}d_{\gamma+1})\tau_n(t_{\alpha+1},s_\beta,l_\gamma)\tau_n(t_\alpha,s_{\beta+1},l_{\gamma+1}) = 0,
\end{aligned}
$$

and

$$
\begin{aligned}
(3.24)\quad & b_{\beta+1}d_{\gamma+1}\tau_{n-1}(t_\alpha,s_\beta,l_{\gamma+1})\tau_{n+1}(t_\alpha,s_{\beta+1},l_\gamma) \\
&\qquad + (1-b_{\beta+1}d_{\gamma+1})\tau_n(t_\alpha,s_\beta,l_\gamma)\tau_n(t_\alpha,s_{\beta+1},l_{\gamma+1}) \\
&\qquad\qquad - \tau_n(t_\alpha,s_{\beta+1},l_\gamma)\tau_n(t_\alpha,s_\beta,l_{\gamma+1}) = 0,
\end{aligned}
$$

from the identities,

$$
(3.25)\qquad \left|\begin{array}{c|c|c|c}
n_{l_{\gamma+1}} & n\dots n+N-2 & & n+N-1_{s_{\beta+1}}\ \ n+N-1_{t_{\alpha+1}} \\ \hline
n_{l_{\gamma+1}} & & n+1\dots n+N-2 & n+N-1_{s_{\beta+1}}\ \ n+N-1_{t_{\alpha+1}}
\end{array}\right| = 0,
$$

and

$$
(3.26)\qquad \left|\begin{array}{c|c|c|c}
n_{l_{\gamma+1}} & n\dots n+N-2 & & n+N-1\ \ n+N-1_{s_{\beta+1}} \\ \hline
n_{l_{\gamma+1}} & & n+1\dots n+N-2 & n+N-1\ \ n+N-1_{s_{\beta+1}}
\end{array}\right| = 0,
$$

and respectively.

Next, we apply the reduction procedure. If we impose the condition,

$$
\begin{aligned}
\tau_n(t_\alpha,s_{\beta+1},l_{\beta+1}) &= f(\beta+1)\tau_n(t_\alpha,s_\beta,l_\beta), \\
\tau_n(t_\alpha,s_\beta,l_{\beta+1}) &= \hat{f}(\beta+1)\tau_n(t_\alpha,s_{\beta-1},l_\beta),
\end{aligned}
$$

then, the shift in l is absorbed in s and hence we may drop the l dependence. This condition is realized by putting the parameters in the entries of the τ function (3.14) as

$$
(3.27a)\qquad (1-b_{\beta+1}p_i)(1-d_{\beta+1}/p_i) = (1-b_{\beta+1}q_i)(1-d_{\beta+1}/q_i),
$$

$$
(3.27b)\qquad (1-b_\beta p_i)(1-d_{\beta+1}/p_i) = (1-b_\beta q_i)(1-d_{\beta+1}/q_i),
$$

which yields

$$
\begin{aligned}
f(\beta+1) &= \prod_{i=1}^{N}(1-b_{\beta+1}p_i)^{-1}\left(1-\frac{d_{\beta+1}}{p_i}\right)^{-1}, \\
\hat{f}(\beta+1) &= \prod_{i=1}^{N}(1-b_{\beta}p_i)^{-1}\left(1-\frac{d_{\beta+1}}{p_i}\right)^{-1}.
\end{aligned}
$$

Concretely, we must choose parameters

$$
q_i = \frac{1}{p_i}, \qquad b_\beta = d_\beta \equiv b = \text{constant}. \tag{3.28}
$$

Applying this procedure on the bilinear difference equations (3.22)–(3.24), we get

$$
\begin{aligned}
&(a_{\alpha+1}-b)\tau_n(t_\alpha,s)\tau_{n+1}(t_{\alpha+1},s+1) - a_{\alpha+1}\tau_n(t_\alpha,s+1)\tau_{n+1}(t_{\alpha+1},s) \\
&\qquad + b\tau_n(t_{\alpha+1},s)\tau_{n+1}(t_\alpha,s+1) = 0,
\end{aligned} \tag{3.29a}
$$

$$
\begin{aligned}
&(a_{\alpha+1}-b)b\tau_{n-1}(t_\alpha,s-1)\tau_{n+1}(t_{\alpha+1},s+1) \\
&\qquad + (1-a_{\alpha+1}b)\tau_n(t_\alpha,s+1)\tau_n(t_{\alpha+1},s-1) \\
&\qquad - (1-b^2)\tau_n(t_{\alpha+1},s)\tau_n(t_\alpha,s) = 0,
\end{aligned} \tag{3.29b}
$$

$$
\begin{aligned}
&b^2\tau_{n-1}(t_\alpha,s-1)\tau_{n+1}(t_\alpha,s+1) + (1-b^2)\tau_n(t_\alpha,s)\tau_n(t_\alpha,s) \\
&\qquad - \tau_n(t_\alpha,s+1)\tau_n(t_\alpha,s-1) = 0,
\end{aligned} \tag{3.29c}
$$

respectively (we rewrite s_β as s, $s_{\beta+1}$ as $s+1$, $s_{\beta-1}$ as $s-1$). The τ function,

$$
\tau_n(t_\alpha,s) = \begin{vmatrix}
\phi_1^{(n)}(t_\alpha,s) & \phi_1^{(n+1)}(t_\alpha,s) & \cdots & \phi_1^{(n+N-1)}(t_\alpha,s) \\
\phi_2^{(n)}(t_\alpha,s) & \phi_2^{(n+1)}(t_\alpha,s) & \cdots & \phi_2^{(n+N-1)}(t_\alpha,s) \\
\vdots & \vdots & \cdots & \vdots \\
\phi_N^{(n)}(t_\alpha,s) & \phi_N^{(n+1)}(t_\alpha,s) & \cdots & \phi_N^{(n+N-1)}(t_\alpha,s)
\end{vmatrix}, \tag{3.30}
$$

with

$$
\begin{aligned}
\phi_i^{(n+j-1)}(t_\alpha,s) &= p_i^{n+j-1}\prod_{\zeta}^{\alpha}(1-a_\zeta p_i)^{-1}(1-bp_i)^{-s} \\
&\qquad + \left(\frac{1}{p_i}\right)^{n+j-1}\prod_{\zeta}^{\alpha}(1-a_\zeta/p_i)^{-1}(1-b/p_i)^{-s},
\end{aligned} \tag{3.31}
$$

satisfies the bilinear equations (3.29a).

Finally, letting

$$
F_n^t = \tau_n(t_\alpha,n), \quad G_n^t = \tau_n(t_\alpha,n-1), \quad \overline{G}_n^t = \tau_n(t_\alpha,n+1), \tag{3.32}
$$

the bilinear equations (3.29a) with $s=n$ become the bilinear equations of the rRTL, and Eq. (3.30) gives their solution (3.5) and (3.9a). We note from this result that the random space-time RTL equation cannot be constructed because of reduction condition (3.28).

3.2. Ultradiscrete relativistic Toda lattice equation. Takahashi and Satsuma presented a simple cellular automaton which shows soliton-like behaviour [**17**]. Tokihiro et al found a connection between the soliton cellular automaton (SCA) and discrete soliton equations [**18**]. We present SCA of the relativistic Toda lattice equation(ultradiscrete relativistic Toda lattice equation) here. We first transform

$$I_n^t V_{n+1}^{t+1} = I_{n+1}^{t+1} V_{n+1}^t, \tag{3.33a}$$

$$\frac{\delta - I_n^{t+1}}{\delta - I_n^t} = \frac{I_n^{t+1} + g^2 V_n^{t+1}}{I_n^t + g^2 V_{n+1}^t} \frac{\delta + g^2 V_{n+1}^{t+1}}{\delta + g^2 V_n^t}, \tag{3.33b}$$

into

$$I_n^t V_{n+1}^{t+1} = I_{n+1}^{t+1} V_{n+1}^t, \tag{3.34a}$$

$$\frac{J_n^t}{J_n^{t+1}} = \frac{I_n^{t+1} + g^2 V_n^{t+1}}{I_n^t + g^2 V_{n+1}^t} \frac{\delta + g^2 V_{n+1}^{t+1}}{\delta + g^2 V_n^t}, \tag{3.34b}$$

$$\delta - I_n^t = -\frac{1}{J_n^t}. \tag{3.34c}$$

Setting $I_n^t = \exp(X_n^t/\varepsilon)$, $V_n^t = \exp(Y_n^t/\varepsilon)$, $J_n^t = \exp(Z_n^t/\varepsilon)$, $\delta = \exp(\alpha/\varepsilon)$, $g = \exp(\beta/\varepsilon)$ and using the formula:

$$\lim_{\varepsilon \to +0} \varepsilon \log(e^{A/\varepsilon} + e^{B/\varepsilon}) = \max(A, B), \tag{3.35}$$

we obtain

$$X_n^t + Y_{n+1}^{t+1} = X_{n+1}^{t+1} + Y_{n+1}^t, \tag{3.36a}$$

$$\begin{aligned} Z_n^t - Z_n^{t+1} = {} & \max(X_n^{t+1}, 2\beta + Y_n^{t+1}) - \max(X_n^t, 2\beta + Y_{n+1}^t) \\ & + \max(\alpha, 2\beta + Y_{n+1}^{t+1}) - \max(\alpha, 2\beta + Y_n^t), \end{aligned} \tag{3.36b}$$

$$\max(\alpha, -Z_n^t) = X_n^t, \tag{3.36c}$$

in the limit $\varepsilon \to +0$. We can also construct ultradiscrete rRTL equation similarly.

4. Conclusion

In this contribution, we have obtained bilinear equations for some discrete integrable equations. These results show that the information of SC test is very important in constructing solutions and SC test is useful not only for detecting integrability but also for construction of solutions. Of course, there is no rigorous proof why SC test give us those information. However, at least, we may say that SC test is really a powerful tool for studying discrete integrable systems.

We have also presented two interesting extension of RTL equation, random time RTL and ultradiscrete RTL equation. It is interesting to study algebraic structure and Hamilton structure of these equations.

References

1. N. C. Freeman and J. J. C. Nimmo, *Soliton solutions of the Korteweg-de Vries and Kadomtsev–Petviashvili equations: The Wronskian technique*, Phys. Lett. A **95** (1983), 1–3.
2. B. Grammaticos, A. Ramani, and V. G. Papageorgiou, *Do integrable mapping have the Painlevé property?*, Phys. Rev. Lett. **67** (1991), 1825–1828.
3. R. Hirota, *Conserved quantities of "Random-time Toda equation"*, J. Phys. Soc. Japan **66** (1997), 283–285.

4. R. Hirota, M. Ito, and F. Kako, *Two-dimensional Toda lattice equations*, Progr. Theoret. Phys. Suppl. **94** (1988), 42–58.
5. R. Hirota and S. Tsujimoto, *Conserved quantities of a class of nonlinear difference-difference equations*, J. Phys. Soc. Japan **64** (1995), 3125–3127.
6. M. Jimbo and T. Miwa, *Solitons and infinite dimensional Lie algebras*, Publ. Res. Inst. Math. Sci. **19** (1983), 943–1001.
7. K. Maruno, K. Kajiwara, S. Nakao, and M. Oikawa, *Bilinearization of discrete soliton equations and singularity confinement*, Phys. Lett. A **229** (1997), 173–182.
8. K. Maruno, K. Kajiwara, and M. Oikawa, *Casorati determinant solution for the discrete-time relativistic Toda lattice equation*, Phys. Lett. A **241** (1998), 335–343.
9. A. Nagai, T. Tokihiro, J. Satsuma, R. Willox, and K. Kajiwara, *Two-dimensional soliton cellular automaton of deautonomized Toda-type*, Phys. Lett. A **234** (1997), 301–309.
10. Y. Ohta, R. Hirota, S. Tsujimoto and T. Imai, *Casorati and discrete Gram type determinant representations of solutions to the discrete* KP *hierarchy*, J. Phys. Soc. Japan **62** (1993), 1872–1886.
11. Y. Ohta, K. Kajiwara, J. Matsukidaira and J. Satsuma, *Casorati determinant solution for the relativistic Toda lattice equation*, J. Math. Phys. **34** (1993), 5190–5204.
12. Y. Ohta, J. Satsuma, D. Takahashi, and T. Tokihiro, *An elementary introduction to Sato theory*, Progr. Theoret. Phys. Suppl. **94** (1988), 210–241.
13. A. Ramani, B. Grammaticos, and J. Satsuma, *Bilinear discrete Painlevé equations*, J. Phys. A **28** (1995), 4655–4665.
14. S. N. M. Ruijsenaars, *Relativistic Toda systems*, Comm. Math. Phys. **133** (1990), 217–247.
15. Y. B. Suris, *On some integrable systems related to the Toda lattice*, J. Phys. A **30** (1997), 2235–2249.
16. ———, *A discrete-time relativistic Toda lattice*, J. Phys. A **29** (1996), 451–465.
17. D. Takahashi and J. Satsuma, *A soliton cellular automaton*, J. Phys. Soc. Japan **59** (1990), 3514–3519.
18. T. Tokihiro, D. Takahashi, J. Matsukidaira, and J. Satsuma, *From soliton equations to integrable cellular automata through a limiting procedure*, Phys. Rev. Lett. **29** (1996), 3247–3250.

Department of Earth System Science and Technology, Kyushu University, Kasuga, Fukuoka 816-8580, Japan.
E-mail address: `maruno@riam.kyushu-u.ac.jp`

Department of Electrical Engineering, Doshisha University, Kyotanabe, Kyoto 610-0321, Japan.
E-mail address: `kaji@elrond.doshosha.ac.jp`

Research Institute for Applied Mechanics, Kyushu University, Kasuga, Fukuoka 816-8580, Japan.
E-mail address: `oikawa@riam.kyushu-u.ac.jp`

Centre de Recherches Mathématiques
CRM Proceedings and Lecture Notes
Volume **25**, 2000

The Integral Formula for the Solutions of the Quantum Knizhnik–Zamolodchikov Equation Associated with $U_q(\widehat{\mathrm{sl}}_n)$ for $|q| = 1$

Tetsuji Miwa and Yoshihiro Takeyama

Dedicated to the memory of Moshe Flato

ABSTRACT. We write the integral formula of Tarasov–Varchenko type for the solutions to the quantum Knizhnik–Zamolodchikov equation associated with a tensor product of the vector representations of sl_n. We consider the case where the deformation parameter q satisfies $|q| = 1$. We use the bosonization of the type II vertex operators in order to find the hypergeometric pairing in this setting.

1. Introduction

In this paper we construct a family of solutions to the quantum Knizhnik–Zamolodchikov equation [**3**] associated with the vector representation of sl_n. To be more precise, we consider the case

$$\lambda = \frac{4\pi}{n}, \tag{1.1}$$

(called level 0) of the following difference equation:

$$\begin{aligned} f(\beta_1, \dots, \beta_r - \lambda i, \dots, \beta_N) &= R_{r,r-1}(\beta_r - \lambda i, \beta_{r-1}) \dots R_{r,1}(\beta_r - \lambda i, \beta_1) \\ &\quad \times D_r R_{r,N}(\beta_r, \beta_N) \dots R_{r,r+1}(\beta_r, \beta_{r+1}) f(\beta_1, \dots, \beta_N). \end{aligned} \tag{1.2}$$

In this equation, the unknown function $f(\beta_1, \dots, \beta_N)$ $(\beta_1, \dots, \beta_N \in \mathbf{C})$ takes its value in $\underbrace{V \otimes \cdots \otimes V}_{N}$ where

$$V = \oplus_{j=0}^{n-1} \mathbf{C} v_j, \tag{1.3}$$

2000 *Mathematics Subject Classification.* Primary: 33D80; Secondary: 17B37, 81R50.

The authors thank A. Nakayashiki, F. Smirnov, and V. Tarasov for valuable discussions. One of the authors (T.M.) thanks the organizers of the SIDE III for invitation and giving the opportunity for writing this paper. He also wishes to express his sorrow at the sudden death of Moshe Flato. He shares the warmest memories of the days with Moshe in Kyoto and Paris.

This is the final form of the paper.

is the vector representation of sl_n. The matrix $R_{rs}(\beta_r, \beta_s)$ acts on the rth and sth components of the tensor product. It is the R-matrix for $U_q(\widehat{\mathrm{sl}}_n)$. The explicit formula of the R-matrix is given by (2.7). In this paper, we restrict to the case where

$$q = e^{-(2\pi^2 i)/(\rho n)} \ (\rho \in \mathbf{R}_{>0}). \tag{1.4}$$

The matrix D_r is a diagonal matrix acting on the rth component. In this paper, we restrict to the case

$$D_r = 1. \tag{1.5}$$

For $n = 2$, Smirnov [**14**] constructed a family of solutions to (1.2) which he identified with the form factors of the quantum sine-Gordon model. Later, Lukyanov [**8**] gave a construction of the form factors by using the bosonic vertex operators. In the sl_n case, we adapt Lukyanov's approach with the modification given in [**4**]. In this paper, however, we only construct the so-called type II vertex operators. This construction gives us the hypergeometric pairing for two functions w and W in the terminology of Tarasov and Varchenko [**11, 15, 16**]:

$$I(w, W) = \prod_{j=1}^{n-1} \prod_{1 \le r \le \nu_j} \int \frac{d\gamma_{j,r}}{2\pi i} K(\{\gamma_{j,r}\}) w(\{\gamma_{j,r}\}) W(\{\gamma_{j,r}\}). \tag{1.6}$$

In this formula, the dependence on $\beta_1, \ldots, \beta_N$ in the right hand side is implicit in the kernel K as well as w and W. We write the integral formula of Tarasov–Varchenko type for the solutions to (1.2) with $\lambda = (4\pi)/n$ by using this pairing:

$$f(\beta_1, \ldots, \beta_N) = \sum_{j_1, \ldots, j_r} I(w_{j_1, \ldots, j_r}, W) v_{j_1} \otimes \cdots \otimes v_{j_N}. \tag{1.7}$$

We prove the convergence of the integral (1.6) in the positive Weyl chamber, and show that the equation (1.2) (with $\lambda = (4\pi)/n$) is indeed satisfied.

Finally we remark that the construction for sl_n case with generic highest weights is considered in [**17**] using the Jackson integral, and in [**18**] in the setting of the hypergeometric pairing (see [**7, 10, 12, 13**] for the conformal case, i.e., $q = 1$).

2. Operator Construction

In this section, we construct the vertex operators

$$\Psi_j^*(\beta) \ (j = 0, 1, \ldots, n-1; \beta \in \mathbf{R}), \tag{2.1}$$

satisfying the commutation relations with the R-matrix for the quantum affine algebra $U_q(\widehat{\mathrm{sl}}_n)$, i.e.,

$$\Psi_j^*(\beta_1)\Psi_j^*(\beta_2) = s(\beta_1 - \beta_2)\Psi_j^*(\beta_2)\Psi_j^*(\beta_1), \tag{2.2}$$

and for $j \ne k$

$$\begin{aligned} \Psi_j^*(\beta_1)\Psi_k^*(\beta_2) = s(\beta_1 - \beta_2)\{&\overline{R}(\beta_1, \beta_2)_{jk}^{jk} \Psi_k^*(\beta_2)\Psi_j^*(\beta_1) \\ &+ \overline{R}(\beta_1, \beta_2)_{jk}^{kj} \Psi_j^*(\beta_2)\Psi_k^*(\beta_1)\}, \end{aligned} \tag{2.3}$$

where

$$s(\beta) = \frac{S_2(-i\beta|\rho, 2\pi) S_2(i\beta + (2(n-1)\pi)/n|\rho, 2\pi)}{S_2(i\beta|\rho, 2\pi) S_2(-i\beta + (2(n-1)\pi)/n|\rho, 2\pi)}, \tag{2.4}$$

$$\overline{R}(\beta_1,\beta_2)_{jk}^{jk} = -\frac{\operatorname{sh}\pi/\rho(\beta_1-\beta_2)}{\operatorname{sh}\pi/\rho(\beta_1-\beta_2-(2\pi i)/n)} \quad (j\neq k), \tag{2.5}$$

$$\overline{R}(\beta_1,\beta_2)_{jk}^{kj} = \begin{cases} -\dfrac{e^{\pi/\rho(\beta_1-\beta_2)}\operatorname{sh}(2\pi^2 i)/(\rho n)}{\operatorname{sh}\pi/\rho(\beta_1-\beta_2-(2\pi i)/n)} & (j>k); \\ -\dfrac{e^{\pi/\rho(\beta_2-\beta_1)}\operatorname{sh}(2\pi^2 i)/n}{\operatorname{sh}\pi/\rho(\beta_1-\beta_2-(2\pi i)/n)} & (j<k). \end{cases} \tag{2.6}$$

We define $\overline{R}(\beta_1,\beta_2)\in \operatorname{End}(V\otimes V)$ by

$$\overline{R}(\beta_1,\beta_2)v_j\otimes v_k = \sum_{j'k'}\overline{R}(\beta_1,\beta_2)_{j'k'}^{jk}v_{j'}\otimes v_{k'}, \tag{2.7}$$

where $\overline{R}(\beta_1,\beta_2)_{j'k'}^{jk}=0$ except for (2.5), (2.6) and $\overline{R}(\beta_1,\beta_2)_{jj}^{jj}=1$.

We refer the reader to [**5**] for the double sine function $S_2(x|\omega_1,\omega_2)$. The deformation parameters q and ρ are identified by the relation (1.4). The construction of the vertex operators forces us to choose the normalization factor $s(\beta_1-\beta_2)$.

The bosonization of the level 1 highest weight representations for $U_q(\widehat{\mathrm{sl}}_n)$ and the vertex operators acting on them are given in [**2**] and [**6**], respectively. However, the present case is not connected to those works in which $|q|<1$. The case $|q|=1$ is obtained in the limit of the elliptic case considered in [**1, 9**]. We adapt the construction in [**1**] by using the method developed in [**4, 5, 8**].

Let $a_j(t)$ $(1\le j\le n-1; t\in \mathbf{R})$ be the free bose fields satisfying the commutation relations

$$\left[a_j(t),a_k(t')\right] = A_{jk}(t)\delta(t+t'), \tag{2.8}$$

$$A_{jk}(t) = -\frac{1}{t}\frac{\operatorname{sh}(a_{jk}\pi t)/n\ \operatorname{sh}(\rho/2+\pi/n)t}{\operatorname{sh}(\pi t)/n\ \operatorname{sh}(\rho t)/2}. \tag{2.9}$$

Here $(a_{jk})_{1\le j,k\le n-1}$ is the Cartan matrix of type A_n.

We consider the Fock space $\mathcal{F}$ generated by the vacuum vector $|\mathrm{vac}\rangle$ satisfying

$$a_j(t)|\mathrm{vac}\rangle = 0 \quad \text{if} \quad t>0. \tag{2.10}$$

Set

$$a_1^*(t) = -\sum_{j=1}^{n-1} a_j(t)\frac{\operatorname{sh}((n-j)\pi t)/n}{\operatorname{sh}\pi t}. \tag{2.11}$$

We have

$$\left[a_1^*(t),a_j(t')\right] = \delta_{1j}\frac{1}{t}\frac{\operatorname{sh}(\rho/2+\pi/n)t}{\operatorname{sh}(\rho t)/2}\delta(t+t'). \tag{2.12}$$

We introduce the currents

$$V_j(\beta) = :\!\exp\left(\int_{-\infty}^{\infty} a_j(t)e^{i\beta t}dt\right)\!: \ (1\le j\le n-1). \tag{2.13}$$

They satisfy the following commutation relations.

$$V_j(\beta_1)V_k(\beta_2) = V_k(\beta_2)V_j(\beta_1) \quad \left(|j-k|\ge 2\right), \tag{2.14}$$

$$V_j(\beta_1)V_{j-1}(\beta_2) = \frac{\mathrm{sh}\,\pi/\rho(\beta_1-\beta_2+(\pi i)/n)}{\mathrm{sh}\,\pi/\rho(\beta_2-\beta_1+(\pi i)/n)} V_{j-1}(\beta_2)V_j(\beta_1) \qquad (2\le j\le n-1), \tag{2.15}$$

$$V_j(\beta_1)V_j(\beta_2) = -\frac{\mathrm{sh}\,\pi/\rho(\beta_1-\beta_2-(2\pi i)/n)}{\mathrm{sh}\,\pi/\rho(\beta_2-\beta_1-(2\pi i)/n)} V_j(\beta_2)V_j(\beta_1) \qquad (1\le j\le n-1). \tag{2.16}$$

Now we define the vertex operators.

$$\Psi_j^*(\beta) = \prod_{1\le k\le j}\int_{-\infty}^{\infty}\frac{d\alpha_k}{2\pi i}\frac{e^{\pi/\rho(\beta-\alpha_j)}V_0(\beta)V_1(\alpha_1)\dots V_j(\alpha_j)}{\prod_{k=1}^{j}\mathrm{sh}\,\pi/\rho(\alpha_{k-1}-\alpha_k+(\pi i)/n)}. \tag{2.17}$$

Here we use $\alpha_0=\beta$ and

$$V_0(\beta) = :\!\exp\Big(\int_{-\infty}^{\infty} a_1^*(t)e^{i\beta t}dt\Big)\!: \tag{2.18}$$

The relations (2.14) and (2.15) are also valid for $V_0(\beta)$. However, the relation (2.16) is modified to

$$V_0(\beta_1)V_0(\beta_2) = s(\beta_1-\beta_2)V_0(\beta_2)V_0(\beta_1). \tag{2.19}$$

We give the proof of (2.2) and (2.3) later, which is the same thing with the proof of Lemma (1) in Section 3.

For a parameter $\lambda\in\mathbf{R}_{>0}$ we define the λ-expectation value by

$$\big\langle a_j(t)a_k(t')\big\rangle_\lambda = \frac{e^{\lambda t}}{e^{\lambda t}-1}A_{jk}(t)\delta(t+t'). \tag{2.20}$$

If $\lambda=\infty$ this reduces to the vacuum expectation value. The λ-expectation values of products of vertex operators can be calculated by using Wick's theorem. We recall the following formula (see [**4**]) for the two point function:

$$\Big\langle :\!e^{\int_{-\infty}^{\infty}a(t)e^{i\beta_1 t}}\!:\,:\!e^{\int_{-\infty}^{\infty}b(t)e^{i\beta_2 t}}\!:\Big\rangle_\lambda = \exp\Big(\int_0^\infty A(t)\frac{\mathrm{ch}(i(\beta_1-\beta_2)+\lambda/2)t}{\mathrm{sh}(\lambda t)/2}dt\Big), \tag{2.21}$$

where $a(t)$ and $b(t)$ are bosons satisfying $\big[a(t),b(t')\big]=A(t)\delta(t+t')$ and $A(t)=-A(-t)$. We also list the two point functions of $V_j(\beta)$. In the formulas below, const. means a constant independent of the spectral parameters.

$$\big\langle V_0(\beta_1)V_0(\beta_2)\big\rangle_\lambda = E_\lambda(\beta_1-\beta_2), \tag{2.22}$$

$$E_\lambda(\beta) = \text{const.}\frac{S_3(-i\beta)S_3(i\beta+\lambda)}{S_3((2\pi)/n+\rho-i\beta)S_3((2\pi)/n+\rho+i\beta+\lambda)}, \quad \text{where} \tag{2.23}$$

$$S_3(\beta) = S_3\big(\beta|\rho,\lambda,2\pi\big). \tag{2.24}$$

We have

$$\frac{E_\lambda(\beta)}{E_\lambda(-\beta)} = s(\beta), \quad E_\lambda(\lambda i-\beta) = E_\lambda(\beta). \tag{2.25}$$

For $1 \leq j \leq n-1$ we have

$$\begin{aligned}\langle V_j(\beta_1)V_{j-1}(\beta_2)\rangle_\lambda &= \langle V_{j-1}(\beta_1)V_j(\beta_2)\rangle_\lambda \\ &= \text{const. sh}\,\frac{\pi}{\rho}\left(\beta_1-\beta_2+\frac{\pi i}{n}\right)\varphi(\beta_1-\beta_2),\end{aligned} \tag{2.26}$$

$$\varphi(\beta) = \frac{1}{S_2(i\beta-\pi/n|\rho,\lambda)S_2(-i\beta-\pi/n|\rho,\lambda)}, \tag{2.27}$$

$$\langle V_j(\beta_1)V_j(\beta_2)\rangle_\lambda = \text{const.}\,\psi(\beta_1-\beta_2)\,\text{sh}\,\frac{\pi}{\rho}\left(\beta_1-\beta_2-\frac{2\pi i}{n}\right)h(\beta_1-\beta_2,\lambda), \tag{2.28}$$

$$h(\beta,\lambda) = \text{sh}\,\frac{\pi}{\lambda}\beta\,\text{sh}\,\frac{\pi}{\lambda}\left(\beta-\frac{2\pi i}{n}\right)\text{sh}\,\frac{\pi}{\lambda}\left(\beta+\frac{2\pi i}{n}\right), \tag{2.29}$$

$$\psi(\beta) = \frac{1}{S_2(i\beta+(2\pi)/n|\rho,\lambda)S_2(-i\beta+(2\pi)/n|\rho,\lambda)}. \tag{2.30}$$

The rest of the two point functions of $V_j(\beta)$'s are 1. These formulas are to be understood as analytic continuations from the regions where the integrals (2.21) are convergent. Because of the existence of poles, if they are used as integrands, we must pay a special attention to the choice of integration contours.

To see this point closely, let us compute

$$\langle \Psi^*_{j_N}(\beta_N)\ldots\Psi^*_{j_1}(\beta_1)\rangle = E(\beta_1,\ldots,\beta_N)\prod_{j,r}\int_C \frac{d\alpha_{j,r}}{2\pi i} I_{j_1,\ldots,j_N}(\{\alpha_{j,r}\}). \tag{2.31}$$

We associated the integration variables $\alpha_{j,r}$ $(1 \leq r \leq N; 1 \leq j \leq j_r)$ to the current $V_j(\alpha_{j,r})$ contained in the vertex operator $\Psi^*_{j_r}(\beta_r)$ (see (2.17)). We also set

$$\alpha_{0,r} = \beta_r, \tag{2.32}$$

$$\mathcal{N}_j = \{r; j_r \geq j\}. \tag{2.33}$$

The factor $E(\beta_1,\ldots,\beta_N)$ is given by the pair product

$$E(\beta_1,\ldots,\beta_N) = \text{const.}\prod_{1\leq r<s\leq N} E_\lambda(\beta_s-\beta_r). \tag{2.34}$$

The integrand consists of three parts,

$$I_{j_1,\ldots,j_N}(\beta_1,\ldots,\beta_N) = K(\{\alpha_{j,r}\})g(\{\alpha_{j,r}\})W(\{\alpha_{j,r}\}), \tag{2.35}$$

where

$$K(\{\alpha_{j,r}\}) = \prod_{j=1}^{n-1}\left\{\prod_{\substack{r\in\mathcal{N}_j\\ s\in\mathcal{N}_{j-1}}}\varphi(\alpha_{j,r}-\alpha_{j-1,s})\prod_{\substack{1\leq r<s\leq N\\ r,s\in\mathcal{N}_j}}\psi(\alpha_{j,s}-\alpha_{j,r})\right\}, \tag{2.36}$$

$$(2.37)\quad g(\{\alpha_{j,r}\}) = \prod_{j=1}^{n-1}\Bigg\{\prod_{r\in\mathcal{N}_j} e^{\pi/\rho(\alpha_{j-1,r}-\alpha_{j,r})}\prod_{\substack{r\in\mathcal{N}_j\\ s\in\mathcal{N}_{j-1}\\ r>s}}\operatorname{sh}\frac{\pi}{\rho}\Big(\alpha_{j,r}-\alpha_{j-1,s}+\frac{\pi i}{n}\Big)$$
$$\times\prod_{\substack{r\in\mathcal{N}_j\\ s\in\mathcal{N}_{j-1}\\ r<s}}\operatorname{sh}\frac{\pi}{\rho}\Big(\alpha_{j-1,s}-\alpha_{j,r}+\frac{\pi i}{n}\Big)$$
$$\times\prod_{\substack{r,s\in\mathcal{N}_j\\ r<s}}\operatorname{sh}\frac{\pi}{\rho}\Big(\alpha_{j,s}-\alpha_{j,r}-\frac{2\pi i}{n}\Big)\Bigg\},$$

$$(2.38)\quad W(\{\alpha_{j,r}\}) = \prod_{j=1}^{n-1}\prod_{\substack{r,s\in\mathcal{N}_j\\ r<s}} h(\alpha_{j,s}-\alpha_{j,r},\lambda).$$

The poles of the integrand come from those of $K(\{\alpha_{j,r}\})$. They are located at

$$(2.39)\quad \alpha_{j,r}-\alpha_{j-1,s} = \pm\Big(\frac{\pi i}{n}-\rho i\mathbf{Z}_{\geq 0}-\lambda i\mathbf{Z}_{\geq 0}\Big),$$

$$(2.40)\quad \alpha_{j,s}-\alpha_{j,r} = \pm\Big(\frac{2\pi i}{n}+\rho i\mathbf{Z}_{\geq 0}+\lambda i\mathbf{Z}_{\geq 0}\Big).$$

The contour $C_{j,r}$ for $\alpha_{j,r}$ is chosen so that

- the poles at $\alpha_{j-1,s}+(\pi i)/n-\rho_i\mathbf{Z}_{\geq 0}-\lambda_i\mathbf{Z}_{\geq 0}$ are below $C_{j,r}$,
- the poles at $\alpha_{j-1,s}-(\pi i)/n+\rho_i\mathbf{Z}_{\geq 0}+\lambda_i\mathbf{Z}_{\geq 0}$ are above $C_{j,r}$,
- the poles at $\alpha_{j,s}+(2\pi i)/n+\rho_i\mathbf{Z}_{\geq 0}+\lambda_i\mathbf{Z}_{\geq 0}$ are above $C_{j,r}$,
- the poles at $\alpha_{j,s}-(2\pi i)/n-\rho_i\mathbf{Z}_{\geq 0}-\lambda_i\mathbf{Z}_{\geq 0}$ are below $C_{j,r}$.

These conditions are not compatible if all the poles are really existent. However, in the actual situation for our matrix element $\langle\Psi^*_{j_N}(\beta_N)\dots\Psi^*_{j_1}(\beta_1)\rangle_\lambda$, this is not the case: the poles at

$$(2.41)\quad \alpha_{j,s}-\alpha_{j,r} = \pm\frac{2\pi i}{n},$$

are canceled by the zeros of $W(\{\alpha_{j,r}\})$. It is easy to see that under this cancellation, a consistent choice of the contours is possible.

In this paper we finish the operator theory at this point. We will not discuss the type I vertex operators and their form factors [**4, 8**]. In the next section, however, following the idea of Tarasov and Varchenko (see [**11, 15, 16**]), we will modify $W(\{\alpha_{j,r}\})$, and thereby construct a family of solutions to the quantum Knizhnik–Zamolodchiov equation in the special case where

$$(2.42)\quad \lambda = \frac{4\pi}{n}.$$

Note that, in this case, the function $\psi(\beta)$ simplifies to

$$(2.43)\quad \psi(\beta) = \frac{1}{2i\operatorname{sh} n/4(\beta-(2\pi i)/n)}.$$

3. Integral Formula

For nonnegative integers $\nu_1, \dots, \nu_{n-1}$ satisfying

$$N = \nu_0 \geq \nu_1 \geq \nu_2 \geq \cdots \geq \nu_{n-1} \geq \nu_n = 0, \tag{3.1}$$

we denote by $\mathcal{Z}_{\nu_1,\dots,\nu_{n-1}}$ the set of all the N-tuples $J = (j_1, \dots, j_N) \in (\mathbf{Z}_{\geq 0})^N$ such that

$$\nu_j = \sharp \mathcal{N}_j. \tag{3.2}$$

Define $r_{j,m}$ $(0 \leq j \leq n-1; 1 \leq m \leq \nu_j)$ as follows:

$$\mathcal{N}_j = \{r_{j,1}, \dots, r_{j,\nu_j}\}, \ r_{j,1} < \cdots < r_{j,\nu_j}. \tag{3.3}$$

We have, in particular, $r_{0,m} = m$. We make a correspondence between two sets of the integration variables:

$$\alpha^J_{j,r_{j,m}} = \gamma_{j,m}. \tag{3.4}$$

We set

$$w_J(\{\gamma_{j,m}\}) = \mathrm{Skew}_{n-1} \circ \cdots \circ \mathrm{Skew}_1 g_J(\{\gamma_{j,m}\}), \tag{3.5}$$

$$\begin{aligned} g_J(\{\gamma_{j,m}\}) = \prod_{j=1}^{n-1} \Bigg\{ \prod_{m=1}^{\nu_j} \Bigg\{ \prod_{r \in \mathcal{N}_j} & e^{\pi/\rho(\alpha^J_{j-1,r} - \alpha^J_{j,r})} \\ & \times \prod_{\substack{m' \\ r_{j-1,m'} < r_{j,m}}} \mathrm{sh} \frac{\pi}{\rho}\left(\gamma_{j,m} - \gamma_{j-1,m'} + \frac{\pi i}{n}\right) \\ & \times \prod_{\substack{m' \\ r_{j,m} < r_{j-1,m'}}} \mathrm{sh} \frac{\pi}{\rho}\left(\gamma_{j,m} - \gamma_{j-1,m'} - \frac{\pi i}{n}\right) \Bigg\} \\ & \times \prod_{1 \leq m < m' \leq \nu_j} \mathrm{sh} \frac{\pi}{\rho}\left(\gamma_{j,m'} - \gamma_{j,m} - \frac{2\pi i}{n}\right) \Bigg\}, \end{aligned} \tag{3.6}$$

where $\gamma_{0,m} = \beta_m$ and Skew_j is the skew-symmetrization with respect to the variables $\{\gamma_{j,m}\}_{m=1,\dots,\nu_j}$:

$$\mathrm{Skew}_j X(\gamma_{j,1}, \dots, \gamma_{j,\nu_j}) = \sum_{\sigma \in S_{\nu_j}} \mathrm{sgn}(\sigma) X(\gamma_{j,\sigma(1)}, \dots, \gamma_{j,\sigma(\nu_j)}). \tag{3.7}$$

Note that the above definition differs from (2.37) by sign. This change is necessary because we will choose different W in the below. Accordingly, we change the R matrix:

$$R(\beta_1, \beta_2)^{j'k'}_{jk} = \begin{cases} -\overline{R}(\beta_1, \beta_2)^{jk}_{jk} & \text{if } (j', k') = (j, k) \quad \text{and} \quad j \neq k; \\ \overline{R}(\beta_1, \beta_2)^{j'k'}_{jk} & \text{otherwise.} \end{cases} \tag{3.8}$$

In the following, we abbreviate

$$w_J(\{\gamma_{j,r}\}) = w_J(\beta_{k_1}, \dots, \beta_{k_N}) \tag{3.9}$$

when the dependence on $\{\gamma_{j,m}\}$ $(j \neq 0)$ is irrelevant.

For the function w_J, the following equality holds.

LEMMA 1.

$$w_{(j_1,\dots,j_{k+1},j_k,\dots,j_N)}(\beta_1,\dots,\beta_{k+1},\beta_k,\dots,\beta_N) = \sum_{j'_k,j'_{k+1}} R(\beta_k,\beta_{k+1})_{j_k j_{k+1}}^{j'_k j'_{k+1}} w_{(j_1,\dots,j'_k,j'_{k+1},\dots,j_N)}(\beta_1,\dots,\beta_k,\beta_{k+1},\dots,\beta_N). \tag{3.10}$$

PROOF. It is enough to prove (3.10) for $N=2$. First, we prove

$$w_{(l,l)}(\beta_1,\beta_2) = w_{(l,l)}(\beta_2,\beta_1), \quad l=0,1,\dots,n-1. \tag{3.11}$$

Note that

$$g_{(l,l)}(\beta_1,\beta_2) = e^{\pi/\rho(\beta_1+\beta_2-\gamma_{l,1}-\gamma_{l,2})} \times \prod_{j=1}^{l}\left\{ \operatorname{sh}\frac{\pi}{\rho}\left(\gamma_{j,2}-\gamma_{j-1,1}+\frac{\pi i}{n}\right) \times \operatorname{sh}\frac{\pi}{\rho}\left(\gamma_{j,1}-\gamma_{j-1,2}-\frac{\pi i}{n}\right) \times \operatorname{sh}\frac{\pi}{\rho}\left(\gamma_{j,2}-\gamma_{j,1}-\frac{2\pi i}{n}\right)\right\}. \tag{3.12}$$

By using

$$\operatorname{sh}\frac{\pi}{\rho}\left(\gamma_2-\beta_1+\frac{\pi i}{n}\right)\operatorname{sh}\frac{\pi}{\rho}\left(\gamma_1-\beta_2-\frac{\pi i}{n}\right) - (\beta_1\leftrightarrow\beta_2) = \operatorname{sh}\frac{\pi}{\rho}(\beta_1-\beta_2)\operatorname{sh}\frac{\pi}{\rho}\left(\gamma_2-\gamma_1+\frac{2\pi i}{n}\right), \tag{3.13}$$

repeatedly, we obtain

$$\operatorname{Skew}_{l-1}\circ\cdots\circ\operatorname{Skew}_0 g_{(l,l)}(\beta_1,\beta_2) = e^{(\pi i)/\rho(\beta_1+\beta_2-\gamma_{l,1}-\gamma_{l,2})}\operatorname{sh}\pi/\rho(\beta_1-\beta_2)\operatorname{sh}\frac{\pi}{\rho}\left(\gamma_{l,2}-\gamma_{l,1}+\frac{2\pi i}{n}\right) \times \operatorname{sh}\frac{\pi}{\rho}\left(\gamma_{l,2}-\gamma_{l,1}-\frac{2\pi i}{n}\right)\prod_{j=1}^{l-1} h(\gamma_{j,1}-\gamma_{j,2}). \tag{3.14}$$

This is symmetric with respect to $\gamma_{l,1},\gamma_{l,2}$, and therefore, we have (3.11).

Next, we show that

$$\operatorname{Skew}_l\circ\cdots\circ\operatorname{Skew}_1 X_{(l+1,l)}(\beta_2,\beta_1) = 0, \tag{3.15}$$

where

$$X_{(l+1,l)}(\beta_2,\beta_1) = g_{(l+1,l)}(\beta_2,\beta_1) - R(\beta_1,\beta_2)_{l,l+1}^{l+1,l} g_{(l+1,l)}(\beta_1,\beta_2) - R(\beta_1,\beta_2)_{l,l+1}^{l,l+1} g_{(l,l+1)}(\beta_1,\beta_2). \tag{3.16}$$

The proof for $g_{(l,l+1)}(\beta_2,\beta_1)$ is similar. The proof for $g_{(l,k)}$ for general j,k reduces to the case $k=l\pm1$.

For $l=0$, one can check (3.15) directly. For $l\geq 1$ we proceed as follows. Set $\gamma_{l,1}=\gamma_1$, $\gamma_{l,2}=\gamma_2$, $\gamma_{l+1,1}=\gamma$. Noting that

$$g_{(l+1,l)}(\beta_1,\beta_2) = g_{(l,l)}(\beta_1,\beta_2)e^{(\pi i)/\rho(\gamma_1-\gamma)}\operatorname{sh}\frac{\pi}{\rho}\left(\gamma-\gamma_2-\frac{\pi i}{n}\right), \tag{3.17}$$

$$g_{(l,l+1)}(\beta_1,\beta_2) = g_{(l,l)}(\beta_1,\beta_2)e^{(\pi i)/\rho(\gamma_2-\gamma)}\operatorname{sh}\frac{\pi}{\rho}\left(\gamma-\gamma_1+\frac{\pi i}{n}\right), \tag{3.18}$$

we have

$$(3.19)\quad X_{l+1,l}(\beta_2,\beta_1) = \left(\frac{1}{2}e^{-(2\pi\gamma)/\rho} - e^{\pi/\rho(\gamma_1-\gamma_2-(\pi i)/n)}\right)\mathrm{Skew}_0 g_{(l,l)}(\beta_1,\beta_2)$$
$$+\frac{2\,\mathrm{sh}\,\pi/\rho(\beta_1-\beta_2)\,\mathrm{sh}\,\pi/\rho(\gamma_1-\gamma_2-(2\pi i)/n)e^{-(\pi^2 i)/(\rho n)}}{\mathrm{sh}\,\pi/\rho(\beta_1-\beta_2-(2\pi i)/n)} g_{(l,l)}(\beta_1,\beta_2).$$

Note that

$$(3.20)\quad \mathrm{Skew}_1\circ\cdots\circ\mathrm{Skew}_l\left\{\mathrm{sh}\,\pi/\rho\left(\gamma_1-\gamma_2-\frac{2\pi i}{n}\right)g_{(l,l)}(\beta_1,\beta_2)\right\}$$
$$=\mathrm{sh}\,\pi/\rho\left(\beta_1-\beta_2-\frac{2\pi i}{n}\right)e^{(\pi i)/\rho(\beta_1+\beta_2-\gamma_1-\gamma_2)}\prod_{j=1}^{l}h(\gamma_{j,1}-\gamma_{j,2}).$$

Using (3.14) and (3.20), we obtain (3.15). □

Following [**11, 15, 16**], we define the hypergeometric pairing. Set

$$(3.21)\quad \mathcal{F}^{(\rho)}_{\nu_1,\dots,\nu_{n-1}} = \sum_{J\in\mathcal{Z}_{\nu_1,\dots,\nu_{n-1}}} \mathbf{C}w_J(\{\gamma_{j,m}\}).$$

In our setting, the hypergeometric pairing gives rise to the pairing given by (1.6) between $w\in\mathcal{F}^{(\rho)}_{\nu_1,\dots,\nu_{n-1}}$ and $W\in\mathcal{F}^{(\lambda)}_{\nu_1,\dots,\nu_{n-1}}$, where the kernel K is given by

$$(3.22)\quad \prod_{j=1}^{n-1}\left\{\prod_{m=1}^{\nu_j}\prod_{m'=1}^{\nu_{j-1}}\varphi(\gamma_{j,m}-\gamma_{j-1,m'})\prod_{1\le m<m'\le\nu_j}\psi(\gamma_{j,m}-\gamma_{j,m'})\right\}.$$

In this paper, we restrict to the level 0 case, i.e., $\lambda=(4\pi)/n$, where we have (2.43). In this case, it is convenient to move the ψ-part of the kernel to the space of W. Therefore, we set

$$(3.23)\quad \mathcal{W}_{\nu_1,\dots,\nu_{n-1}} = \sum_{J\in\mathcal{Z}_{\nu_1,\dots,\nu_{n-1}}} \mathbf{C}W_J(\{\gamma_{j,m}\}),$$

where

$$(3.24)\quad W_J(\{\gamma_{j,m}\}) = \mathrm{Skew}_{n-1}\circ\cdots\circ\mathrm{Skew}_1 G_J(\{\gamma_{j,m}\}),$$

$$(3.25)\quad G_J(\{\gamma_{j,m}\}) = \prod_{j=1}^{n-1}\left\{\prod_{r\in\mathcal{N}_j} e^{n/4(\alpha^J_{j-1,r}-\alpha^J_{j,r})}\right.$$
$$\times\prod_{m=1}^{\nu_j}\prod_{\substack{m'\\ r_{j-1,m'}<r_{j,m}}}\mathrm{sh}\,\frac{n}{4}\left(\gamma_{j,m}-\gamma_{j-1,m'}+\frac{\pi i}{n}\right)$$
$$\left.\times\prod_{\substack{m'\\ r_{j,m}<r_{j-1,m'}}}\mathrm{sh}\,\frac{n}{4}\left(\gamma_{j,m}-\gamma_{j-1,m'}-\frac{\pi i}{n}\right)\right\}.$$

We define the pairing as follows. For $w \in \mathcal{F}^{(\rho)}_{\nu_1,\ldots,\nu_{n-1}}$ and $W \in \mathcal{W}_{\nu_1,\ldots,\nu_{n-1}}$, we set

$$I(w,W)(\beta_1,\ldots,\beta_N)=\left(\prod_{j=1}^{n-1}\prod_{m=1}^{\nu_j}\int_{C_j}d\gamma_{j,m}\right)\prod_{j=1}^{n-1}\prod_{m=1}^{\nu_j}\prod_{m'=1}^{\nu_{j-1}}\varphi(\gamma_{j,m}-\gamma_{j-1,m'}) \times w\big(\{\gamma_{j,m}\}\big)W\big(\{\gamma_{j,m}\}\big). \tag{3.26}$$

In the above formula, φ is given by (2.27) with $\lambda = (4\pi)/n$. The contour C_j for $\gamma_{j,m}$ is $(-\infty,\infty)$ except that the poles at

$$\gamma_{j\pm1,m'} + \frac{\pi i}{n} - \rho_i \mathbf{Z}_{\geq 0} - \frac{4\pi_i}{n}\mathbf{Z}_{\geq 0}, \tag{3.27}$$

are below C_j and the poles at

$$\gamma_{j\pm1,m'} - \frac{\pi i}{n} + \rho_i \mathbf{Z}_{\geq 0} + \frac{4\pi_i}{n}\mathbf{Z}_{\geq 0}, \tag{3.28}$$

are above C_j.

The weight of the function $f(\beta_1,\ldots,\beta_N)$ given by (1.7) is

$$\sum_{j=0}^{n-1}(\nu_j-\nu_{j+1})\varepsilon_j. \tag{3.29}$$

The positive Weyl chamber is given by

$$\nu_{j-1}+\nu_{j+1}\geq 2\nu_j, \quad \text{for all } j=1,\ldots,n-1. \tag{3.30}$$

PROPOSITION 2. *Suppose that $\nu_1,\ldots,\nu_{n-1}$ satisfy* (3.1) *and* (3.30). *Then the integral* (3.26) *is absolutely convergent.*

PROOF. It is easy to see that

$$\begin{aligned}&\big|w\big(\{\gamma_{j,m}\}\big)\big|\\ &\quad\leq \text{const.}\exp\left\{\frac{\pi}{\rho}\sum_{j=1}^{n-1}\left(\sum_{m=1}^{\nu_j}\sum_{m'=1}^{\nu_{j-1}}|\gamma_{j,m}-\gamma_{j-1,m'}| + \sum_{1\leq m<m'\leq\nu_j}|\gamma_{j,m}-\gamma_{j,m'}|\right)\right\},\\ &\big|W\big(\{\gamma_{j,m}\}\big)\big|\\ &\quad\leq \text{const.}\exp\left\{\frac{n}{4}\sum_{j=1}^{n-1}\left(\sum_{m=1}^{\nu_j}\sum_{m'=1}^{\nu_{j-1}}|\gamma_{j,m}-\gamma_{j-1,m'}|\right)\right\},\end{aligned}$$

where const. is a constant independent of $\{\gamma_{j,m}\}$.

The asymptotic behaviour of φ is as follows (see [**5**]):

$$\varphi(\beta)\sim\exp\left\{-\left(\frac{n}{4}+\frac{3\pi}{2\rho}\right)|\beta|\right\}, \quad \beta\to\pm\infty. \tag{3.31}$$

Therefore, we have

$$\big|\text{the integrand of } I(w,W)\big| \leq \text{const.}\exp\left(-\frac{\pi}{2\rho}H_{\nu_1,\ldots,\nu_{n-1}}\big(\{\gamma_{j,m}\}\big)\right), \tag{3.32}$$

where

$$H_{\nu_1,\ldots,\nu_{n-1}}(\{\gamma_{j,m}\}) = \sum_{j=1}^{n-1}\left(\sum_{m=1}^{\nu_j}\sum_{m'=1}^{\nu_{j-1}}|\gamma_{j,m}-\gamma_{j-1,m'}| - 2\sum_{1\le m<m'\le\nu_j}|\gamma_{j,m}-\gamma_{j,m'}|\right). \tag{3.33}$$

Set

$$\nu = \sum_{j=0}^{n-1}\nu_j. \tag{3.34}$$

Let $\gamma_1,\ldots,\gamma_\nu$ be a renumbering of the variables $\{\gamma_{j,m}\}_{\substack{0\le j\le n-1\\ 1\le m\le\nu_j}}$. We consider the integral in the region

$$\gamma_1 < \cdots < \gamma_\nu. \tag{3.35}$$

Set

$$\gamma_{k_1} = \min(\beta_1,\ldots,\beta_N), \tag{3.36}$$

$$\gamma_{k_2} = \max(\beta_1,\ldots,\beta_N). \tag{3.37}$$

One can rewrite (3.33) as

$$H_{\nu_1,\ldots,\nu_{n-1}}(\{\gamma_{j,m}\}) = \sum_{l=1}^{\nu-1} M_l(\gamma_{l+1}-\gamma_l). \tag{3.38}$$

Let

$$\tilde\gamma_1 < \cdots < \tilde\gamma_{\nu-N}, \tag{3.39}$$

be the renumbering of the variables $\{\gamma_{j,m}\}_{\substack{1\le j\le n-1\\ 1\le m\le\nu_j}}$. We change the integration variables from $\tilde\gamma_k$ to t_k by setting

$$t_k = \begin{cases} \gamma_{k+1}-\gamma_k & \text{if } 1\le k<k_1;\\ \gamma_{k+N}-\gamma_{k+N-1} & \text{if } k_2-N<k\le\nu-N;\\ \tilde\gamma_k & \text{otherwise.}\end{cases} \tag{3.40}$$

For the convergence, it is enough to show that

$$M_k > 0 \ (1\le k<k_1), \tag{3.41}$$

and

$$M_{k+N-1} > 0 \ (k_2-N<k\le\nu-N). \tag{3.42}$$

We will show (3.41). The proof of (3.42) is similar.

Set

$$x_j = \sharp\{m;\gamma_{j,m}\le\gamma_k\} \ (0\le j\le n-1). \tag{3.43}$$

Note that $x_0 = 0$ because $k<k_1$. Note also that not all x_j $(1\le j\le n-1)$ are zero. We have

$$M_k = \sum_{j=1}^{n-1}(\nu_{j-1}+\nu_{j+1}-2\nu_j)x_j + 2\left(\sum_{j=1}^{n-1}x_j^2 - \sum_{j=1}^{n-2}x_jx_{j+1}\right). \tag{3.44}$$

The second term of the right hand side of the above formula is positive. Therefore, $M_k>0$ if the condition (3.30) is satisfied. □

The hypergeometric pairing has the following property.

LEMMA 3.

$$I\big(g_{\bar{J}}(\beta_N,\beta_1,\dots,\beta_{N-1}),W(\beta_1,\dots,\beta_{N-1},\beta_N)\big)\big|_{\beta_N\to\beta_N-(4\pi i)/n} = I\big(g_J(\beta_1,\dots,\beta_N),W(\beta_1,\dots,\beta_N)\big), \tag{3.45}$$

where $\bar{J}=(j_N,j_1,\dots,j_{N-1})$ for $J=(j_1,\dots,j_N)$, and the left hand side is understood as the analytic continuation of the integral.

PROOF. We prove (3.45) in the form

$$I\big(g_{\bar{J}}(\beta_N,\beta_1,\dots,\beta_{N-1}),W(\beta_1,\dots,\beta_{N-1},\beta_N)\big) = I\big(g_J(\beta_1,\dots,\beta_N),W(\beta_1,\dots,\beta_N)\big)\big|_{\beta_N\to\beta_N+(4\pi i)/n}. \tag{3.46}$$

Note that the integration variables γ_{j,ν_j} $(j=1,\dots,j_N)$ are associated with the Nth component in the sense of the operator construction in Section 2. In other words, we have

$$r_{j,\nu_j}=N. \tag{3.47}$$

Therefore, it is natural to shift the contours for these variables at the same time when we change β_N. In fact, by this simultaneous shift, no crossing of contours by poles occurs. To see this, it is enough to observe that the function g_J has zeros at $\gamma_{j\pm1,m}=\gamma_{j,\nu_j}+(\pi i)/n$ if $r_{j\pm1,m}<N$.

We have

$$W\big(\{\gamma_{j,m}\}_{m\neq N}\big)\big|_{\substack{\gamma_{j,\nu_j}\to\gamma_{j,\nu_j}+(4\pi i)/n\\(0\le j\le j_N)}} = (-1)^{N+\nu_{j_N}+\nu_{j_N+1}-1}W\big(\{\gamma_{j,m}\}\big). \tag{3.48}$$

Note also that (see [**5**])

$$\frac{\varphi(\beta+(4\pi i)/n)}{\varphi(\beta)} = -\frac{\operatorname{sh}\pi/\rho(\beta-(\pi i)/n)}{\operatorname{sh}\pi/\rho(\beta+(5\pi i)/n)}. \tag{3.49}$$

We shift the contours in the integral (3.26) (with $w=g_J$), and then change the variables $\gamma_{j,r}$ $(1\le j\le j_N)$: first by

$$\gamma_{j,\nu_j}\to\gamma_{j,\nu_j}+\frac{4\pi i}{n}, \tag{3.50}$$

and second by

$$\gamma_{j,r}\to\gamma_{j,r+1}\ (1\le r\le \nu_j-1),\quad \gamma_{j,\nu_j}\to\gamma_{j,1}. \tag{3.51}$$

Rewriting the integrand by using (3.48) and (3.49), we get (3.46). □

From Lemma 1 and Lemma 3, it is easy to show the following theorem.

THEOREM 1. *We assume the condition* (3.30). *For* $W\in\mathcal{W}_{\nu_1,\dots,\nu_{n-1}}$, *we set*

$$f_W(\beta_1,\dots,\beta_N)=\sum_{J\in\mathcal{Z}_{\nu_1,\dots,\nu_{n-1}}} I(w_J,W)(\beta_1,\dots,\beta_N)v_J, \tag{3.52}$$

where $v_J=v_{j_1}\otimes\cdots\otimes v_{j_N}$ *for* $J=(j_1,\dots,j_N)$. *Then* f_W *is a solution to* (1.2) *with the restriction* (1.1) *and* (1.5).

References

1. Y. Asai, M. Jimbo, T. Miwa, and Y. Pugai, *Bosonization of vertex operators for the $A_{n-1}^{(1)}$ face models*, J. Phys. A **29** (1996), 6595–6616.
2. I. Frenkel and N. Jing, *Vertex representations of quantum affine algebras*, Proc. Nat. Acad. Sci. U.S.A **85** (1988), 9373–9377.
3. I. Frenkel and N. Reshetikhin, *Quantum affine algebras and holonomic difference equations*, Comm. Math. Phys. **146** (1992), 1–60.
4. M. Jimbo, H. Konno, and T. Miwa, *Massless* XXZ *model and degeneration of the elliptic algebra $\mathcal{A}_{q,p}(\widehat{\mathrm{sl}}_2)$*, Deformation Theory and Symplectic Geometry (Ascona, 1996), Math. Phys. Stud., vol. 20, Kluwer Acad. Publ., Dordrecht, 1997, pp. 117–138.
5. M. Jimbo and T. Miwa, *Quantum KZ equation with $|q| = 1$ and correlation functions of the XXZ model in the gapless regime*, J. Phys. A **29** (1996), 2923–2958.
6. Y. Koyama, *Staggered polarization of vertex models with $U_q(\widehat{SL}_n)$ symmetry*, Comm. Math. Phys. **164** (1994), 277–291.
7. G. Kuroki, *Fock space representations of affine Lie algebras and integral representations in the Wess–Zumino–Witten models*, Comm. Math. Phys. **142** (1991), 511–542.
8. S. Lukyanov, *Free field representation for massive integrable models*, Comm. Math. Phys. **167** (1995), 183–226.
9. S. Lukyanov and Y. Pugai, *Multipoint local height probabilities in the integrable* RSOS *model*, Nucl. Phys. B **473** (1996) 631–658.
10. A. Matsuo, *An application of Aomoto–Gelfand hypergeometric functions to the* SU(n) *Knizhnik–Zamolodchikov equation*, Comm. Math. Phys. **134** (1990), 65–77.
11. T. Miwa and Y. Takeyama, *Determinant formula for the solutions of the quantum Knizhnik–Zamolodchikov equation with $|q| = 1$*, math. QA/9812096.
12. V. Schechtman and A. Varchenko, *Hypergeometric solutions of Knizhnik–Zamolodchikov equations*, Lett. Math. Phys. **20** (1990), 279–283.
13. ———, *Arrangements of hyperplanes and Lie algebra homology*, Invent. Math. **106** (1991), 139–194.
14. F. Smirnov, *Form factors in completely integrable models of quantum field theory*, Advanced Series in Mathematics, vol. 14, World Scientific Publ. Co., Inc., River Edge, NJ, 1992.
15. V. Tarasov and A. Varchenko, *Geometry of q–hypergeometric functions as a bridge between Yangians and quantum affine algebras*, Invent. Math. **128** (1997), no.3, 501–588.
16. ———, *Geometry of q–hypergeometric functions, quantum affine algebras and elliptic quantum groups*, Astérisque **246** (1997), 1–135.
17. ———, *Jackson integral representations for solutions to the quantized Knizhnik–Zamolodchikov equation*, St. Petersburg Math J. **6** (1994), 275–313.
18. ———, private communication. (in preparation)

Research Institute for Mathematical Sciences, Kyoto University, Kyoto 606-8502, Japan.

E-mail address, T. Miwa: `miwa@kurims.kyoto-u.ac.jp`

E-mail address, Y. Takeyama: `minihuni@kurims.kyoto-u.ac.jp`

Centre de Recherches Mathématiques
CRM Proceedings and Lecture Notes
Volume **25**, 2000

Symmetries of the Discrete Schrödinger Equation and Its Quantum Deformations

Luis M. Nieto, Javier Negro, Francisco J. Herranz, and Angel Ballesteros

ABSTRACT. After a short revision of the symmetries appearing in the free Schrödinger equation, the nonstandard quantum deformations of the $(1+1)-$ Schrödinger algebra are considered. Two of them are identified with the symmetry algebras of either a space or time uniform lattice discretization of the free Schrödinger equation. For both cases, the deformation parameter of the corresponding Hopf algebra can be interpreted as the step of the lattice. The problem of finding a quantum algebra linked to the full space-time discretization is also discussed.

1. Introduction

In recent works **[13]** new techniques have been developed in order to deal with the symmetries of difference or differential-difference equations. These studies have tried to adapt in this field the standard methods that have been so successful when used for analyzing differential equations.

On the other side, since they were introduced, quantum algebras have been connected with different versions of space-time lattices through several algebraic constructions, independently of the usual Lie symmetry theory **[3, 11, 14]**. An exhaustive study for the discretization on q-lattices of classical linear differential equations has shown that their symmetries obeyed to q-deformed commutation relations with respect to the Lie algebra structure of the continuous symmetries **[7,9,10]**. However, not all the Hopf algebra structures underlying these q-symmetry algebras have been discussed in detail.

When the discretization of linear equations is made on uniform lattices it is also well known that the relevant symmetries preserve the Lie algebra structure that appears in the continuous case **[8, 15]**. Perhaps, this is the reason why the symmetry approach to these equations has never been related to quantum algebras.

We will address this question now for the discrete $(1+1)$-Schrödinger equation. In this case we will show that Lie algebras of (discrete) symmetries can be put in

2000 *Mathematics Subject Classification.* Primary: 39A10, 81R50, 16W30.

This work has been partially supported by DGES (Projects PB94-1115 and PB95-0719) from Ministerio de Educación y Cultura (Spain), and also by Junta de Castilla y León (Projects CO1/396, CO2/197, and CO2/297).

This is the final form of the paper.

correspondence to nontrivial quantum Hopf algebra structures of the nonstandard type. Indeed, two quantum deformations of the Schrödinger algebra have been recently obtained [**4, 5**], both of them endowed with a triangular Hopf structure. The former was derived by starting from the nonstandard quantum deformation of its harmonic oscillator subalgebra h_4, while the latter was constructed by means of the nonstandard quantum deformation of the extended $\overline{\mathrm{sl}}(2,\mathbb{R})$ subalgebra. We will show later that the symmetry algebra [**8**] of the space discretization of the Schrödinger equation obtained from (2.1) by taking the limit $\tau \to 0$ is just the quantum Schrödinger algebra of [**4**], and the deformation parameter z is related with the space lattice constant σ. Likewise, we also show that the time discretization of the Schrödinger equation obtained from (2.1) by means of the limit $\sigma \to 0$ has the quantum Schrödinger algebra of [**5**] as its symmetry algebra; in this case, the time lattice step τ plays the role of the deformation parameter. As a consequence, a relationship between nonstandard deformations and regular lattice discretizations can be established.

2. Symmetries of the Discrete Heat or Schrödinger Equation

To begin with, let us consider the discrete version of the heat or (time imaginary) Schrödinger equation on a two-dimensional uniform lattice introduced in [**8**]

$$(\Delta_x^2 - 2m\Delta_t)\phi(x,t) = 0. \tag{2.1}$$

The difference operators Δ_x and Δ_t which appear in (2.1) can be expressed in terms of shift operators $T_x = e^{\sigma\partial_x}$ and $T_t = e^{\tau\partial_t}$ as

$$\Delta_x = \frac{T_x - 1}{\sigma}, \quad \Delta_t = \frac{T_t - 1}{\tau}, \tag{2.2}$$

where the parameters σ and τ are the lattice constants in the space x and time t directions, respectively. The action of Δ_x or Δ_t on a function $\phi(x,t)$ consist in a discrete derivative, which in the limit $\sigma \to 0$ and $\tau \to 0$ come into ∂_x and ∂_t, respectively.

We will say that a given operator $\mathcal{O}$ is a symmetry of the linear equation $E\phi(x,t) = 0$ if $\mathcal{O}$ transforms solutions into solutions, that is, if $\mathcal{O}$ is such that

$$E\mathcal{O} = \Lambda E, \tag{2.3}$$

where Λ is another operator. In this way, the symmetries of the equation (2.1) were computed in [**8**] showing that they realized the Schrödinger algebra, which is exactly the same result as for the continuous case [**12**]. To be more precise, we look for the symmetries of the following form

$$\begin{aligned} \mathcal{O}(t,x,T_t,T_x) = {} & \mathcal{O}_{[0]}(T_t,T_x) + t\mathcal{O}_{[t]}(T_t,T_x) + x\mathcal{O}_{[x]}(T_t,T_x) \\ & + t^2\mathcal{O}_{[tt]}(T_t,T_x) + tx\mathcal{O}_{[tx]}(T_t,T_x) + x^2\mathcal{O}_{[xx]}(T_t,T_x), \end{aligned} \tag{2.4}$$

that is, the symmetries are supposed to be polynomials up to the second order in the discrete variables $t = k\tau$, $x = n\sigma$, k, $n \in \mathbb{Z}$. After some calculations, the following symmetry operators are found (for the particular case $m = 1/2$):

$$
\begin{aligned}
&\mathbb{P}_t = \Delta_t,\\
&\mathbb{P}_x = \Delta_x,\\
&\mathbb{D} = 2t\Delta_t T_t^{-1} + x\Delta_x T_x^{-1} - \frac{1}{2}T_x^{-1} + 1,\\
&\mathbb{K} = 2t\Delta_x T_t^{-1} + \left(x + \frac{\sigma}{2}\right)T_x^{-1},\\
&\mathbb{C} = t^2\Delta_t T_t^{-2} + tx\Delta_x T_t^{-1}T_x^{-1}\\
&\qquad + \frac{x^2}{4}T_x^{-2} + t\left(T_t^{-2} - \frac{1}{2}T_t^{-1}T_x^{-1}\right) - \frac{\sigma^2}{16}T_x^{-2}.
\end{aligned}
\tag{2.5}
$$

The nonzero commutators are

$$
\begin{aligned}
&[\mathbb{P}_t, \mathbb{K}] = 2\mathbb{P}_x; \qquad &[\mathbb{P}_x, \mathbb{K}] = 1;\\
&[\mathbb{P}_t, \mathbb{D}] = 2\mathbb{P}_t; \qquad &[\mathbb{P}_x, \mathbb{D}] = \mathbb{P}_x;\\
&[\mathbb{P}_t, \mathbb{C}] = \mathbb{D}; \qquad &[\mathbb{P}_x, \mathbb{C}] = \mathbb{K}/2;\\
&[\mathbb{D}, \mathbb{K}] = \mathbb{K}; \qquad &[\mathbb{D}, \mathbb{C}] = 2\mathbb{C}.
\end{aligned}
\tag{2.6}
$$

They are (σ, τ)–independent, and define the Schrödinger algebra. When $\sigma, \tau \to 0$ the symmetry operators become:

$$
\begin{aligned}
\mathbb{P}_t = \partial_t, \quad \mathbb{P}_x = \partial_x, \quad \mathbb{D} = 2t\partial_t + x\partial_x + 1/2,\\
\mathbb{K} = 2t\partial_x + x, \quad \mathbb{C} = t^2\partial_t + tx\partial_x + x^2/4 + t/2.
\end{aligned}
\tag{2.7}
$$

Using the results of (2.5), it is possible to find eigenfunctions of the discrete operators $\mathbb{P}_t$ and $\mathbb{P}_x$ which are at the same time solutions of the discrete heat equation:

$$
\begin{aligned}
&\phi_\lambda(k,n) = (1+\lambda^2\tau)^k(1+\lambda\sigma)^n, \quad \lambda \in \mathbb{R},\ t = k\tau,\ x = n\sigma,\ k,n \in \mathbb{Z},\\
&\mathbb{P}_t\phi_\lambda(k,n) = \lambda^2\phi_\lambda(k,n), \quad \mathbb{P}_x\phi_\lambda(k,n) = \lambda\phi_\lambda(k,n).
\end{aligned}
\tag{2.8}
$$

In the limit τ, $\sigma \to 0$, we recover the solutions of the continuous heat equation where the variables t and x are separated:

$$\lim_{\tau,\sigma\to 0} \phi_\lambda(t,x;\tau,\sigma) = e^{\lambda^2 t}\, e^{\lambda x}. \tag{2.9}$$

Different solutions are obtained when we look for the eigenfunctions $\varphi_\ell(k,n)$ of $\mathbb{D}$, with eigenvalue $n + \frac{1}{2}$:

$$\varphi_\ell(k,n) = (-\sigma)^\ell \sum_{a=0}^{[[\ell/2]]} \frac{(-\ell)_{2a}}{a!}(-k)_a\left(-n-\frac{1}{2}\right)_{\ell-2a}\left(-\frac{\tau}{\sigma^2}\right)^a, \tag{2.10}$$

where $[[\ell/2]]$ signifies the integral part of $\ell/2$ and $(-\ell)_{2a}$ is a Pochhammer's symbol. In the continuum limit we get Hermite polynomials

$$\lim_{\tau,\sigma\to 0} \varphi_\ell(k,n) = (-t)^{n/2} H_\ell\left(\frac{x}{\sqrt{-t}}\right). \tag{2.11}$$

The functions $\phi_\lambda(k,n)$ can be expanded in terms of $\varphi_\ell(k,n)$ as follows

$$(1+\lambda^2\tau)^{t/\tau}(1+\lambda\sigma)^{x/\sigma+1/2} = \sum_{n=0}^{\infty} \frac{\lambda^n}{n!}\varphi_n(k,n)\ . \tag{2.12}$$

The case of having a discrete and a continuous variable is straightforwardly obtained by taking the limits $\tau \to 0$ or $\sigma \to 0$ in the last expression.

The main conclusion of this study is that the symmetry algebra of the discrete, semidiscrete and continuous cases is exactly the same, and leads to similar properties of the corresponding eigenfunctions.

3. A Discrete Space Schrödinger Equation

We are going to consider now a different approach to the discretization of the Schrödinger equation (2.1), using quantum deformations. First of all, let us introduce the more familiar notation for the Schrödinger generators: time translation H, space translation P, Galilean boost K, dilation D, conformal transformation C, and the central generator M [**12**]. Let us consider now the quantum Schrödinger algebra $U_z^{(s)}(\mathcal{S})$ studied in [**4**], whose underlying Lie bialgebra is generated by the nonstandard classical r-matrix

$$r^{(s)} = zP \wedge \left(D + \frac{1}{2}M\right). \tag{3.1}$$

The coproduct of $U_z^{(s)}(\mathcal{S})$ is

$$\begin{aligned}
&\Delta(M) = 1 \otimes M + M \otimes 1,\\
&\Delta(P) = 1 \otimes P + P \otimes 1,\\
&\Delta(H) = 1 \otimes H + H \otimes e^{-2zP},\\
&\Delta(K) = 1 \otimes K + K \otimes e^{zP} - zD' \otimes e^{zP}M,\\
&\Delta(D) = 1 \otimes D + D \otimes e^{zP} + \frac{1}{2}M \otimes (e^{zP} - 1),\\
&\Delta(C) = 1 \otimes C + C \otimes e^{2zP} + \frac{z}{2}K \otimes e^{zP}D' - \frac{z}{2}D' \otimes e^{zP}(K + zD'M),
\end{aligned} \tag{3.2}$$

and its commutation rules are

$$\begin{gathered}
[D,P] = \frac{1}{z}(1 - e^{zP}), \quad [D,K] = K, \quad [K,P] = Me^{zP}, \quad [M,\cdot] = 0,\\
[D,H] = -2H, \quad [D,C] = 2C - \frac{z}{2}KD', \quad [H,P] = 0,\\
[H,C] = \frac{1}{2}(1 + e^{-zP})D' - \frac{1}{2}M + zKH, \quad [K,C] = -\frac{z}{2}K^2,\\
[P,C] = -\frac{1}{2}(1 + e^{zP})K - \frac{z}{2}e^{zP}MD', \quad [K,H] = \frac{1}{z}(1 - e^{-zP}),
\end{gathered} \tag{3.3}$$

where hereafter we shall use the notation $D' = D + \frac{1}{2}M$ in order to simplify some expressions. The following differential-difference realization of (3.3) can be found:

$$\begin{gathered}
P = \partial_x, \quad H = \partial_t, \quad M = m,\\
K = -t\frac{1 - e^{-z\partial_x}}{z} - mxe^{z\partial_x}, \quad D = 2t\partial_t + x\frac{e^{z\partial_x} - 1}{z} + \frac{1}{2},\\
C = t^2\partial_t e^{-z\partial_x} + tx\left(\frac{\sinh z\partial_x}{z} - zm\partial_t e^{z\partial_x}\right) + \frac{1}{2}mx^2 e^{z\partial_x}\\
- \frac{1}{4}t\{1 - 3e^{-z\partial_x} + m(1 - e^{-z\partial_x})\} + \frac{1}{4}zm(1 - m)xe^{z\partial_x}.
\end{gathered} \tag{3.4}$$

Note that the Galilei generators $\{K, H, P, M\}$ close a deformed subalgebra (but not a Hopf subalgebra) whose Casimir is

$$E_z^{(s)} = \frac{1}{z^2}(1 - e^{-zP})^2 - 2MH. \tag{3.5}$$

The action of $E_z^{(s)}$ on $\phi(x,t)$ (choosing for the Casimir the value zero) provides a space discretization of the Schrödinger equation:

$$\left(\left(\frac{1 - e^{-z\partial_x}}{z}\right)^2 - 2m\partial_t\right)\phi(x,t) = 0. \tag{3.6}$$

Furthermore, the quantum Hopf algebra $U_z^{(s)}(\mathcal{S})$ is a symmetry algebra of this equation since

$$\begin{gathered}[E_z^{(s)}, X] = 0 \quad \text{for} \quad X \in \{K, H, P, M\}, \quad [E_z^{(s)}, D] = 2E_z^{(s)},\\ [E_z^{(s)}, C] = \{t(e^{-z\partial_x} + 1) - zmxe^{z\partial_x}\}E_z^{(s)}.\end{gathered} \tag{3.7}$$

Now we will relate these results with the discretizations of the Schrödinger equation studied in [**8**] (and reproduced with some detail in Section 2) through the nonlinear change of basis defined by

$$\begin{gathered}\mathcal{H} = H, \quad \mathcal{P} = \frac{1 - e^{-zP}}{z}, \quad \mathcal{M} = M,\\ \mathcal{D} = D + \frac{1}{2}(1 - e^{zP}), \quad \mathcal{K} = -2K - zMe^{zP},\\ \mathcal{C} = C - \frac{z}{2}KD' + \frac{z}{2}Ke^{zP} - \frac{z^2}{8}Me^{2zP}, \quad \sigma = -z.\end{gathered} \tag{3.8}$$

In this new basis the commutation rules (3.3) of the Hopf algebra $U_\sigma^{(s)}(\mathcal{S})$ are

$$\begin{array}{llll}[\mathcal{D}, \mathcal{P}] = -\mathcal{P}, & [\mathcal{D}, \mathcal{K}] = \mathcal{K}, & [\mathcal{K}, \mathcal{P}] = -2\mathcal{M}, & [\mathcal{M}, \cdot] = 0,\\ [\mathcal{D}, \mathcal{H}] = -2\mathcal{H}, & [\mathcal{D}, \mathcal{C}] = 2\mathcal{C}, & [\mathcal{H}, \mathcal{C}] = \mathcal{D}, & [\mathcal{H}, \mathcal{P}] = 0,\\ [\mathcal{P}, \mathcal{C}] = \frac{1}{2}\mathcal{K}, & [\mathcal{K}, \mathcal{H}] = -2\mathcal{P}, & [\mathcal{K}, \mathcal{C}] = 0, & \end{array} \tag{3.9}$$

that is, they come into the nondeformed Schrödinger Lie algebra. Hence the deformation parameter σ appears only within the coproduct, which now reads

$$\begin{aligned}\Delta(\mathcal{M}) &= 1 \otimes \mathcal{M} + \mathcal{M} \otimes 1,\\ \Delta(\mathcal{P}) &= 1 \otimes \mathcal{P} + \mathcal{P} \otimes 1 + \sigma\mathcal{P} \otimes \mathcal{P},\\ \Delta(\mathcal{H}) &= 1 \otimes \mathcal{H} + \mathcal{H} \otimes (1 + \sigma\mathcal{P})^2,\\ \Delta(\mathcal{K}) &= 1 \otimes \mathcal{K} + \mathcal{K} \otimes \frac{1}{1 + \sigma\mathcal{P}} - 2\sigma\mathcal{D}' \otimes \frac{\mathcal{M}}{1 + \sigma\mathcal{P}},\\ \Delta(\mathcal{D}) &= 1 \otimes \mathcal{D} + \mathcal{D} \otimes \frac{1}{1 + \sigma\mathcal{P}} - \frac{1}{2}\mathcal{M} \otimes \frac{\sigma\mathcal{P}}{1 + \sigma\mathcal{P}},\\ \Delta(\mathcal{C}) &= 1 \otimes \mathcal{C} + \mathcal{C} \otimes \frac{1}{(1 + \sigma\mathcal{P})^2} - \frac{\sigma}{2}\mathcal{D}' \otimes \frac{1}{1 + \sigma\mathcal{P}}\mathcal{K}\\ &\qquad + \frac{\sigma^2}{2}\mathcal{D}'(\mathcal{D}' - 1) \otimes \frac{\mathcal{M}}{(1 + \sigma\mathcal{P})^2},\end{aligned} \tag{3.10}$$

where $\mathcal{D}' = \mathcal{D} + \frac{1}{2}\mathcal{M}$. Note that although the new generator $\mathcal{P}$ is nonprimitive, its coproduct satisfies

$$\Delta\big((1+\sigma\mathcal{P})^a\big) = (1+\sigma\mathcal{P})^a \otimes (1+\sigma\mathcal{P})^a, \tag{3.11}$$

since the old generator P fulfils $\Delta(e^{azP}) = e^{azP} \otimes e^{azP}$ for any real number a.

On the other hand, the mapping (3.8) transforms the differential-difference realization (3.4) into

$$\begin{gathered}
\mathcal{P} = \Delta_x, \quad \mathcal{H} = \partial_t, \quad \mathcal{M} = m, \\
\mathcal{K} = 2t\Delta_x + m(2x+\sigma)T_x^{-1}, \quad \mathcal{D} = 2t\partial_t + x\Delta_x T_x^{-1} - \frac{1}{2}T_x^{-1} + 1, \\
\mathcal{C} = t^2\partial_t + tx\Delta_x T_x^{-1} + \frac{1}{2}mx^2 T_x^{-2} + t\left(1 - \frac{1}{2}T_x^{-1}\right) - \frac{m\sigma^2}{8}T_x^{-2}.
\end{gathered} \tag{3.12}$$

We emphasize that these are just the symmetry operators of the equation (2.1) obtained in [**8**], provided the continuum time limit $\tau \to 0$ is performed and $m = \frac{1}{2}$. This connection is consistent with the fact that (3.8) transforms the Casimir of the Galilei subalgebra $E_z^{(s)}$ into the nondeformed one $E = \mathcal{P}^2 - 2\mathcal{M}\mathcal{H}$. Therefore, the discretized Schrödinger equation

$$(\Delta_x^2 - 2m\partial_t)\phi(x,t) = 0, \tag{3.13}$$

is obtained as the realization (3.12) of E, and the operators (3.12) are symmetries of this equation satisfying

$$[E, X] = 0, \quad X \in \{\mathcal{K}, \mathcal{H}, \mathcal{P}, \mathcal{M}\}, \quad [E, \mathcal{D}] = 2E, \quad [E, \mathcal{C}] = 2tE. \tag{3.14}$$

Thus, we have shown that the space differential-difference Schrödinger equation introduced in [**8**] has $U_\sigma^{(s)}(\mathcal{S})$ as its quantum Hopf symmetry algebra. The deformation parameter $\sigma = -z$ is interpreted as the lattice step in the x coordinate, meanwhile the time t remains a continuum variable.

4. A Discrete Time Schrödinger Equation

A similar procedure can be applied to the quantum Schrödinger algebra $U_z^{(t)}(\mathcal{S})$ coming from the nonstandard classical r-matrix

$$r^{(t)} = 2zH \wedge \left(D + \frac{1}{2}M\right) \equiv 2zH \wedge D'. \tag{4.1}$$

The coproduct and commutation rules of $U_z^{(t)}(\mathcal{S})$ are given by [**5**]

$$\begin{aligned}
\Delta(M) &= 1 \otimes M + M \otimes 1, \\
\Delta(H) &= 1 \otimes H + H \otimes 1, \\
\Delta(P) &= 1 \otimes P + P \otimes e^{-2zH}, \\
\Delta(K) &= 1 \otimes K + K \otimes e^{2zH} - 2zD' \otimes e^{4zH}P, \\
\Delta(D) &= 1 \otimes D + D \otimes e^{4zH} + \frac{1}{2}M \otimes (e^{4zH} - 1), \\
\Delta(C) &= 1 \otimes C + C \otimes e^{4zH} - zD' \otimes e^{4zH}M,
\end{aligned} \tag{4.2}$$

$$
\begin{aligned}
&[D,P]=-P, \quad [D,K]=K, \quad [K,P]=M, \quad [M,\cdot]=0,\\
&[D,H]=\frac{1}{2z}(1-e^{4zH}), \quad [D,C]=2C+2z(D')^2, \quad [H,P]=0,\\
&[H,C]=D'-\frac{1}{2}Me^{4zH}, \quad [K,C]=z(D'K+KD'),\\
&[P,C]=-K-z(D'P+PD'), \quad [K,H]=e^{4zH}P.
\end{aligned} \tag{4.3}
$$

A realization of (4.3) reads [5]

$$
\begin{aligned}
&H=\partial_t, \quad P=\partial_x, \quad M=m,\\
&K=-(t+4z)e^{4z\partial_t}\partial_x-mx, \quad D=2(t+4z)\frac{e^{4z\partial_t}-1}{4z}+x\partial_x+\frac{1}{2},\\
&C=(t^2-4zbt)\frac{e^{4z\partial_t}-1}{4z}+tx\partial_x+\frac{1}{2}t+\frac{1}{2}mx^2-4z(b+1)e^{4z\partial_t}\\
&\qquad -zx^2\partial_x^2-2z(b+1)x\partial_x-z\left(b+\frac{1}{2}\right)^2,
\end{aligned} \tag{4.4}
$$

where $b=m/2-2$. Now, a time discretization of the Schrödinger equation is obtained by considering the deformed Casimir of the Galilei subalgebra

$$
E_z^{(t)}=P^2-\frac{1}{2z}M(1-e^{-4zH}), \tag{4.5}
$$

written in terms of the realization (4.4):

$$
\left(\partial_x^2-2m\frac{1-e^{-4z\partial_t}}{4z}\right)\phi(x,t)=0. \tag{4.6}
$$

Under the realization (4.4) the generators of $U_z^{(t)}(\mathcal{S})$ are symmetry operators of this equation as they satisfy

$$
\begin{aligned}
&[E_z^{(t)},X]=0, \quad X\in\{K,H,P,M\}, \quad [E_z^{(t)},D]=2E_z^{(t)},\\
&[E_z^{(t)},C]=2\{t+z(1-m-2x\partial_x)\}E_z^{(t)}.
\end{aligned} \tag{4.7}
$$

The relationship with the time discretization of the Schrödinger equation analysed in [8] is provided by the nonlinear map defined by

$$
\begin{aligned}
&\mathcal{H}=\frac{1-e^{-4zH}}{4z}, \quad \mathcal{P}=P, \quad \mathcal{M}=M,\\
&\mathcal{D}=D+2(1-e^{4zH}), \quad \mathcal{K}=-2K-8zPe^{4zH},\\
&\mathcal{C}=C+z(D')^2-4zDe^{4zH}, \quad \tau=-4z.
\end{aligned} \tag{4.8}
$$

If we apply (4.8) to (4.3) we find again the classical commutation rules of the Schrödinger algebra (3.9) while the coproduct is now given by

$$
\begin{aligned}
\Delta(\mathcal{M}) &= 1\otimes\mathcal{M}+\mathcal{M}\otimes 1,\\
\Delta(\mathcal{H}) &= 1\otimes\mathcal{H}+\mathcal{H}\otimes 1+\tau\mathcal{H}\otimes\mathcal{H},\\
\Delta(\mathcal{P}) &= 1\otimes\mathcal{P}+\mathcal{P}\otimes(1+\tau\mathcal{H})^{1/2},\\
\Delta(\mathcal{K}) &= 1\otimes\mathcal{K}+\mathcal{K}\otimes\frac{1}{(1+\tau\mathcal{H})^{1/2}}-\tau\mathcal{D}'\otimes\frac{\mathcal{P}}{1+\tau\mathcal{H}},\\
\Delta(\mathcal{D}) &= 1\otimes\mathcal{D}+\mathcal{D}\otimes\frac{1}{1+\tau\mathcal{H}}-\frac{1}{2}\mathcal{M}\otimes\frac{\tau\mathcal{H}}{1+\tau\mathcal{H}},\\
\Delta(\mathcal{C}) &= 1\otimes\mathcal{C}+\mathcal{C}\otimes\frac{1}{1+\tau\mathcal{H}}-\frac{\tau}{2}\mathcal{D}'\otimes\frac{1}{1+\tau\mathcal{H}}\mathcal{D}\\
&\qquad\qquad +\frac{\tau}{4}\mathcal{D}'(\mathcal{D}'-2)\otimes\frac{\tau\mathcal{H}}{(1+\tau\mathcal{H})^2}.
\end{aligned}
\tag{4.9}
$$

Note that the generators $\{\mathcal{D},\mathcal{H},\mathcal{C},\mathcal{M}\}$ close a quantum extended $\mathrm{sl}(2,\mathbb{R})$ algebra. As a byproduct, if we take $\mathcal{M}=0$, we obtain a new expression for the coproduct of the nonstandard quantum $\mathrm{sl}(2,\mathbb{R})$ with "classical" commutation rules (compare this result with a different solution given in [**1**]). Notice also that the generator $\mathcal{H}$ fulfils a similar property to (3.11).

In this new basis the realization (4.4) turns out to be

$$
\begin{gathered}
\mathcal{H}=\Delta_t,\quad \mathcal{P}=\partial_x,\quad \mathcal{M}=m,\\
\mathcal{K}=2t\partial_x T_t^{-1}+2mx,\quad \mathcal{D}=2t\Delta_t T_t^{-1}+x\partial_x+\frac{1}{2},\\
\mathcal{C}=t^2\Delta_t T_t^{-2}+tx\partial_x T_t^{-1}+\frac{1}{2}mx^2+t\left(T_t^{-2}-\frac{1}{2}T_t^{-1}\right),
\end{gathered}
\tag{4.10}
$$

which is again the limit $\sigma\to 0$ of the symmetry operators of [**8**] when $m=\frac{1}{2}$. The corresponding discretized Schrödinger equation is provided by the (nondeformed) Casimir E of the Galilei subalgebra written through (4.10):

$$
(\partial_x^2-2m\Delta_t)\phi(x,t)=0. \tag{4.11}
$$

The new operators (4.10) are symmetries of this equation satisfying

$$
[E,X]=0,\quad X\in\{\mathcal{K},\mathcal{H},\mathcal{P},\mathcal{M}\},\quad [E,\mathcal{D}]=2E,\quad [E,\mathcal{C}]=2tT_t^{-1}E. \tag{4.12}
$$

Thus we have obtained a discrete version of the Schrödinger equation on a uniform lattice such that its symmetry algebra is the quantum Hopf algebra $U_\tau^{(t)}(\mathcal{S})$, and the time lattice step is related to the deformation parameter in the form $\tau=-4z$.

5. Space-Time Discretization

It would be certainly interesting to analyse whether there exists a (at least two parameter) quantum Schrödinger algebra giving rise to the symmetries of the full space-time uniform discretization provided by (2.1). In general, the search for quantum deformations of a given Lie algebra can be guided by the study of its Lie bialgebra structures, since they are just the first order deformations of the coproduct (see, for instance, [**2**]). Therefore, in our case, one could try to find the most general Schrödinger bialgebra $(\mathcal{S},\delta)$ such that it includes among its components the two Lie bialgebras linked to the previous deformations; hence $(\mathcal{S},\delta)$ should depend on σ, τ and perhaps on some additional parameters λ_i. Under such conditions and by imposing δ to fulfil the cocycle condition and the dual map δ^* to define a Lie bracket, we obtain a unique cocommutator family depending only on three parameters: $\tau = -4z_1$, $\sigma = -z_2$ and λ. Furthermore, it can be shown that this Schrödinger bialgebra is a coboundary one generated by the classical r-matrix:

$$
\begin{aligned}
r = & 2z_1 H \wedge D' + z_2 P \wedge D' - \lambda H \wedge C - \frac{\lambda^2}{8z_1} C \wedge D - \frac{2z_1 z_2}{\lambda} P \wedge H \\
& + \frac{\lambda z_2}{8z_1}\left(K \wedge D - \frac{\lambda}{2z_1} K \wedge C - 3P \wedge C\right) + \left(\frac{\lambda}{4} + \frac{3z_2^2}{16z_1}\right) K \wedge P \\
& + \left(\frac{3\lambda z_2}{16z_1} + \frac{z_2^3}{32z_1^2}\right) K \wedge M - \left(\frac{\lambda^2}{16z_1} + \frac{\lambda z_2^2}{32z_1^2}\right) M \wedge C \\
& + \left(\frac{\lambda}{4} + \frac{z_2^2}{16z_1}\right) M \wedge D.
\end{aligned} \tag{5.1}
$$

Note that the first two terms correspond to the time and space classical r-matrices $r^{(t)}$ and $r^{(s)}$, but both deformations do not fulfil a simple "superposition principle" and extra contributions have to be added. The full quantization of such Lie bialgebra seems to be a difficult task and we shall not address this problem here. The "additional" deformation parameter λ allows us to distinguish between nonstandard (when $\lambda = -(z_2^2)/(4z_1)$) and standard solutions (otherwise). On the other hand, although the limits $z_2 \to 0$ and $\lambda \to 0$ lead to the time classical r-matrix, unfortunately the limit $z_1 \to 0$ leads to divergences in (5.1).

6. Final Remarks

We would like to point out that the existence of a Hopf structure for the symmetries of a given equation associated to an elementary system allows us to write equations of composed systems keeping the same symmetry algebra [**6**]. In order to use this property for the two cases here discussed, we see that only the last commutator in either (3.14) or (4.12) involving the conformal generator $\mathcal{C}$ is not algebraic, but depends explicitly on the chosen representation (the same happens at the continuum level). Consequently the composed systems characterized by the equation $\Delta(E)\phi = 0$ will have, by construction, $\Delta(\mathcal{H})$, $\Delta(\mathcal{P})$, $\Delta(\mathcal{K})$, $\Delta(\mathcal{D})$, and $\Delta(\mathcal{M})$ as symmetry operators (they close a Hopf subalgebra). However, in general this will not be the case for $\Delta(\mathcal{C})$, and a further study on the behaviour of this operator is needed in order to construct coupled equations with full Schrödinger algebra symmetry.

The applicability of the constructive approach presented here is not limited to the cases analyzed before; in fact it can be directly extended to other quantum algebras by means of their corresponding difference realizations. Work on this line is nowadays in progress.

References

1. N. Aizawa, *Irreducible decomposition for tensor product representations of Jordanian quantum algebras*, J. Phys. A **30** (1997), 5981–5992.
2. A. Ballesteros and F. J. Herranz, *Lie bialgebra quantizations of the oscillator algebra and their universal R-matrices*, J. Phys. A **29** (1996), 4307–4320.
3. A. Ballesteros, F. J. Herranz, M. A. del Olmo, and M. Santander, *A new "null-plane" quantum Poincaré algebra*, Phys. Lett. B **351** (1995), 137–145.

4. A. Ballesteros, F. J. Herranz, and P. Parashar, *A Jordanian quantum two-photon/Schrödinger algebra*, J. Phys. A **30** (1997), 8587–8597.
5. ———, *Quantum two-photon algebras from nonstandard $U_z(\mathrm{sl}(2,\mathbb{R}))$ and discrete time Schrödinger equation*, Modern Phys. Lett. A **13** (1998), 1241–1252.
6. F. Bonechi, E. Celeghini, R. Giachetti, E. Sorace, and M. Tarlini, *Inhomogeneous quantum groups as symmetries of phonons*, Phys Rev Lett **68** (1992), 3718–3720; *Quantum Galilei group as symmetry of magnons*, Phys. Rev. B **46** (1992), 5727–5730.
7. V. K. Dobrev, H. D. Doebner, and C. Mrugalla, *A q-Schrödinger algebra, its lowest-weight representations and generalized q-deformed heat/Schrödinger equations*, J. Phys. A **29** (1996), 5909–5918.
8. R. Floreanini, J. Negro, L. M. Nieto, and L. Vinet, *Symmetries of the heat equation on the lattice*, Lett. Math. Phys. **36** (1996), 351–355.
9. R. Floreanini and L. Vinet, *Symmetries of the q-difference heat equation*, Lett. Math. Phys. **32** (1994), 37–44.
10. ———, *Quantum symmetries of q-difference equations*, J. Math. Phys. **36** (1995), 3134–3156.
11. L. Frappat and A. Sciarrino, *Lattice space-time from Poincaré and κ-Poincaré algebras*, Phys. Lett. B **347**, (1995), 28–32.
12. C. R. Hagen, *Scale and conformal transformations in Galilean-invariant field theory*, Phys. Rev. D **5** (1972), 377–388.
13. D. Levi, L. Vinet, and P. Winternitz, *Lie group formalism for difference equations*, J. Phys. A **30** (1997), 633–649.
14. J. Lukierski, A. Nowicki, and H. Ruegg, *New quantum Poincaré algebra and κ-deformed field theory*, Phys. Lett. B **293** (1992), 344–352.
15. J. Negro and L. M. Nieto, *Symmetries of the wave equation in a uniform lattice*, J. Phys. A **29** (1996), 1107–1114.

Departamento de Física Teórica, Facultad de Ciencias, Universidad de Valladolid, 47011 Valladolid, Spain.
E-mail address: luismi@metodos.fam.cie.uva.es

Departamento de Física Teórica, Facultad de Ciencias, Universidad de Valladolid, 47011 Valladolid, Spain.
E-mail address: jnegro@fta.uva.es

Departamento de Física, Facultad de Ciencias Químicas, Universidad de Burgos, 09001 Burgos, Spain.
E-mail address: fteorica@cpd.uva.es

Departamento de Física, Facultad de Ciencias Químicas, Universidad de Burgos, 09001 Burgos, Spain.
E-mail address: fteorica@cpd.uva.es

Centre de Recherches Mathématiques
CRM Proceedings and Lecture Notes
Volume **25**, 2000

Determinant and Pfaffian Solutions for Discrete Soliton Equations

Y. Ohta

ABSTRACT. The Pfaffian solution for the discrete coupled modified KdV equation with arbitrary coupling constants is discussed. In the single component case, discrete modified KdV equation, the solution is given by determinant, while the multicomponentization naturally induces the Pfaffian solution. There are two types of Pfaffian representions, Wronski type Pfaffian and Gram Pfaffian. The Gram one gives a simple expression of the most general soliton solutions.

1. Introduction

The solutions of the soliton equations are usually given by the τ functions which are expressible in terms of determinants or Pfaffians with Wronski or Gram structure, and in most cases the bilinear forms of the equations are simply reduced to the algebraic identities of determinants or Pfaffians, so-called Plücker relations [**7**]. However there are some exceptions in which the bilinear equations are not Plücker. In this paper, we consider such an example, the discrete coupled modified KdV (cmKdV) equation [**2, 9**],

$$\begin{aligned} \frac{1}{a}(v^\lambda_{n,t+1} - v^\lambda_{nt}) &= \Gamma_{nt}\left(1 + \sum_{\mu=1}^{M}\sum_{\nu=1}^{M} a^{\mu\nu} v^\mu_{nt}\overline{v}^\nu_{nt}\right)(v^\lambda_{n+1,t+1} - v^\lambda_{n-1,t}), \\ &\qquad 1 \le \lambda \le M, \\ \frac{1}{a}(\overline{v}^\lambda_{n,t+1} - \overline{v}^\lambda_{nt}) &= \Gamma_{nt}\left(1 + \sum_{\mu=1}^{M}\sum_{\nu=1}^{M} a^{\mu\nu} v^\mu_{nt}\overline{v}^\nu_{nt}\right)(\overline{v}^\lambda_{n+1,t+1} - \overline{v}^\lambda_{n-1,t}), \\ &\qquad 1 \le \lambda \le M, \\ \frac{\Gamma_{nt}}{\Gamma_{n-1,t}} &= \frac{1 + \sum_{\mu=1}^{M}\sum_{\nu=1}^{M} a^{\mu\nu} v^\mu_{n,t+1}\overline{v}^\nu_{n,t+1}}{1 + \sum_{\mu=1}^{M}\sum_{\nu=1}^{M} a^{\mu\nu} v^\mu_{nt}\overline{v}^\nu_{nt}}, \end{aligned} \tag{1.1}$$

2000 *Mathematics Subject Classification.* 35Q51, 58F07.

The author would like to thank Prof. Ablowitz, Dr. Trubatch, Prof. Hirota, Dr. Tsujimoto, Dr. Iwao, and Dr. Tsuchida for their discussions and interests.

This is the final form of the paper.

where $a^{\mu\nu}$ is the coupling constant and a is the difference interval of t. We note that this is equivalent with the discrete coupled nonlinear Schrödinger equation. In the single component case ($M = 1$), the solutions are given by determinants and the equations are the usual Plücker relations [**1**] (see Appendix). However in the multicomponent case, the soliton solutions are written in terms of Pfaffians rather than determinants. We should note that the simple vectorization of the single component equation leads to another more complicated coupled system [**5**]. The continuous limits of (1.1) give the following semi-discrete cmKdV and continuous cmKdV equations.

$$\frac{\mathrm{d}}{\mathrm{d}t}v_n^\lambda = \left(1+\sum_{\mu=1}^{M}\sum_{\nu=1}^{M}a^{\mu\nu}v_n^\mu\overline{v}_n^\nu\right)(v_{n+1}^\lambda - v_{n-1}^\lambda),$$
$$\frac{\mathrm{d}}{\mathrm{d}t}\overline{v}_n^\lambda = \left(1+\sum_{\mu=1}^{M}\sum_{\nu=1}^{M}a^{\mu\nu}v_n^\mu\overline{v}_n^\nu\right)(\overline{v}_{n+1}^\lambda - \overline{v}_{n-1}^\lambda),$$
$$\partial_t v^\lambda = 6\sum_{\mu=1}^{M}\sum_{\nu=1}^{M}a^{\mu\nu}v^\mu\overline{v}^\nu\partial_x v^\lambda + \partial_x^3 v^\lambda,$$
$$\partial_t \overline{v}^\lambda = 6\sum_{\mu=1}^{M}\sum_{\nu=1}^{M}a^{\mu\nu}v^\mu\overline{v}^\nu\partial_x \overline{v}^\lambda + \partial_x^3 \overline{v}^\lambda.$$

The stationary case of the above systems was studied by Suris as a symplectic map and shown to be integrable [**8**]. Tsuchida, Ujino and Wadati have analysed these cmKdV equations by using the inverse scattering method and solved the initial value problems [**9, 10**]. The wronki type Pfaffian solution for the semi-discrete case was proposed in [**6**]. Iwao and Hirota investigated the Gram Pfaffian solutions for the above systems [**2, 4**]. In this paper, we propose another Gram Pfaffian solution of (1.1) which gives a compact representation for the most general soliton solutions, and show how the bilinear equation arises from a non Plücker algebraic identity.

2. Bilinear Form and Pfaffian Solution

By applying the variable transformation,

$$v_{nt}^\mu = \frac{g_{nt}^\mu}{f_{nt}}, \quad \overline{v}_{nt}^\nu = \frac{\overline{g}_{nt}^\nu}{f_{nt}}, \quad \Gamma_{nt} = \frac{f_{n+1,t+1}f_{nt}}{f_{n,t+1}f_{n+1,t}},$$

we obtain the bilinear form of (1.1),

(2.1)
$$\frac{1}{a}(g_{n,t+1}^\mu f_{nt} - g_{nt}^\mu f_{n,t+1}) = g_{n+1,t+1}^\mu f_{n-1,t} - g_{n-1,t}^\mu f_{n+1,t+1},$$
$$\frac{1}{a}(\overline{g}_{n,t+1}^\nu f_{nt} - \overline{g}_{nt}^\nu f_{n,t+1}) = \overline{g}_{n+1,t+1}^\nu f_{n-1,t} - \overline{g}_{n-1,t}^\nu f_{n+1,t+1},$$

(2.2)
$$f_{n+1,t}f_{n-1,t} - f_{nt}f_{nt} = \sum_{\mu=1}^{M}\sum_{\nu=1}^{M}a^{\mu\nu}g_{nt}^\mu\overline{g}_{nt}^\nu.$$

The Pfaffian solution of the above bilinear equations is given as

$$\begin{aligned} f_{nt} &= (1,2,\ldots,2N)_{nt}, \\ g^{\mu}_{nt} &= (d_{00}, c^{\mu}, 1,2,\ldots,2N)_{nt}, \\ \overline{g}^{\nu}_{nt} &= (d_{00}, \overline{c}^{\nu}, 1,2,\ldots,2N)_{nt} \end{aligned}$$

where the elements of Pfaffian are defined by

$$(i,j)_{nt} = c_{ij} + \frac{p_i - p_j}{p_i p_j - 1}\alpha_i p_i^n \left(\frac{1 - a\,1/(p_j)}{1 - ap_i}\right)^t \alpha_j p_j^n \left(\frac{1 - a\,1/(p_j)}{1 - ap_j}\right)^t, \quad i < j,$$

$$(d_{00}, i)_{nt} = \alpha_i p_i^n \left(\frac{1 - a\,1/(p_i)}{1 - ap_i}\right)^t, \quad (d_{00}, c^{\mu})_{nt} = 0, \quad (d_{00}, \overline{c}^{\nu})_{nt} = 0$$

$$(c^{\mu}, i)_{nt} = c^{\mu}_i, \quad 1 \le i \le m, \quad (c^{\mu}, j)_{nt} = 0, \quad m+1 \le j \le 2N,$$

$$(\overline{c}^{\nu}, i)_{nt} = 0, \quad 1 \le i \le m, \quad (\overline{c}^{\nu}, j)_{nt} = \overline{c}^{\nu}_j, \quad m+1 \le j \le 2N,$$

$$c_{ij} = \begin{cases} -\frac{1}{(p_i + 1/(p_i)) - (p_j + 1/(p_j))} \sum_{\mu=1}^{M} \sum_{\nu=1}^{M} a^{\mu\nu} c^{\mu}_i \overline{c}^{\nu}_j & 1 \le i \le m,\ m+1 \le j \le 2N, \\ 0 & 1 \le i < j \le m \quad \text{or} \\ & m+1 \le i < j \le 2N, \end{cases}$$

where p_i is the complex wave number of ith soliton, α_i is the complex phase parameter of ith soliton and c^{μ}_i, $\overline{c}^{\mu}_i$ give the complex amplitude of ith soliton in μth channel.

The time evolution part of bilinear equations (2.1) is just a Plücker relation. In fact, we have

$$\begin{aligned} &\frac{1}{a}(d_{00}, c, 1,2,\ldots,2N)_{n,t+1}\ (1,2,\ldots,2N)_{nt} \\ &\qquad -(d_{00}, c, 1,2,\ldots,2N)_{nt}\frac{1}{a}(1,2,\ldots,2N)_{n,t+1} \\ &\qquad - (d_{00}, c, 1,2,\ldots,2N)_{n+1,t+1}\ (1,2,\ldots,2N)_{n-1,t} \\ &\qquad + (d_{00}, c, 1,2,\ldots,2N)_{n-1,t}\ (1,2,\ldots,2N)_{n+1,t+1} \\ &= (d_{00}, d_{-10}, d_{11}, c, 1,2,\ldots,2N)_{nt}\ (1,2,\ldots,2N)_{nt} \\ &\qquad - (d_{00}, c, 1,2,\ldots,2N)_{nt}\ (d_{-10}, d_{11}, 1,2,\ldots,2N)_{nt} \\ &\qquad - (d_{11}, c, 1,2,\ldots,2N)_{nt}\ (d_{00}, d_{-10}, 1,2,\ldots,2N)_{nt} \\ &\qquad + (d_{-10}, c, 1,2,\ldots,2N)_{nt}\ (d_{00}, d_{11}, 1,2,\ldots,2N)_{nt} \\ &= 0, \end{aligned}$$

where the Pfaffian elements are defined by

$$(d_{n't'}, i)_{nt} = \alpha_i p_i^{n+n'} \left(\frac{1 - a\,1/(p_i)}{1 - ap_i}\right)^{t+t'}, \quad (d_{n't'}, c)_{nt} = 0, \quad (c, i)_{nt} = c_i,$$

$$(d_{00}, d_{-10})_{nt} = 1, \quad (d_{00}, d_{11})_{nt} = 1, \quad (d_{-10}, d_{11})_{nt} = \frac{1}{a}.$$

On the other hand, the space part of bilinear equations (2.2) is not a Plücker relation but a direct consequence of the following identity for the Pfaffian,

$$(1,2,\dots,2N)_{n+1,t}\,(1,2,\dots,2N)_{n-1,t}-(1,2,\dots,2N)_{nt}\,(1,2,\dots,2N)_{nt}$$
$$=\sum_{1\le i<j\le 2N}(-1)^{i+j-1}\Big(\Big(p_i+\frac{1}{p_i}\Big)-\Big(p_j+\frac{1}{p_j}\Big)\Big)$$
$$\times c_{ij}(d_{00},1,2,\dots,\widehat{i},\dots,2N)_{nt}(d_{00},1,2,\dots,\widehat{j},\dots,2N)_{nt}.$$

We note that the continuous counterpart of this identity is

$$D_x^2(1,2,\dots,2N)\cdot(1,2,\dots,2N)$$
$$=2\sum_{1\le i<j\le 2N}(-1)^{i+j-1}(p_i^2-p_j^2)c_{ij}(d_0,1,2,\dots,\widehat{i},\dots,2N)(d_0,1,2,\dots,\widehat{j},\dots,2N),$$

where D is the bilinear differential operator and

$$(i,j)=c_{ij}+\frac{p_i-p_j}{p_i+p_j}e^{\xi_i+\xi_j},\ i<j,\quad (d_0,i)=e^{\xi_i},\quad \xi_i=p_ix+\xi_{i0}.$$

3. Example of Soliton Solution

Here we explicitly give the 2-soliton solution in 2-channel system with changing the notation appropriately. The equation is

$$\frac{1}{a}(v^1_{n,t+1}-v^1_{nt})=\Gamma_{nt}(1+a^{11}v^1_{nt}\overline{v}^1_{nt}+a^{12}v^1_{nt}\overline{v}^2_{nt}+a^{21}v^2_{nt}\overline{v}^1_{nt}+a^{22}v^2_{nt}\overline{v}^2_{nt})$$
$$(v^1_{n+1,t+1}-v^1_{n-1,t}),$$
$$\frac{1}{a}(v^2_{n,t+1}-v^2_{nt})=\Gamma_{nt}(1+a^{11}v^1_{nt}\overline{v}^1_{nt}+a^{12}v^1_{nt}\overline{v}^2_{nt}+a^{21}v^2_{nt}\overline{v}^1_{nt}+a^{22}v^2_{nt}\overline{v}^2_{nt})$$
$$(v^2_{n+1,t+1}-v^2_{n-1,t}),$$
$$\frac{1}{a}(\overline{v}^1_{n,t+1}-\overline{v}^1_{nt})=\Gamma_{nt}(1+a^{11}v^1_{nt}\overline{v}^1_{nt}+a^{12}v^1_{nt}\overline{v}^2_{nt}+a^{21}v^2_{nt}\overline{v}^1_{nt}+a^{22}v^2_{nt}\overline{v}^2_{nt})$$
$$(\overline{v}^1_{n+1,t+1}-\overline{v}^1_{n-1,t}),$$
$$\frac{1}{a}(\overline{v}^2_{n,t+1}-\overline{v}^2_{nt})=\Gamma_{nt}(1+a^{11}v^1_{nt}\overline{v}^1_{nt}+a^{12}v^1_{nt}\overline{v}^2_{nt}+a^{21}v^2_{nt}\overline{v}^1_{nt}+a^{22}v^2_{nt}\overline{v}^2_{nt})$$
$$(\overline{v}^2_{n+1,t+1}-\overline{v}^2_{n-1,t}),$$
$$\frac{\Gamma_{nt}}{\Gamma_{n-1,t}}$$
$$=\frac{1+a^{11}v^1_{n,t+1}\overline{v}^1_{n,t+1}+a^{12}v^1_{n,t+1}\overline{v}^2_{n,t+1}+a^{21}v^2_{n,t+1}\overline{v}^1_{n,t+1}+a^{22}v^2_{n,t+1}\overline{v}^2_{n,t+1}}{1+a^{11}v^1_{nt}\overline{v}^1_{nt}+a^{12}v^1_{nt}\overline{v}^2_{nt}+a^{21}v^2_{nt}\overline{v}^1_{nt}+a^{22}v^2_{nt}\overline{v}^2_{nt}}.$$

Under the variable transformation,

$$v^1_{nt}=\frac{g^1_{nt}}{f_{nt}},\quad v^2_{nt}=\frac{g^2_{nt}}{f_{nt}},\quad \overline{v}^1_{nt}=\frac{\overline{g}^1_{nt}}{f_{nt}},\quad \overline{v}^2_{nt}=\frac{\overline{g}^2_{nt}}{f_{nt}},\quad \Gamma_{nt}=\frac{f_{n+1,t+1}f_{nt}}{f_{n,t+1}f_{n+1,t}},$$

the Pfaffian solution is give as

$$f_{nt}=\mathrm{Pf}(P),$$

$$g_{nt}^{\mu} = \mathrm{Pf}\begin{pmatrix} 0 & 0 & \varphi_1^{nt} & \varphi_2^{nt} & \overline{\varphi}_2^{nt} & \overline{\varphi}_1^{nt} \\ 0 & 0 & c_1^{\mu} & c_2^{\mu} & 0 & 0 \\ -\varphi_1^{nt} & -c_1^{\mu} & & & & \\ -\varphi_2^{nt} & -c_2^{\mu} & & & & \\ -\overline{\varphi}_2^{nt} & 0 & & \multicolumn{2}{c}{P} & \\ -\overline{\varphi}_1^{nt} & 0 & & & & \end{pmatrix},$$

$$\overline{g}_{nt}^{\nu} = \mathrm{Pf}\begin{pmatrix} 0 & 0 & \varphi_1^{nt} & \varphi_2^{nt} & \overline{\varphi}_2^{nt} & \overline{\varphi}_1^{nt} \\ 0 & 0 & 0 & 0 & \overline{c}_2^{\nu} & \overline{c}_1^{\nu} \\ -\varphi_1^{nt} & 0 & & & & \\ -\varphi_2^{nt} & 0 & & & & \\ -\overline{\varphi}_2^{nt} & -\overline{c}_2^{\nu} & & \multicolumn{2}{c}{P} & \\ -\overline{\varphi}_1^{nt} & -\overline{c}_1^{\nu} & & & & \end{pmatrix},$$

where P is the following 4×4 antisymmetric matrix,

$$\begin{pmatrix} 0 & \frac{p_1-p_2}{p_1p_2-1}\varphi_1^{nt}\varphi_2^{nt} & c_{1\overline{2}}+\frac{p_1-\overline{p}_2}{p_1\overline{p}_2-1}\varphi_1^{nt}\overline{\varphi}_2^{nt} & c_{1\overline{1}}+\frac{p_1-\overline{p}_1}{p_1\overline{p}_1-1}\varphi_1^{nt}\overline{\varphi}_1^{nt} \\ -\frac{p_1-p_2}{p_1p_2-1}\varphi_1^{nt}\varphi_2^{nt} & 0 & c_{2\overline{2}}+\frac{p_2-\overline{p}_2}{p_2\overline{p}_2-1}\varphi_2^{nt}\overline{\varphi}_2^{nt} & c_{2\overline{1}}+\frac{p_2-\overline{p}_1}{p_2\overline{p}_1-1}\varphi_2^{nt}\overline{\varphi}_1^{nt} \\ -c_{1\overline{2}}-\frac{p_1-\overline{p}_2}{p_1\overline{p}_2-1}\varphi_1^{nt}\overline{\varphi}_2^{nt} & -c_{2\overline{2}}-\frac{p_2-\overline{p}_2}{p_2\overline{p}_2-1}\varphi_2^{nt}\overline{\varphi}_2^{nt} & 0 & \frac{\overline{p}_2-\overline{p}_1}{\overline{p}_2\overline{p}_1-1}\overline{\varphi}_2^{nt}\overline{\varphi}_1^{nt} \\ -c_{1\overline{1}}-\frac{p_1-\overline{p}_1}{p_1\overline{p}_1-1}\varphi_1^{nt}\overline{\varphi}_1^{nt} & -c_{2\overline{1}}-\frac{p_2-\overline{p}_1}{p_2\overline{p}_1-1}\varphi_2^{nt}\overline{\varphi}_1^{nt} & -\frac{\overline{p}_2-\overline{p}_1}{\overline{p}_2\overline{p}_1-1}\overline{\varphi}_2^{nt}\overline{\varphi}_1^{nt} & 0 \end{pmatrix},$$

where

$$\varphi_i^{nt} = \alpha_i p_i^n \left(\frac{1-a\,1/(p_i)}{1-ap_i}\right)^t, \quad \overline{\varphi}_j^{nt} = \overline{\alpha}_j \overline{p}_j^n \left(\frac{1-a\,1/(\overline{p}_j)}{1-a\overline{p}_j}\right)^t,$$

$$c_{i\overline{j}} = -\frac{1}{(p_i+1/(p_i))-(\overline{p}_j+1/(\overline{p}_j))}(a^{11}c_i^1\overline{c}_j^1 + a^{12}c_i^1\overline{c}_j^2 + a^{21}c_i^2\overline{c}_j^1 + a^{22}c_i^2\overline{c}_j^2).$$

4. Lax Pair

Since we obtained the Pfaffian solution, we can construct the Lax pair using the standard technique [**3**]. Let

$$\begin{aligned} \Psi_{nt} &= (\psi_{nt}^1, \dots, \psi_{nt}^M)^T, & \overline{\Psi}_{nt} &= (\overline{\psi}_{nt}^1, \dots, \overline{\psi}_{nt}^M)^T, \\ V_{nt} &= (v_{nt}^1, \dots, v_{nt}^M)^T, & \overline{V}_{nt} &= (\overline{v}_{nt}^1, \dots, \overline{v}_{nt}^M)^T, \\ A &= (a^{\mu\nu})_{1\le\mu,\nu\le M}, & u_{nt} &= 1 + V_{nt}^T A \overline{V}_{nt}, \end{aligned}$$

where ^{T} denotes transposition. Then

$$\begin{aligned} \Psi_{n+1,t} - \Psi_{nt} &= V_{n+1,t}\theta_{nt} - V_{nt}\theta_{n+1,t}, \\ \overline{\Psi}_{n+1,t} - \overline{\Psi}_{nt} &= \overline{V}_{n+1,t}\theta_{nt} - \overline{V}_{nt}\theta_{n+1,t}, \\ u_{nt}(\theta_{n+1,t} + \theta_{n-1,t}) &= \lambda\theta_{nt} + \Psi_{nt}^T A \overline{V}_{nt} + V_{nt}^T A \overline{\Psi}_{nt}, \end{aligned}$$

and

$$\begin{aligned} \Psi_{n+1,t+1} - \Psi_{nt} &= V_{n+1,t+1}\theta_{nt} - V_{nt}\theta_{n+1,t+1}, \\ \overline{\Psi}_{n+1,t+1} - \overline{\Psi}_{nt} &= \overline{V}_{n+1,t+1}\theta_{nt} - \overline{V}_{nt}\theta_{n+1,t+1}, \\ \frac{1}{a}(\theta_{n,t+1} - \theta_{nt}) &= \Gamma_{nt} u_{nt}(\theta_{n+1,t+1} - \theta_{n-1,t}), \end{aligned}$$

give the Lax pair of (1.1). Here Ψ, $\overline{\Psi}$ and θ are the eigenfunctions and λ is the eigenvalue. The part of Ψ and $\overline{\Psi}$ is the usual Lax pair of discrete BKP. We should note that Tsuchida, Ujino and Wadati derived another Lax pair from the discrete AKNS formulation by using a redundant form [**9, 10**].

5. Concluding Remarks

We have shown that the most general soliton solutions of the discrete cmKdV equation (1.1) have the simple expression in terms of Gram Pfaffian. As is seen in the Pfaffian solution, we can regard that the system is derived by applying an extra constraint to the usual BKP hierarchy. In the Lax pair, the part of θ corresponds to the constraint. This suggests that imposing some extra conditions may help us to create new integrable equations especially in the case of coupled systems.

Appendix

We give the determinant solution for the single component case. The discrete modified KdV equation

$$\frac{1}{a}(v_{n,t+1}-v_{nt})=\Gamma_{nt}(1+v_{nt}\overline{v}_{nt})(v_{n+1,t+1}-v_{n-1,t}),$$
$$\frac{1}{a}(\overline{v}_{n,t+1}-\overline{v}_{nt})=\Gamma_{nt}(1+v_{nt}\overline{v}_{nt})(\overline{v}_{n+1,t+1}-\overline{v}_{n-1,t}),$$
$$\frac{\Gamma_{nt}}{\Gamma_{n-1,t}}=\frac{1+v_{n,t+1}\overline{v}_{n,t+1}}{1+v_{nt}\overline{v}_{nt}},$$

is transformed into the bilinear form

$$\frac{1}{a}(g_{n,t+1}f_{nt}-g_{nt}f_{n,t+1})=g_{n+1,t+1}f_{n-1,t}-g_{n-1,t}f_{n+1,t+1},$$
$$\frac{1}{a}(\overline{g}_{n,t+1}f_{nt}-\overline{g}_{nt}f_{n,t+1})=\overline{g}_{n+1,t+1}f_{n-1,t}-\overline{g}_{n-1,t}f_{n+1,t+1},$$
$$f_{n+1,t}f_{n-1,t}-f_{nt}f_{nt}=g_{nt}\overline{g}_{nt},$$

through the variable transformation

$$v_{nt}=\frac{g_{nt}}{f_{nt}},\quad \overline{v}_{nt}=\frac{\overline{g}_{nt}}{f_{nt}},\quad \Gamma_{nt}=\frac{f_{n+1,t+1}f_{nt}}{f_{n,t+1}f_{n+1,t}}.$$

The determinant solution is given as

$$f_{nt}=\tau_N^n(t,-t),\quad g_{nt}=\tau_{N+1}^n(t,-t),\quad \overline{g}_{nt}=\tau_{N-1}^n(t,-t),$$

$$\tau_N^n(k,l)=\begin{vmatrix}\varphi_{n-N+1}(k,l) & \cdots & \cdots & \varphi_{n-1}(k,l) & \varphi_n(k,l)\\ \iddots & & \iddots & \varphi_n(k,l) & \varphi_{n+1}(k,l)\\ \cdots & \iddots & \iddots & \cdots & \iddots\\ \varphi_{n-1}(k,l) & \varphi_n(k,l) & \iddots & & \cdots\\ \varphi_n(k,l) & \varphi_{n+1}(k,l) & \cdots & \cdots & \varphi_{n+N-1}(k,l)\end{vmatrix},$$

where φ is an arbitrary function satisfying

$$\frac{1}{a}\big(\varphi_n(k,l)-\varphi_n(k-1,l)\big)=\varphi_{n+1}(k,l),$$
$$\frac{1}{a}\big(\varphi_n(k,l)-\varphi_n(k,l-1)\big)=\varphi_{n-1}(k,l).$$

References

1. M. J. Ablowitz and F. J. Ladik, *A nonlinear difference scheme and inverse scattering*, Stud. Appl. Math. **55** (1976), 213–229; R. Hirota, unpublished work.
2. R. Hirota, *Discretization of coupled modified* KdV *equations*, Publ. Res. Inst. Math. Sci. **1020** (1997), 143–159.
3. ———, *Direct method in soliton theory*, Iwanami, Tokyo, 1992.

4. M. Iwao and R. Hirota, *Soliton solutions of a coupled modified* KdV *equations*, J. Phys. Soc. Japan **66** (1997), 577–588; private communication; R. Hirota, *Molecule solutions of coupled modified* KdV *equations*, J. Phys. Soc. Japan **66** (1997), 2530–2532.
5. S. V. Manakov, *On the theory of two-dimensional stationary self-focusing of electromagnetic waves*, Soviet Physics JETP **38** (1974), 248–253; V. E. Zakharov and A. B. Shabat, *A plan for integrating the nonlinear equations of mathematical physics by the method of the inverse scattering problem.* I, Functional Anal. Appl. **8** (1974), 43–53; N. Yajima and M. Oikawa, *A class of exactly solvable nonlinear evolution equations*, Progr. Theoret. Phys. **54** (1975), 1576–1577; C. Athorne and A. Fordy, *Generalised* KdV *and* MKdV *equations associated with symmetric spaces*, J. Phys. A **20** (1987) 1377–1386; M. J. Ablowitz, Y. Ohta, and A. D. Trubatch, *On discretizations of the vector nonlinear Schrödinger equation*, Phys. Lett. A. (submitted).
6. Y. Ohta, *Pfaffian solution for coupled discrete nonlinear Schrödinger equation*, Chaos, Solitons and Fractals. (to appear)
7. M. Sato, *Soliton equations as dynamical systems on a infinite dimensional Grassmann manifolds*, Publ. Res. Inst. Math. Sci. **439** (1981), 30–46; M. Jimbo and T. Miwa, *Solitons and infinite dimensional Lie algebras*, Publ. Res. Inst. Math. Sci. **19** (1983), 943–1001; J. Satsuma, *A Wronskian representation of* N*-soliton solutions of nonlinear evolution equations*, J. Phys. Soc. Japan **46** (1979), 359–360; N. C. Freeman and J. J. C. Nimmo, *Soliton solutions of the Korteweg-de Vries and Kadomtsev-Petviashvili equations*: *The Wronskian technique*, Phys. Lett. A **95** (1983), 1–3; A. Nakamura, *A bilinear* N*-soliton formula for the* KP *equation*, J. Phys. Soc. Japan **58** (1989), 412–422; R. Hirota, *Soliton solutions to the* BKP *equations.* I. *The Pfaffian technique*, J. Phys. Soc. Japan **58** (1989), 2285–2296; R. Hirota and Y. Ohta, *Hierarchies of coupled soliton equations.* I, J. Phys. Soc. Japan **60** (1991), 798–809.
8. Yu. B. Suris, *A discrete-time Garnier system*, Phys. Lett. A **189** (1994), 281–289; *A family of integrable symplectic standard-like maps related to symmetric spaces*, Phys. Lett. A **192** (1994), 9–16.
9. T. Tsuchida, *Inverse scattering method for coupled modified* KdV *equations*, talk at the Kyushu Symposium, 1998.
10. T. Tsuchida, H. Ujino, and M. Wadati, *Integrable semi-discretization of the coupled modified* KdV *equations*, J. Math. Phys. **39** (1998), 4785–4813; T. Tsuchida and M. Wadati, *The coupled modified Korteweg-de Vries equations*, J. Phys. Soc. Japan **67** (1998), 1175–1187.

DEPARTMENT OF APPLIED MATHEMATICS, FACULTY OF ENGINEERING, HIROSHIMA UNIVERSITY, 1-4-1 KAGAMIYAMA, HIGASHI-HIROSHIMA 739–8527, JAPAN.
E-mail address: `ohta@kurims.kyoto-u.ac.jp`

Centre de Recherches Mathématiques
CRM Proceedings and Lecture Notes
Volume **25**, 2000

Difference Isomonodromy Problems and Discrete Painlevé Equations

Vassilios G. Papageorgiou

ABSTRACT. Discrete time isomonodromy problems of difference equations are presented. The relation of the resulting nonautonomous nonlinear mappings to discrete Painlevé equations is discussed.

Consider the best known chain of dressing transformations, i.e. that for the Schrödinger equation [**8,12,14**], and Refs. therein), which reads:

$$\frac{d}{dx}\big(f_{n+1}(x)+f_n(x)\big)=f_{n+1}^2(x)-f_n^2(x)+a_n, \tag{1}$$

where a_n does not depend on the continuous variable x. If one starts with the dressing chain (1) and imposes periodicity conditions in the n-direction, one can recover Painlevé equations [**14,1**]. Similarly, from discrete analogs of dressing chains for the difference Schrödinger operator, with proper periodicity conditions, one obtains discrete Painlevé equations (d-P's). Indeed starting with the linear problem (related to the Quotient-Difference (QD) algorithm, see e.g. [**10**]):

$$\Phi(n+1)=U(n)\Phi(n),\quad L(n)\Phi(n+1)=\lambda\Phi(n), \tag{2}$$

where U, L are operators which depend explicitly on n (while the assumption leading to the standard QD algorithm is that the eigenvalue λ is independent of n), allowing an additive increment of λ, i.e. $\lambda_{n+1}-\lambda_n=\alpha_n$, the compatibility condition of (2) reads:

$$L(n+1)U(n+1)=U(n)L(n)+\alpha(n)I. \tag{3}$$

Next introducing the ansatz:

$$L(n)=I+\Sigma^*E(n),\quad U=\Sigma+Q(n), \tag{4}$$

where I is the identity and $E(n)$, $Q(n)$ are infinite, diagonal, matrices with elements $e_{n,\mu}\delta_{\mu\nu}$, $q_{n,\mu}\delta_{\mu\nu}$ respectively, the operators Σ, Σ^* are represented by infinite

2000 *Mathematics Subject Classification.* Primary 33E30, 81U40; Secondary 58F19, 39A10.

The author acknowledges support by the C. Caratheodory Research Program of the University of Patras under project No. 1938.

This is the final form of the paper.

matrices $\Sigma_{\mu\nu} = \delta_{\mu(\nu-1)}$, $\Sigma^*_{\mu\nu} = \delta_{(\mu-1)\nu}$, one gets from the compatibility condition (3) the equations of the QD algorithm with shifts:

$$q_{n+1,m} + e_{n+1,m-1} = q_{n,m} + e_{n,m} + \alpha_n, \quad e_{n+1,m}q_{n+1,m} = e_{n,m}q_{n,m+1}. \tag{5}$$

It is interesting to notice that the system (5) passes the singularity confinement criterion [**6**] due to the fact that α is independent of m. Equations (5) form a discrete analogue of the nonlinear dressing chain (1). Indeed the continuum limit with respect to the m direction can be performed by putting $e = 1 + \varepsilon f$ and $q = 1 - \varepsilon f + \varepsilon^2 h$, $\alpha = \varepsilon^2 a$ and taking the limit $\varepsilon \to 0$. Then equations (7) yield the following pair of equations

$$f'_{n+1} = h_{n+1} - h_n - a_n, \quad h_{n+1} - h_n = f^2_{n+1} - f^2_n - f'_n, \tag{6}$$

from which one recovers the nonlinear chain (1) by eliminating h.

We present also the Zakharov–Shabat (nonisospectral) zero-curvature representation of (5) in terms of 2×2 matrices (of difference-difference type cf. [**7**]) that one is going to use in order to derive the isomonodromy problems. We reformulate (2) (with the ansatz (4)) to obtain the linear problems

$$\Psi_{n+1,m}(\lambda + \alpha_n) = M_{n,m}(\lambda)\Psi_{n,m}(\lambda), \quad \Psi_{n,m+1}(\lambda) = N_{n,m}(\lambda)\Psi_{n,m}(\lambda), \tag{7}$$

where

$$M_{n,m}(\lambda) = \begin{bmatrix} q_{n+1,m-1} - e_{n,m-1} & \lambda \\ -e_{n,m-1} & \lambda \end{bmatrix}, \quad N_{n,m}(\lambda) = \begin{bmatrix} -e_{n,m-1} & \lambda \\ -e_{n,m-1} & \lambda - q_{n,m} \end{bmatrix}, \tag{8}$$

(with spectral parameter λ), the compatibility of which (that encodes (5)) reads:

$$M_{n,m+1}(\lambda)N_{n,m}(\lambda) = N_{n+1,m}(\lambda + \alpha_n)M_{n,m}(\lambda). \tag{9}$$

The periodic reduction in the discrete case is performed by imposing periodicity p in the n-direction allowing a shift s in the m-direction i.e.:

$$q_{n+p,m-s} = q_{n,m}, \quad e_{n+p,m-s} = e_{n,m}, \alpha_{n+p} = \alpha_n, \tag{10}$$

as it is implemented in partial difference equations (see e.g. [**2**]). In the sequel we will keep in mind that the obtaining mappings are indeed nonautonomous provided $\delta \equiv \alpha_0 + \alpha_2 + \cdots + \alpha_{n+p-1}$ is not equal to zero.

For the case $p = 1$, $s = 2$, the resulting equations read:

$$q_{m+2} + e_{m+1} = q_m + e_m + \alpha, \quad e_{m+2}q_{m+2} = e_m q_{m+1}, \tag{11}$$

(where the first index disappears as fixed). Each one of equations (11) integrates to obtain $q_m + q_{m+1} + e_{m+1} = c + m\alpha$, $q_m = c_1/e_m e_{m-1}$ (c, c_1 a constants) and elimination of the q's, results in a single equation which in terms of $x = 1/e$ reads:

$$x_{m+1} + x_{m-1} = \frac{1}{c_1}\left(\frac{c + m\alpha}{x_m} - \frac{1}{x_m^2}\right). \tag{12}$$

Equation (12) is a form of d-$\mathrm{P_I}$ (see [**3,11**]).

For the case $p = 2$, $s = 0$, the equations (5) read:

$$q_{1,m} + e_{1,m-1} = q_{0,m} + e_{0,m} + \alpha_0, \quad q_{0,m} + e_{0,m-1} = q_{1,m} + e_{1,m} + \alpha_1, \tag{13a}$$

$$e_{1,m}q_{1,m} = e_{0,m}q_{0,m+1}, \quad e_{0,m}q_{0,m} = e_{1,m}q_{1,m+1}. \tag{13b}$$

From equations (13b) of "quotient type" by taking their product and integrating we obtain $q_{1,m} = \kappa/q_{0,m}$ and $e_{1,m} = q_{0,m}q_{0,m+1}e_{0,m}/\kappa$ by substituting back into the first. Also the sum of equations (13a) of "difference type" integrates once to $e_{0,m} + e_{1,m} = z_m$ and substitution to the first of them yields also $z_m + \alpha_1 = e_{0,m} + e_{0,m-1} + q_{0,m} - q_{1,m}$, where $z \equiv c - m(\alpha_0 + \alpha_1)$, c the constant of integration. Finally substitution of the $q_{1,m}$, $e_{1,m}$ into the previous two equations results in a 2-dimensional nonautonomous mapping $(e_{0,m-1}, q_{0,m}) \mapsto (e_{0,m}, q_{0,m+1})$ that can be written in terms of the variable $x \equiv q_0$ as the following equation:

$$\frac{\kappa z_m}{\kappa + x_m x_{m-1}} + \frac{\kappa z_{m+1}}{\kappa + x_m x_{m+1}} = \frac{c}{x_m} - x_m + z_m + \alpha_1. \tag{14}$$

This is precisely the equation known as alternate d-$\mathrm{P_{II}}$ [**3**,**9**], and it is a discrete form of the Painlevé II equation.

For the case of the periodic conditions (10) with $p = 3, s = 1$, we write the corresponding equations (5)

$$\begin{aligned}
&(15a) \qquad q_{1,m} + e_{1,m-1} = q_{0,m} + e_{0,m} + \alpha_0, \quad e_{1,m}q_{1,m} = e_{0,m}q_{0,m+1},\\
&(15b) \qquad q_{2,m} + e_{2,m-1} = q_{1,m} + e_{1,m} + \alpha_1, \quad e_{2,m}q_{2,m} = e_{1,m}q_{1,m+1},\\
&(15c) \qquad q_{0,m+1} + e_{0,m} = q_{2,m} + e_{2,m} + \alpha_2, \quad e_{0,m+1}q_{0,m+1} = e_{2,m}q_{2,m+1}.
\end{aligned}$$

We proceed as before by integrating once the sum of equations (15) of "difference type" and the product of the three equations (15) of "quotient type". Then equations (15) can be replaced by the following equations

$$\begin{aligned}
&(16a) \qquad z_m = e_{1,m-1} + e_{2,m-1} - q_{0,m}, \qquad e_{0,m} = \kappa q_{1,m}q_{2,m},\\
&(16b) \qquad z_m - \alpha_0 = e_{2,m-1} + e_{0,m} - q_{1,m}, \quad e_{1,m} = \kappa q_{2,m}q_{0,m+1},\\
&(16c) \qquad z_{m+1} - \alpha_2 = e_{0,m} + e_{1,m} - q_{2,m}, \quad e_{2,m} = \kappa q_{0,m+1}q_{1,m+1},
\end{aligned}$$

where $z_m \equiv c - m(\alpha_0 + \alpha_1 + \alpha_2)$ and c, κ costants. Finally eliminating the e's between (16), we obtain a nonautonomous mapping $(q_{1,m-1}, q_{2,m-1}) \mapsto (q_{1,m}, q_{2,m})$

$$\begin{aligned}
&(17a) \qquad \kappa(q_{0,m} + q_{1,m-1}) = 1 + (z_m + \alpha_2)/q_{2,m-1},\\
&(17b) \qquad \kappa(q_{2,m-1} + q_{1,m}) = 1 + z_m/q_{0,m},\\
&(17c) \qquad \kappa(q_{2,m} + q_{0,m}) = 1 + (z_m - \alpha_0)/q_{1,m},
\end{aligned}$$

which is a form of d-$\mathrm{P_{IV}}$ (see [**5**]).

We consider now again (3) but, instead of (4), the operators L, U are of the form

$$L(n) = \Sigma + E(n), \quad U = \Sigma + Q(n). \tag{18}$$

The compatibility condition (3) translates into the following equations:

$$q_{n+1,m} + e_{n+1,m-1} = e_{n,m} + q_{n,m-1}, \quad e_{n+1,m}q_{n+1,m} = e_{n,m}q_{n,m} + \alpha_n. \tag{19}$$

The $p = 1$, $s = 2$ case of periodic conditions gives the equations

$$q_{m+2} + e_{m+1} = e_m + q_{m-1}, \quad e_{m+2}q_{m+2} = e_m q_m + \alpha. \tag{20}$$

Equations (20) integrate as $e_m+q_{m-1}+q_m+q_{m+1}=c$, $e_mq_m=c_0+c_1(-1)^m+m\alpha/2$ with c, c_0, c_1 constants, and in terms of the variable q the final equations form a nonautonomous mapping $(q_{2m-1},q_{2m})\mapsto(q_{2m+1},q_{2m+2})$:

$$q_{2m+1}+q_{2m} = c-q_{2m-1}-(c_0+c_1+m\alpha)/q_{2m}, \tag{21a}$$

$$q_{2m+2}+q_{2m+1} = c-q_{2m}-(c_0-c_1+\alpha/2+m\alpha)/q_{2m+1}, \tag{21b}$$

which is an odd-even coupling of two d-P$_{\text{I}}$ equations (see [**3**,**11**]).

Another instance of an odd-even coupling of two discrete Painlevé equations, this time with d-P$_{\text{II}}$, occures for the case of $p=2$, $s=2$, periodic conditions for equations (22). Starting from the equations

$$q_{1,m}+e_{1,m-1}=e_{0,m}+e_{0,m-1}, \qquad q_{0,m+1}+e_{0,m}=q_{1,m-2}+e_{1,m-1}, \tag{22a}$$

$$e_{1,m}q_{1,m}=e_{0,m}q_{0,m}+\alpha_1, \qquad e_{0,m+2}q_{0,m+2}=e_{1,m}q_{1,m}+\alpha_2, \tag{22b}$$

the sum of equations (22a) as well as the sum of equations (22b) integrate to give respectively:

$$q_{0,m}+q_{1,m-1}=c_1+(-1)^m c_2, \quad e_{0,m}q_{0,m}=c_3+c_4(-1)^m+m(\alpha_1+\alpha_2)/2. \tag{23}$$

Then substituting into the odd and even versions of the first equation (22a), taking into account the first equation (22b) also and using the notation $x_m\equiv q_{1,2m}-c$, $y_m\equiv q_{1,2m+1}-d$, $c\equiv(c_1+c_2)/2$, $d\equiv(c_1-c_2)/2$, $\delta=\alpha_1+\alpha_2$, the nonautonomous mapping $(x_{m-1},y_{m-1})\mapsto(x_m,y_m)$

$$x_m+x_{m-1} = \frac{(2c_3+2m\delta+(\alpha_1-\alpha_2)/2)y_{m-1}+c(2c_4-(\alpha_1-\alpha_2)/2)}{c^2-y_{m-1}^2}, \tag{24a}$$

$$y_m+y_{m-1} = \frac{(2c_3+2(m+1/2)\delta+(\alpha_1-\alpha_2)/2)x_m-d(2c_4+(\alpha_1-\alpha_2)/2)}{d^2-x_m^2}, \tag{24b}$$

is obtained which is a d-P$_{\text{III}}$ as an odd-even coupling of the d-P$_{\text{II}}$ equation (see [**4**]).

The case of higher order operators L, U give rise also to discrete Painlevé equations. Indeed, using (3) with $L=I+\Sigma^*E$, $U=\Sigma^2+\Sigma Q+R$, one obtains the following equations

$$q_{n+1,m}+e_{n+1,m-2}=e_{n,m}+q_{n,m}, \tag{25a}$$

$$r_{n+1,m}+e_{n+1,m_1}q_{n+1,m}=q_{n,m+1}e_{n,m}+r_{n,m}+\alpha_n, \tag{25b}$$

$$e_{n+1,m}r_{n+1,m}=r_{n,m+1}e_{n,m}, \tag{25c}$$

which in the case $p=1$, $s=2$, simplify as

$$q_{m+2}+e_m=e_m+q_m, \quad r_{m+2}+e_{m+1}q_{m+2}=q_{m+1}e_m+r_m+\alpha, \tag{26a}$$

$$e_{m+2}r_{m+2}=r_{m+1}e_m. \tag{26b}$$

Equations (26) integrate to give

$$q_{2k}=q_0, \quad q_{2k+1}=q_1, \quad r_m=c/(e_me_{m-1}), \tag{27a}$$

$$r_{m+1}+r_m+e_mq_{m+1}=c_0+m\alpha, \tag{27b}$$

and in terms of the variable $x = 1/e$ the following equations

$$(28)\qquad \begin{aligned} x_{2k+1} + x_{2k-1} &= \frac{c_0 + 2k\alpha}{cx_{2k}} - \frac{q_1}{cx_{2k}^2}, \\ x_{2k+2} + x_{2k} &= \frac{c_0 + (2k+1)\alpha}{cx_{2k+1}} - \frac{q_0}{cx_{2k+1}^2}, \end{aligned}$$

which are an odd-even coupling of equation (12). Other forms of operators with associated periodic conditions give rise to 3-dimensional nonautonomous mappings the relation of which with discrete Painlevé equations is under investigation.

The derivation of difference isomonodromy problems is most easily illustrated for the case of equation (14): from (7), because of the periodicity with zero shift s we obtain

$$(29)\qquad \begin{aligned} \Psi_{0,m}(\lambda + \delta) &= M_{1,m}(\lambda + \alpha_0) M_{0,m}(\lambda) \Psi_{0,m}(\lambda), \\ \Psi_{0,m+1}(\lambda) &= N_{0,m}(\lambda) \Psi_{0,m}(\lambda). \end{aligned}$$

In case s is nonzero then some M-type factors have to be shifted w.r.t. m and N-type factors to be inserted in the product of the first equation (29) following a "staircase " construction on the (n, m)-lattice (cf. [**2**]).

All nonautonomous mappings related to discrete Painlevé equations that are presented here have the common feature that they are derived in a unified way from linear problems of shifted QD type (cf. (3)) with operators L, U graded with powers of Σ supplemented by periodic conditions. It is this feature that gives them a symmetric form and some common patterns for the derivation of the 2-dimensional final reduction. It has to be noticed that the same technique was used in [**5,13**]. In [**13**], where emphasis was given on the factorization of the discrete Schrödinger operator and the nature of the spectra, equations (12), (14) where derived (and linear ones related to orthogonal polynomials) but system (15) was not reduced to the final form of d-$\mathrm{P_{IV}}$. The latter was done in [**5**] (as well as the identification of (14) as alternate d-$\mathrm{P_{II}}$) but the derivation was slightly "disturbed" by the (a posteriori unnecessary) global integration of the "quotient type" equations (5) (since at the end the variables q, e are used). Nevertheless the integration variables seem to be closer related to structured determinants (Hankel type in the $\alpha_n \equiv 0$ case) so it would be interesting to see if one can use the same approach adapted to the shifted case in order to obtain similar representations of solutions for the discrete Painlevé equations presented above.

References

1. V. E. Adler, *Nonlinear chains and Painlevé equations*, Physica D **73** (1994), 335–351.
2. H. W. Capel, F. W. Nijhoff, and V. G. Papageorgiou, *Complete integrability of Lagrangian lattices and mappings of* KdV *type*, Phys. Lett. A **155** (1991), 337–387.
3. A. S. Fokas, B. Grammaticos, and A. Ramani, *From continuous to discrete Painlevé equations*, J. Math. Anal. Appl. **180** (1993), 342–360.
4. B. Grammaticos, F. W. Nijhoff, V. Papageorgiou, A. Ramani, and J. Satsuma, *Linearization and solutions of the discrete Painlevé* III *equation*, Phys. Lett. A **185** (1994), 446–452.
5. B. Grammaticos, V. Papageorgiou, and A. Ramani, *Discrete dressing transformations and Painlevé equations*, Phys. Lett. A **235** (1997), 475–479.
6. B. Grammaticos, A. Ramani, and V. G. Papageorgiou, *Do integrable mappings have the Painlevé property?*, Phys. Rev. Lett. **67** (1991), 1825–1828.

7. D. Levi, O. Ragnisco, and M. Bruschi, *Extension of the Zakharov–Shabat generalized inverse method to solve differential-difference and difference-difference equations*, Nuovo Cimento A **14** (1980), 56-66.
8. V. B. Matveev and M. A. Salle, *Differential-difference evolution equations.* II. *Darboux transformations for the Toda lattice*, Lett. Math. Phys **3** (1979), 425–429.
9. F. Nijhoff, J. Satsuma, K. Kajiwara, B. Grammaticos, and A. Ramani, *A study of the alternate discrete Painlevé* II *equation*, Inverse Problems **12** (1996), 697–716.
10. V. Papageorgiou, B. Grammaticos, and A. Ramani, *Orthogonal polynomial approach to discrete Lax pairs for initial-boundary value problems of the* QD *algorithm*, Lett. Math. Phys. **34** (1995), 91–101.
11. A. Ramani, B. Grammaticos, *Discrete Painlevé equations: Coalescences, limits and degeneracies*, Physica A **228** (1996), 160–171.
12. A. B. Shabat and R. I. Yamilov, *Symmetries of nonlinear lattices*, Leningrad Math. J. **2** (1991), 377–381.
13. V. Spiridonov, L. Vinet, and A. Zhedanov, *Difference Shrödinger operators with linear and exponentially discrete spectra*, Lett. Math. Phys. **29** (1993), 63–73.
14. A. P. Veselov and A. B. Shabat, *Dressing chains and the spectral theory of the Schrödinger operator*, Funct. Anal. Appl. **27** (1993), 1–21.

Department of Mathematics, University of Patras, GR-265 00, Patras, Greece.
E-mail address: vassilis@math.upatras.gr

Centre de Recherches Mathématiques
CRM Proceedings and Lecture Notes
Volume **25**, 2000

Discrete Equations of Motion for A_{k-1} Algebra

A. P. Protogenov and V. A. Verbus

ABSTRACT. We study integrals of motion for Hirota bilinear difference equation that is satisfied by the eigenvalues of the transfer-matrix. The combinations of the eigenvalues of the transfer-matrix are found, which are integrals of motion for integrable discrete models for the A_{k-1} algebra with zero and quasiperiodic boundary conditions. Discrete analogs of the equations of motion for the Bullough–Dood model and non-Abelian generalization of Liouville model are obtained.

1. Introduction

The thermodynamical Bethe–Ansatz (TBA) [**25**] is one of the main methods of studying statistical properties of integrable low-dimensional systems. Its formulation yields the functional equations [**10, 14, 17**] which have the form of multidimensional recurrent relations. The same equations (identical in form but different in the analytical properties of solutions) are known in the theory of nonlinear waves as the Hirota equations. Discrete Hirota equations with specified boundary conditions may yield all the known exactly integrated $(1+1)$D equations at the continuous limit. The problem of integrability of discrete equations was discussed in recent papers [**2, 13, 15, 16, 19, 21, 22**]. In [**13, 15, 21, 22**] a significant progress was achieved in applying Bethe–Ansatz (BA) for integrable discrete $(1+1)$D systems with internal degrees of freedom (see also [**11**]). It was found out thereby that the discrete time variable and the variable corresponding to the internal degrees of freedom, as well as discrete rapidities equally enter the Hirota equation as the arguments of the searched function. The limit of continuous time in this case corresponds to large spin values [**4**].

The eigenvalues $T_s^a(u)$ of the transfer matrix, satisfying Eq. (2.1) with certain boundary conditions (see below), are characterized by analytical properties of solutions specific for the problem under consideration. Thus, for example, the functional recurrent relations of a thermodynamical Bethe–Ansatz [**17**], being outwardly similar in form with equations for A_1-case with zero boundary condition, are satisfied by functions with different analytical properties. The distinction in

2000 *Mathematics Subject Classification.* Primary: 39A10, 81R50.

We want to thank A. I. Bobenko and W. K. Schief for useful discussion. This work was supported in part by the Russian Foundation for Basic Research under Grant No. 98-02-16237.

This is the final form of the paper.

analytical properties consists in the fact that for recurrent TBA equations the shift of rapidities occurs along the imaginary axis.

In the present paper we consider the case of real shifts along the axis of rapidities. The main goal of the paper is study of the integrals of motion following from the analysis of the Hirota classical equation for zero and quasiperiodic boundary conditions in the case of A_{k-1} algebra. The integrals of motion are useful for the investigation of the solutions of BA equations. As an example of the equations that require integrals of motion to be analyzed we consider equations in the discrete and continuous limits for the $k = 3$ case. The general approach for studying the integrals of motion for discrete sine-Gordon model was used in [**3**]. The representation of the zero curvature [**5, 24**] for the classical discrete sine-Gordon model via the R-matrix was studied in [**23**].

2. Hirota Equation

The Hirota bilinear difference equation may be written down in different forms. One of the presentation of this equation for the function $T_s^a(u)$ depending on the discrete variables a, s and u is

$$T_s^a(u+1)T_s^a(u-1) - T_{s+1}^a(u)T_{s-1}^a(u) = T_s^{a+1}(u)T_s^{a-1}(u). \tag{2.1}$$

The discrete index $a = 0, 1, \ldots, k$ for the A_{k-1} algebra and the discrete time s in this equation are, respectively, the length and the height of the Young rectangular diagram, to which the eigenvalue $T_s^a(u)$ of the transfer matrix corresponds; u in Eq. (2.1) is the discrete spectral parameter. Equation (2.1) preserves its form if the function $T_s^a(u)$ is multiplied by the product of four arbitrary functions χ_i depending on one variable with the following arguments:

$$T_s^a(u) \to \chi_1(a+u+s)\chi_2(a-u+s)\chi_3(a+u-s)\chi_4(a-u-s)T_s^a(u). \tag{2.2}$$

For the function

$$Y_s^a(u) = \frac{T_{s+1}^a(u)T_{s-1}^a(u)}{T_s^{a+1}(u)T_s^{a-1}(u)}, \tag{2.3}$$

which is invariant with respect to the transform (2.2), Hirota equation (2.1) has the form

$$Y_s^a(u+1)Y_s^a(u-1) = \frac{(1+Y_{s+1}^a(u))(1+Y_{s-1}^a(u))}{(1+(Y_s^{a+1}(u))^{-1})(1+(Y_s^{a-1}(u))^{-1})}. \tag{2.4}$$

The zero boundary conditions for it are as follows

$$\begin{gathered} T_s^a(u) = 0 \text{ for } a < 0 \quad \text{and} \quad a > k, \\ T_s^k(u) = \phi(u-s-k)\,,\, T_s^0(u) = \phi(u+s), \end{gathered} \tag{2.5}$$

where $\phi(u) = \prod_{k=1}^N \sigma(\eta(u - y_k))$, σ is the Weirshtrass function, y_k being its roots; N is the size of the system and η is a parameter.

The quasiperiodic boundary conditions for the function

$$T_s^a(u) = e^\alpha \lambda^{-2a} T_s^{a+2}(u+2) \tag{2.6}$$

in the A_1-algebra case lead to the discrete sine-Gordon equation[**12**].

We consider discrete equations of motion for the A_2 algebra for the zero and quasiperiodic boundary conditions, and their continuous limit. Applying the general approach of the paper [**12**] in the case of zero boundary conditions for gauge-invariant functions $Y_s^a(u)$ with $a = 1, 2$ we obtain the following system of discrete equations:

$$(1+Y_s^2(u))Y_s^1(u+1)Y_s^1(u+1) = (1+Y_{s+1}^1(u))(1+Y_{s-1}^1(u))Y_s^2(u), \tag{2.7}$$

$$(1+Y_s^1(u))Y_s^2(u+1)Y_s^2(u+1) = (1+Y_{s+1}^2(u))(1+Y_{s-1}^2(u))Y_s^1(u). \tag{2.8}$$

This system of equations is one of the results of our paper.

We introduce the functions $\phi_l(x,t)$ with $l = 1, 2$ in such a way that $Y_s^l(u) = \delta^{-2}\exp(-\phi_l(x,t))$. Then the continuous limit $\delta \to 0$ with $u = x/\delta$, $s = t/\delta$ yields a non-Abelian analog [**9**] of the Liouville equation:

$$\frac{\partial^2\phi_1}{\partial t^2} - \frac{\partial^2\phi_1}{\partial x^2} = 2e^{\phi_1} - e^{\phi_2}, \tag{2.9}$$

$$\frac{\partial^2\phi_2}{\partial t^2} - \frac{\partial^2\phi_2}{\partial x^2} = 2e^{\phi_2} - e^{\phi_1}, \tag{2.10}$$

which corresponds to SU(3) Toda equation.

In the case of non-Abelian discrete sine-Gordon model the quasiperiodic boundary condition (2.6) should be generalized for the A_{k-1} algebra in the following way:

$$T_s^a(u) = e^{\alpha}\lambda^{-ka}T_s^{a+k}(u+k). \tag{2.11}$$

Here the functions $X_s^a(u)$ satisfy the relation

$$X_s^a(u)X_s^{a+1}(u+1)\dots X_s^{a+k-1}(u+k-1) = (-1)^k. \tag{2.12}$$

Consider the case $k = 3$. The Hirota equation for gauge-invariant functions $X_s^a(u) = -\lambda(1+Y_s^a(u))$ [**6, 8, 20**] for the A_2-algebra with boundary conditions (2.11) can be written as

$$\begin{aligned}(\lambda + X_s^2(u))X_{s+1}^1(u)X_{s-1}^1(u)\\ = \frac{(\lambda+X_s^1(u-1))(\lambda+X_s^1(u+1))}{(1-\lambda X_s^1(u+1)X_s^2(u+2))}X_s^2(u),\end{aligned} \tag{2.13}$$

$$\begin{aligned}(\lambda + X_s^1(u))X_{s+1}^2(u)X_{s-1}^2(u)\\ = \frac{(\lambda+X_s^2(u-1))(\lambda+X_s^2(u+1))}{(1-\lambda X_s^1(u-2)X_s^2(u-1))}X_s^1(u).\end{aligned} \tag{2.14}$$

The continuous limit in these equations can be achieved as follows. Assume that $\lambda \to 0$ and $s \to \infty$ in such a way that $s = t/\sqrt{\lambda}$ for a fixed t. Similarly, assume that $\lambda \to 0$ and $u \to \infty$ with $u = x/\sqrt{\lambda}$ for a fixed x. Once parametrized with the aid of the function $X^l(x,t) = -\exp(-\phi_l(x,t))$, Eqs. (2.13) and (2.14) transform in the continuous limit to the following equations of motion

$$\frac{\partial^2\phi_1}{\partial t^2} - \frac{\partial^2\phi_1}{\partial x^2} = 2e^{\phi_1} - e^{\phi_2} - e^{-(\phi_1+\phi_2)}, \tag{2.15}$$

$$\frac{\partial^2\phi_2}{\partial t^2} - \frac{\partial^2\phi_2}{\partial x^2} = 2e^{\phi_2} - e^{\phi_1} - e^{-(\phi_1+\phi_2)}, \tag{2.16}$$

which correspond to SU(3)-case of affine Toda equation. Assuming $\phi = \phi_1 = \phi_2$ in Eqs. (2.15) or (2.16) yields

$$\frac{\partial^2\phi}{\partial t^2} - \frac{\partial^2\phi}{\partial x^2} = e^{\phi} - e^{-2\phi}, \tag{2.17}$$

which is the Bullough–Dodd equation. A discrete analog of this equation is derived from (2.13) on assumption that $X_s(u) = X_s^1(u) = X_s^2(u)$:

$$\frac{X_{s+1}(u)X_{s-1}(u)}{X_s(u)} = \frac{(\lambda + X_s(u-1))(\lambda + X_s(u+1))}{(1 - \lambda X_s(u-2)X_s(u-1))(\lambda + X_s(u))}, \tag{2.18}$$

or from (2.14) with the same assumption

$$\frac{X_{s+1}(u)X_{s-1}(u)}{X_s(u)} = \frac{(\lambda + X_s(u-1))(\lambda + X_s(u+1))}{(1 - \lambda X_s(u+2)X_s(u+1))(\lambda + X_s(u))}. \tag{2.19}$$

We have two versions of discrete Bullough–Dodd equation. These equations are symmetrical to each other with respect to changing the variable $u \to -u$. It means that the discrete version of the Bullough–Dodd equation broke the symmetry with respect to u and it is connected with selection of the direction of motion. One can symmetrize these equations adding together Eqs. (3.6), (3.7)

$$\begin{aligned} &\frac{X_s(u)}{X_{s+1}(u)X_{s-1}(u)} \\ &= \frac{(\lambda + X_s(u))(1 - (\lambda/2)X_s(u-2)X_s(u-1) - (\lambda/2)X_s(u+2)X_s(u+1))}{(\lambda + X_s(u-1))(\lambda + X_s(u+1))}. \end{aligned} \tag{2.20}$$

Another discrete variant of the Bullough–Dodd equation was found from geometrical point of view in [**1**].

3. Integrals of Motion

Integrals of motion are known to play a key role in the study of this types of motion. In [**12**] integrals of motion for discrete dynamics were obtained for the A_1 algebra and zero boundary conditions. A general determinant representation of integrals of motion was also written out for the A_k algebra with zero boundary conditions. In this section we shall show

(i) how to get integrals of motion in the A_2 case from the known integrals for the A_1 algebra. In fact, we use this example to show how integrals of motion are transformed in the Bäcklund flow. Besides,

(ii) we write down some integrals of motion for quasiperiodic boundary conditions.

We shall use the equations of the linear problem, which follow from the gauge and dual transformations on the lattice [**18**] of the variables in question. These equations are equivalent to those resulting from the zero curvature condition [**5**, **24**]. In the terms of the functions $T_s^a(u)$ and $G_s^a(u)$, the direct equations of the linear problem are written as follows

$$G_s^a(u)T_{s+1}^a(u+1) - G_{s+1}^a(u+1)T_s^a(u) = c^{-1}G_s^{a-1}(u+1)T_{s+1}^{a+1}(u), \tag{3.1}$$

$$G_s^{a-1}(u+1)T_{s+1}^a(u-1) - G_{s+1}^{a-1}(u)T_s^a(u) = cG_s^a(u)T_{s+1}^{a-1}(u). \tag{3.2}$$

Similarly, the inverse equations are written as

$$G_s^a(u)T_{s-1}^a(u-1) - G_{s-1}^a(u-1)T_s^a(u) = -c^{-1}G_{s-1}^{a-1}(u)T_s^{a+1}(u-1), \tag{3.3}$$

$$G_s^a(u)T_{s-1}^a(u) - G_{s-1}^a(u+1)T_s^{a+1}(u-1) = -cG_{s-1}^{a+1}(u)T_s^a(u). \tag{3.4}$$

To make the comparison of the approaches and notations of various papers easier, we write out the equations of the linear problems having expressed the functions included in them via the eigenvalues of the transfer matrix, used in [**12**]. We note that the complete equivalence is achieved at $c=-1$ and the substitution $s \to -s$.

Let us focus our attention on the symmetry of pairs of Eqs. (3.1), (3.2) and (3.3), (3.4). It is seen that after the substitution $T_s^a(u) \to G_{-s}^{-a}(-u)$ and $G_s^a(u) \to T_{-s}^{-a}(-u)$ Eqs. (3.1), (3.2) transform to (3.3), (3.4).

In the case of zero boundary conditions (2.5) for the function $T_s^a(u)$, the function $G_s^a(u)$ satisfies also the same boundary conditions, but with the maximum values a less by a unity, i.e., k should be replaced by $(k-1)$. This feature expresses Backlund transform which realizes the transform $A_k \to A_{k-1}$. It was used in the paper [**12**] to get nested Bethe–Ansatz equation. This transform for $k=3$ case can be schmetically presented in the following form:

$$\begin{array}{cccccc}
0 & 1 & 0 & & & \\
0 & Q_1(u+s) & \overline{Q}_1(u-s) & 0 & & \\
0 & Q_2(u+s) & G_s^1(u) & \overline{Q}_2(u-s) & 0 & \\
0 & \phi(u+s) & T_s^1(u) & T_s^2(u) & \bar{\phi}(u-s) & 0
\end{array} \tag{3.5}$$

Here $Q_\alpha(u)$ are some boundary functions expressed via Weierstrass functions $\sigma(\eta u)$, $\overline{Q}_\alpha(u) = Q_\alpha(u-\alpha)$, $e\bar{\phi}(u) = \phi(u-k)$. It follows also from the boundary condition (2.5) that $Q_k(u) = \phi(u)$. The functions at each level satisfy Hirota equation. There is a relation between two adjacent levels, defined by the equations of the linear problem. The function $G_s^1(u)$ corresponds to the case of A_1 algebra. The equations of the linear problem written for it allow to obtain integrals of motion, which for u-dynamics [**12**] are:

$$I_1(s) = \frac{Q_2(s-2)G_{s-u+1}^1(u) + Q_2(s)G_{s-u-1}^1(u-2)}{G_{s-u}^1(u-1)}, \tag{3.6}$$

$$I_2(s) = \frac{\overline{Q}_2(s+2)G_{u-s+1}^1(u) + \overline{Q}_2(s)G_{u-s-1}^1(u+2)}{G_{u-s}^1(u+1)}. \tag{3.7}$$

For A_2 case Eq. (3.1) of the linear problem for $a=1$ and $a=2$ may be written in the following way

$$Q_2(u+s)T_{s+1}^1(u) - Q_2(u+s+2)T_s^1(u-1) = c^{-1}G_{s+1}^1(u)\phi(u+s), \tag{3.8}$$

$$G_s^1(u)T_{s+1}^2(u) - G_{s+1}^1(u+2)T_s^2(u-1) = c^{-1}\overline{Q}_2(u-s-1)T_s^1(u). \tag{3.9}$$

Similarly Eq. (3.2) is written down

$$G_s^1(u)T_{s+1}^1(u) - G_{s+1}^1(u-1)T_s^1(u) = cQ_2(u+s+1)T_s^2(u-1), \tag{3.10}$$

$$\overline{Q}_2(u-s)T_{s+1}^2(u-1) - \overline{Q}_2(u-s-2)T_s^2(u) = cG_{s+1}^1(u)\phi(u+s). \tag{3.11}$$

Having expressed from (3.8) the necessary values $G_s^1(u)$ and substituting them in (3.6) we get the first integral of motion

$$(3.12)\quad I_1(s) = \phi(s-2)\left[\frac{Q_2(s-2)[T^1_{s-u+1}(u)Q_2(s) - T^1_{s-u}(u-1)Q_2(s+2)]}{\phi(s)\left[T^1_{s-u}(u-1)Q_2(s-2) - T^1_{s-u-1}(u-2)Q_2(s)\right]}\right.$$
$$\left. + \frac{Q_2(s)[T^1_{s-u-1}(u-2)Q_2(s-4) - T^1_{s-u-2}(u-3)Q_2(s-2)]}{\phi(s-4)\left[T^1_{s-u}(u-1)Q_2(s-2) - T^1_{s-u-1}(u-2)Q_2(s)\right]}\right].$$

Similarly having expressed from (3.9) the corresponding values $G_s^1(u)$ and substituting them in (3.7) we get the second integral of motion for u-dynamics

$$(3.13)\quad I_2(s) = \phi(-s-2)$$
$$\times\left[\frac{\overline{Q}_2(2-s)[T^2_{s+u+1}(u-1)\overline{Q}_2(-s) - T^2_{s+u}(u)\overline{Q}_2(-s-2)]}{\phi(s)[T^2_{s+u}(u)\overline{Q}_2(2-s) - T^2_{s+u-1}(u+1)\overline{Q}_2(-s)]}\right.$$
$$\left. - \frac{\overline{Q}_2(-s)[T^2_{s+u-1}(u+1)\overline{Q}_2(4-s) - T^1_{s+u-2}(u+2)\overline{Q}_2(s-2)]}{\phi(s-4)[T^2_{s+u}(u)\overline{Q}_2(2-s) - T^2_{s+u-1}(u+1)\overline{Q}_2(-s)]}\right].$$

In order to obtain equations for s-dynamics one should use the other pair of Eqs. (3.9), (3.11). The integrals of motion are related with this form (2.5) of boundary conditions. Here one of them is associated with the boundary condition at the left end of the segment $[0,k]$ and the other—at the right end. The chosen example $k=3$ illustrates the method of obtaining integrals of motion from the integrals of motion for k less by a unity.

Now we consider quasiperiodic boundary conditions. In order Eqs. (3.1), (3.2) and (3.3), (3.4), are fulfilled, the function $G_s^a(u)$ in the case of the quasiperiodic boudary conditions (2.11) should also satisfy the same boundary conditions. Therefore there are no Backlund flows in this case and the transition from A_k algebra to A_{k-1} algebra does not take place. It follows from the condition that two functions $T_s^a(u)$ and $G_s^a(u)$ which satisfy one and the same Hirota equation with equal boundary conditions are proportional to each other. The relations between the indices in these functions are defined by the law of the transformation of direct equations of the linear problem (3.1), (3.2) into the inverse ones (3.3), (3.4). The coefficient of proportionality should retain Hirota equation invariant, i.e. it should coincide with the gauge transformation (2.2). Thus we may write down

$$(3.14)\quad G_s^a(u) = \chi_1(a+u+s)\chi_2(a-u+s)\chi_3(a+u-s)\chi_4(a-u-s)T_{-s}^{-a}(-u).$$

We use Eq. (3.14) in (3.1), (3.2). As the result we obtain

$$(3.15)\quad \frac{\chi_4(a-u-s)}{\chi_4(a-u-s-2)}T_{-s}^{-a}(-u)T_{s+1}^{a}(u+1)$$
$$- \frac{\chi_1(a+u+s+2)}{\chi_1(a+u+s)}T_{-s-1}^{-a}(-u-1)T_s^a(u)$$
$$= c^{-1}\frac{\chi_2(a-u+s-2)}{\chi_2(a-u+s)}T_{-s}^{-a+1}(-u-1)T_{s+1}^{a+1}(u),$$

$$(3.16)\quad \frac{\chi_2(a-u+s-2)}{\chi_2(a-u+s)}T_{-s}^{-a-1}(-u-1)T_{s+1}^{a}(u-1)$$
$$- \frac{\chi_3(a+u-s-2)}{\chi_3(a+u-s)}T_{-s-1}^{-a+1}(-u)T_s^a(u)$$

$$= c\frac{\chi_4(a-u-s)}{\chi_4(a-u-s-2)}T^{-a}_{-s}(-u)T^{a-1}_{s+1}(u).$$

We use this form of equations to get integrals of motion. Here we take into account that quasiperiodic boundary conditions limit the type of the functions χ_1 and χ_3. They should be periodic with the period $2k$, while the functions χ_2 and χ_4 may be arbitrary. We single out the functions χ_1 and χ_3 from Eqs. (3.15), (3.16), having constructed the relations

$$(3.17)\quad \frac{\chi_1(a+u+s+2)}{\chi_1(a+u+s)} = \frac{\chi_4(a-u-s)T^{-a}_{-s}(-u)T^{a}_{s+1}(u+1)}{\chi_4(a-u-s-2)T^{-a}_{-s-1}(-u-1)T^{a}_{s}(u)} - c^{-1}\frac{\chi_2(a-u+s-2)T^{-a+1}_{-s}(-u-1)T^{a+1}_{s+1}(u)}{\chi_2(a-u+s)T^{-a}_{-s-1}(-u-1)T^{a}_{s}(u)},$$

$$(3.18)\quad \frac{\chi_3(a+u-s-2)}{\chi_3(a+u-s)} = \frac{\chi_2(a-u+s-2)T^{-a+1}_{-s}(-u-1)T^{a}_{s+1}(u-1)}{\chi_2(a-u+s)T^{-a+1}_{-s-1}(-u)T^{a}_{s}(u)} - c\frac{\chi_4(a-u-s)T^{-a}_{-s}(-u)T^{a-1}_{s+1}(u)}{\chi_4(a-u-s-2)T^{-a+1}_{-s-1}(-u)T^{a}_{s}(u)}.$$

It follows from the periodicity of the functions χ_1 and χ_3 that the r.h.s. of Eqs. (3.17), (3.18) are periodic with the period $2k$ with respect to three indices. After the replacement $u \to u-s$ in Eq. (3.17) and the replacement $u \to u+s$ in (3.18) the right part of these equations does not depend on the time s. Consequently, after this replacement of variables the right hand size of the equations is the integral of motion for s-dynamics. Similarly we may obtain the preserving values of u-dynamics, substituting $s \to s-u$ in Eq. (3.17) and $s \to s+u$ in Eq. (3.18). For simplicity we assume that $\chi_2 = \chi_4 = 1$ and we consider the case of A_1-algebra. For $a = 0$ and $a = 1$ from Eq. (3.17) we may get two invariant combinations of functions:

$$(3.19)\quad \frac{\chi_1(u+s+2)}{\chi_1(u+s)} = \frac{T^{0}_{-s}(-u)T^{0}_{s+1}(u+1)}{T^{0}_{-s-1}(-u-1)T^{0}_{s}(u)} - c^{-1}\frac{T^{1}_{-s}(-u-1)T^{1}_{s+1}(u)}{T^{0}_{-s-1}(-u-1)T^{0}_{s}(u)},$$

$$(3.20)\quad \frac{\chi_1(u+s+3)}{\chi_1(u+s+1)} = \frac{T^{1}_{-s}(-u+2)T^{1}_{s+1}(u+1)}{T^{1}_{-s-1}(-u+1)T^{1}_{s}(u)} - c^{-1}\frac{T^{0}_{-s}(-u-1)T^{0}_{s+1}(u-2)(\lambda e^{\alpha})^2}{T^{1}_{-s-1}(-u+1)T^{1}_{s}(u)}.$$

The combinations made of the functions χ_3, which follow from (3.18), have the form

$$(3.21)\quad \frac{\chi_3(u-s-2)}{\chi_3(u-s)} = \frac{T^{1}_{-s}(-u-1)T^{0}_{s+1}(u-1)}{T^{1}_{-s-1}(-u)T^{0}_{s}(u)} - c\frac{T^{0}_{-s}(-u)T^{1}_{s+2}(u+2)}{T^{1}_{-s-1}(-u)T^{0}_{s}(u)},$$

$$(3.22)\quad \frac{\chi_3(u-s-1)}{\chi_3(u-s+1)} = \frac{T^{0}_{-s}(-u-1)T^{1}_{s+1}(u-1)}{T^{0}_{-s-1}(-u)T^{0}_{s}(u)} - c\frac{T^{1}_{-s}(-u+2)T^{0}_{s+1}(u)}{T^{0}_{-s-1}(-u)T^{0}_{s}(u)}.$$

It is seen directly from these equations that after the substitution $u \to u - s$ $(u \to u + s)$ or $s \to s - u$ $(s \to s + u)$ in Eqs. (3.19), (3.20), (3.21), (3.22) we obtain in the r.h.s. of these equations the magnitudes which do not depend on s (on u) after the change of these discrete variables.

In conclusion, we found the integrals of motion for s- and u-dynamics for A_k-algebra with quasiperiodic boundary conditions. The case $k = 2$ corresponds to the discrete sine-Gordon model. It is shown that for A_2- algebra the discrete equations of motion at zero boundary conditions convert in the continuous limit to the equations of motion of non-Abelian generalization of Liouville equation and at the quasiperiodic boundary conditions—to the equation of motion of the SU(3) affine Toda equation and to the equation of motion of the Bullough–Dodd model.

References

1. A. Bobenko and W. Schief, *Discrete indefinite affine spheres*, preprint SFB288, Technische Universität Berlin **263**, (1997); *Affine spheres: Discretization via duality relations*, preprint SFB288, Technische Universität Berlin **297**, (1997).
2. H. W. Capel and F. W. Nijhoff, *Integrable lattice equations*, Important Developments in Soliton Theory (A. S. Fokas and V. E. Zakharov, eds.), Springer Ser. Nonlinear Dynam., Springer, Berlin, 1993.
3. B. Enriquez and B. L. Feigin, *Integrals of motion of classical lattice sine-Gordon system*, Theoret. Math. Phys. **103** (1995), 738–756.
4. L. D. Faddeev, *Les Houches lectures: "How Algebraic Bethe–Ansatz work for integrable model"*, hep-th/9605187.
5. L. Faddeev and L. Takhtadjan, *Hamiltonian methods in the theory of solitons*, Springer, Berlin, 1987.
6. L. Faddeev and A. Volkov, *Hirota equation as an example of integrable symplectic map*, Lett. Math. Phys. **32** (1994), 125–136.
7. R. Hirota, *Discrete two-dimensional Toda molecule equation*, J. Phys. Soc. Japan **56** (1987), 4285–4288.
8. ———, *Nonlinear partial difference equation.* III. *Discrete sine-Gordon equation*, J. Phys. Soc. Japan **43** (1977), 2079–2086.
9. R. Jackiw and So-Young Pi, *Self-dual Chern-Simons solitons*, Progr. Theoret. Phys. Suppl. **107** (1992), 1–39.
10. A. Klümper and P. Pearce, *Conformal weights of* RSOS *lattice models and their fusion hierarchies*, Physica A **183** (1992), 304–350.
11. I. G. Korepanov, *Algebraic integrable dynamical systems,* $2+1$*-dimensional model in wholly discrete space-time, and inhomogeneous models in* 2*-dimensional statistical physics*, hep-th/9312197; *Vacuum curves, classical integrable systems in discrete space-time and statistical physics*, solv-int/9506003.
12. I. Krichever, O. Lipan, P. Wiegmann, and A. Zabrodin, *Quantum integrable systems and elliptic solutions of classical discrete nonlinear equations*, Comm. Math. Phys. **188** (1997), 267–304.
13. I. Krichever, P. Wiegmann, and A. Zabrodin, *Elliptic solutions to difference nonlinear equation and related many-body problems*, Comm. Math. Phys. **193** (1998), 373–396.
14. A. Kuniba, T. Nakanishi, and Suzuki, *Functional relations in solvable lattice model.* I. *Functional relations and representations theory.*II. *Application*, Internat. J. Modern Phys. A **9** (1994), 5215–5312.
15. O. Lipan, P. Wiegmann, and A. Zabrodin, *Fusion rules for quantum transfer matrices as a dynamical system on Grassmann manifolds*, solv-int/9704015.
16. F. W. Nijhoff, O. Ragnisco, and V. B. Kuznetsov, *Integrable time-discretization of the Ruijsenaars-Schneider model*, Comm. Math. Phys. **176** (1996), 681-700.
17. F. Ravanini, A. Valleriani, and R. Tateo, *Dynkin* TBA, Internat. J. Modern Phys. A **8** (1993), 1707–1727.

18. S. Saito and N. Saitoh, *Lineriazation of bilinear difference equations*, Phys. Lett. A **120** (1987), 322–326; *Gauge and dual symmetries and linearization of Hirota's bilinear equations*, J. Math. Phys. **28**, (1987), 1052–1055 .
19. Y. B. Suris, *A discrete time relativistic Toda lattice*, J. Phys. A **29** (1996), 451–465.
20. A. Volkov, *Quantum lattice* KdV *equation*, Lett. Math. Phys. **39** (1997), 313–329.
21. P. B. Wiegmann, *Bethe–Ansatz and classical Hirota equation*, Internat. J. Modern Phys.B **11** (1997), 75–90.
22. A. V. Zabrodin, *Bethe–Ansatz and classical Hirota equation*, preprint ITEP/TH-28/96, hep-th/9607162; *A survey of Hirota's difference equations*, preprint ITEP/TH-10/97, solv-int/9704001; *Discrete Hirota's equation in quantum integrable model*, preprint ITEP/TH-44/96, hep-th/9610039.
23. ———, *Zero curvature representation for classical lattice sine-Gordon model via quantum R-matrix*, JETP Lett. **66** (1997), no. 9, 653–659. (Russian)
24. V. Zakharov, S. Manakov, S. Novikov, and L. Pitaevskiy, *Theory of solitons. The inverse problem method*, Moscow, Nauka, 1980.
25. Al. B. Zamolodchikov, *Thermodynamic Bethe–Ansatz in relativistic model: Scaling 3-state Potts and Lee-Yang models*, Nuclear Phys. B **342** (1990), 695–720.

Institute of Applied Physics, Russian Academy of Sciences, 46 Ul'yanov Street, Nizhny Novgorod 603600, Russia.

E-mail address: alprot@iap.sci-nnov.ru

Institute for Physics of Microstructures, Russian Academy of Sciences, GSP-105, Nizhny Novgorod 603600, Russia.

E-mail address: verbus@ipm.sci-nnov.ru

Centre de Recherches Mathématiques
CRM Proceedings and Lecture Notes
Volume **25**, 2000

Discrete Painlevé Equations from Nonisospectral Soliton Equations

G. R. W. Quispel and D. Levi

1. Introduction

The standard method to obtain Painlevé equations from soliton equations is to apply a (point) symmetry reduction [**1,2,9**]. When one or more of the independent variables that are involved is discrete, however, this method does not work, and for a long time it was not clear in what way a discrete analog of the above construction might be found.

The solution to this problem was found in [**7**]. There, for the example of the Korteweg-de Vries equation, it was shown that symmetry reduction of an isospectral equation is equivalent to stationary reduction of a corresponding nonisospectral equation.

In this paper, the method of [**7**] is applied to the Nonlinear Schrödinger (NLS) equation. In Section 2.1, a symmetry reduction is applied to reduce the continuous NLS to an equation known as Ince XXXIV, which is equivalent to the second Painlevé equation [**5**]. In Section 2.2, we show that exactly the same equation is obtained by stationary reduction of a nonisospectral NLS equation. Finally, in Section 3 it is shown that the latter method (but not the former) has a straightforward generalization to the discrete case, and we obtain a discrete version of Ince XXXIV from the stationary reduction of the discrete (differential-difference) NLS equation ($D\Delta$NLS).

1.1. The isospectral NLS and its symmetry reduction. The NLS equation is the first member of a hierarchy of isospectral equations.

$$\sigma_3 \frac{d}{dt}\begin{pmatrix} r \\ q \end{pmatrix} = \omega(\mathcal{L}_1)\begin{pmatrix} r \\ q \end{pmatrix}, \tag{1.1}$$

2000 *Mathematics Subject Classification.* 34K05, 34M55, 39A05.

DL is grateful to the Australian Research Council for supporting his visit to La Trobe University during which the research for this paper was done. GRWQ is grateful to the Mathematical Sciences Research Institute, Berkeley (funded in part by the National Science Foundation through grant no. DMS-9701755) where this work was written up. We are grateful to Basil Grammaticos for valuable comments.

This is the final form of the paper.

where ω is an arbitrary polynomial,

$$\sigma_3 := \begin{pmatrix} -1 & 0 \\ 0 & 1 \end{pmatrix}, \tag{1.2}$$

and the recursion operator $\mathcal{L}_1$ is defined by

$$2i\mathcal{L}_1 \begin{pmatrix} a \\ b \end{pmatrix} := \begin{pmatrix} -a_x \\ b_x \end{pmatrix} + 2 \begin{pmatrix} r \\ q \end{pmatrix} \int_x^\infty dx\big(q(x)a(x) - r(x)b(x)\big), \tag{1.3}$$

(here and below partial derivatives are denoted by subscripts, e.g. $a_x := (\partial a)/(\partial x)$).

Taking

$$r(x) = -q^*(x), \quad \omega(\mathcal{L}_1) = -4i\mathcal{L}_1^2, \tag{1.4}$$

(the asterisk denoting complex conjugation), we obtain the NLS equation:

$$-iq_t = q_{xx} + 2|q|^2 q. \tag{1.5}$$

In [**6**], Johnson et al give a list of infinitesimal Lie point symmetries of the NLS. One of the symmetries in their list is:

$$\begin{aligned} x &\to x + \varepsilon bt \\ t &\to t + \varepsilon d \\ q &\to 9 + \varepsilon \frac{1}{2} ibxq. \end{aligned} \tag{1.6}$$

By the usual methods [**1, 9**] this symmetry leads to the following reduction:

$$q(x,t) = e^{i(-\gamma xt - 2/3\gamma^2 t^3)} r(\eta), \tag{1.7}$$

where

$$\eta := x + \gamma t^2, \quad \gamma := -\frac{b}{2d}. \tag{1.8}$$

Substituting the similarity solution (1.7) in the NLS, we obtain the ODE

$$r_{\eta\eta} + 2|r|^2 r + \gamma\eta r = 0. \tag{1.9}$$

Defining

$$c := r^* r \tag{1.10}$$

we obtain Ince XXXIV:

$$c_{\eta\eta} = \frac{1}{2}\frac{c_\eta^2}{c} + 4c^2 - 2\gamma\eta c + \frac{\alpha^2}{c} \tag{1.11}$$

(where α is an integration constant). Note that Eq. (1.11) is related to Painlevé II.

1.2. The nonisospectral NLS and its stationary reduction. There is also a nonisospectral NLS hierarchy [**4**]

$$\sigma_3 \frac{d}{dt}\begin{pmatrix} r \\ q \end{pmatrix} = \omega(\mathcal{L}_1)\begin{pmatrix} r \\ q \end{pmatrix} + \widehat{\omega}(\mathcal{L}_1)x\begin{pmatrix} r \\ q \end{pmatrix}, \tag{1.12}$$

where ω and $\widehat{\omega}$ are 2 arbitrary polynomials.

The simplest nonlinear member of the double hierarchy (1.12) is given by:

$$\omega(\mathcal{L}_1) = -4i(\mathcal{L}_1^2), \quad \widehat{\omega}(\mathcal{L}_1) = i\mu, \tag{1.13}$$

where μ is an arbitrary constant. This gives the nonisospectral NLS:

$$-iq_t = q_{xx} + 2|q|^2 q + \mu xq. \tag{1.14}$$

Taking the stationary reduction $q_t = 0$, we get:

$$q_{xx} + 2|q|^2 q + \mu x q = 0. \tag{1.15}$$

Note that Eq. (1.15) is identical to Eq. (1.9)!

So putting $s = q^* q$ we obtain again Ince XXXIV:

$$s_{xx} = \frac{1}{2}\frac{s_x^2}{s} + 4s^2 - 2\mu x s + \frac{\alpha^2}{s}. \tag{1.16}$$

2. The Nonisospectral $D\Delta$NLS and Its Stationary Reduction

Whereas the isospectral point symmetry reduction of Section 2.1 has no straightforward generalization to the discrete case, the nonisospectral stationary reduction does.

The nonisospectral discrete NLS hierarchy [**3**] is given by

$$\sigma_3 \frac{d}{dt}\begin{pmatrix} R_n \\ Q_n \end{pmatrix} = \omega(\mathcal{L}_2)\begin{pmatrix} R_n \\ Q_n \end{pmatrix} + \widehat{\omega}(\mathcal{L}_2)\Delta\left(n + \frac{1}{2}\right)\begin{pmatrix} R_n \\ Q_n \end{pmatrix}, \tag{2.1}$$

where Δ denotes the lattice spacing, and the discrete recursion operator $\mathcal{L}_2$ is defined by:

$$\mathcal{L}_2 \begin{pmatrix} A_n \\ B_n \end{pmatrix} := \begin{pmatrix} A_{n-1} + R_{n-1}(1 - R_n Q_n) D_n + R_n E_n \\ B_{n+1} + Q_{n+1}(1 - R_n Q_n) D_{n+1} + Q_n E_{n+1} \end{pmatrix}, \tag{2.2}$$

with

$$D_n := \sum_{j=n}^{\infty} \frac{R_j B_j - Q_j A_j}{1 - R_j Q_j}, \tag{2.3}$$

and

$$E_n := \sum_{j=n}^{\infty} (R_j B_{j+1} - Q_j A_{j-1}). \tag{2.4}$$

Taking $R_n = -Q_n^*$, and

$$\omega(\mathcal{L}_2) = -\frac{i}{\Delta^2}(\mathcal{L}_2 - 2 + \mathcal{L}_2^{-1}), \quad \widehat{\omega}(\mathcal{L}_2) = i\mu, \tag{2.5}$$

we get

$$-i\frac{d}{dt}Q_n = \frac{Q_{n+1} - 2Q_n + Q_{n-1}}{\Delta^2} + \frac{|Q_n|^2(Q_{n+1} + Q_{n-1})}{\Delta^2} + \mu\Delta\left(n + \frac{1}{2}\right)Q_n. \tag{2.6}$$

We take the stationary reduction $(dQ_n)/(dt) = 0$, and define

$$C_n := Q_{n+1}Q_n^* + Q_{n+1}^* Q_n.$$

This way we obtain a discrete analog of Ince XXXIV:[1]

$$\left(\frac{\zeta(n)}{C_n + C_{n-1}} - 1\right)\left(\frac{\zeta(n+1)}{C_n + C_{n+1}} - 1\right) = \frac{4}{C_n^2 + \beta^2}, \tag{2.7}$$

where β is an integration constant, and

$$\zeta(n) := 4 - \mu\Delta^3(2n + 1). \tag{2.8}$$

[1]Similar equations are studied in [**8**].

Equation (2.7) reduces to Ince XXXIV in the continuum limit:

$$C_n = \Delta^2 C(x), \quad \beta = \alpha\Delta^3, \ x = n\Delta. \tag{2.9}$$

References

1. M. J. Ablowitz and P. A. Clarkson, *Solitons, nonlinear evolution equations and inverse scattering*, Cambridge Univ. Press, Cambridge, 1991.
2. M. J. Ablowitz, A. Ramani and H. Segur, *Nonlinear evolution equations and ordinary differential equations of Painlevé type*, Lett. Nuovo Cim. **23** (1978) 333–338.
3. M. Bruschi, D. Levi, and O. Ragnisco, *Discrete version of the Nonlinear Schrödinger equation with linearly x-dependent coefficients*, Nuovo Cimento A **53** (1979), 21–30.
4. F. Calogero and A. Degasperis, *Exact solution via the spectral transform of a generalization with linearly x-dependent coefficients of the nonlinear Schrödinger equation*, Lett. Nuovo Cimento **22** (1978), 270–273.
5. E. L. Ince, *Ordinary differential equations*, Dover, New York, 1956.
6. S. F. Johnson, K. E. Lonngren, and D. R. Nicholson, *On the invariants of the Nonlinear Schrödinger equation*, Phys. Lett. A **74** (1979), 393–394.
7. D. Levi, O. Ragnisco, and M. A. Rodrigues, *On nonisospectral flows, Painlevé equations and symmetries of differential and difference equations*, Theoret. and Math. Phys. **93** (1992), 473–480.
8. F. Nijhoff, J. Satsuma, K. Kajiwara, B. Grammaticos, and A. Ramani, *A study of the alternate discrete Painlevé. II. Equation*, Inverse Problems **12** (1996), 697–716.
9. P. J. Olver, *Applications of Lie groups to differential equations*, Springer, Berlin, 1986.

Department of Mathematics, Latrobe University, Bundoora, Victoria 3083, Australia.

E-mail address: R.Quispel@latrobe.edu.au.

Dipartimento di Fisica "E.Amaldi", Universita di Roma Tre, Via della Vasca Navale 84, 00146 Roma, Italy.

E-mail address: Levi@Amaldi.fis.uniroma3.it

Centre de Recherches Mathématiques
CRM Proceedings and Lecture Notes
Volume **25**, 2000

A Study of the Continuous and Discrete Gambier Systems

A. Ramani, B. Grammaticos, and S. Lafortune

ABSTRACT. We present a systematic study of the Gambier systems, which in the continuous case is given by two Riccati equations in cascade. We derive the condition for its integrability and show that the generic Gambier system contains one free function. We also derive the Schlesinger transformations for this system which allows in principle the systematic construction of the integrable cases. The above procedure is carried over to a discrete setting. We show thus how the discrete Gambier system can be expressed as a system of two homographic mappings in cascade. The integrable cases are obtained through the singularity confinement discrete integrability criterion. Finally the discrete Schlesinger transformations are also derived giving a handle to the construction of the integrable Gambier mapping.

1. Introduction

Linearizable equations play a very particular role in the domain of nonlinear systems. Their nonlinearity is not a genuine one in the sense that elementary transformations can reduce them to a linear system. Calogero coined the term C-integrability (which applies to both ordinary and partial differential equations) in order to describe this situation and to distinguish it from the more complicated situations of systems integrable through IST methods (S-integrability). The simplest example of a linearizable system is the Riccati equation [**8**]:

$$x' = ax^2 + bx + c, \tag{1.1}$$

which can be transformed through the Cole–Hopf transformation $x = -u'/(au)$ to the linear second-order equation:

$$u'' = (b + a'/a)u' - cau. \tag{1.2}$$

Higher degree, first-order linearisable equations also exist [**8**]. Already limiting ourselves to equations of binomial type we find

$$(x')^2 = (x-a)^2(x-k_1)(x-k_2), \tag{1.3}$$

2000 *Mathematics Subject Classification.* Primary: 39A10; Secondary: 34A34.

S. Lafortune acknowledges two scholarships: one from NSERC (National Science and Engineering Research Council of Canada) for his Ph.D. and one from "Programme de Soutien de Cotutelle de Thèse de doctorat du Gouvernement du Québec" for his stay in Paris.

This is the final form of the paper.

where a is a function of the independent variable and k_1, k_2 are constants. Putting $u^2 = (x - k_1)/(x - k_2)$ equation (1.3) reduces to the Riccati:

$$u' = \frac{1}{2}\big(k_1 - a - (k_2 - a)u^2\big). \tag{1.4}$$

When we consider second-order equations, the situation becomes immediately richer, in the sense that there exist several types of linearizable equation [**8**]. The first (and simplest) one is the equation obtained if one computes the derivative of both sides of the Riccati (1.1). It is usually given in canonical form and reads:

$$x'' = -2xx' + bx' + b'x. \tag{1.5}$$

In the same spirit, one can compute the derivative of (1.1) after having divided both sides of the equation by x. We find thus the equation (again given in canonical form):

$$x'' = \frac{x'^2}{x} + \left(ax - \frac{c}{x}\right)x' + a'x^2 + c'. \tag{1.6}$$

Another linearizable equation does exist which can be linearized by a Cole–Hopf transformation just like the Riccati. Its form is:

$$x'' = -3xx' - x^3 + q(x' + x^2). \tag{1.7}$$

Indeed, putting $x = u'/u$ we can reduce it to the linear third order equation:

$$u''' = qu''. \tag{1.8}$$

But the most interesting linearizable equation discovered at second-order is the one obtained by Gambier in his classification of second-order equations having the Painlevé property [**4**]. The Gambier equation is usually given as a second-order equation for a single variable:

$$x'' = \frac{n-1}{n}\frac{x'^2}{x} + a\frac{n+2}{n}xx' + bx' - \frac{n-2}{n}\frac{x'}{x}\sigma - \frac{a^2}{n}x^3 + (a' - ab)x^2 + \left(cn - \frac{2a\sigma}{n}\right)x - b\sigma - \frac{\sigma^2}{nx}, \tag{1.9}$$

where n is integer, a, b, c are functions of the independent variable and σ is equal to 0 or 1. Its structure becomes clearer when one writes it as a system. We have in this case a system of two Riccati's:

$$y' = -y^2 + by + c, \tag{1.10a}$$

$$x' = ax^2 + nxy + \sigma. \tag{1.10b}$$

Although the Gambier equation is always linearizable this does not mean it is always integrable. Indeed, in a system such as (1.10) of two equations in cascade we can always solve the first equation for y, obtain $y(t)$ and inject it into the second equation. The latter can always be written as a linear, second-order differential equation for x. So, in principle, the problem can always be solved formally. The difficulty comes when one wishes to actually compute x, in terms of contour integrals, while y has an analytic structure that interferes badly with that of x. This is where the Painlevé property comes into play. If we require that the system possess the Painlevé property the integration can be performed and we can indeed obtain the solution for $x(t)$ over the complex t-plane. Thus, the integrability of the Gambier equation will be closely related to its singularity structure.

The fact that the integrability of the Gambier equation is related to the Painlevé property allows us to obtain another interesting result. Just as in the case of the Painlevé equations, it is possible to introduce transformations relating the solution of an equation with parameter n to one with parameter $n+1$ or $n+2$ (two different transformations do exist). Thus the Gambier system possesses Schlesinger transformations [**10**].

Another feature of the Gambier equation is that it can be integrable discretized [**5, 6**]. Our approach follows closely the spirit of Gambier based on coupled Riccati equations. In perfect analogy to the continuous case it is possible to introduce a system of two homographic mappings in cascade which represent the discretization of the Gambier system. The general form of this system is the following:

$$y_{n+1} = \frac{ay_n + b}{cy_n + d}, \tag{1.11a}$$

$$x_{n+1} = \frac{(\alpha y_n + \beta)x_n + \gamma y_n + \delta}{(\varepsilon y_n + \zeta)x_n + \eta y_n + \theta}. \tag{1.11b}$$

The form (1.11) can be simplified through homographic transformations and the integrable cases can be obtained through the application of the singularity confinement discrete integrability criterion. As in the continuous case, although (1.11) is always linearizable it is not automatically integrable. The difficulty arises when one tries to compute x_n in terms of matrix products the elements of which contain y_n. Some of these matrices are singular in such a way that degrees of freedom are irretrievably lost when y_n has the wrong properties. Singularity confinement precisely means that these degrees of freedom are in fact recovered at some later stage.

Quite expectedly, the Gambier mapping has Schlesinger transformations. Their derivation is based on a study of the singularities of the mapping (1.11) and their confinement. Thus the parallel between the discrete and continuous cases is perfect.

In what follows, we shall present the singularity analysis of the Gambier equation and derive its integrable cases. Next, we study the discrete case and use the singularity structure in order to derive the Gambier mapping. Finally, we present the Schlesinger transformations of both the continuous and discrete Gambier systems which allow us, starting from some elementary case (n=0 or n=1), to construct recursively the Gambier systems for higher n's.

2. The Continuous Gambier Equation

The Gambier equation is given as a system of two Riccati equations in cascade. This means that we start with a first Riccati for some variable y

$$y' = -y^2 + by + c, \tag{2.1}$$

and then couple its solution to a second Riccati by making the coefficients of the latter depend explicitly on y:

$$x' = ax^2 + nxy + \sigma. \tag{2.2}$$

The precise form of the coupling introduced in (2.2) is due to integrability requirements. In fact, the application of singularity analysis shows that the Gambier system cannot be integrable unless the coefficient of the xy term in (2.2) is an integer n. This is not the only integrability requirement. Depending on the value of

n one can find constraints on the a, b, c, σ (where the latter is traditionally taken to be constant 1 or 0) which are necessary for integrability.

The common lore [**8**] is that out of the functions a, b, c two are free. This turns out not to be the case. The reason for this is that the system (2.1)–(2.2) is not exactly canonical i.e. we have not used all possible transformations in order to reduce its form. We introduce a change of independent variable from t to T through $dt = gdT$ where g is given by $1/gdg/dt = bn/2 - n$, a gauge through $x = gX$ and also $Y = gy - 1/ndg/dt$. The net result is that system (2.1)–(2.2) reduces to one where $b = 0$ while σ remains equal to 0 or 1. It is clear from the equations above that n must be different from 2. On the other hand when $n = 2$ the integrability conditions, if $\sigma = 1$, is precisely $b = 0$. So we can always take $b = 0$. (As a matter of fact in the case $\sigma = 0$ an additional gauge freedom allows us to take both b and c to zero for all n, even for $n = 2$). Thus the Gambier system can be written in full generality

$$y' = -y^2 + c, \tag{2.3a}$$

$$x' = ax^2 + nxy + \sigma. \tag{2.3b}$$

One further remark is in order here. The system (2.3) retains its form under the transformation $x \to 1/x$. In this case $n \to -n$ and σ and $-a$ are exchanged. Thus in some cases it will be interesting to consider a Gambier system where σ is not constant but rather a function of t. Still, it is possible to show that we can always reduce this case to one where $\sigma = 1$, while preserving the form of (2.3a) i.e. $b = 0$. To this end we introduce the change of variables $dt = hdT$, $x = gX$ and $Y = hy - \frac{1}{2}dh/dt$ with $h = \sigma^{2/(n-2)}$, $g = \sigma^{n/(n-2)}$. With these transformations system (2.3) reduces to one with $\sigma = 1$ and $b = 0$. (In the special case $n = 2$, with $b = 0$, integrability implies σ =constant, whereupon its value can always be reduced to 1).

In order to study the movable singularities of the coupled Riccati system we start from the observation that from (2.3a) the dominant behaviour of y can only be $y \approx 1/(t - t_0)$. The next terms in the expansion of y can be easily obtained, and involve the function c and its derivatives. In order to study the structure of the singularities of (2.3b), we first remark that since the latter is a Riccati, its movable singularities are poles. However, (2.3b) also has singularities that are due to the singular behaviour of the coefficients of the r.h.s of (2.3b), namely y. Now, the locations of the singularities of the coefficients are 'fixed' as far as (2.3b) is concerned. However, from the point of view of the full system (2.3), these singularities are *movable* and thus should be studied. The 'fixed' character reflects itself in the fact -1 is *not* a resonance. (The terms 'resonance' is used here following the ARS terminology [**1**] and means the order, in the expansion, where a free coefficient enters. A resonance -1 is related to the arbitrariness of the location of the singularity, and is thus absent when the location of the singularity is determined from the 'outside' rather than by the initial conditions). Because of the pole in y, x has a singular expansion with a resonance different from -1 which may introduce a compatibility condition to be satisfied.

We consider below the case of the full Riccati (2.3b) with $a \neq 0$, $\sigma = 1$. (The analysis of the case $a\sigma = 0$ was given in [**5**]). As we explained above, only the singularity due to y can lead to trouble. Rewriting (2.3b) as $x'/x = ax + ny + \sigma/x$ for $y = 1/(t - t_0) + \dots$ we remark that unless $n = \pm 1$ a behaviour of the form

$x \sim (t-t_0)^n$ is impossible when $a\sigma \neq 0$. For $n = 1$, a logarithmic leading behaviour will be present for $\sigma \neq 0$. (Note that the condition $\sigma = 0$ is sufficient for the absence of a critical singularity for $n = 1$ irrespective of the value of a. Similarly, in a dual way, for $n = -1$ the necessary and sufficient condition for the absence of a critical singularity is $a = 0$, whether σ is 1 or 0).

Next we assume $n \neq \pm 1$, in which case it suffices to study the singularities $x \approx \lambda(t - t_0)$, $\big(\lambda = \sigma/(1-n)\big)$ and $x \approx \mu/(t - t_0)$, $\big(\mu = -(n+1)/a\big)$. The first singular behaviour $\big(x \approx \lambda(t - t_0)\big)$ has a resonance at $n - 1$, which is negative for $n < 1$ and thus does not introduce any further condition. For $n > 1$, the resonance condition can be studied at least for the first few values of n. (In fact, $\sigma = 0$ suffices for the resonance condition to be satisfied even for $a \neq 0$). In the particular case $n = 2$, we have already mentioned that the integrability condition for $\sigma = 1$ is precisely $b = 0$. For higher n (and $\sigma = 1$) we find the further possibilities:

$$\begin{aligned} n &= 3, \quad 2c - a = 0, \\ n &= 4, \quad 3c' - a' = 0. \end{aligned} \tag{2.4}$$

The second singular behaviour, $x \approx \mu/(t-t_0)$, has a resonance at $-1-n$. Thus for $n > 0$ this resonance is negative and does not introduce any further condition, while for $n < 0$ a compatibility condition must be satisfied. (We find that for every case $n < 0$, $a = 0$ is a sufficient condition for the absence of critical singularity. This is not in the least astonishing given the duality of a and σ). On the other hand if we demand $a \neq 0$ then a different resonance condition is obtained, at each value of n. For $n = -1$, whenever $a \neq 0$, a logarithmic singularity of the form $(t - t_0)^{-1} \log(t - t_0)$ appears irrespective of the value of σ. For $n < -1$, we find:

$$\begin{gathered} n = -2, \quad a' = 0, \quad n = -3, \quad 2ac - a^2\sigma - 2a'' = 0, \\ n = -4, \quad 2ac' + a'(4c - 2a\sigma) - a''' = 0. \end{gathered} \tag{2.5}$$

The integrability condition for higher values of n can be obtained through the use of computer algebra.

3. The Discrete Gambier Equation

The discretisation of the Gambier equation is based on the idea of two Riccati equations in cascade. The discrete form of the first is simply:

$$\overline{y} = \frac{ay + b}{cy + d}, \tag{3.1}$$

where $y \equiv y_n$ and $\overline{y} \equiv y_{n+1}$. The second equation which contains the coupling can be discretised in several, not necessarily equivalent, ways. In [**6**] we have considered the generic coupling of the form:

$$\alpha x\overline{x}y + \beta x\overline{x} + \gamma\overline{x}y + \delta\overline{x} + \varepsilon xy + \zeta x + \eta y + \theta = 0. \tag{3.2}$$

Implementing a homographic transformation on x and y we can generically bring (3.2) under the form:

$$x\overline{x} + \gamma\overline{x}y - \varepsilon xy - \theta = 0. \tag{3.3}$$

A choice of different transformations can bring (3.2) also to the form

$$\overline{x} - x = -fx\overline{x} + (gx + h\overline{x})y + k. \tag{3.4}$$

Note that (3.2) contains an 'additive' type coupling $x\overline{x} + \delta\overline{x} + \zeta x + \eta y + \theta = 0$ for special values of its parameters, but the generic form (3.3) is that of a 'multiplicative' coupling where γ, ε do not vanish. Solving (3.3) for $\overline{x}$ we obtained the second equation of the discrete Gambier system in the form:

$$\overline{x} = \frac{\varepsilon x y + \theta}{x + \gamma y}. \tag{3.5}$$

Clearly, a scaling freedom remains in equation (3.5). We can use it in order to bring it to the final form:

$$\overline{x} = \frac{xy/d + c^2}{x + dy}. \tag{3.6}$$

Eliminating y and $\overline{y}$ from (3.1), (3.6) and its upshift, we can obtain a 3-point mapping for x alone but the analysis is clearer if we deal with both y and x.

In what follows we shall present a different derivation based on the study of singularities of the system. As we explained in [**10**] we can use the homographic freedom in order to bring the mapping for y to the form:

$$\overline{y} = \frac{y + c}{y + 1}, \tag{3.7}$$

instead of (3.1) where c is a function of n. Next, we turn to the equation for x. This equation is homographic in x. However we require that when y takes the value 0, the resulting value of x be ∞. Thus the denominator must be proportional to y, and since we can freely translate x, we can reduce its form to just xy. The remaining overall gauge factor is chosen so as to put the coefficient of xy of the numerator to unity resulting to the following mapping:

$$\overline{x} = \frac{x(y - r) + q(y - s)}{xy}. \tag{3.8}$$

The system (3.7)–(3.8) is a discrete form of the Gambier system. In order to study the confinement of the singularity induced by $y = 0$ we introduce the auxiliary quantity ψ_N which is the N'th iterate of $y = 0$ in equation (3.7), N times downshifted. Thus $\psi_0 = 0$, $\psi_1 = \underline{c}$, $\psi_2 = (\underline{\underline{c}} + \underline{c})/(\underline{\underline{c}} + 1)$, etc. The confinement requirement is that after N steps x becomes 0 in such a way as to lead to 0/0 at the next step. Thus the mapping (3.8) has in fact the form:

$$\overline{x} = \frac{x(y - r) + q(y - \psi_N)}{xy}. \tag{3.9}$$

Thus when at some step N we have $y = \psi_N$ and $x = 0$, on the view of (3.9) $\overline{x}$ will then be indeterminate of the form 0/0. However it turns out that in fact this value is well-determined and finite. Let us take a closer look at the conditions for confinement. The generic patterns for x and y are:

$$y : \left\{ \quad 0 \qquad \overline{\psi_1} \qquad \overline{\overline{\psi_2}} \qquad \cdots \qquad \overline{\overline{\overline{\vdots}}}\,\psi_N \quad \right\},$$

$$x : \left\{ \text{free} \qquad \infty \qquad \frac{\overline{\psi_1} - \overline{r}}{\overline{\psi_1}} \qquad \cdots \qquad 0 \qquad \text{free} \right\}.$$

At $N = 1$ it is clearly impossible to confine with a form (3.9) since we do not have enough steps. In this case the only integrable form of the x-equation is a linear one. This case must be studied separately. For an arbitrary N, the general form of the linear x-mapping can be obtained using confinement arguments in a way similar to what we did for the generic, nonlinear, case. We obtain

$$\overline{x} = \frac{x(y - \psi_N) + g}{y}, \tag{3.10}$$

where g is free.

The first genuinely confining case of the form (3.9) is $N = 2$. From the requirement $\overline{\overline{x}} = 0$ we have $r = \psi_1$ and q free: this is indeed the only integrability condition. For higher N's we can similarly obtain the confinement condition which takes the form of an equation for r in terms of q.

At this point it is natural to ask whether the mapping (3.7)–(3.9) does indeed correspond to the Gambier equation (2.3). In order to do this we construct its continuous limit. We first introduce:

$$c = \varepsilon^2 D, \quad y = \frac{\varepsilon D}{Y + H}, \tag{3.11}$$

with $H \approx D'/(2D)$ and obtain the continuous limit of (3.7) for $\varepsilon \to 0$. We find as expected

$$Y' = -Y^2 + C, \tag{3.12}$$

(i.e. Eq. (2.3a)) where $C = D - (D'')/(2D) + 3/4(D'^2)/(D^2)$. Using (3.7) and (3.11) we can also compute ψ_N and we find at lowest order:

$$\psi_N = \varepsilon^2 \Psi_N \quad \text{with } \Psi_N \approx N\left(D - \varepsilon \frac{N+1}{2} D'\right) + \varepsilon^2 \Phi_N, \tag{3.13}$$

where Φ_N is an explicit function of D depending on N.

Next we turn to the equation for x and introduce:

$$r = \varepsilon^2 R, \quad x \approx \frac{1}{2} + \frac{\varepsilon}{2X} - \varepsilon \frac{RD'}{4D^2}, \quad q \approx -\frac{1}{4} + \varepsilon^2 Q, \tag{3.14}$$

and for the continuous limit of the form (2.3b) to exist in canonical form (i.e. b=0, σ=1) we find that we must have

$$R \approx \frac{ND}{2} - \varepsilon(N + 2)\frac{ND'}{8}. \tag{3.15}$$

This leads to the equation for x:

$$X' = AX^2 + NXY + 1, \tag{3.16}$$

with $A = N/4(N/4+1)(D'^2)/(D^2) - (ND'')/(4D) - 4Q$. Moreover the confinement constraint implies a differential relation between D and Q which depends on N. We can verify explicitly in the first few cases that this is indeed the integrability constraint obtained in the continuous case. In the case $N = 2$, the condition for confinement is

$$\overline{r} = c. \tag{3.17}$$

Using the approximate expansion in ε (3.15) which gives r up to the third order in ε and putting $N = 2$, we get that equation (3.17) is automatically satisfied for the two first nonzero orders in ε (ε^2 and ε^3). This reflects the fact that for the continuous Gambier equation, in the case $N = 2$, the only integrability condition is $b = 0$. Thus no condition has to be imposed on D and Q.

The confinement condition for $N = 3$ is obtained by imposing $\overline{\overline{\overline{x}}} = 0$ when $y = 0$. This condition reads:

$$(\overline{c} + c - (\overline{c}+1)\overline{\overline{\psi}}_3) = (\overline{r} - c)(\overline{c} + c - \overline{\overline{r}}(\overline{c}+1)). \tag{3.18}$$

Putting $N = 3$, we find that equation (3.18) is identically satisfied for the first two nonzero orders in ε (ε^4 and ε^5). To get the integrability condition, we must calculate (3.18) at order ε^6. To do this, we need the value of Φ_3 in (3.13). We easily find that $\Phi_3 = 7D'' - 8D^2$. Satisfying (3.18) in ε^5 gives a relation giving explicitly Q in terms of D. Implementing this relation, we find that $A = 2C$ which is the continuous integrability condition for $N = 3$.

4. Schlesinger Transformations for the Continuous Gambier Equation

The theory of auto-Bäcklund transformations of Painlevé equations is well established. As was shown in [**3**] the general form of auto-Bäcklund transformations for most Painlevé equations is of the form:

$$\widetilde{x} = \frac{\alpha x' + \beta x^2 + \gamma x + \delta}{\varepsilon x' + \zeta x^2 + \eta x + \theta}. \tag{4.1}$$

In the case of the Gambier equation considered as a coupled system of two Riccati's it is more convenient to look for an auto-Bäcklund of the form:

$$\widetilde{x} = \frac{\alpha x y + \beta x + \gamma y + \delta}{(\zeta y + \eta)(\theta x + \kappa)}, \tag{4.2}$$

with a factorized denominator, with hindsight from the discrete case. We require that the equation satisfied by $\widetilde{x}$ do not comprise terms nonlinear in y. We examine first the case $\zeta \neq 0$ and reach easily the conclusion that there exists no solution. So we take $\zeta = 0$, $\eta = 1$ which implies that α and γ do not both vanish (otherwise (4.2) would have been independent of y). We find in this case $\alpha = 0$ and thus the general form of the auto-Bäcklund can be written as:

$$\widetilde{x} = \frac{\beta x + \gamma y + \delta}{\theta x + \kappa}. \tag{4.3}$$

From (4.3) we can obtain the two possible forms of the Gambier system auto-Bäcklund:

$$\widetilde{x} = \beta x + \gamma y + \delta, \tag{4.4}$$

$$\widetilde{x} = \frac{\beta x + \gamma y + \delta}{x + \kappa}. \tag{4.5}$$

As we shall see in what follows both forms lead to Schlesinger transformations.

Let us first work with form (4.4). Our approach is straightforward. We assume (4.4) and require that $\widetilde{x}$ satisfy an equation of the form (2.3b) while y is always the same solution of (2.3a). The calculation is easily performed leading to:

$$\widetilde{x} = \gamma y + \frac{a\gamma}{n+1} x + \frac{\gamma'}{n}, \tag{4.6}$$

where γ satisfies:

$$\frac{\gamma'}{\gamma} = \frac{n}{n+2}\frac{a'}{a}. \tag{4.7}$$

Here we have assumed $a \neq 0$; otherwise $\widetilde{x}$ does not depend on x and (4.6) does not define a Schlesinger. The parameters of the equation satisfied by $\widetilde{x}$ are given (in obvious notations) by:

$$\widetilde{n} + n + 2 = 0, \tag{4.8a}$$

$$\widetilde{a} = \frac{n+1}{\gamma}, \quad \text{and} \tag{4.8b}$$

$$\widetilde{\sigma} = \gamma\left(c + \frac{a\sigma}{n+1} + \frac{1}{n+2}\frac{a''}{a} - \frac{n+3}{(n+2)^2}\frac{a'^2}{a^2}\right). \tag{4.8c}$$

Thus (4.6) is indeed a Schlesinger transformation since it takes us from a Gambier system with parameter n to one with parameter $\widetilde{n} = -n-2$. It suffices now to invert $\widetilde{x}$ in order to obtain an equation with parameter $N = n+2$. Expressions (4.6) and (4.8) can be written in a more symmetric way:

$$\widetilde{a}\widetilde{x} - ax = (n+1)\left(y - \frac{a'}{\widetilde{n}a}\right), \tag{4.9}$$

and

$$\begin{aligned} \widetilde{n} + 1 &= -(n+1), \\ \widetilde{n}\frac{\widetilde{a}'}{\widetilde{a}} &= n\frac{a'}{a}, \\ \widetilde{a}\widetilde{\sigma} - a\sigma &= (n+1)\left(c - \frac{1}{\widetilde{n}}\left(\frac{a'}{a}\right)' - \frac{1}{\widetilde{n}^2}\frac{a'^2}{a^2}\right). \end{aligned} \tag{4.10}$$

The inverse transformation can be easily obtained if we introduce $\widetilde{\gamma}$ such that $a\widetilde{\gamma} = -(n+1) = -\widetilde{a}\gamma$. We thus find

$$x = y\widetilde{\gamma} + \frac{\widetilde{a}\widetilde{\gamma}}{\widetilde{n}+1}\widetilde{x} + \frac{\widetilde{\gamma}'}{\widetilde{n}}, \tag{4.11}$$

and the relations (4.10) are still valid.

Iterating the Schlesinger transformations one can construct the integrable Gambier systems for higher n's and obtain by construction the functions which appear in them. However it may happen that when we implement the Schlesinger we find $\widetilde{\sigma} = 0$. If we invert x we get a system with $N = -\widetilde{n} = n+2$ but $A = 0$ for which one cannot iterate the Schlesinger.

Let us give example of the application of this Schlesinger transformation. Let us start from $n = 0$, in which case we find $\widetilde{n} = -2$ and, after inversion, $N = 2$. For $n = 0$ we start from $a = -1$ and $\sigma = 0$ or 1 (always possible through the appropriate changes of variable). This leads to $\widetilde{a} = -1$, $\widetilde{\sigma} = -c + \sigma$ and the Schlesinger reads: $\widetilde{x} = -y + x$. Next we invert $\widetilde{x}$ and have $X = 1/(x-y)$. We find thus that the Schlesinger takes us from

$$y' = -y^2 + c, \quad x' = -x^2 + \sigma, \tag{4.12}$$

to the system

$$y' = -y^2 + c, \quad X' = AX^2 + 2Xy + \Sigma, \tag{4.13}$$

with $A = c - \sigma$, $\Sigma = 1$. In the particular case $n = 2$, a change of variables exists which allows us to put $A = -1$ (unless $A = 0$), without introducing b in the equation for y, while keeping $\Sigma = 1$ and changing only the value of c. Thus the generic case of the Gambier equation for $n = 2$ can be written with $A = -1$. Eliminating y between the two equations we find:

$$x'' = \frac{x'^2}{2x} - 2xx' - \frac{x^3}{2} - \frac{1}{2x} + (2c+1)x. \tag{4.14}$$

This is the generic form of the $n = 2$, Gambier equation [8] and it contains just one free function.

We turn now to the second Schlesinger transformation corresponding to the form (4.5). As we shall show, a Schlesinger transformation of this form does indeed exist and corresponds to changes in n with $\Delta n = 1$. Let us start from the basic equations (2.3). Next we ask that $\widetilde{x}$ defined by (4.5) indeed satisfy a system like (2.3). We find thus that and $\kappa = -x_0$ and γ must be given by:

$$\frac{\gamma'}{\gamma} = y_0 + \frac{2ax_0}{n+1}, \tag{4.15}$$

where y_0 is a solution of the Riccati (2.3a) and a solution x_0 of (2.3b), obtained with y replaced by y_0. We introduce the quantities $\widetilde{x}_0 = (a\gamma)/(n+1)$, $\widetilde{a} = -(nx_0)/\gamma$. In this case (4.15) becomes:

$$\frac{\gamma'}{\gamma} = y_0 + \frac{2\widetilde{a}\widetilde{x}_0}{\widetilde{n}+1} = y_0 + \frac{2x_0\widetilde{x}_0}{\gamma}, \tag{4.16}$$

where

$$\widetilde{n} + n + 1 = 0. \tag{4.17}$$

We have thus, starting from a generic solution x, y of (2.3) for some n, the Schlesinger:

$$\widetilde{x} = \widetilde{x}_0 + \frac{\gamma(y - y_0)}{x - x_0}, \tag{4.18}$$

where $\widetilde{x}$ is indeed a solution of (2.3) for $\widetilde{n} = -n - 1$ for the *same* y

$$\widetilde{x}' = \widetilde{a}\widetilde{x}^2 + \widetilde{n}\widetilde{x}y + \widetilde{\sigma}, \tag{4.19}$$

where $\widetilde{a}$ has been defined as $-nx_0/\gamma$ and

$$\widetilde{\sigma} = \frac{\gamma}{n+1}\left(a' + a^2x_0\frac{n+2}{n+1} + ay_0(n+2)\right). \tag{4.20}$$

Note that $\widetilde{x}_0$ is a solution of the same equation with y replaced by y_0. As in the previous case if we invert $\widetilde{x}$ we obtain an equation corresponding to $N = n + 1$.

It is worth pointing out here that the Schlesinger transformation corresponding to $\Delta n = 2$ was known to Gambier himself. As a matter of fact when faced with the problem of determining the functions appearing in his equation so as to satisfy the integrability requirement, Gambier proposed a recursive method which is essentially the Schlesinger $\Delta n = 2$. On the other hand the Schlesinger $\Delta n = 1$ is quite new.

5. Schlesinger Transformations for the Discrete Gambier Equation

Once the singularity pattern of the Gambier mapping is established we can use it in order to construct the Schlesinger transformation. Let us first look for a transformation that corresponds to $\Delta N = 2$. The idea is that given the N-steps singularity pattern of the equation for x we introduce a variable w with $N + 2$

singularity steps where we enter the singularity one step before x and exit it one step later. The general form of the Schlesinger transformation, which defines w, is:

$$w = X\frac{y - \psi_{N+1}}{y}, \tag{5.1}$$

where X is homographic in x. The presence of the y and $y - \psi_{N+1}$ terms is clear: they ensure that w becomes infinite one step before x, and vanishes one step after x. Next we turn to the determination of X. Since X is homographic in x we can rewrite (5.1) as:

$$w = \frac{\alpha x + \beta}{y}\frac{y - \psi_{N+1}}{\gamma x + \delta}. \tag{5.2}$$

Our requirement is that w becomes infinite when $y = 0$ for every value of x. This statement must be qualified. The numerator $\alpha x + \beta$ *will* vanish for some x (namely $x = -\beta/\alpha$) so this value of x must be the only one which should *not* occur in the confined singularity. Indeed there is a unique value of x where instead of being confined, the singularity extends to infinity in *both* directions of the independent variable n, while the only nonsingular values of the dependent variable occur in a finite range. The value of x such that $\overline{x}$ is finite and free even though y is zero is such that the numerator $-xr - q\psi_N$ of $\overline{x}$ vanishes. For this value of x, the values of the dependent variable are fixed for $n \leq 0$ and $n \geq N+1$ and the value can be considered as 'forbidden'. Thus $\alpha x + \beta = xr + q\psi_N$ up to a multiplicative constant. Similarly when $y = \psi_{N+1}$, w must vanish. Thus $\gamma x + \delta$ must not be zero except for the unique value of x that does not occur in the confined singularity. Note that $y = \psi_{N+1}$ means $\underline{y} = \underline{\psi}_N$ and the only value of x that comes from a nonzero $\underline{x}$ in that case is $x = (\underline{\psi}_N - \underline{r})/\underline{\psi}_N$. In that case the values of the dependent variable are fixed for $n \geq 0$ and $n \leq -N-1$. This value of x being 'forbidden', $\gamma x + \delta$ must be proportional to $\underline{\psi}_N x - (\underline{\psi}_N - \underline{r})$. We now have the first form of the Schlesinger:

$$w = \frac{xr + q\psi_N}{y}\frac{y - \psi_{N+1}}{\underline{\psi}_N x + \underline{r} - \underline{\psi}_N}, \tag{5.3}$$

where the proportionality constant has been taken equal to 1 (but any other value would have been equally acceptable). Here w effectively depends on x unless $r(\underline{r} - \underline{\psi}_N) = q\underline{\psi}_N\psi_N$. But in this case the mapping (3.8) is in fact linear in the variable $\xi = (x - 1 + \overline{r}/\overline{\psi}_N)^{-1}$. This case is the analog of the case $a = 0$ in the continuous case where the Schlesinger does not exist. Let us give an application of the Schlesinger transformation by obtaining the $N = 2$ equation starting from $N = 0$. We have always the equation for y which reads:

$$\overline{y} = \frac{y + c}{y + 1}, \tag{5.4}$$

and $\psi_0 = 0$, $\psi_1 = \underline{c}$. For $N = 0$ the equation for x reads:

$$\overline{x} = \frac{x(y - r) + qy}{xy} = \frac{x + q}{x}, \tag{5.5}$$

since for integrability $r = 0$ and indeed $N = 0$ means that the x equation does *not* depend on y. We introduce the Schlesinger:

$$w = x\frac{y - \underline{c}}{y}. \tag{5.6}$$

Using (5.5) and (5.6) to eliminate x we obtain the equation for w:

$$\overline{w} = (1-c)\frac{yw + q(y-\underline{c})}{(y+c)w}. \tag{5.7}$$

This equation is of the form (3.8) but not quite canonical. We can transform it to canonical form simply by introducing $\overline{y}$ instead of y because indeed w is infinite one step before x, so $w = \infty$ means $\overline{x} = \infty$ i.e. $\overline{y} = 0$. We obtain thus:

$$\overline{w} = \frac{w(\overline{y}-c) + q(1+\underline{c})(\overline{y}-\overline{\psi}_2)}{\overline{y}w}, \tag{5.8}$$

with $\overline{\psi}_2 = (c+\underline{c})/(1+\underline{c})$ which coupled to (5.4) is indeed a $N = 2$ Gambier mapping.

As we pointed out in Section 3, the linear case (3.10) necessitates a special treatment. The Schlesinger transformation is again given by:

$$w = X\frac{y-\psi_{N+1}}{y}, \tag{5.9}$$

and arguments similar to those of the nonlinear case allow us to determine the form of the homographic object X leading to:

$$w = \frac{x\psi_N - g}{y}\,\frac{y-\psi_{N+1}}{x\underline{\psi}_N - \underline{g}}. \tag{5.10}$$

Thus one can also perform a Schlesinger in the linear case. This is not in disagreement with the continuous case. It is, in fact, the analog of the case where $\sigma = 0$ but $a \neq 0$ (which is linear in $1/x$) for which the Schlesinger *can* be performed. The analog of the case $\sigma = 0$ and $a = 0$ is the situation when $g = k\psi_N$ with constant k in which case the mapping rewrites $\overline{\xi} = \xi(y-\psi_N)/y$ with $\xi = x - k$. Then w does not depend on ξ (or x) and (5.10) does not define a Schlesinger in analogy to the case $r(\underline{r} - \underline{\psi}_N) = q\underline{\psi}_N\psi_N$ in the nonlinear case.

Finally, we examine the possibility of the existence of a $\Delta N = 1$ Schlesinger. In this case, the structure of the transformation will be obtained by asking that the $N+1$ case enter the singularity one step before the N case but exit at the same point. The general structure is thus:

$$w = \frac{rx + q\psi_N}{y}\,\frac{y-\eta}{x-\xi}, \tag{5.11}$$

where η and ξ must be determined. We do this by requiring that the equation for w contain no coefficients nonlinear in y. As a result we find that η must satisfy the equation (3.7) for y:

$$\overline{\eta} = \frac{\eta + c}{\eta + 1}, \tag{5.12}$$

and ξ the equation (3.9) for x with η instead of y:

$$\overline{\xi} = \frac{\xi(\eta - r) + q(\eta - \psi_N)}{\xi\eta}. \tag{5.13}$$

We remark here the perfect parallel to the continuous case (and as we pointed out the discrete case led the investigation back to the continuous one). Let us point out here that the w obtained through (5.11) does not lead to $w = 0$ at the exit of the singularity (i.e. when $x = 0$, $y = \psi_N$) and a translation is needed. One has in principle to define a new variable

$$\omega = w - w(x = 0, y = \psi_N) = w + \frac{q}{\xi}(\psi_N - \eta).$$

Finally we derive the $\Delta N = 1$ Schlesinger for the case of a linear mapping (5.9). We start from:

$$w = \frac{x\psi_N - g}{y}\frac{y - \eta}{x - \xi}, \tag{5.14}$$

and again require for w an equation with coefficients linear in y. We find that η must again be a solution of the equation for y i.e. it must satisfy (5.12) and moreover ξ is a solution of (3.10) with $y = \eta$:

$$\overline{\xi} = \frac{\xi(\eta - \psi_N) + g}{\eta}. \tag{5.15}$$

Thus the list of the Schlesinger transformations of the Gambier mapping is complete.

6. Conclusion

In this work we have reviewed the Gambier system in both its continuous and discrete forms. We have shown that the singularity analysis can be used in order to obtain the integrable subcases of this equation and moreover we have derived their Schlesinger transformations. Several open questions appear at this point. The Gambier equation is the generic linearizable second-order equation of first degree. If one relaxes this last constraint, one can already obtain further linearizable equations. Cosgrove and Scoufis [**2**] have presented two such examples:

$$(x'')^2 = (ax' + a'x + c')^2 x', \tag{6.1}$$

$$\begin{aligned}(x'')^2 = A(x)\big(c_1(tx' - x) + c_2 x'(tx' - x) + c_3(x')^2 \\ + c_4(tx' - x) + c_5 x' + c_6\big).\end{aligned} \tag{6.2}$$

It would be interesting to study closely the properties of these equations and in particular their discretization.

Moving to higher orders one can wonder how the Gambier approach can be extended. In [**9**] we have presented a first exploration of this problem, in both continuous and discrete settings, for third-order systems.

Finally, the general problem of n-th-order linearizable systems is far from being solved. Only a special class of such systems is known, based on projective constructions. Recently, we have obtained the discretization of these projective Riccati systems [**7**]. Clearly, this is not the last word as far as linearizability is concerned.

References

1. M. J. Ablowitz, A. Ramani, and H. Segur, *Nonlinear evolution equations and ordinary differential equations of Painlevé type*, Lett. Nuovo Cimento **23** (1978), 333–338.
2. C. M. Cosgrove and G. Scoufis, *Painlevé classification of a class of differential equations of the second order and second degree*, Stud. Appl. Math. **88** (1993), 25–87.
3. A. S. Fokas and M. J. Ablowitz, *On a unified approach to transformations and elementary solutions of Painlevé equations*, J. Math. Phys. **23** (1982), 2033–2042.

4. B. Gambier, *Sur les équations différentielles du second order et du premier degré dont l'intégrale générale est à points critiques fixes*, Acta Math. **33** (1910), 1–55.
5. B. Grammaticos and A. Ramani, *The Gambier mapping*, Physica A **223** (1996), 125–136.
6. B. Grammaticos, A. Ramani, and S. Lafortune, *The Gambier mapping, revisited*, Physica A **253** (1998), 260–270.
7. B. Grammaticos, A. Ramani, and P. Winternitz, *Discretising families of linearizable equations*, Phys. Lett. A **245** (1998), 382–388.
8. E. L. Ince, *Ordinary differential equations*, Dover, New York, 1956.
9. S. Lafortune, B. Grammaticos, and A. Ramani, *Constructing integrable third-order systems: The Gambier approach*, Inverse Problems **14** (1998), 287–298.
10. B. Ramani, A. Grammaticos, and S. Lafortune, *Schlesinger transformations for linearizable equations*, Lett. Math. Phys. **46** (1998), 131–145.

CPT, École Polytechnique, CNRS, UPR 14, 91128 Palaiseau, France.
E-mail address: ramani@cpht.polytechnique.fr

LPN, Université Paris VII, Tour 24-14, 5^{e}étage, 75251 Paris, France.
E-mail address: grammati@paris7.jussieu.fr

Current address: Centre de recherches mathématiques, Université de Montréal, Montréal, Qué., H3C 3J7, Canada.

LPTM et GMPIB, Université Paris VII, Tour 24-14, 5e étage, 75251 Paris, France.
E-mail address: lafortus@crm.umontreal.ca

Centre de Recherches Mathématiques
CRM Proceedings and Lecture Notes
Volume **25**, 2000

Discrete Surfaces of Revolution

W. K. Schief and B. G. Konopelchenko

ABSTRACT. Geodesics and parallels on surfaces of revolution are discretized in a purely geometric manner. Discrete principal curvatures are defined and it is shown that the natural discrete Gauss equation is given by the standard discrete Schrödinger equation with the discrete Gaussian curvature as its potential. Algebraic and geometric properties which are reminiscent of those known in the continuous case are established. Important connections with integrable discrete equations are also recorded.

1. Introduction

The study of coordinate systems which are canonical on a given class of surfaces constitutes an integral part of the differential geometry of surfaces. The most important coordinate systems to be found in the classical literature are asymptotic, conjugate, curvature and geodesics coordinates. In the past few years, the discretizations of asymptotic and conjugate coordinates proposed by Sauer in the forties [**29, 30**] have been exploited to construct integrable discrete versions of pseudospherical surfaces [**3**], affine spheres [**6, 7**] and conjugate coordinate systems [**8, 12, 31**]. Recently, lattices composed of cyclic quadrilaterals, which are relevant in computer aided surface design [**22, 26**], have been identified as canonical discretizations of curvature coordinates. They have been used to obtain integrable discrete isothermic surfaces [**4**], surfaces of constant positive Gaussian curvature [**5**], orthogonal coordinate systems [**2,9,20**] and coordinate systems of Egorov type [**31**]. In the first part of the paper, we briefly summarize some of the above-mentioned results.

It appears that geodesics and their corresponding orthogonal trajectories (parallels), which were introduced in Gauss' *Disquisitiones* [**16**], have not been considered in the context of integrable discrete geometry. The present paper addresses this issue. As a first step towards the discretization of geodesic coordinates on surfaces, we here consider geodesics and parallels on surfaces of revolution. We give natural discretizations of the geodesics and parallels and define discrete principal curvatures in a purely geometric manner. In the classical case of surfaces of revolution, the Gauss equation reduces to a Schrödinger equation. Remarkably, it turns out that the canonical discrete analogue of the Gauss equation is given by the

2000 *Mathematics Subject Classification.* Primary: 52C99; Secondary: 53A99.
This is the final form of the paper.

standard discrete Schrödinger equation with the product of the discrete principal curvatures being the potential.

We discuss several specializations of surfaces of revolution in order to test the validity of our definitions in the discrete case. Thus, we investigate in detail surfaces of constant Gaussian curvature, their Darboux transform and the class of Weingarten surfaces which is obtained by assuming that the principal curvatures are proportional. We establish important algebraic and geometric analogies between the discrete and continuous cases. In particular, connections with an integrable discrete pendulum equation, the integrable differential-difference Heisenberg spin equation [**15, 19**] and a multi-parameter class of integrable mappings [**27, 28**] are recorded.

2. Integrable Discretizations of Coordinate Systems

This section is concerned with a review of canonical (integrable) discretizations of particular coordinate systems on surfaces in $\mathbb{R}^3$ and their applications with regard to modelling classes of parametrized surfaces. To this end, it is assumed that 'discrete coordinate lines' constitute coordinate polygons $\boldsymbol{r} = \boldsymbol{r}(n_i = \text{const.})$, $i = 1, 2$ on a $\mathbb{Z}^2$ lattice

$$\boldsymbol{r} : \mathbb{Z}^2 \to \mathbb{R}^3, \quad (n_1, n_2) \mapsto \boldsymbol{r}(n_1, n_2). \tag{2.1}$$

In the sequel, we refer to this lattice as a *discrete surface*. Moreover, the arguments of any vertex $\boldsymbol{r} = \boldsymbol{r}(n_1, n_2)$ will be suppressed and its four neighbours denoted by

$$\boldsymbol{r}_1 = \boldsymbol{r}(n_1 + 1, n_2), \quad \boldsymbol{r}_2 = \boldsymbol{r}(n_1, n_2 + 1), \tag{2.2}$$

$$\boldsymbol{r}_{\bar{1}} = \boldsymbol{r}(n_1 - 1, n_2), \quad \boldsymbol{r}_{\bar{2}} = \boldsymbol{r}(n_1, n_2 - 1). \tag{2.3}$$

Other vertices are represented by multiple indices. For instance, $(\boldsymbol{r}, \boldsymbol{r}_1, \boldsymbol{r}_2, \boldsymbol{r}_{12})$ designates an elementary quadrilateral. The symbols Δ_i and Δ_{12} denote the usual difference operators of first and second order respectively, that is $\Delta_i f = f_i - f$ and $\Delta_{12} f = f_{12} - f_1 - f_2 + f$.

2.1. Discrete asymptotic coordinates. A surface $\Sigma : \boldsymbol{r} = \boldsymbol{r}(u, v)$ in Euclidean space $\mathbb{R}^3$ is said to be parametrized in terms of *asymptotic* coordinates if the 2nd fundamental form $\mathrm{II} = -d\boldsymbol{r} \cdot d\boldsymbol{N}$ is off-diagonal, that is

$$\boldsymbol{N}_u \cdot \boldsymbol{r}_u = \boldsymbol{N}_v \cdot \boldsymbol{r}_v = 0, \tag{2.4}$$

where $\boldsymbol{N} = (\boldsymbol{r}_u \times \boldsymbol{r}_v)/|\boldsymbol{r}_u \times \boldsymbol{r}_v|$ denotes the unit normal to Σ. Consider a point P on Σ and four infinitesimally close points P_u, $P_{\bar{u}}$ and P_v, $P_{\bar{v}}$ lying on the asymptotic lines $v =$ const. and $u =$ const. respectively which pass through P. The corresponding position vectors may be written as $\boldsymbol{r}(P) = \boldsymbol{r}(u, v)$, $\boldsymbol{r}(P_u) = \boldsymbol{r}(u + du, v)$, $\boldsymbol{r}(P_{\bar{u}}) = \boldsymbol{r}(u - du, v)$ and $\boldsymbol{r}(P_v) = \boldsymbol{r}(u, v + dv)$, $\boldsymbol{r}(P_{\bar{v}}) = \boldsymbol{r}(u, v - dv)$. If we introduce the difference vectors

$$\Delta_u \boldsymbol{r} = \boldsymbol{r}(P_u) - \boldsymbol{r}(P), \quad \Delta_{\bar{u}} \boldsymbol{r} = \boldsymbol{r}(P_{\bar{u}}) - \boldsymbol{r}(P), \tag{2.5}$$

$$\Delta_v \boldsymbol{r} = \boldsymbol{r}(P_v) - \boldsymbol{r}(P), \quad \Delta_{\bar{v}} \boldsymbol{r} = \boldsymbol{r}(P_{\bar{v}}) - \boldsymbol{r}(P), \tag{2.6}$$

then Taylor's theorem delivers

$$\begin{aligned} |\Delta_{\bar{u}} \boldsymbol{r}, \Delta_u \boldsymbol{r}, \Delta_v \boldsymbol{r}| &\approx 4|\boldsymbol{r}_u \times \boldsymbol{r}_v|(\boldsymbol{r}_{uu} \cdot \boldsymbol{N}) du^3\, dv = 0, \\ |\Delta_{\bar{v}} \boldsymbol{r}, \Delta_u \boldsymbol{r}, \Delta_v \boldsymbol{r}| &\approx 4|\boldsymbol{r}_u \times \boldsymbol{r}_v|(\boldsymbol{r}_{vv} \cdot \boldsymbol{N}) du\, dv^3 = 0, \end{aligned} \tag{2.7}$$

by virtue of (2.4). This implies that, in the above approximation, the points P, $P_u, P_{\bar{u}}$ and P_v, $P_{\bar{v}}$ lie on a plane. The latter observation forms the basis of the following definition of discrete asymptotic nets:

The coordinate polygons of a discrete surface $\boldsymbol{r} : \mathbb{Z}^2 \to \mathbb{R}^3$ form a *discrete asymptotic net* if any vertex $\boldsymbol{r}$ and its four neighbours $\boldsymbol{r}_1, \boldsymbol{r}_{\bar{1}}, \boldsymbol{r}_2, \boldsymbol{r}_{\bar{2}}$ lie on a plane $\mathcal{P}$ (c.f. Fig. 1).[1]

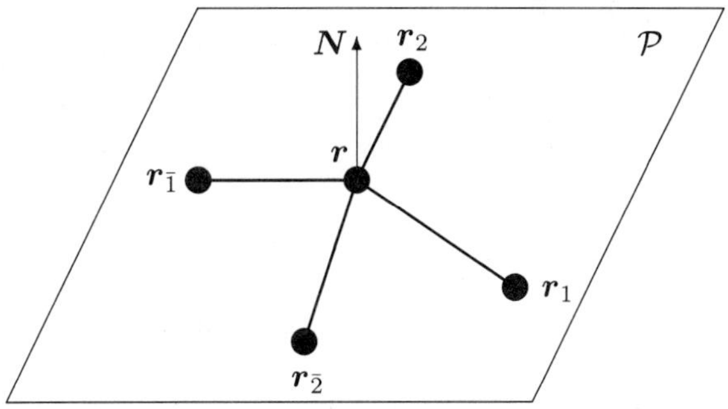

FIGURE 1. Discrete asymptotic nets.

Remarkably, the unit normal $\boldsymbol{N}$ to the plane $\mathcal{P}$, which may be regarded as 'normal' to the discrete surface, obeys the relations

$$\Delta_1 \boldsymbol{N} \cdot \Delta_1 \boldsymbol{r} = \Delta_2 \boldsymbol{N} \cdot \Delta_2 \boldsymbol{r} = 0, \tag{2.8}$$

which are nothing but the 'naive' discretization of the defining relations (2.4) for asymptotic coordinates.

The above definition of discrete asymptotic nets is due to Sauer [**29,30**] and has been used to construct integrable discretizations of pseudospherical surfaces [**3**] and affine spheres [**6**]. The corresponding underlying integrable discrete equations turn out to be Hirota's discrete sine-Gordon equation [**18**] and the discrete Tzitzeica equation [**32**].

2.2. Discrete conjugate coordinates. If the 2nd fundamental form associated with a surface $\Sigma : \boldsymbol{r} = \boldsymbol{r}(u,v) \in \mathbb{R}^3$ is diagonal then the coordinates (u,v) are *conjugate*. In this case, it is readily shown that the position vector $\boldsymbol{r}$ satisfies a hyperbolic equation of the form [**14**]

$$\boldsymbol{r}_{uv} = a(u,v)\boldsymbol{r}_u + b(u,v)\boldsymbol{r}_v. \tag{2.9}$$

Hence, if we consider four infinitesimally close points P, P_u, P_v and P_{uv} on Σ associated with the position vectors $\boldsymbol{r}(u,v)$, $\boldsymbol{r}(u+du,v)$, $\boldsymbol{r}(u,v+dv)$ and $\boldsymbol{r}(u+du, v+dv)$ and define the difference vectors

$$\Delta_u \boldsymbol{r} = \boldsymbol{r}(P_u) - \boldsymbol{r}(P), \quad \Delta_v \boldsymbol{r} = \boldsymbol{r}(P_v) - \boldsymbol{r}(P), \quad \Delta_{uv} \boldsymbol{r} = \boldsymbol{r}(P_{uv}) - \boldsymbol{r}(P), \tag{2.10}$$

then

$$|\Delta_u \boldsymbol{r}, \Delta_v \boldsymbol{r}, \Delta_{uv}| \approx |\boldsymbol{r}_u, \boldsymbol{r}_v, \boldsymbol{r}_{uv}| du^2\, dv^2 = 0, \tag{2.11}$$

on use of (2.9). Accordingly, the points P, P_u, P_v and P_{uv} approximately lie on a plane. This gives rise to the following definition:

[1]We usually commit the impropriety of referring to $\boldsymbol{r}$ as a vertex even though, strictly speaking, $\boldsymbol{r}$ is the position vector associated with a vertex.

A discrete surface $\boldsymbol{r} : \mathbb{Z}^2 \to \mathbb{R}^3$ constitutes a *discrete conjugate net* [**30**] if the elementary quadrilaterals $(\boldsymbol{r}, \boldsymbol{r}_1, \boldsymbol{r}_2, \boldsymbol{r}_{12})$ are planar as indicated in Fig. 2.

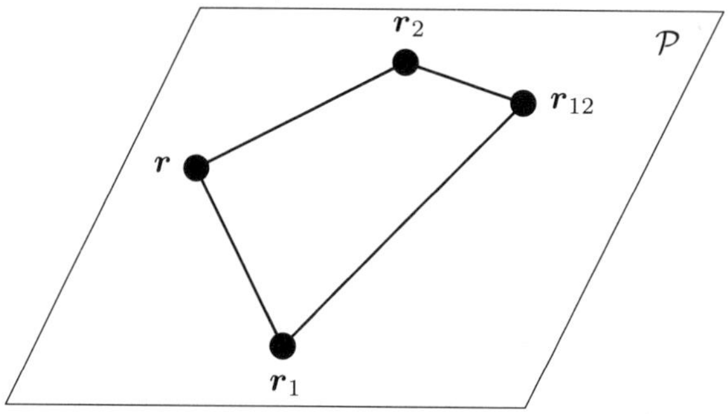

FIGURE 2. A planar elementary quadrilateral.

The position vector $\boldsymbol{r}$ then satisfies a linear equation of the form

$$\Delta_{12}\boldsymbol{r} = a(n_1, n_2)\Delta_1\boldsymbol{r} + b(n_1, n_2)\Delta_2\boldsymbol{r}, \tag{2.12}$$

which represents the natural discretization of the hyperbolic equation (2.9).

The discrete equation (2.12) has been discussed in [**11**] in connection with a discrete Laplace–Darboux transformation and the associated discrete Toda lattice. Discrete conjugate nets have been used in [**7**] to construct natural discretizations of affine spheres with definite Blaschke metric and the underlying elliptic Tzitzeica equation. Multi-dimensional discrete conjugate coordinate systems have been the subject of [**12**]. In particular, the integrable discrete Darboux system [**8**] has been shown to govern discrete conjugate coordinate systems in $\mathbb{R}^3$.

2.3. Discrete curvature coordinates. A canonical discretization of orthogonal coordinates on surfaces in the context of integrability has not yet been found. However, if the coordinate system on a surface $\Sigma : \boldsymbol{r} = \boldsymbol{r}(u, v) \in \mathbb{R}^3$ is orthogonal *and* conjugate, that is both fundamental forms are diagonal, then the coordinate lines constitute *lines of curvature.* In this case, a natural discretization is known:

The elementary quadrilaterals of *discrete 'curvature' nets* are inscribed in circles. In analytic terms, 'cyclic' quadrilaterals are described by the relation

$$\boldsymbol{e}_1 \cdot \boldsymbol{e}_3 + \boldsymbol{e}_2 \cdot \boldsymbol{e}_4 = 0, \tag{2.13}$$

where the vectors $\boldsymbol{e}_i$ are the unit vectors in 'tangential' direction as illustrated in Fig. 3. It is evident that the constraint (2.13) reduces to the orthogonality condition

$$\boldsymbol{r}_u \cdot \boldsymbol{r}_v = 0, \tag{2.14}$$

in the continuum limit.

Cyclic quadrilateral lattices which model surfaces parametrized in terms of curvature coordinates have been used in computer aided surface design [**22, 26**]. Several additional assumptions lead to important (integrable) subcases. Firstly, the additional constraint $a/d = b/c$ on the edges of the quadrilaterals (c.f. Fig. 3) corresponds to the condition that the 1st fundamental form be conformal. This has been exploited to construct discrete isothermic surfaces in [**4**]. Secondly, if one of the angles in the quadrilaterals, β, say, is $\pi/2$ then a discrete analogue of

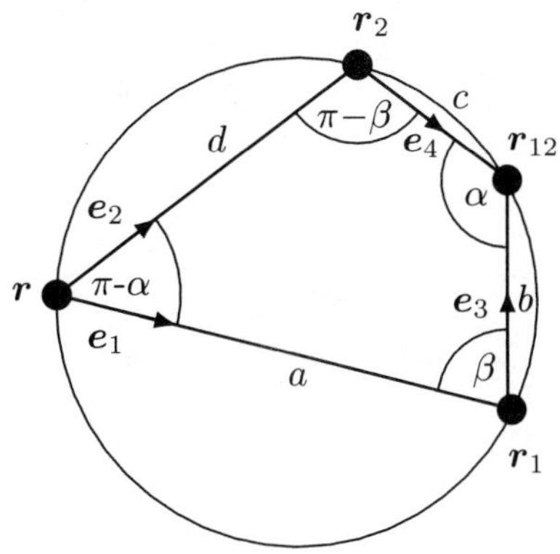

FIGURE 3. A cyclic elementary quadrilateral.

classical Egorov coordinates is obtained [**31**]. Thirdly, the quadrilaterals of the discrete surfaces of revolution considered here are trapezoidal and cyclic. Multi-dimensional discrete curvature coordinates have been extensively studied in [**2, 9, 20**]. In particular, cyclic lattices on the line, plane and in $\mathbb{R}^3$ have been shown to be governed by the discrete BKP and 2+1-dimensional sine-Gordon equations [**24, 25**].

3. Surfaces of Revolution

3.1. The classical case. If a surface $\Sigma : \boldsymbol{r} = \boldsymbol{r}(s,t)$ in Euclidean space $\mathbb{R}^3$ is parametrized in terms of geodesic coordinates, that is a family of geodesics and their orthogonal trajectories (parallels), then the 1st fundamental form $\mathrm{I} = d\boldsymbol{r} \cdot d\boldsymbol{r}$ takes the simple form

$$\mathrm{I} = ds^2 + \phi^2\, dt^2. \tag{3.1}$$

Here, the geodesics $t =$ const. are parametrized in terms of arclength, that is $\boldsymbol{r}_s^2 = 1$, and the lines $s =$ const. form a family of parallels on the surface. According to the *Theorema egregium*, the Gaussian curvature of the surface may be expressed entirely in terms of the coefficients of the 1st fundamental form. In particular, in the present case, the Gauss equation takes the form of a Schrödinger equation with the Gaussian curvature $\mathcal{K}$ as potential [**14**]:

$$\phi_{ss} + \mathcal{K}\phi = 0. \tag{3.2}$$

In the sequel, we focus on the subclass of surfaces of revolution. A surface of revolution is generated by a planar curve (generator) which is rotated around an axis lying in the plane of the generator. The meridians of the surface of revolution are the geodesics $t =$ const. and the parallels $s =$ const. consist of circles. In this case, the geodesics and parallels form lines of curvature so that the normal curvatures $\kappa^{(s)}$ and $\kappa^{(t)}$ of the s- and t-lines respectively coincide with the principal curvatures. Accordingly, the Gaussian curvature is given by

$$\mathcal{K} = \kappa^{(s)}\kappa^{(t)}. \tag{3.3}$$

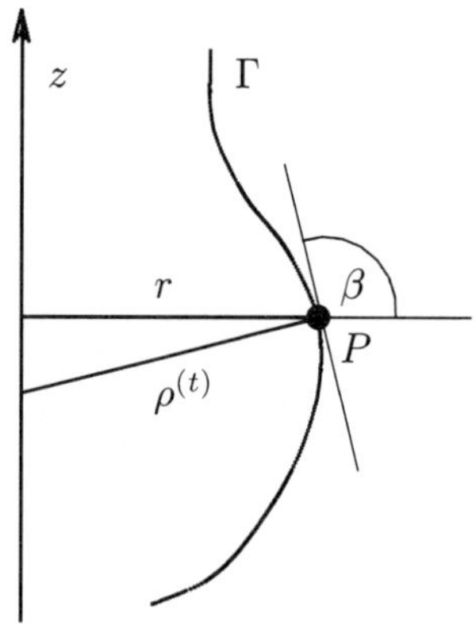

FIGURE 4. The generator of a surface of revolution.

Without loss of generality, the position vector $\boldsymbol{r} = \boldsymbol{r}(s,t)$ of a surface of revolution may be taken to be

$$\boldsymbol{r} = \begin{pmatrix} r(s)\cos(\omega t) \\ r(s)\sin(\omega t) \\ z(s) \end{pmatrix}, \tag{3.4}$$

where the functions $r(s)$ and $z(s)$ are related by the arclength condition

$$r_s^2 + z_s^2 = 1. \tag{3.5}$$

In order to be able to compare the continuous case with the discrete one, the constant angular velocity ω has not been scaled to unity. The curve $\Gamma : \boldsymbol{r} = \boldsymbol{r}(s,0)$ constitutes the generator of the surface of revolution. The constraint (3.5) may be solved identically by introducing an angle $\beta(s)$ according to

$$r_s = \cos\beta, \quad z_s = \sin\beta. \tag{3.6}$$

The geometric relation between r, z and β is illustrated in Fig. 4.

Now, in view of a canonical discretization procedure, it turns out convenient to focus on the standard geometric interpretation of the principal curvatures of surfaces of revolution. Thus, given a point P on the surface of revolution, the osculating circle associated with the meridian passing through P lies in the plane spanned by the axis of rotation and the meridian. If we denote its radius by ρ, the principal curvature $\kappa^{(s)}$ takes the form

$$\kappa^{(s)} = \kappa = \frac{1}{\rho} = \beta_s. \tag{3.7}$$

On the other hand, it is readily verified that

$$\kappa^{(t)} = \frac{1}{\rho^{(t)}} = \frac{\sin\beta}{r}, \tag{3.8}$$

where $\rho^{(t)}$ is the distance between the point P and the axis in the direction of the (principal) normal (c.f. Fig. 4). It is important to note that the Gauss equation (3.2) constitutes a Schrödinger equation for the radial distance r since $\phi = \omega r$ and whence

$$r_{ss} + \mathcal{K} r = 0. \tag{3.9}$$

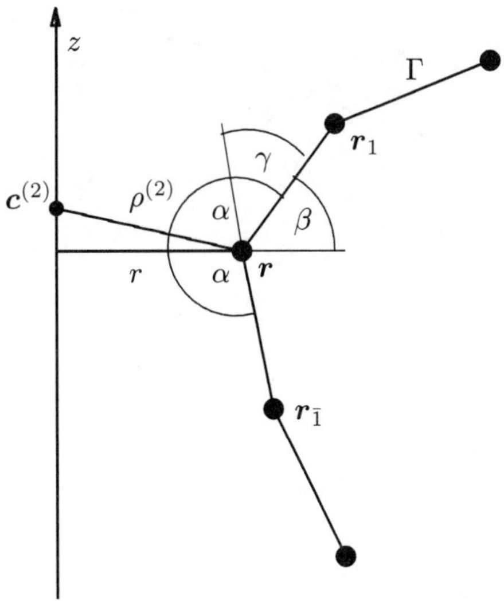

FIGURE 5. The generator of a discrete surface of revolution.

The latter may be derived directly via differentiation of $(3.6)_1$ and use of (3.7), (3.8).

3.2. The discrete case. Here, we are concerned with a canonical discretization of surfaces of revolution. More precisely, the geodesics and parallels on surfaces of revolution are discretized. Indeed, it is natural to consider a planar polygon $\Gamma : \boldsymbol{r} = \boldsymbol{r}(n_1)$ whose vertices are equidistant, that is $|\boldsymbol{r}_1 - \boldsymbol{r}| = 1$ without loss of generality (cf. Fig. 5).

If we now successively rotate Γ around an axis which lies in the same plane by a finite angle $\omega = 2\pi/N$, $N \in \boldsymbol{N}\!\boldsymbol{N}$, then a discrete surface $\Sigma : \boldsymbol{r} = \boldsymbol{r}(n_1, n_2)$ is generated which possesses a $\mathbb{Z}_N$ rotational symmetry. It is evident that the discrete meridians n_2 = const. represented by Γ may be regarded as discrete geodesics while the discrete parallels n_1 = const. consist of planar regular polygons ('discrete circles').

In order to construct a discrete counterpart of the Gauss equation (3.9) for these 'discrete surfaces of revolution', it is necessary to give natural geometric analogues of the principal curvatures $\kappa^{(s)}$ and $\kappa^{(t)}$. For this purpose, we first recall the standard definition of the radius of curvature for a curve in space. Consider a point P on a curve and two sufficiently close points P_1 and P_2 on this curve on either side of P. The circle passing through these three points is uniquely defined and the limiting circle as P_1, $P_2 \to P$ is called the osculating circle of the curve at P. Its radius ρ is termed the radius of curvature at P and $\kappa = 1/\rho$ is the corresponding curvature.

In the case of the discrete meridians, we define the circle passing through three adjacent vertices $\boldsymbol{r}_{\bar{1}}$, $\boldsymbol{r}$ and $\boldsymbol{r}_1$ as the osculating circle at $\boldsymbol{r}$. We regard the radius ρ of the osculating circle as the radius of curvature and its inverse $\kappa = 1/\rho$ as the curvature of the meridian at $\boldsymbol{r}$. In analogy with the continuous case, we therefore

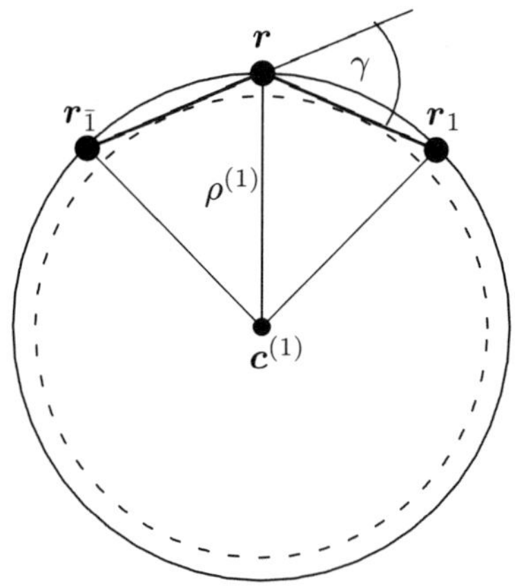

FIGURE 6. The discrete osculating circle.

define the discrete principal curvature $\kappa^{(1)}$ associated with the discrete meridian by

$$\kappa^{(1)} = \kappa = \frac{1}{\rho}, \tag{3.10}$$

and make the identification $\rho^{(1)} = \rho$. In terms of the angle between two adjacent edges, this means that

$$\kappa^{(1)} = 2\sin\frac{\gamma}{2}, \quad \cos\gamma = (\boldsymbol{r}_1 - \boldsymbol{r})\cdot(\boldsymbol{r} - \boldsymbol{r}_{\bar{1}}), \tag{3.11}$$

as indicated in Fig. 6.

On the other hand, the bisector of the polygon $(\boldsymbol{r}_{\bar{1}}, \boldsymbol{r}, \boldsymbol{r}_1)$ and the axis of revolution intersects at a point $\boldsymbol{c}^{(2)}$ which may be regarded as the centre of normal curvature associated with the discrete parallels (c.f. Fig. 5). Hence, if we denote the distance between $\boldsymbol{r}$ and $\boldsymbol{c}^{(2)}$ by $\rho^{(2)}$, then it is natural to define the principal curvature associated with the discrete parallels by

$$\kappa^{(2)} = \frac{1}{\rho^{(2)}}. \tag{3.12}$$

Now, without loss of generality, the position vector of a discrete surface of revolution may be taken to be

$$\boldsymbol{r} = \begin{pmatrix} r(n_1)\cos(\omega n_2) \\ r(n_1)\sin(\omega n_2) \\ z(n_1) \end{pmatrix}, \tag{3.13}$$

where the functions $r(n_1)$ and $z(n_1)$ are related by the 'discrete arclength condition'

$$(\Delta_1 r)^2 + (\Delta_1 z)^2 = 1. \tag{3.14}$$

The latter may be solved identically by setting

$$\Delta_1 r = \cos\beta, \quad \Delta_1 z = \sin\beta, \tag{3.15}$$

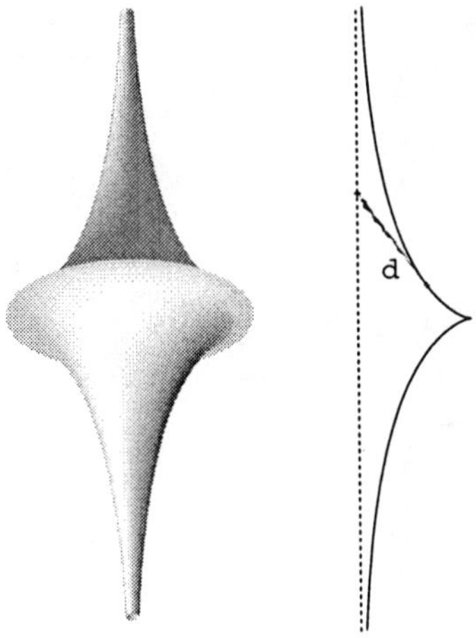

FIGURE 7. The pseudosphere and its generator (tractrix).

as indicated in Fig. 5. Consequently, the second radius of curvature at the vertex $\boldsymbol{r}$ reads

$$\rho^{(2)} = \frac{r}{\sin((\beta + \beta_{\bar{1}})/2)}, \tag{3.16}$$

so that the principal curvatures become

$$\kappa^{(1)} = 2\sin\left(\frac{\beta - \beta_{\bar{1}}}{2}\right), \quad \kappa^{(2)} = \frac{\sin((\beta + \beta_{\bar{1}})/2)}{r}. \tag{3.17}$$

Finally, combination of the relations $(3.15)_1$, (3.17) and application of a trigonometric identity produce

$$\Delta_{\bar{1}1} r + \mathcal{K} r = 0, \tag{3.18}$$

which is the discrete analogue of the Schrödinger equation (3.9).

4. The Pseudosphere

4.1. The classical case. In the case of pseudospherical surfaces, that is, without loss of generality, $\mathcal{K} = -1$, the Schrödinger equation (3.9) admits the first-order reduction

$$r_s = \pm r. \tag{4.1}$$

This implies that the distance of a point on the generator to the axis in tangential direction is

$$d = \left|\frac{r}{\cos\beta}\right| = \left|\frac{r}{r_s}\right| = 1 = \text{const}. \tag{4.2}$$

The latter property defines the classical *tractrix* which is displayed in Fig. 7. The associated surface of revolution constitutes the *pseudopshere.*

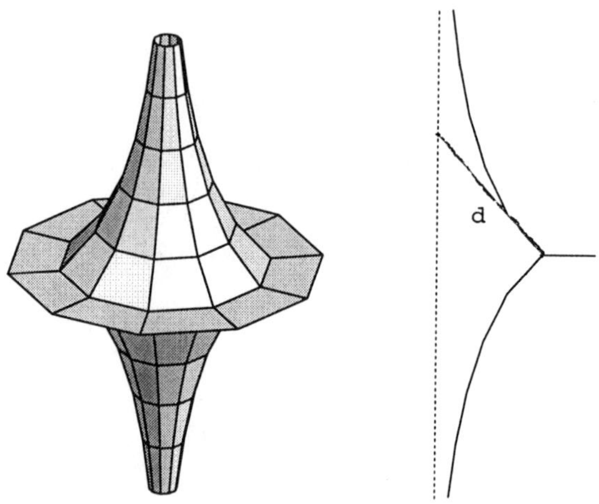

FIGURE 8. The discrete tractrix as the generator of the discrete pseudosphere.

4.2. The discrete case. As a first test of our definition of the discrete principal curvatures $\kappa^{(i)}$, we consider the case of constant discrete Gaussian curvature $\mathcal{K} = -\nu^2$. Since (3.18) now constitutes a linear difference equation with constant coefficients, it admits the first-order reduction

$$r_1 = \mu r, \tag{4.3}$$

where μ and $1/\mu$ are the roots of the characteristic equation $\mu^2 - (2+\nu^2)\mu + 1 = 0$. If $\mu < 1$, then we denote by d the distance between the vertex $\boldsymbol{r}$ and the axis in the ('tangential') direction $\Delta_1 \boldsymbol{r}$ (cf. Fig. 8).

Since $|\Delta_1 \boldsymbol{r}| = 1$, the relation

$$\frac{d-1}{d} = \frac{r_1}{r} = \mu, \tag{4.4}$$

holds and hence d is constant as in the continuous case. For $\mu > 1$, a similar argument is applicable. We may therefore regard the polygon $\Gamma : \boldsymbol{r} = \boldsymbol{r}(n_1, 0)$ as a discrete tractrix which generates a discrete pseudosphere. Remarkably, in his book on difference geometry, Sauer [**30**] constructed the same discrete pseudosphere in a purely geometric manner by taking the condition $d =$ const. as a natural discretization of that defining the classical tractrix. The discrete pseudosphere is displayed in Fig. 8.

5. Hasimoto Surfaces

In this section, we first review the (semi-)discretization of a class of surfaces governed by the important Heisenberg spin equation and then focus on the subcase of surfaces of revolution.

5.1. The Heisenberg spin equation. In a key paper [**17**], Hasimoto demonstrated that the intrinsic equation governing the curvature κ and torsion τ of an isolated thin vortex filament without stretching in an incompressible inviscid fluid may be approximated by the nonlinear Schrödinger equation

$$iq_t + q_{ss} + \frac{1}{2}|q|^2 q = 0, \tag{5.1}$$

where s denotes arclength measured along the filament, t is time and q is related to κ and τ by the 'Hasimoto transformation'

$$q = \kappa e^{\mathrm{i}\int \tau\, ds}. \tag{5.2}$$

His derivation was based on the fact that the vortex filament evolves in binormal direction with speed proportional to the curvature κ. Thus, if $\boldsymbol{r}(s,t)$ denotes the position of the vortex filament Γ in space, then its motion is constrained by

$$\boldsymbol{r}_t = \kappa \boldsymbol{b}, \quad \boldsymbol{r}_s^2 = 1, \tag{5.3}$$

where $\boldsymbol{b}$ designates the binormal to Γ. The latter condition encodes the assumption of a nonstretching filament.

Since the binormal $\boldsymbol{b}$ is orthogonal to both the unit tangent $\boldsymbol{t} = \boldsymbol{r}_s$ and the principal normal $\boldsymbol{n}$, the Serret–Frenet equations [**14**]

$$\begin{pmatrix} \boldsymbol{t} \\ \boldsymbol{n} \\ \boldsymbol{b} \end{pmatrix}_s = \begin{pmatrix} 0 & \kappa & 0 \\ -\kappa & 0 & \tau \\ 0 & -\tau & 0 \end{pmatrix} \begin{pmatrix} \boldsymbol{t} \\ \boldsymbol{n} \\ \boldsymbol{b} \end{pmatrix}, \tag{5.4}$$

deliver

$$\kappa \boldsymbol{b} = \boldsymbol{t} \times \boldsymbol{t}_s. \tag{5.5}$$

Consequently, cross-differentiation of $(5.3)_1$ and $\boldsymbol{r}_s = \boldsymbol{t}$ results in the integrable Heisenberg spin equation

$$\boldsymbol{t}_t = \boldsymbol{t} \times \boldsymbol{t}_{ss}. \tag{5.6}$$

In the present context, the latter was set down by Betchov [**1**] and related to the nonlinear Schrödinger equation by Hasimoto [**17**]. Later, in terms of integrability, it was pointed out that the Heisenberg spin equation is linked to the nonlinear Schrödinger equation via a gauge (Miura-type) transformation [**21, 34**].

5.2. The differential-difference Heisenberg spin equation. Consider a polygon $\Gamma : \boldsymbol{r} = \boldsymbol{r}(n_1)$ with $|\Delta_1 \boldsymbol{r}| = 1$ so that

$$\boldsymbol{t} = \Delta_1 \boldsymbol{r}, \tag{5.7}$$

may be regarded as a discrete unit 'tangent'. It is then natural to interpret the unit vector which is perpendicular to the edges $\boldsymbol{t}$ and $\boldsymbol{t}_{\bar{1}}$, that is

$$\boldsymbol{b} = \frac{\boldsymbol{t}_{\bar{1}} \times \boldsymbol{t}}{|\boldsymbol{t}_{\bar{1}} \times \boldsymbol{t}|} = \frac{\boldsymbol{t}_{\bar{1}} \times \boldsymbol{t}}{\sin\gamma} = \cot\left(\frac{\gamma}{2}\right) \frac{\boldsymbol{t}_{\bar{1}} \times \boldsymbol{t}}{1 + \boldsymbol{t} \cdot \boldsymbol{t}_{\bar{1}}}, \tag{5.8}$$

as the discrete binormal attached to the vertex $\boldsymbol{r}$, where γ is, as usual, defined by

$$\cos\gamma = \boldsymbol{t} \cdot \boldsymbol{t}_{\bar{1}}. \tag{5.9}$$

If we now assume in analogy with the Hasimoto vortex motion that the discrete curve Γ evolves in binormal direction with speed

$$\tilde{\kappa} = 2\tan\left(\frac{\gamma}{2}\right), \tag{5.10}$$

in such a way that the discrete arclength condition $|\Delta_1 \boldsymbol{r}| = 1$ is preserved, then compatibility of

$$\dot{\boldsymbol{r}} = \tilde{\kappa} \boldsymbol{b} = 2\tan\left(\frac{\gamma}{2}\right) \boldsymbol{b}, \tag{5.11}$$

and (5.7) produces the integrable differential-difference Heisenberg spin equation [**13, 15, 19**]

$$\dot{\boldsymbol{t}} = 2\Delta_1\left(\frac{\boldsymbol{t}_{\bar{1}} \times \boldsymbol{t}}{1 + \boldsymbol{t} \cdot \boldsymbol{t}_{\bar{1}}}\right). \tag{5.12}$$

Here, the overdot indicates differentiation with respect to time t. It is emphasized that the quantity $\tilde{\kappa}$ is another measure for the curvature of the polygon Γ. In fact, $1/\tilde{\kappa}$ is the radius of the dashed circle in Fig. 6 which touches two adjacent edges of Γ. It is noted that the centres of the 'osculating' circles associated with κ and $\tilde{\kappa}$ coincide.

5.3. A Hasimoto surface of revolution. The classical Darboux transformation [**10**] provides a purely algebraic means of generating a sequence of potentials in a Schrödinger equation along with their corresponding eigenfunctions if the eigenfunctions associated with a seed potential are known. It is therefore natural to investigate the action of the Darboux transformation on (3.9). Here, we consider a degenerate Darboux transformation which reads as follows: If r and $\widehat{r}$ are two solutions of the Schrödinger equation (3.9) with seed potential $\mathcal{K}$, then another solution of the Schrödinger equation is given by

$$r' = \widehat{r}_s - \frac{r_s}{r}\widehat{r}, \quad \mathcal{K}' = \mathcal{K} - 2(\ln r)_{ss}. \tag{5.13}$$

However, since the Wronskian $w(r, \widehat{r}) = r\widehat{r}_s - r_s\widehat{r}$ is constant, the above transformation for r reduces to an 'inversion'

$$r' = \frac{c}{r}, \tag{5.14}$$

where c is an arbitrary constant.

Once again, it is instructive to consider the simplest case $\mathcal{K} = -1$. Hence,

$$r = \cosh s, \quad r' = \frac{2}{\cosh s}, \tag{5.15}$$

form a corresponding pair of eigenfunctions with associated new potential

$$\mathcal{K}' = -1 - \frac{1}{\cosh^2 s}, \tag{5.16}$$

and integration of the primed version of the relations $(3.6)_2$ yields

$$z' = -s + 2\tanh s, \tag{5.17}$$

without loss of generality. The corresponding generator Γ' is depicted in Fig. 9. It contains a loop and therefore generates a self-intersecting surface of revolution.

The appearance of a one-soliton sech^2-potential and a 'loop soliton' as generator is not coincidental. Indeed, the connection with soliton theory is readily established if one makes use of the fact that the new principal curvature $\kappa^{(s)\prime}$ reads

$$\kappa^{(s)\prime} = \kappa' = \beta_s' = r'. \tag{5.18}$$

Thus, on the one hand, it is natural to consider the general case of a surface of revolution subject to the constraint

$$r = \tilde{c}\kappa = \tilde{c}\beta_s, \tag{5.19}$$

where $\tilde{c}$ is an arbitrary constant. Insertion of this condition into $(3.6)_1$ produces the pendulum equation

$$\tilde{c}\beta_{ss} = \cos\beta. \tag{5.20}$$

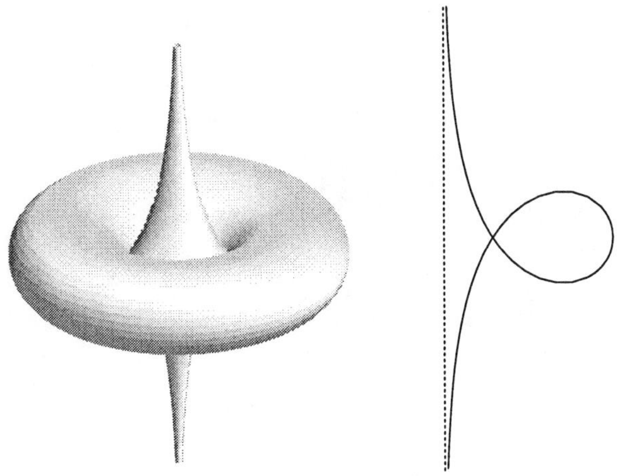

FIGURE 9. A Hasimoto surface of revolution and its generator.

It turns out that the well-known kink solution of the pendulum equation gives rise to the above surface of revolution.

On the other hand, combination of (5.18) and the primed version of (3.4) shows that $\boldsymbol{r}'_t = \omega\kappa'\boldsymbol{b}'$ which implies that the generator Γ' evolves in binormal direction with speed proportional to the curvature κ'. Accordingly, Γ' is governed by the integrable Hasimoto evolution (5.3). The latter is linked to the nonlinear Schrödinger equation and may also be related to a complex loop soliton equation by employing a reciprocal transformation. It turns out that the stationary one-soliton solution of the NLS equation induces a motion of the filament in such a way that the above surface of revolution is swept out. Moreover, the corresponding solution of the loop soliton equation at any fixed time t is represented by the generator of that surface [**33**].

5.4. A discrete Hasimoto surface of revolution. It is well-known that the discrete Schrödinger equation (3.18) is form-invariant under a discrete version of the classical Darboux transformation [**23**]. In parallel with the previous subsection, we now investigate the action of a degenerate discrete Darboux transformation on (3.18). Thus, if r and $\widehat{r}$ are two solutions of the discrete Schrödinger equation (3.18) with seed potential $\mathcal{K}$, then another solution of the discrete Schrödinger equation is given by

$$r' = \widehat{r}_1 - \frac{r_1}{r}\widehat{r}, \quad \mathcal{K}' - 2 = \frac{r^2}{r_1 r_{\bar{1}}}(\mathcal{K} - 2). \tag{5.21}$$

Once again, the Wronskian $w(r, \widehat{r}) = r\widehat{r}_1 - r_1\widehat{r}$ is constant so that the above transformation reduces to

$$r' = \frac{c}{r}, \tag{5.22}$$

where c is an arbitrary constant.

As a seed potential, we now choose $\mathcal{K} = -\nu^2$ so that a particular solution of the discrete Schrödinger equation (3.18) is given by the 'discrete hyperbolic cosine'

$$r = \mu^{n_1} + \mu^{-n_1}. \tag{5.23}$$

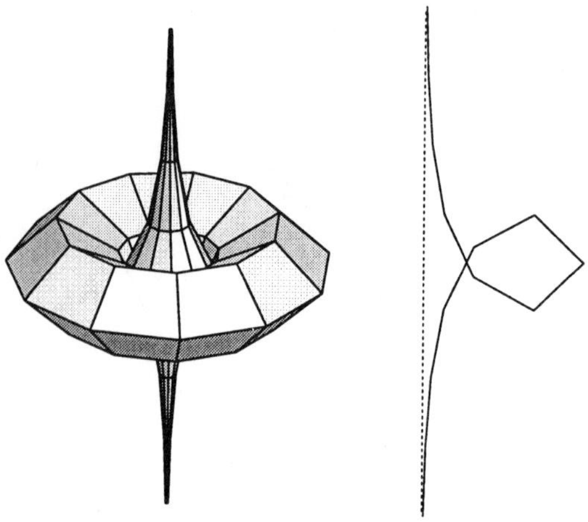

FIGURE 10. A discrete Hasimoto surface of revolution and its generator.

It is readily shown that the new angle β' represents a discrete kink if the constant c in the transformation law (5.22) is chosen to be

$$c = 2\frac{\mu+1}{\mu-1}. \tag{5.24}$$

Indeed, for this value of c, the primed version of the relation $(3.15)_1$ delivers

$$\beta' = 2\arctan\left(\frac{\mu^{n_1+1} - \mu^{-n_1}}{\mu+1}\right) + \frac{\pi}{2}, \tag{5.25}$$

and the new z-component of the position vector reads

$$z' = -n_1 + \left(\frac{\mu+1}{\mu-1}\right)\left(\frac{\mu^{n_1} - \mu^{-n_1}}{\mu^{n_1} + \mu^{-n_1}}\right). \tag{5.26}$$

The corresponding generator Γ' is depicted in Fig. 10. It contains a discrete loop and generates a discrete version of the Hasimoto surface of revolution displayed in Fig. 9.

The discrete Hasimoto surface of revolution as shown in Fig. 10 may be considered 'integrable' in the following sense. It is straightforward to show that

$$r' = \tilde{c}\tilde{\kappa}', \quad \tilde{c} = \frac{1}{(\mu^{1/2} - \mu^{-1/2})^2}, \quad \tilde{\kappa}' = 2\tan\left(\frac{\beta' - \beta'_{\bar{1}}}{2}\right). \tag{5.27}$$

It is therefore natural to investigate the class of discrete surfaces of revolution defined by the relation

$$r = \tilde{c}\tilde{\kappa}, \quad \tilde{\kappa} = 2\tan\left(\frac{\beta - \beta_{\bar{1}}}{2}\right), \tag{5.28}$$

where $\tilde{c}$ is an arbitrary constant. Now, insertion of the constraint (5.28) into $(3.15)_1$ results in the discrete pendulum equation

$$\tan\left(\frac{\beta_1 - \beta}{2}\right) - \tan\left(\frac{\beta - \beta_{\bar{1}}}{2}\right) = \frac{1}{2\tilde{c}}\cos\beta. \tag{5.29}$$

Its kink solution is given by (5.25) if we parametrize $\tilde{c}$ according to $(5.27)_2$. Interestingly, the kink solutions (5.25) and

$$q = 4\arctan\mu^{n_1}, \tag{5.30}$$

of the standard integrable discrete pendulum equation [**18**]

$$\sin\left(\frac{q_1 - 2q + q_{\bar{1}}}{4}\right) = \left(\frac{\mu-1}{\mu+1}\right)^2 \sin\left(\frac{q_1 + 2q + q_{\bar{1}}}{4}\right), \tag{5.31}$$

are related by

$$\beta' = \frac{q_1 + q}{2} - \frac{\pi}{2}. \tag{5.32}$$

However, this connection between the discrete pendulum equations (5.29) and (5.31) is only valid in the case of the kink solutions. Thus, the constraint (5.32) leads to a *special* first integral which is common to both discrete pendulum equations. This particular integral of motion is then parametrized by the kink solution (5.25) (or (5.30)).

It turns out that the discrete pendulum equation (5.29) is indeed integrable. In fact, Quispel *et al.* [**27**] have noted that the ansatz

$$\boldsymbol{t} = \begin{pmatrix} \cos\beta\cos 2\sigma t \\ \cos\beta\sin 2\sigma t \\ \sin\beta \end{pmatrix}, \tag{5.33}$$

reduces the differential-difference Heisenberg spin equation (5.12) to a second-order difference equation for $x = \tan\beta/2$ which is contained in an 18-parameter class of integrable reversible mappings of the plane that may be parametrized in terms of elliptic functions. In the present context, the reduction (5.33) has a distinct geometric meaning. Indeed, the relation (5.7) shows that the position vector of Γ takes the form

$$\boldsymbol{r} = \begin{pmatrix} r(n_1)\cos 2\sigma t \\ r(n_1)\sin 2\sigma t \\ z(n_1) \end{pmatrix}, \tag{5.34}$$

where $r(n_1)$ and $z(n_1)$ are related to β by (3.15). Thus, the polygon Γ sweeps out a (semi-discrete) surface of revolution . Now, on the one hand, it is readily verified that

$$\dot{\boldsymbol{r}} = -2\sigma r\boldsymbol{b}, \tag{5.35}$$

so that comparison with the evolution (5.11) reveals that the condition (5.28) is satisfied with $2\tilde{c}\sigma = -1$. On the other hand, the surface of revolution represented by (5.34) may be obtained from the discrete surfaces of revolution considered here in the limit $N \to \infty$. Hence, the constraint (5.33) reduces the differential-difference Heisenberg spin equation (5.12) to the discrete pendulum equation (5.29) since the latter does not depend on N. We therefore conclude that the class of discrete surfaces of revolution defined by (5.28) is governed by the integrable discrete pendulum equation (5.29).

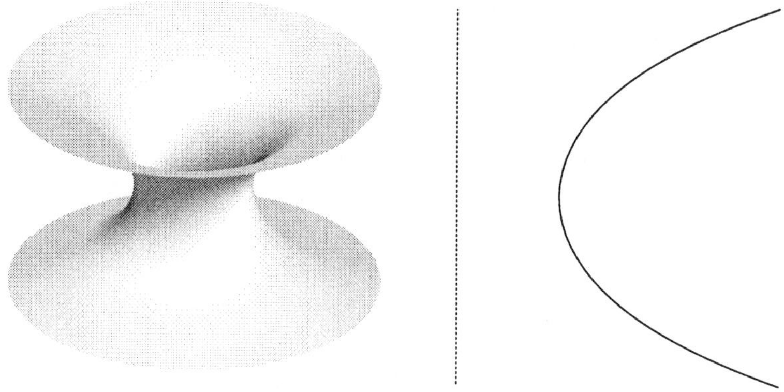

FIGURE 11. The catenoid and its generator (catenary).

6. Particular Weingarten Surfaces

6.1. The classical case. A class of surfaces which contains surfaces of constant Gaussian curvature, constant mean curvature and minimal surfaces was first studied by Weingarten [**14**] in the last century. Weingarten surfaces are defined by the property that the principal curvatures be functionally dependent. Surfaces of revolution constitute a subclass of Weingarten surfaces since the principal curvatures depend only on the arclength parameter s and hence $\kappa^{(t)} = F(\kappa^{(s)})$ in the generic case. Here, we briefly mention (Weingarten) surfaces of revolution for which a relation of the form

$$\kappa^{(s)} = c\kappa^{(t)}, \quad c = \text{const.}, \tag{6.1}$$

holds. In the case $c = 1$, it is readily shown that the principal curvatures are constant and the meridians are circles. Hence, the corresponding surfaces of revolution constitute spheres. On the other hand, if the surface is minimal, that is

$$\mathcal{H} = \kappa^{(s)} + \kappa^{(t)} = 0, \tag{6.2}$$

then the classical *catenoid* is obtained (Fig. 11).

It is natural to enquire as to whether there exists some intersection between the classes (5.19) and (6.1). Thus, insertion of r as given by (5.19) into the constraint (6.1) produces the relation

$$\tilde{c}\beta_s^2 - c\sin\beta = I = 0. \tag{6.3}$$

Differentiation shows that I is a first integral of the pendulum equation (5.20) if and only if $c = 2$. Since $\tilde{c}$ is arbitrary, the complete class of surfaces of revolution with

$$\kappa^{(s)} = 2\kappa^{(t)}, \tag{6.4}$$

is therefore contained in the 'Hasimoto' class (5.19). The discrete analogue of this result is presented below.

6.2. The discrete case. In the preceding, we have made assumptions about the discrete Gaussian curvature $\mathcal{K} = \kappa^{(1)}\kappa^{(2)}$ in order to establish further analogies between the discrete and continuous cases. Here, as a first test of the appropriateness of the indvidual discrete principal curvatures $\kappa^{(1)}$ and $\kappa^{(2)}$, we consider the 'discrete Weingarten surfaces' of revolution defined by the relation

$$\kappa^{(1)} = c\kappa^{(2)}, \quad c = \text{const.} \tag{6.5}$$

Insertion of the definitions (3.17) into the above constraint and elimination of β by means of $(3.15)_1$ produce the second-order difference equation

$$\left(\frac{2r-c}{2r+c}\right)^2 = \left(\frac{1+r_1-r}{1-r_1+r}\right)\left(\frac{1-r+r_{\bar{1}}}{1+r-r_{\bar{1}}}\right). \tag{6.6}$$

Alternatively, elimination of r in favour of $x = \tan\beta/2$ leads to the second-order difference equation

$$\frac{2}{c}\left(\frac{1-x^2}{1+x^2}\right) = \frac{x_1+x}{x_1-x} - \frac{x+x_{\bar{1}}}{x-x_{\bar{1}}}. \tag{6.7}$$

Both equations may be brought into the compact form

$$X_1 = \frac{f^1(X) - X_{\bar{1}}f^2(X)}{f^2(X) - X_{\bar{1}}f^3(X)}. \tag{6.8}$$

The functions $f^i(X)$ may easily be calculated, depending on whether $X = r$ or $X = x$.

Within the class of difference equations (6.8), Quispel *et al.* [**28**] have isolated a 12-parameter sub-class of equations which admit first integrals that are biquadratic and symmetric in X and X_1. Furthermore, these 'invariant curves' may be parametrized in terms of elliptic functions so that the corresponding two-dimensional mappings (6.8) are considered integrable. The 12-parameter class is given by

$$\boldsymbol{f}(X) = \begin{pmatrix} f^1(X) \\ f^2(X) \\ f^3(X) \end{pmatrix} = (A\boldsymbol{X}) \times (B\boldsymbol{X}), \quad \boldsymbol{X} = \begin{pmatrix} X^2 \\ X \\ 1 \end{pmatrix}, \tag{6.9}$$

where A and B are symmetric but otherwise arbitrary constant matrices, and the corresponding first integral reads

$$\frac{\boldsymbol{X}\cdot(A\boldsymbol{X}_1)}{\boldsymbol{X}\cdot(B\boldsymbol{x}_1)} = \text{const.} \tag{6.10}$$

In the case $c = 1$, that is $\kappa^{(1)} = \kappa^{(2)}$, the difference equation (6.6) is indeed a member of the above 12-parameter class. This does *not* apply to the difference equation (6.7). However, it is not difficult to show that the latter admits a first integral which turns out to be a discrete Riccati equation and hence may be linearized. In fact, it may be directly verified that

$$\beta = \gamma n_1 + \delta, \quad r = \frac{\sin(\beta-\gamma/2)}{2\sin(\gamma/2)}, \quad z = z_0 - \frac{\cos(\beta-\gamma/2)}{2\sin(\gamma/2)}, \tag{6.11}$$

so that

$$\kappa^{(1)} = \kappa^{(2)} = \text{const.}, \quad r^2 + (z-z_0)^2 = \text{const.} \tag{6.12}$$

Accordingly, as in the continuous case, the principal curvatures are constant and the vertices of the discrete surface of revolution lie on a sphere.

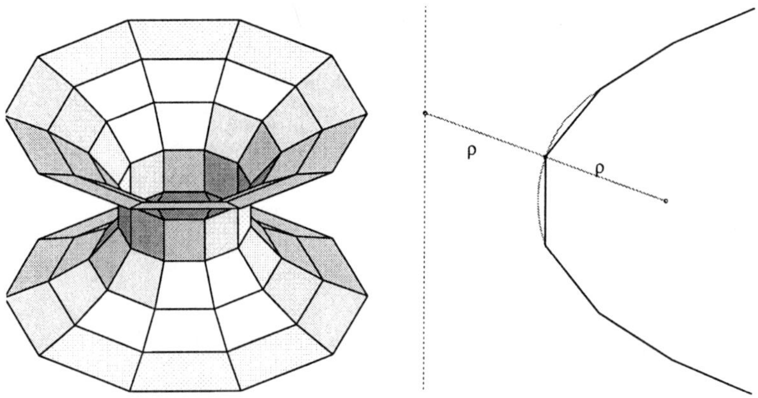

FIGURE 12. The discrete catenoid and its generator.

The case of vanishing 'discrete mean curvature'

$$\mathcal{H} = \kappa^{(1)} + \kappa^{(2)} = 0, \tag{6.13}$$

that is $c = -1$, is of particular interest since the corresponding discrete surface of revolution may be interpreted as a 'discrete catenoid'. It turns out that neither the difference equation (6.6) nor (6.7), which may be written as

$$x^2 \Delta_{\bar{1}1} x = x_{\bar{1}} x_1 \Delta_{\bar{1}1} \frac{1}{x}, \tag{6.14}$$

may be located in the 12-parameter class. In fact, the functions f^i associated with (6.14) are given by (6.9) with

$$A = \begin{pmatrix} 0 & -1 & 0 \\ 2 & 0 & 2 \\ 0 & -1 & 0 \end{pmatrix}, \quad B = \begin{pmatrix} 0 & 0 & -1 \\ 0 & 2 & 0 \\ -1 & 0 & 0 \end{pmatrix}, \tag{6.15}$$

but the first integral (6.10) no longer holds since the matrix A is not symmetric. It it is therefore desirable to classify the matrices A and B which allow integration of the mappings (6.8), (6.9). Thus far, we have not been able to find nontrivial solutions to the difference equation (6.14).

The discrete catenoid along with its generator (discrete catenary) is displayed in Fig. 12. Note that, by construction, the distances from the vertex $\boldsymbol{r}$ to the centre of the corresponding osculating circle and the axis of revolution in 'normal' direction are the same.[2]

In the continuous setting, the case $c = 2$ turned out to be governed by the pendulum equation. Remarkably, in the discrete case, the functions $f^i(X = x)$ have the form (6.9) with *symmetric* matrices

$$A = \begin{pmatrix} 0 & 0 & 1 \\ 0 & -2 & 0 \\ 1 & 0 & 0 \end{pmatrix}, \quad B = \begin{pmatrix} 0 & 1 & 0 \\ 1 & 0 & 1 \\ 0 & 1 & 0 \end{pmatrix}. \tag{6.16}$$

---

[2]It is evident that the discrete surfaces of revolution considered here are discrete isothermic in the sense of Bobenko and Pinkall [**4**] since the cross-ratio q of isosceles trapezoids is real. However, q depends on n_1 so that our discrete catenoid is different from that presented in [**4**].

Hence, the corresponding first integral reads

$$\tilde{K}(x - x_1)^2 = (x + x_1)(xx_1 + 1). \tag{6.17}$$

On the other hand, as pointed out earlier, the discrete pendulum equation (5.29) constitutes a symmetry reduction of the differential-difference Heisenberg spin equation with first integral [**28**]

$$\begin{aligned}(1+2K)x^2x_1^2 + \sigma(1+K)(x^2x_1 + xx_1^2) + (1+K)(x^2+x_1^2) \\ + 2Kxx_1 + \sigma(1+K)(x+x_1) + 1 + 2K = 0,\end{aligned} \tag{6.18}$$

where $x = \tan\beta/2$ and $2\check{c}\sigma = -1$. Comparison with (6.17) shows that the latter is a specialization ($K = -1/2$) of (6.18) if the identification $\tilde{K} = -1/\sigma$ is made. Thus, as in the continuous case, discrete Weingarten surfaces of revolution with $c = 2$ are contained in the 'discrete Hasimoto' class given by (5.28).

References

1. R. Betchov, *On the curvature and torsion of an isolated vortex filament*, J. Fluid Mech. **22** (1965), 471–479.
2. A. Bobenko, *Discrete conformal maps and surfaces*, Symmetries and Integrability of Difference Equations. II (P. Clarkson and F. Nijhoff, eds.), Cambridge Univ. Press, 1999, Cambridge, pp. 97–108.
3. A. Bobenko and U. Pinkall, *Discrete surfaces with constant negative Gaussian curvature and the Hirota equation*, J. Differential Geom. **43** (1996), 527–611.
4. ———, *Discrete isothermic surfaces*, J. Reine Angew. Math. **475** (1996), 187–208.
5. ———, *Discretization of surfaces and integrable systems*, Discrete Integrable Geometry and Physics (A. Bobenko and R. Seiler, eds.), Oxford Univ. Press, 1999, New York, pp. 3–58.
6. A. I. Bobenko and W. K. Schief, *Discrete indefinite affine spheres*, *Discrete Integrable Geometry and Physics* (A. Bobenko and R. Seiler, eds.), Oxford Univ. Press, 1999, New York, pp. 113–138.
7. ———, *Affine spheres: Discretization via duality relations*, J. Exp. Math. (to appear)
8. L. V. Bogdanov and B. G. Konopelchenko, *Lattice and q-difference Darboux–Zakharov–Manakov systems via $\bar{\delta}$-dressing method*, J. Phys. A **28** (1995,) L173–L178.
9. J. Cieśliński, A. Doliwa, and P. M. Santini, *The integrable discrete analogues of orthogonal coordinate systems are multidimensional circular lattices*, Phys. Lett. A **235** (1997), 480–487.
10. G. Darboux, *Sur une proposition relative des équations linéaires*, C. R. Acad. Sci. Paris Sér. I Math. **96** (1882), 1456–1459.
11. A. Doliwa, *Geometric discretisation of the Toda system*, Phys. Lett. A **234** (1997), 187–192.
12. A. Doliwa and P. M. Santini, *Multidimensional quadrilateral lattices are integrable*, Phys. Lett. A **233** (1997), 365–372.
13. ———, *Geometry of discrete curves and lattices and integrable difference equations*, (A. Bobenko and R. Seiler, eds.), Discrete Integrable Geometry and Physics, Oxford Univ. Press., New York, 1999, pp. 139–154.
14. L. P. Eisenhart, *A treatise on the differential geometry of curves and surfaces*, Dover, New York, 1960.
15. L. D. Faddeev, Lectures delivered at Les Houches Summer School, preprint CEN-Saclay S. Ph.T./82/76.
16. C. F. Gauss, *Disquisitiones Generales Circa Superficies Curvas*, 1827; *General Investigations on Curved Surfaces*, Princeton, 1902.
17. H. Hasimoto, *A soliton on a vortex filament*, J. Fluid Mech. **51** (1972), 477–485.
18. R. Hirota, *Nonlinear partial difference equations.* III. *Discrete sine-Gordon equation*, J. Phys. Japan **43** (1977), 2079–2086.
19. Y. Ishimori, *An integrable classical spin chain*, J. Phys. Soc. Japan **51** (1982), 3417–3418.
20. B. G. Konopelchenko and W. K. Schief, *Three-dimensional integrable lattices in Euclidean spaces: Conjugacy and orthogonality*, Proc. Roy. Soc. London Ser. A **454** (1998), 3075–3104.
21. M. Lakshmanan, *Continuum spin system as an exactly solvable dynamical system*, Phys. Lett. A **61** (1977), 53–54.

22. R. R. Martin, J. de Pont, and T. J. Sharrock, *Cyclide surfaces in computer aided design*, The Mathematics of Surfaces (J. A. Gregory, ed.), Clarendon Press, Oxford, 1986, pp. 253–268.
23. V. B. Matveev and M. Salle, *Darboux transformations and solitons*, Springer Series in Nonlinear Dynamics, Springer-Verlag, Berlin, 1991.
24. J. J. C. Nimmo and W. K. Schief, *Superposition principles associated with the Moutard transformation: An integrable discretization of a* $(2+1)$*-dimensional sine-Gordon system*, Proc. Roy. Soc. London Ser. A **453** (1997), 255–279.
25. ———, *An integrable discretization of a* $2+1$*-dimensional sine-Gordon equation*, Stud. Appl. Math. **100** (1998), 295–309.
26. A. W. Nutbourne, *The solution of the frame matching equation*, The Mathematics of Surfaces (J. A. Gregory, ed.), Clarendon Press, Oxford, 1986, pp. 233–252.
27. G. R. W. Quispel, J. A. G. Roberts, and C. J. Thompson, *Integrable mappings and soliton equations*, Phys. Lett. A **126** (1988), 419–421.
28. ———, *Integrable mappings and soliton equations.* II, Physica D **34** (1989), 183–192.
29. R. Sauer, *Finite Analoga zur Differentialgeometrie der Asymptotennetze*, Ber. Math.-Tagung Tübingen **1946** (1947), 127–129.
30. ———, *Differenzengeometrie*, Springer-Verlag, Berlin-New York, 1970. (German)
31. W. K. Schief, *Finite analogues of Egorov-type coordinates*, Nonlinear Systems, Solitons and Geometry, Mathematisches Forschungsinstitut Oberwolfach, Germany, 1997. (to appear)
32. ———, *Self-dual Einstein spaces and a discrete Tzitzeica equation. A permutability theorem link*, Symmetries and Integrability of Difference Equations. II (P. Clarkson and F. Nijhoff, eds.) Cambridge Univ. Press, 1999, Cambridge, pp. 137–148.
33. A. Sym, *Soliton surfaces.* VI. *Gauge invariance and final formulation of the approach*, Lett. Nuovo Cimento **41** (1984), 353–360.
34. V. E. Zakharov and L. A. Takhtadzhyan, *Equivalence of the nonlinear Schrödinger equation and the equation of a Heisenberg ferromagnet*, Theoret. Math. Phys. **38** (1979), 17–23.

School of Mathematics, The University of New South Wales, Sydney, NSW 2052, Australia.
E-mail address: `schief@maths.unsw.EDU.AU`

Dipartimento di Fisica, Università di Lecce, Via Arnesano, 73100 Lecce, Italy.
E-mail address: `konopel@le.infn.it`

Centre de Recherches Mathématiques
CRM Proceedings and Lecture Notes
Volume **25**, 2000

Kac–Moody–Virasoro Algebras and Integrability of Certain Higher Dimensional Nonlinear Evolutionary Equations

M. Senthilvelan

ABSTRACT. We study the existence of Virasoro type subalgebras in certain higher dimensional nonlinear evolutionary equations. In contrast with the common belief, our investigations reveal that not all the integrable equations in $(2+1)$-dimensions admit Virasoro type subalgebras. Our examples include

(i) breaking soliton equation and
(ii) a nonlinear Schrödinger type equation. We also explore certain similarity reductions.

Recently the invariance properties of certain physically important $(2+1)$-dimensional *integrable* nonlinear evolutionary equations (NLEEs) have been investigated through Lie group analysis and it has been shown that all of them admit infinite dimensional symmetry groups with a specific Virasoro type subalgebras. For example the invariance properties of the following integrable equations have been investigated by different authors and a specific Virasoro type subalgebras has been constructed for all of them:

(i) Kadomtsev–Petviashvili (KP) equation [**3**],
(ii) Davey–Stewartson equation [**2**],
(iii) the three wave interaction problem [**10**],
(iv) cylindrical KP equation [**9**],
(v) stimulated Raman scattering equation [**8**],
(vi) Nizhnik–Novikov–Veselov equation and
(vii) $(2+1)$-dimensional sine-Gordon equation [**7**].

The same investigation made on certain *nonintegrable* higher dimensional NLEEs shows that it does not admit Virasoro type subalgebras. Specific examples of this category are

(i) Infeld–Rowlands equation [**4**] and
(ii) a nonintegrable dispersive long-wave equation [**11**].

2000 *Mathematics Subject Classification.* Primary: 54C40, 14E20; Secondary: 46E25, 20C20.

The author wishes to thank Professor M. Lakshmanan, Professor P. M. Santini and Professor P. Winternitz for their helpful discussions. He also wishes to thank Organisers of the Meeting, Prof. M. Torrisi and Consiglio Nazionale delle Ricerche (CNR), Italy, for inviting him to participate this meeting. The work is partially supported by FAPESP, Brazil.

This is the final form of the paper.

So it is commonly believed in the current literature that all integrable higher dimensional NLEEs will admit Virasoro type subalgebras while nonintegrable equations will not. However, we wish to mention that *not all the integrable* higher dimensional NLEEs admit Virasoro type subalgebras. Our counter examples are

(i) $(2+1)$-dimensional breaking soliton equation and
(ii) $(2+1)$-dimensional nonlinear Schrödinger type equation introduced by Zakharov and studied recently by Strachan.

Our investigations on these examples show that they do not admit Virasoro type subalgebras even though it has been shown in the literature that they are integrable.

The plan of the paper is as follows. In Section 1 we study the invariance analysis of a $(2+1)$-dimensional breaking soliton equation and show that it does not admit a Virasoro type subalgebra. We also investigate general similarity reductions and explore some particular solutions. In Section 2 we present our conclusions.

1. Breaking Soliton Equation

Let us consider an asymmetric generalisation of the KdV equation in $(2+1)$-dimensions of the form [**1**]

$$u_t + B\,u_{xxy} + 4B\,uv_x + 2B\,vu_x = 0,\; u_y = v_x, \tag{1.1}$$

which describes the interaction of a Riemann wave propagating along the y-axis with a long wave propagating along the x-axis. Eq. (1.1) admits Lax representation [**1**] and Painlevé (P)-property.

1.1. Lie symmetries. The invariance of Eq. (1.1) under the infinitesimal point transformations,

$$\begin{aligned} &x \to X = x + \varepsilon\xi_1(t,x,y,u,v), \quad y \to Y = y + \varepsilon\xi_2(t,x,y,u,v),\\ &t \to T = t + \varepsilon\xi_3(t,x,y,u,v), \quad u \to U = u + \varepsilon\phi_1(t,x,y,u,v),\\ &v \to V = v + \varepsilon\phi_2(t,x,y,u,v), \quad \varepsilon \ll 1, \end{aligned} \tag{1.2}$$

leads to the expressions for the infinitesimals [**5**]

$$\begin{aligned} \xi_1 &= \frac{-c_1}{3}xt - \frac{c_2}{2}x + f(t),\\ \xi_2 &= \frac{-2c_1}{3}yt + \left(\frac{c_2}{2} - c_4\right)y - 4B\,c_3 t + c_5,\\ \xi_3 &= \frac{-2c_1}{3}t^2 - \left(\frac{c_2}{2} + c_4\right)t - c_6,\\ \phi_1 &= \frac{2c_1}{3}ut + c_2 u - \frac{c_1}{6B}y - c_3, \qquad \phi_2 = c_1 vt + c_4 v - \frac{c_1}{6B}x + \frac{\dot f(t)}{2B}, \end{aligned} \tag{1.3}$$

where c_1, c_2, c_3, c_4, c_5, c_6 are arbitrary constants, $f(t)$ is an arbitrary function of t and the dot denotes differentiation with respect to t.

1.2. Lie algebra of symmetry vector fields. A general element of this Lie algebra can be written as

$$V = V_1(f) + V_2 + V_3 + V_4 + V_5 + V_6 + V_7, \tag{1.4}$$

where

$$
(1.5)\quad
\begin{aligned}
V_1(f) &= f(t)\frac{\partial}{\partial x} + \frac{\dot f(t)}{2B}\frac{\partial}{\partial v},\\
V_2 &= -\frac{xt}{3}\frac{\partial}{\partial x} - \frac{2yt}{3}\frac{\partial}{\partial y} - \frac{2t^2}{3}\frac{\partial}{\partial t} + \left(\frac{2}{3}ut - \frac{y}{6B}\right)\frac{\partial}{\partial u} + (vt - \frac{x}{6B})\frac{\partial}{\partial v},\\
V_3 &= -\frac{1}{2}x\frac{\partial}{\partial x} + \frac{1}{2}y\frac{\partial}{\partial y} - \frac{1}{2}t\frac{\partial}{\partial t} + u\frac{\partial}{\partial u}, \qquad V_4 = -4Bt\frac{\partial}{\partial y} - \frac{\partial}{\partial u},\\
V_5 &= -y\frac{\partial}{\partial y} - t\frac{\partial}{\partial t} + v\frac{\partial}{\partial v}, \qquad V_6 = \frac{\partial}{\partial y}, \qquad V_7 = \frac{\partial}{\partial t}.
\end{aligned}
$$

The associated Lie algebra between these vector fields becomes

$$
(1.6)\quad
\begin{array}{lll}
[V_1, V_2] = V_1\left(\frac{2t^2\dot f}{3} - \frac{tf}{3}\right), & [V_1, V_3] = V_1\left(\frac{f}{2} + \frac{t\dot f}{2}\right), & [V_1, V_4] = 0,\\
[V_1, V_5] = V_1(t\dot f), & [V_1, V_6] = 0, & [V_1, V_7] = -V_1(\dot f),\\
[V_2, V_3] = \frac{V_2}{2}, & [V_2, V_4] = 0, & [V_2, V_5] = V_2,\\
[V_2, V_6] = -\frac{V_4}{6B}, & [V_2, V_7] = -\frac{2V_3}{3} - V_5, & [V_3, V_4] = -V_4,\\
[V_3, V_5] = 0, & [V_3, V_6] = -\frac{V_6}{2}, & [V_3, V_7] = \frac{1}{2}V_7,\\
[V_4, V_5] = 0, & [V_4, V_6] = 0, & [V_4, V_7] = 4B\,V_6,\\
[V_5, V_6] = V_6, & [V_5, V_7] = V_7, & [V_6, V_7] = 0,
\end{array}
$$

whereas the commutation relation between $V_1(f_1)$, $V_1(f_2)$ turns out to be

$$
(1.7)\qquad \big[V_1(f_1), V_1(f_2)\big] = 0,
$$

which is not of *Virasoro type* typically appearing in most of the integrable systems mentioned in the introduction.

1.3. Similarity variables and similarity reductions. The associated characteristic equation reads

$$
(1.8)\quad
\begin{aligned}
\frac{dx}{-c_1/3\,xt - c_2/2\,x + f(t)} &= \frac{dy}{-2/3c_1yt + (c_2/2 - c_4)y - 4B\,c_3t + c_5},\\
&= \frac{dt}{-2/3\,c_1t^2 - (c_2/2 + c_4)t - c_6},\\
&= \frac{du}{2/3c_1\,ut + c_2u - 1/6B\,c_1y - c_3},\\
&= \frac{dv}{(-c_1)/6B\,x + 1/2B\,\dot f(t) + c_1vt + c_4v}.
\end{aligned}
$$

Solving Eq. (1.8) we obtain similarity variables of the form

$$
\begin{aligned}
\tau_1 &= \frac{x(t+k_1+k_2)^{n-(1/4)}}{(t+k_1-k_2)^{n+(1/4)}} \\
&\quad + \frac{3}{2c_1}\int^t \frac{f(t')(t'+k_1+k_2)^{n-(5/4)}}{(t'+k_1-k_2)^{n+(5/4)}}\,dt', \\
\tau_2 &= \frac{y(t+k_1+k_2)^{-2n-(1/2)}}{(t+k_1-k_2)^{-2n+(1/2)}} \\
&\quad - \frac{3}{2c_1}\int^t \frac{(4B\,c_3t'-c_5)(t'+k_1+k_2)^{-2n-(3/2)}}{(t'+k_1-k_2)^{-2n+(3/2)}}\,dt', \\
F &= \frac{u(t+k_1+k_2)^{2n+(1/2)}}{(t+k_1-k_2)^{2n-(1/2)}} \\
&\quad - \frac{3}{8B\,c_1}\int^t\left[\int^t \frac{(4B\,c_3t-c_5)(t+k_1+k_2)^{-2n-(3/2)}}{(t+k_1-k_2)^{-2n+(3/2)}}\,dt\right]dt \\
&\quad - \frac{\tau_2 t}{4B} - \frac{3c_3}{2c_1}\int^t \frac{(t'+k_1-k_2)^{2n-(1/2)}}{(t'+k_1+k_2)^{2n+(1/2)}}\,dt', \\
G &= \frac{v(t+k_1+k_2)^{n+(3/4)}}{(t+k_1-k_2)^{n-(3/4)}} \\
&\quad + \frac{3}{8B\,c_1}\int^t\left[\int^{t'} \frac{f(t'')(t''+k_1+k_2)^{n-(5/4)}}{(t''+k_1-k_2)^{n+(5/4)}}\,dt''\right]dt' \\
&\quad - \frac{\tau_1 t}{4B} + \frac{3}{4B\,c_1}\int^t \frac{\dot f(t')(t'+k_1-k_2)^{-(n+(1/4))}}{(t'+k_1+k_2)^{-n+(1/4)}}\,dt',
\end{aligned}
\tag{1.9}
$$

where F and G are functions of τ_1 and τ_2. Using the similarity transformation one can reduce Eq. (1.1) to the following system of partial differential equation (PDE) in (1+1) dimensions

$$
\begin{aligned}
F_{\tau_1\tau_1\tau_2} + 4FG_{\tau_1} + 2GF_{\tau_1} + \frac{(3c_2-6c_4)}{4B\,c_1}\tau_2F_{\tau_2} - \frac{3c_2}{4B\,c_1}\tau_1F_{\tau_1}& \\
- \frac{3c_2}{2c_1}F + \frac{3c_6}{8B\,c_1}\tau_2 = 0&, \\
F_{\tau_2} = G_{\tau_1}&.
\end{aligned}
\tag{1.10}
$$

In addition to the above general similarity reduction one can also look into the subcases by assuming one or more of the vector fields to be zero. The detailed analysis is reported elsewhere [**14**].

Now the reduced PDE (1.10) in two independent variables can itself be further analysed for its symmetry properties by looking at its invariance property under the classical Lie algorithm. The invariance of the Eq. (1.10) leads to the following infinitesimal symmetries

$$
\begin{aligned}
\xi_1 &= c_7\tau_1 + c_8, & \xi_2 &= -2c_7\tau_2, \\
\eta_1 &= -2c_7F, & \eta_2 &= c_7G + \frac{3c_2c_8}{8B\,c_1},
\end{aligned}
\tag{1.11}
$$

where c_7 and c_8 are arbitrary constants. The associated reduced ordinary differential equation takes the form

$$\begin{aligned} w_1''' - \frac{2}{z}w_1'' + \frac{2}{z^2}w_1' + \frac{8}{c_1^2}w_1w_2' + \frac{4}{c_1^2}w_1'w_2 - \left(\frac{3c_2+6c_4}{4B\,c_1^3}\right)w_1' \\ + \frac{3c_6}{2B^2c_1^3}z = 0, \end{aligned} \tag{1.12}$$

$$w_1' = 2z(w_2 + zw_2').$$

While the exact solution for Eq. (1.12) has not been found for the general two parameter case, particular solutions can be obtained for the special one parameter case. For example, by choosing $c_7 = 0$ and redoing the calculations one gets the following solution

$$F = \frac{3c_2}{8B\,c_1}\tau_2, \quad G = \frac{3c_2}{8B\,c_1}\tau_1 + w_2(\tau_2), \tag{1.13}$$

where $w(\tau_2)$ is an arbitrary function of τ_2. Now rewriting Eq. (1.13) in terms of old variables one can get a solution for the PDE (1.1). Similar analysis can be performed for the other subcases also.

2. Conclusions

In this paper we have shown that Eq. (1.1) does not admit Virasoro type subalgebras even though it has been shown in the literature that it is integrable. We have also carried out a detailed analysis for some more (2+1)-dimensional integrable equations and found that they also do not admit Virasoro type subalgebras. The examples are

(i) Zakharov–Strachan equation:

$$i\psi_t - \psi_{xy} + V\psi = 0, \quad V_x = 2\partial_y|\psi|^2. \tag{2.1}$$

Eq. (2.1) was introduced originally by Zakharov and studied recently by Strachan [**15**]. When $\partial_x = \partial_y$ Eq. (2.1) reduces to the NLS equation. It has been that Eq. (2.1) admits P-property and a new class of localized solutions called "induced localized structures" [**12**].

(ii) Breaking soliton equation:

$$u_t + A\,u_{xxx} + B\,u_{xxy} + 6Auu_x + 4B\,u_x\partial_x^{-1}u_y dx' = 0. \tag{2.2}$$

Eq. (2.2) is another asymmetric generalization of the KdV equation in (x,y) spatial variables. When $B = 0$ Eq. (2.2) reduces to the KdV equation. Eq. (2.2) possesses a Lax pair but the spectral parameter breaks to be a multivalued function and hence the solution may also get multivalued [**6**]. Eq. (2.2) admits P-property for $A = 0$ and admit dromion—like solutions [**13**].

Even though there is no specific reason known for the absence of Virasoro type subalgebras in these systems we offer here a speculative guess. By carefully looking at the Lax operators of Eqs. (1.1) and (2.2), one notices that they are of the form

$$L = -\partial_x^2 + u(x,y,t) - \lambda, \tag{2.3}$$

which is an one-dimensional operator rather than a two-dimensional operator encountered in other integrable equations. Due to this property one may loose the Virasoro type subalgebras in these systems. A detailed analysis between the Lax operators and its connection with the existence of Virasoro type subalgebras deserves further study.

References

1. O. I. Bogoyavlenskii, *Overturning solitons in two-dimensional integrable equations*, Russian Math. Surveys **45** (1990), 1–86. (Russian)
2. B. Champagne and P. Winternitz, *On the infinite dimensional symmetry groups of the Davey–Stewartson equations*, J. Math. Phys. **29** (1988), 1–8.
3. D. David, N. Kamran, D. Levi, and P. Winternitz, *Symmetry reduction for the Kadomstev–Petviashvili equation using a loop—algebra*, J. Math. Phys. **27** (1986), 1225–1237.
4. M. Faucher and P. Winternitz, *Symmetry analysis of the Infeld–Rowlands equation*, Phys. Rev. E **48** (1993), 3066–3071.
5. A. Head, *LIE: A PC program for Lie analysis of differential equations*, Comput. Phys. Comm. **77** (1993), 241–248.
6. B. G Konopelchenko, *Solitons in multidimensions*, World Scientific, Singapore, 1993.
7. M. Lakshmanan and M. Senthil Velan, *Lie symmetries, infinite dimensional Lie algebras and similarity reductions of certain* $(2+1)$*-dimensional equations*, J. Nonlinear Math. Phys. **3** (1996), 24–39.
8. D. Levi, C. R. Menyuk, and P. Winternitz, *Similarity reduction and perturbation solution of the stimulated—Raman scattering equation in the presence of dissipation*, Phys. Rev. A **49** (1994), 2844–2852.
9. D. Levi and P. Winternitz, *The cylindrical Kadomstev–Petviashvili Equation: Its Kac–Moody–Virasoro algebra and relation to* KP *equation*, Phys. Lett. A **129** (1988), 165–167.
10. L. Martina and P. Winternitz, *Analysis and applications of the symmetry group of the multidimensional three-wave resonant interaction problem*, Ann. Physics **196** (1989), 231–237.
11. G. Paquin and P. Winternitz, *Group theoretical analysis of dispersive long wave equations in two space dimensions*, Physica D **46** (1990), 122–138.
12. R. Radha and M. Lakshmanan, *Singularity structure analysis and bilinear form of a* $(2+1)$*-dimensional nonlinear Schrödinger* (NLS) *equation*, Inverse Problems **10** (1994), L29–33.
13. ———, *Dromion structures in the* $(2+1)$*-dimensional breaking soliton equation*, Phys. Lett. A **197** (1995), 7–12.
14. M. Senthil Velan and M. Lakshmanan, *Lie symmetries, Kac–Moody Virasoro algebras and integrability of certain* $(2+1)$*-dimensional nonlinear evolution equations*, J. Nonlinear Math. Phys. **5** (1998), 190–211.
15. I. A. B. Strachan, *Some integrable hierarchies in* $(2+1)$*-dimensions and their twister description*, J. Math. Phys. **34** (1993), 243–259.

Dipartimento di Mathematica, Via. A. Doria 6, I–95125, Catania, Italy.

Current address: Instituto de Física Teórica, Universidade Estadual Paulista, Rua Pamplona 145, 01405–900, São Paulo, Brazil.

E-mail address: `senthil@ift.unesp.br`

Centre de Recherches Mathématiques
CRM Proceedings and Lecture Notes
Volume **25**, 2000

Different Faces of Harmonic Oscillator

Alexander Turbiner

ABSTRACT. Harmonic oscillator in Fock space is defined. Isospectral as well as polynomiality-of-eigenfunctions preserving, translation-invariant discretization of the harmonic oscillator is presented. Dilatation-invariant and polynomiality-of-eigenfunctions preserving discretization is also given.

Undoubtedly the harmonic oscillator plays a fundamental role in science. The goal of the present note is to give a review of the different representations of the quantum harmonic oscillator and its deformations in terms of the elements of an Heisenberg algebra of differential, finite-difference and discrete operators.

The Hamiltonian of harmonic oscillator is defined by

$$\mathcal{H} = -\frac{\partial^2}{\partial x^2} + \omega^2 x^2, \tag{1}$$

where ω is the oscillator frequency. The eigenfunctions and eigenvalues are given by

$$\Psi_k(x) = H_k\big(\sqrt{\omega}x\big)e^{-\omega(x^2)/2}, \quad E_k = \omega(2k+1), \quad k = 0, 1, \dots, \tag{2}$$

where H_k is the kth Hermite polynomial in standard notation. With no loss of generality one puts all normalization constants equal to 1. The Hamiltonian (1) is Z_2-invariant, $x \to -x$, which leads to two families of eigenstates: even and odd, symmetric and anti-symmetric with respect of reflection, correspondingly. This property is coded in the parity of the Hermite polynomials:

$$H_{2n+p}\big(\sqrt{\omega}x\big) = \omega^{p/2} x^p L_n^{(p-1/2)}(\omega x^2), \quad n = 0, 1, \dots, \tag{3}$$

where $L_n^{(\alpha)}(y)$ is the nth associated Laguerre polynomial in standard notation, and $p = 0,\ 1$ has a meaning of parity. Hereafter we can call a ground state the lowest energy state of parity p:

$$\Psi_0^{(p)}(x) = x^p e^{-\omega\, x^2/2d}. \tag{4}$$

Thus the formula (4) makes an unification of both possible values of parity and for the sake of simplicity we will call (4) the ground state eigenfunctions without specifying parity.

2000 *Mathematics Subject Classification.* Primary: 81R99; Secondary: 17B99, 34A05, 39A05.

This is the final form of the paper.

Now, let us perform a gauge transformation of the Hamiltonian (1), taking as gauge factor the ground state eigenfunction (4), and change the variable x to $y = \omega x^2$, which incorporates the reflection symmetry. After dropping all constant terms we get the operator

$$h(y, \partial_y) = -\frac{1}{\omega}\left(\Psi_0^{(p)}(x)\right)^{-1}\mathcal{H}\Psi_0^{(p)}(x)\big|_{y=\omega x^2} = 4y\partial_y^2 - 4\left(y - p - \frac{1}{2}\right)\partial_y, \tag{5}$$

with the spectrum $(-4n)$, where $n = 0,\ 1,\ 2\ldots.$ The operator (5) simultaneously describes a family of eigenstates of positive parity if $p = 0$ and a family of eigenstates of negative parity if $p = 1$. We will call (5) *the algebraic form* of the Hamiltonian of the harmonic oscillator. The word 'algebraic' reflects the fact that the operator (5) has a form of linear differential operator with polynomial coefficients and furthermore possesses infinitely-many polynomial eigenfunctions. The latter implies that any eigenfunction can be found by algebraic means by solving a system of linear algebraic equations.

The algebraic form (5) admits a generalization of the original Hamiltonian (1) we started with. If we assume that the parameter p can take any real value, $p > -1/2$, we can transform the operator (5) back to an Hamiltonian (1), arriving at

$$\begin{aligned}\omega y^{p/2} e^{-y/2}\left[4y\partial_y^2 - 4\left(y - p - \frac{1}{2}\right)\partial_y\right] y^{-p/2} e^{y/2}\Big|_{x=\sqrt{y/\omega}} \\ = \left[\partial_x^2 - \omega^2 x^2 - \frac{p(p-1)}{x^2}\right] \equiv -\mathcal{H}_k,\end{aligned} \tag{6}$$

which is known in literature as Kratzer Hamiltonian. It is worth to mention that this Hamiltonian coincides also with 2-body Calogero Hamiltonian after separation the center of mass motion. Hereafter we will call the system characterized by the Hamiltonian (6) *the harmonic oscillator.*

The resulting Hamiltonian (6) is characterized by the eigenfunctions

$$\Psi_k(x) = x^p L_n^{-1/2+p}(\omega x^2) e^{-\omega x^2/2}, \tag{7}$$

which coincides with (2) at $p = 0, 1$. The spectrum (6) is still equidistant with energy gap ω and after appropriate shift of the reference point it coincides with the spectrum of the original harmonic oscillator (1). Thus, the deformation of (1) to (6) is isospectral, which is, of course, well-known.

In order to move ahead let us introduce the notion of Fock space. Take two operators a and b obeying the commutation relation

$$[a, b] \equiv ab - ba = I, \tag{8}$$

with the identity operator I on the r.h.s.—they span a three-dimensional Lie algebra which is called the Heisenberg algebra h_3. By definition the universal enveloping algebra of h_3 is the algebra of all normal-ordered polynomials in a, b: any monomial is taken to be of the form $b^k a^m$.[1] If, besides the polynomials, all entire functions in a, b are considered, then the *extended* universal enveloping algebra of the Heisenberg algebra appears or in other words, the extended Heisenberg–Weyl algebra. In the (extended) Heisenberg–Weyl algebra one can find the nontrivial embedding of the Heisenberg algebra: nontrivial elements obeying the commutation relations (8),

[1] Sometimes this is called the Heisenberg–Weyl algebra.

whose can be treated as a certain type of quantum canonical transformations. We say that the (*extended*) Fock space, $\mathcal{F}$ is determined if we take the (extended) universal enveloping algebra of the Heisenberg algebra and attach to it the vacuum state $|0\rangle$ such that

$$a|0\rangle = 0. \tag{9}$$

It is easy to check that the following statement holds: if the operators a, b obey (8), then the operators

$$J_n^+ = b^2 a - nb, \quad J_n^0 = ba - \frac{n}{2}, \quad J_n^- = a, \tag{10}$$

span the sl_2-algebra with the commutation relations:

$$[J^0, J^\pm] = \pm J^\pm, \quad [J^+, J^-] = -2J^0,$$

where $n \in \mathbf{C}$.[2] For the realization (10) the quadratic Casimir operator is a c-number given by:

$$C_2 \equiv \frac{1}{2}\{J_n^+, J_n^-\} - J_n^0 J_n^0 = -\frac{n}{2}\left(\frac{n}{2} + 1\right), \tag{11}$$

where $\{\ ,\ \}$ denotes the anticommutator. If $n \in \mathbb{Z}_+$, then (10) possesses a finite-dimensional, irreducible representation in Fock space leaving invariant the space of polynomials in b:

$$\mathcal{P}_n(b) = \langle 1, b, b^2, \ldots, b^n \rangle |0\rangle, \tag{12}$$

of dimension $\dim \mathcal{P}_n = (n+1)$. The spaces $\mathcal{P}_n$ possess the property that $\mathcal{P}_n \subset \mathcal{P}_{n+1}$ for each $n \in \mathbb{Z}_+$, and form an infinite flag such that

$$\bigcup_{n \in \mathbb{Z}_+} \mathcal{P}_n = \mathcal{P}.$$

It is evident that any polynomial in generators $J_n^{0,-}$ operator preserves the flag $\mathcal{P}$. Such an operator we will call sl_2-*exactly-solvable operator.*

Take as an example the sl_2-exactly-solvable operator of the form

$$h_f(b, a) = 4J^0 J^- - 4J^0 + 4\left(p + \frac{1}{2}\right)J^- = 4ba^2 - 4\left(b - p - \frac{1}{2}\right)a, \tag{13}$$

where p is a parameter and $J^{\pm,0} \equiv J_0^{\pm,0}$ (see (10)). One can demonstrate that the eigenfunctions of h_f are the associated Laguerre polynomials of the argument b, $L_n^{(p-1/2)}(b)$ and their eigenvalues, $E_n = -4n$.

As the next step we consider two different realization of the Heisenberg algebra (8) in terms of differential and finite-difference operators. A traditional realization of (8) appearing in text-books is the coordinate-momentum representation:

$$a = \frac{d}{dy} \equiv \partial_y, \quad b = y, \tag{14}$$

where the operator $b = y$ stands for the multiplication operator on y in a space of functions $f(y)$. In this case the vacuum is a constant and without a loss of generality we put $|0\rangle = 1$. Recently, a finite-difference analogue of (14) has been found [**2, 3**]:

$$a = \mathcal{D}_+, \quad b = y(1 - \delta\mathcal{D}_-), \tag{15}$$

[2]For details and discussion see, for example, [**6**].

where

$$\mathcal{D}_{\pm} f(y) = \frac{f(y \pm \delta) - f(y)}{\pm\delta},$$

is the finite-difference operator, δ is a real number and $\mathcal{D}_{\pm}(-\delta) = \mathcal{D}_{\mp}(\delta)$. A remarkable property of this realization is that the vacuum remains the same for both cases (14)–(15) and it can be written as $|0\rangle = 1$.

Substitution of (14) into (13) leads to the operator (5)—the algebraic form of the Hamiltonian of the harmonic oscillator. Thus, the operator (13) can be called the *algebraic form of the Hamiltonian of the harmonic oscillator in the Fock space*. It is evident that the procedure of realization of the Heisenberg generators a, b by concrete operators (differential, finite-difference, discrete) provided that the vacuum remains unchanged leaves any polynomial operator in a, b isospectral. Now let us study another 'face' of harmonic oscillator by substituting the realization (15) in (13). Finally, we obtain

$$h_d(y, D_{\pm}) = \frac{4}{\delta}\left[y + \delta\left(p + \frac{1}{2}\right)\right] D_+ - 4\left(1 + \frac{1}{\delta}\right) y D_- . \tag{16}$$

Thus, in the realization by finite-difference operators the corresponding spectral problem can be defined as

$$\frac{4}{\delta^2}\left[y + \delta\left(p + \frac{1}{2}\right)\right]\phi(y+\delta) - \frac{4}{\delta}\left[\left(1 + \frac{2}{\delta}\right) y + p + \frac{1}{2}\right]\phi(y) + \frac{4}{\delta}\left(1 + \frac{1}{\delta}\right) y \phi(y - \delta) = E\phi(y). \tag{17}$$

The operator $h_d(y, D_{\pm})$ is a nonlocal, three-point, finite-difference operator. It is illustrated by Fig. 1.

In general, the function $\phi(y)$ in the rhs of (17) can be replaced by $\phi(y+\delta)$ or $\phi(y-\delta)$, or by a linear combination of $\phi(y \pm \delta)$, $\phi(y)$. It does not change the statement that (17) has infinitely-many polynomial eigenfunctions.

It is easy to check that

$$[y e^{-\delta \partial_y}]^n = y^{(n)} e^{-n\delta\partial_y}$$

where $y^{(n+1)} = y(y-\delta)\ldots(y-n\delta)$ is a so-called *quasi-monomial*. Using this relation one can show that the eigenfunctions of (16) remain polynomials and furthermore the solutions of (17) are given by polynomials

$$\widehat{L}_n^{(p+1/2)}(y, \delta) = \sum_{\ell=0}^{n} a_\ell^{(p+1/2)} y^{(\ell)}, \tag{18}$$

where $a_\ell^{(p+1/2)}$ are the coefficients in the expansion of the Laguerre polynomials, $L_n^{(p+1/2)}(y) = \sum_{\ell=0}^{n} a_\ell^{(p+1/2)} y^\ell$. We call these polynomials the *modified associated Laguerre polynomials*. Simultaneously, the eigenvalues of the equation (16) remain equal to $(-4n)$, $n = 0, 1, 2\ldots$ and they are the same as the eigenvalues of the

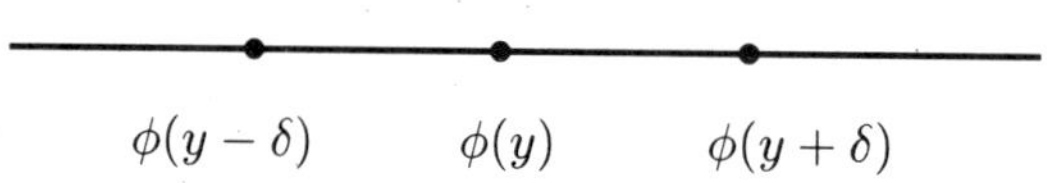

FIGURE 1. Graphical representation of the operator (16).

harmonic oscillator problem (1), (6) and (13). Thus, one can say that the operator (16) defines a *finite-difference form of harmonic oscillator* Hamiltonian.

A natural question can be posed about the most general second-order linear differential operator, which

(i) has infinitely-many polynomial eigenfunctions and
(ii) is isospectral to the harmonic oscillator (1).

In [**4, 5**] it has been shown that this operator has the form

$$(19)\quad h_g(y,\partial_y) = 4(AJ^0 + BJ^-)J^- - 4J^0 + 4\left(p+\frac{1}{2}\right)CJ^- = 4(Ay+B)\partial_y^2 - 4\left[y - \left(p-\frac{1}{2}\right)C\right]\partial_y,$$

where A, B, C are arbitrary constants and the generators (10) are realized by differential operators (14). However, by a linear change of variable, $y \to \alpha y + \beta$ the operator (19) is transformed to (5). Thus, without loss of generality we can put $A = \alpha = 1$ and also $C = 1$. The eigenfunctions of (19) remain the Laguerre polynomials but of a shifted argument, $L_n^{(p-1/2)}(y+\beta)$. It leads to the statement that among the second-order differential operators there exist no nontrivial isospectral deformation of the harmonic oscillator potential preserving polynomiality of the eigenfunctions.

The operator (19) can be rewritten in the Fock space formalism by using (14)

$$(20)\qquad h_g(b,a) = 4(b+B)a^2 - 4\left[b - \left(p-\frac{1}{2}\right)\right]a.$$

It is evident that the operator (20) is the most general second order polynomial in a, which is isospectral to (13) and also preserves the space of polynomials (12). By substitution (15) into the operator (20) it becomes transformed into a finite-difference operator

$$(21)\qquad h_g(y,D_\pm) = 4BD_+^2 + \frac{4}{\delta}\left[y + \delta\left(p+\frac{1}{2}\right)\right]D_+ - 4\left(1+\frac{1}{\delta}\right)yD_-$$

(cf. (16)). Thus, in the realization by finite-difference operators the corresponding spectral problem can be defined as

$$(22)\qquad \begin{gathered} \frac{4B}{\delta^2}\phi(y+2\delta) + \frac{4}{\delta^2}\left[y - 2B + \delta\left(p+\frac{1}{2}\right)\right]\phi(y+\delta), \\ -\frac{4}{\delta}\left[\left(1+\frac{2}{\delta}\right)y - \frac{B}{\delta} + \left(p+\frac{1}{2}\right)\right]\phi(y) + \frac{4}{\delta}\left(1+\frac{1}{\delta}\right)y\phi(y-\delta) = E\phi(y), \end{gathered}$$

and it has infinitely-many polynomial eigenfunctions.

The operator $h_d(y,D_\pm)$ now becomes the four-point finite-difference operator, see Fig. 2.

It is quite surprising that a linear shift of variable y in differential operator (16) (which gives nothing nontrivial, see discussion above) leads to occurrence of

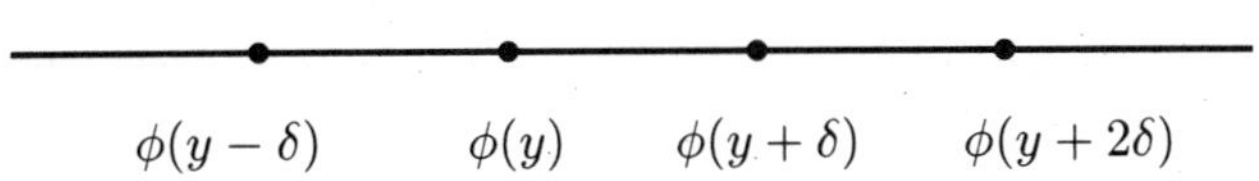

FIGURE 2. Graphical representation of the operator (21).

the extra point in their isospectral finite-difference counterpart changing a type of nonlocality.

Now it is time to ask what would happen if in the expressions (13), (20) the operators a, b are not the generators of the Heisenberg algebra (8) but the generators of the $q-$deformed Heisenberg algebra

$$[a, b]_q \equiv ab - qba = 1, \tag{23}$$

where q is a parameter. Following the Theorem proved in [**4,5**], one can demonstrate that within the q-deformed Fock space built on using q-deformed Heisenberg algebra (23) there exists the flag of linear spaces of polynomials in b, $\mathcal{P}$ (see (12)), which is preserved by the operators (13), (20). By a simple calculation one can find the eigenvalues of (13), (20)

$$E_n^{(q)} = -4\{n\}, \quad n = 0, 1, \ldots, \tag{24}$$

where

$$\{n\} = \frac{1 - q^n}{1 - q},$$

is a so-called $q-$number and $\{n\} \to n$, if $q \to 1$. If the parameter q is real the spectra of (13), (20) are real.

The algebra (23) has a realization in terms of discrete operators (see, for example, [**6**])

$$a = \mathcal{D}_q, \quad b = y, \tag{25}$$

where

$$\mathcal{D}_q f(y) = \frac{f(qy) - f(y)}{y(1 - q)}.$$

This realization has a property that the vacuum remains the same as well as for the cases (14)–(15) and it can be written as $|0\rangle = 1$. Now we can substitute (25) in (13) and the following operator emerges

$$h_q(y, \mathcal{D}_q) = 4\tilde{J}^0 \tilde{J}^- - 4\tilde{J}^0 + 4\left(p + \frac{1}{2}\right)\tilde{J}^- = 4y\mathcal{D}_q^2 - 4\left(y - p - \frac{1}{2}\right)\mathcal{D}_q, \tag{26}$$

where the generators $\tilde{J}^0 = ba$, $\tilde{J}^- = a$ have the same functional form as (10) but obey the $q-$ deformed commutation relation

$$[\tilde{J}^0, \tilde{J}^-]_{1/q} \equiv \tilde{J}^0 \tilde{J}^- - \frac{1}{q} \tilde{J}^0 \tilde{J}^- = -\tilde{J}^-,$$

forming the q-deformed Borel subalgebra $b(2)_q$ of the q-deformed algebra $\mathrm{sl}(2)_q$.

The operator $h_q(y, D_q)$ is a nonlocal, three-point, discrete, dilatation-invariant operator illustrated by Fig. 3.

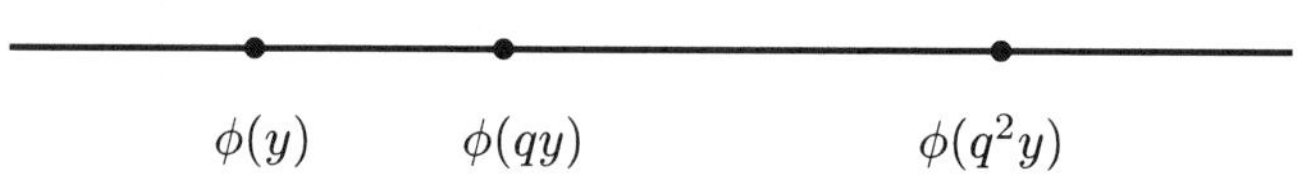

FIGURE 3. Graphical representation of the operator (26).

The spectral problem for the operator (26) has the form

$$
\begin{aligned}
(27)\qquad & 4\frac{\phi(q^2y)}{yq(q-1)^2} - 4\frac{1+q+(y-p-1/2)q(1-q)}{yq(q-1)^2}\phi(qy) \\
& + 4\frac{1+(y-p-1/2)(1-q)}{y(q-1)^2}\phi(y) \\
& = E^{(q)}\phi(y),
\end{aligned}
$$

or, the rhs can be taken as

$$(28)\qquad = E^{(q)}\phi(qy)$$

or as

$$(29)\qquad = E^{(q)}\phi(q^2y).$$

For the case (27) the eigenvalues are given by (24) while for (28), (29) the eigenvalues are equal to

$$(30)\qquad E_n^{(q)} = -4\{-n\}, \qquad n = 0, 1, \ldots,$$

$$(31)\qquad E_n^{(q)} = -4q^{-2n}\{n\}, \quad n = 0, 1, \ldots,$$

correspondingly, while in the limit $q \to 1$ all three expressions coincide. The spectral problems (27)–(29) can be considered as a possible definition of a q-deformed harmonic oscillator. In the literature many other definitions of the $q-$deformed harmonic oscillator are known. Such a situation reflects the existence of an ambiguity appearing when a q-deformation is performed[3] and the absence of clear criteria, which can remove or reduce this ambiguity. For instance, in the literature three types of $q-$Laguerre polynomials (see, for example, an excellent review [**1**])are discussed, but it is not clear why other possible $q-$deformations of Laguerre polynomials are not studied.

Substitution of (25) in (20) gives a slight modification of the expressions (26)–(27). Unlike the translation-invariant case, it does not lead to a change of the number of points in the operator (26) as it happened for the operators (16) and (21).

References

1. R. Koekoek and R. F. Swarttouw, *The Askey-scheme of hypergeometric orthogonal polynomials and its q-analogue*, report 94-05, Delft Univ. of Technology, Delft, 1994.
2. Yu. F. Smirnov and A. V. Turbiner, *Lie-algebraic discretization of differential equations*, Modern Phys. Lett. A **10**, 1795–1802 (1995); Errata: 3139.
3. ———, *Hidden* sl_2*-algebra of finite-difference equations*, Proc. of the IV Wigner Symposium (Guadalajara, 1995) (N. M. Atakishiyev, T. H. Seligman and K.B. Wolf, eds.), World Scientific, River Edge, NJ, 1996, pp. 435–440.
4. A. V. Turbiner, *Lie algebras and linear operators with invariant subspace*, Lie Algebras, Cohomologies and New Findings in Quantum Mechanics (N. Kamran and P. J. Olver, eds.) (Springfield, 1992), Contemp. Math. vol. 160, Amer. Math. Soc., Providence, RI, 1994, pp. 263–310.
5. ———, *Quasi-exactly-solvable differential equations*, Chapter 12, CRC Handbook of Lie Group Analysis of Differential Equations (N. Ibragimov, ed.), vol. 3, New Trends in Theoretical Developments and Computational Methods, CRC Press, Boca Raton, FL, 1995, pp. 331–366.

[3]For example, any term in nondeformed expression can be modified by multipliers of the type q^a and it can be added extra terms with vanishing coefficients in the limit $q \to 1$ like $(1-q)^b$.

6. ———, *Lie algebras in Fock space*, preprint ICN-UNAM/97-14, Mexico, 1997; Proc. of the International Conference Complex Analysis and related topics (Cuernavaca, 1996), Birkhäuser, 1999. (to appear)

Institute for Theoretical and Experimental Physics, Moscow 117259, Russia.

Current address: Instituto de Ciencias Nucleares, UNAM, Apartado Postal 70–543, 04510 Mexico D.F., Mexico.

E-mail address: `turbiner@xochitl.nuclecu.unam.mx`

Centre de Recherches Mathématiques
CRM Proceedings and Lecture Notes
Volume **25**, 2000

Nonsymmetric Linear Difference Equations for Multiple Orthogonal Polynomials

Walter Van Assche

Abstract. We first give a brief survey of some aspects of orthogonal polynomials. The three-term recurrence relation gives a tridiagonal matrix and the corresponding Jacobi operator gives useful information about the orthogonalizing measure and the asymptotic behavior of the zeros of the orthogonal polynomials. The Toda lattice and other similar dynamical systems (Langmuir lattice or Kac-Van Moerbeke lattice) can be solved explicitly using Jacobi operators. Then we present multiple orthogonal polynomials, which are less known. These multiple orthogonal polynomials are defined using orthogonality conditions spread out over r different measures. There is a higher order recurrence relation with $r+2$ terms, which gives a banded Hessenberg matrix and a corresponding operator which is essentially nonsymmetric. We give some examples and indicate how one can start working out a spectral theory for such operators. As an application we show that one can explicitly solve the Bogoyavlenskii lattice using certain multiple orthogonal polynomials.

1. Orthogonal Polynomials

In this paper we will introduce multiple orthogonal polynomials and show how they are related to certain nonsymmetric linear operators that correspond to a finite order linear recurrence relation. In this section we will first recall some relevant facts from orthogonal polynomials (see Szegő [**25**] or [**27**] for a more thorough treatment) and in the next section we will see how some of these facts have an extension to multiple orthogonal polynomials, but that the new setting is richer and still needs further study (see Nikishin and Sorokin [**20**] and Aptekarev [**2**] for more information on multiple orthogonal polynomials).

let μ be a positive measure on the real line for which all the moments exist and for which the support contains infinitely many points. Without loss of generality we will normalize μ so that it is a probability measure. The monic orthogonal polynomials P_n $(n = 0, 1, 2, \dots)$ for the measure μ are such that $P_n(x) = x^n + \dots$

2000 *Mathematics Subject Classification.* Primary: 42C05, 33C45; Secondary: 37L60, 47B36.

Research Director of the Belgian National Fund for Scientific Research (FWO). This research is supported by FWO research project G.0278.97 and INTAS 93-219ext.

This is the final form of the paper.

has degree n and

$$\int P_n(x)x^k\,d\mu(x) = 0, \quad k = 0,1,2,\ldots,n-1. \tag{1.1}$$

These polynomials always exist and are unique. Often one considers the orthonormal polynomials $p_n(x) = \gamma_n P_n(x)$, with $\gamma_n > 0$ and

$$\int p_n(x)p_m(x)\,d\mu(x) = 1, \quad n,m \geq 0. \tag{1.2}$$

1.1. Recurrence relation. A very useful fact is that orthogonal polynomials on the real line always satisfy a three-term recurrence relation of the form

$$P_{n+1}(x) = (x - b_n)P_n(x) - a_n^2 P_{n-1}(x), \quad n \geq 0, \tag{1.3}$$

with initial values $P_0 = 1$ and $P_{-1} = 0$ and with $a_{n+1}^2 > 0$ and $b_n \in \mathbb{R}$ for all n. The recurrence relation for the orthonormal polynomials is given by

$$xp_n(x) = a_{n+1}p_{n+1}(x) + b_n p_n(x) + a_n p_{n-1}(x), \quad n \geq 0, \tag{1.4}$$

with $a_{n+1} > 0$ and $b_n \in \mathbb{R}$ the same coefficients as in (1.3) and $p_0 = 1$, $p_{-1} = 0$. To show that (1.4) holds, one expands $xp_n(x)$ into a Fourier series in the orthonormal polynomials,

$$xp_n(x) = \sum_{k=0}^{n+1} a_{k,n}p_k(x),$$

with Fourier coefficients given by

$$a_{k,n} = \int xp_n(x)p_k(x)\,d\mu(x).$$

Obviously $xp_k(x)$ is a polynomial of degree $k+1$ and by orthogonality it follows that $a_{k,n} = 0$ whenever $k+1 < n$. Therefore the Fourier series only contains three terms and we put

$$a_{n-1,n} = \int xp_n(x)p_{n-1}(x)\,d\mu(x) := a_n,$$

$$a_{n,n} = \int xp_n^2(x)\,d\mu(x) := b_n,$$

$$a_{n+1,n} = \int xp_n(x)p_{n+1}(x)\,d\mu(x) = a_{n+1}.$$

By comparing the coefficient of x^{n+1} in (1.4) we find that $a_{n+1} = \gamma_n/\gamma_{n+1}$, and if we use $p_n(x) = \gamma_n P_n(x)$, then we indeed obtain (1.3).

There is also an important converse of this connection between orthogonal polynomials and three-term recurrence relations, known as Favard's theorem. It states that any sequence of monic polynomials P_n that satisfies a three-term recurrence relation (1.3) with $a_n^2 > 0$ and $b_n \in \mathbb{R}$ for all n and initial conditions $P_0 = 1$ and $P_{-1} - 0$, is always a sequence of monic orthogonal polynomials for some positive measure μ on the real line. This means that the orthogonality conditions (1.1) and the three-term recurrence relation (1.3) are two equivalent ways to characterize orthogonal polynomials. This gives two important problems in the theory of orthogonal polynomials:

(1) Given the recurrence coefficients, how can one obtain the orthogonality measure? (the direct problem)

(2) Given the orthogonality measure, how can we find the recurrence coefficients? (the inverse problem)

A numerical analyst usually knows the measure (or its moments) and wants to compute the recurrence coefficients, hence problem 2 would be the relevant problem. On the other hand, there are many situations when one knows the recurrence coefficients (e.g., they may be birth and death rates in a birth and death process) and then one is interested in obtaining the measure μ, in which case problem 1 is the relevant problem. Later we will see that both the direct and inverse problem are needed for solving the differential equations for the Toda lattice and other similar dynamical systems.

1.2. Jacobi matrix and Jacobi operator. Starting with the recurrence relation (1.3) we can construct a tridiagonal matrix

$$J_n = \begin{pmatrix} b_0 & 1 & & & & \\ a_1^2 & b_1 & 1 & & & \\ & a_2^2 & b_2 & 1 & & \\ & & \ddots & \ddots & \ddots & \\ & & & a_{n-2}^2 & b_{n-2} & 1 \\ & & & & a_{n-1}^2 & b_{n-1} \end{pmatrix},$$

and we can write the recurrence relation (1.3) as

$$\begin{pmatrix} b_0 & 1 & & & & \\ a_1^2 & b_1 & 1 & & & \\ & a_2^2 & b_2 & 1 & & \\ & & \ddots & \ddots & \ddots & \\ & & & a_{n-2}^2 & b_{n-2} & 1 \\ & & & & a_{n-1}^2 & b_{n-1} \end{pmatrix} \begin{pmatrix} P_0(x) \\ P_1(x) \\ P_2(x) \\ \vdots \\ P_{n-2}(x) \\ P_{n-1}(x) \end{pmatrix}$$

$$= x \begin{pmatrix} P_0(x) \\ P_1(x) \\ P_2(x) \\ \vdots \\ P_{n-2}(x) \\ P_{n-1}(x) \end{pmatrix} - P_n(x) \begin{pmatrix} 0 \\ 0 \\ 0 \\ \vdots \\ 0 \\ 1 \end{pmatrix}.$$

This shows that each zero $x_{j,n}$ of P_n is an eigenvalue of J_n. Zeros of orthogonal polynomials can therefore be identified with the spectrum of a tridiagonal matrix. The disadvantage of the matrix J_n is that it is not symmetric. One can easily symmetrize the matrix:

$$R_n J_n R_n^{-1} = \tilde{J}_n = \begin{pmatrix} b_0 & a_1 & & & & \\ a_1 & b_1 & a_2 & & & \\ & a_2 & b_2 & a_3 & & \\ & & \ddots & \ddots & \ddots & \\ & & & a_{n-2} & b_{n-2} & a_{n-1} \\ & & & & a_{n-1} & b_{n-1} \end{pmatrix},$$

where R_n is the diagonal matrix with on the diagonal the vector $(\gamma_0, \gamma_1, \gamma_2, \dots, \gamma_{n-1})$. This symmetric Jacobi matrix corresponds to the recurrence relation (1.4) for the orthonormal polynomials and has the same eigenvalues as J_n. Since $\tilde{J}_n$ is symmetric, it follows immediately that the zeros of p_n are all real.

1.3. Spectral theory and perturbations. More interesting is the infinite Jacobi matrix

$$\tilde{J} = \begin{pmatrix} b_0 & a_1 & & & \\ a_1 & b_1 & a_2 & & \\ & a_2 & b_2 & a_3 & \\ & & \ddots & \ddots & \ddots \end{pmatrix},$$

which acts as a linear operator $\tilde{J} : \ell^2 \to \ell^2$. If the recurrence coefficients are bounded, then $\tilde{J}$ is a bounded symmetric operator, hence selfadjoint. For unbounded recurrence coefficients the operator $\tilde{J}$ is unbounded, but well defined on elements of ℓ^2 with only finitely many nonzero elements. One can extend this matrix to its maximal domain, but there may be several selfadjoint extensions. Let us assume that we are dealing with bounded recurrence coefficients. The tridiagonal structure implies that $e_0 = (1, 0, 0, 0 \ldots) \in \ell^2$ is a cyclic vector, i.e.,

$$\operatorname{span}\{\tilde{J}^n e_0 : n \in \mathbb{N}\} = \ell^2,$$

hence $\tilde{J}$ has a simple spectrum and the spectral theorem tells us that there exists a positive measure μ (the spectral measure) such that ℓ^2 and $L^2(\mu)$ are isometric. The operator $\tilde{J} : \ell^2 \to \ell^2$ corresponds to the multiplication operator $M : L^2(\mu) \to L^2(\mu)$ for which $Mf(x) = xf(x)$ for every $f \in L^2(\mu)$. The isometry maps $\tilde{J}^n e_0$ to x^n and the three-term recurrence relation (1.4) shows that the unit vector $e_n \in \ell^2$ is mapped to the polynomial p_n. Obviously the orthogonality $\langle e_n, e_m \rangle = \delta_{m,n}$ holds in ℓ^2, hence the isometry also gives $\langle p_n, p_m \rangle = \delta_{m,n}$ in $L^2(\mu)$, which gives (1.2). The spectral theorem therefore proves the Favard theorem.

The infinite Jacobi matrix (operator) therefore gives the orthogonality measure (which is the spectral measure of $\tilde{J}$) and the finite sections $\tilde{J}_n$ give the orthonormal polynomials. A very useful class of orthogonal polynomials is the class for which the recurrence coefficients converge

$$\lim_{n\to\infty} a_n = a \geq 0, \quad \lim_{n\to\infty} b_n = b. \tag{1.5}$$

This class is known as $\mathrm{M}(a, b)$ and has been studied intensively by Nevai [**17**] and others. Consider the constant Jacobi operator

$$\tilde{J}_0 = \begin{pmatrix} b & a & & & \\ a & b & a & & \\ & a & b & a & \\ & & \ddots & \ddots & \ddots \end{pmatrix},$$

then in the class $\mathrm{M}(a, b)$ the operator $\tilde{J} - \tilde{J}_0$ has diagonals that converge to 0, which means that $\tilde{J} - \tilde{J}_0$ is a compact operator. Weyl's theorem (see, e.g., [**15**]) then implies that $\tilde{J}$ and $\tilde{J}_0$ have the same essential spectrum. The spectrum of $\tilde{J}_0$ for $a > 0$ is the interval $[b - 2a, b + 2a]$ and the spectral measure has a density

$$\frac{1}{2\pi a^2}\sqrt{4a^2 - (x - b)^2}, \quad x \in [b - 2a, b + 2a].$$

Hence Weyl's theorem implies that the spectrum of $\tilde{J}$ in $\mathrm{M}(a, b)$ is equal to $[b - 2a, b + 2a] \cup E$, where E is bounded, at most denumerable and the accumulation points (if any) are at $b \pm 2a$. One also knows the asymptotic distribution of the zeros $x_{1,n} < x_{2,n} < \cdots < x_{n,n}$ of the orthogonal polynomials p_n in $\mathrm{M}(a,b)$:

$$\lim_{n\to\infty} \frac{1}{n}\sum_{j=1}^{n} f(x_{j,n}) = \frac{1}{\pi}\int_{b-2a}^{b+2a} \frac{f(x)}{\sqrt{4a^2-(x-b)^2}}\,dx,$$

which holds for every bounded and continuous real function f. The limit measure for this asymptotic zero distribution is the equilibrium measure for the set $[b-2a, b+2a]$ in logarithmic potential theory.

1.4. Application: Toda lattice. An important application of Jacobi operators and the direct and inverse problem for orthogonal polynomials is the explicit solution of the differential equations for the Toda lattice, as was observed by Moser [**16**]. Consider the chain of differential equations

$$\begin{cases} \dot a_n = \frac{1}{2} a_n(b_{n-1}-b_n), \\ \dot b_n = a_n^2 - a_{n+1}^2, \end{cases} \qquad n \geq 1, \tag{1.6}$$

with $a_0 \equiv 0$ and initial data $a_{n+1}(0) > 0$, $b_n(0) \in \mathbb{R}$ for $n \in \mathbb{N}$. It is well known that this is an integrable system which can be written as

$$\dot L = ML - LM,$$

where (L, M) is the Lax pair

$$L = \begin{pmatrix} b_0 & a_1 & & & \\ a_1 & b_1 & a_2 & & \\ & a_2 & b_2 & a_3 & \\ & & \ddots & \ddots & \ddots \end{pmatrix}, \qquad M = \frac{1}{2}\begin{pmatrix} 0 & -a_1 & & & \\ a_1 & 0 & -a_2 & & \\ & a_2 & 0 & -a_3 & \\ & & \ddots & \ddots & \ddots \end{pmatrix}.$$

In order to find the solution $a_{n+1}(t)$, $b_n(t)$ for $t > 0$ we first solve the direct problem for the Jacobi operator $L(0)$ with the initial data. This will give the spectral measure μ_0 and the spectrum E, with corresponding orthogonal polynomials $p_n(x;0)$ for which

$$\int_E p_n(x;0)p_m(x;0)\,d\mu_0(x) = \delta_{m,n}.$$

We modify the spectral measure and use the measure μ_t for which $d\mu_t(x) = e^{-xt}\,d\mu_0(x)$ which has the same spectrum as μ_0 (isospectral deformation). Now we have to solve the inverse problem for the measure μ_t to find the orthogonal polynomials $p_n(x;t)$ for which

$$\int_E p_n(x;t)p_m(x;t)\,d\mu_t(x) = \delta_{m,n}.$$

The recurrence coefficients $a_{n+1}(t)$, $b_n(t)$ for these orthonormal polynomials then give the solution of the equations (1.6).

A related dynamical system is the Langmuir lattice (or Kac-Van Moerbeke lattice [**9**]) for which we have the chain of differential equations

$$\dot a_n = \frac{1}{2} a_n(a_{n-1}^2 - a_{n+1}^2), \quad n \geq 1, \tag{1.7}$$

with $a_0 \equiv 0$. This can also be solved using the direct and inverse problem for orthogonal polynomials for initial data $a_n(0) > 0$ whenever $n > 0$. Indeed, consider the Jacobi matrix

$$L(0) = \begin{pmatrix} 0 & a_1(0) & & & \\ a_1(0) & 0 & a_2(0) & & \\ & a_2(0) & 0 & a_3(0) & \\ & & \ddots & \ddots & \ddots \end{pmatrix},$$

then we solve the direct problem to find the spectral measure μ_0 and the corresponding orthogonal polynomials $p_n(x;0)$. We now modify the measure and define a measure μ_t for which

$$d\mu_t(x) = e^{-x^2 t}\, d\mu_0(x),$$

so that μ_t has the same spectrum as μ_0. Now solve the inverse problem for this new measure μ_t to find the orthogonal polynomials $p_n(x;t)$. The Jacobi operator $L(0)$ has zeros on the diagonal, which implies that the spectral measure μ_0 is symmetric with respect to the origin. As a consequence the orthogonal polynomials $p_n(x;0)$ are also symmetric in the sense that $p_n(-x;0) = (-1)^n p_n(x;0)$. The new measure μ_t is also symmetric (since the modification $e^{-x^2 t}$ is symmetric in x), hence the orthogonal polynomials $p_n(x;t)$ are also symmetric and therefore satisfy a recurrence relation of the form

$$xp_n(x;t) = a_{n+1}(t)p_{n+1}(x;t) + a_n(t)p_{n-1}(x;t), \quad n \geq 0.$$

The recurrence coefficients $a_n(t)$ then give the solution of the differential equations (1.7) for the Langmuir lattice.

For the symmetric orthogonal polynomials one can write

$$p_{2n}(x;t) = q_n(x^2;t), \quad p_{2n+1}(x^2;t) = xr_n(x^2;t),$$

where $q_n(x;t)$ ($n \in \mathbb{N}$) and $r_n(x;t)$ ($n \in \mathbb{N}$) are two new sequences of orthogonal polynomials for the measures μ_t^1 and μ_t^2 respectively, which now have a spectrum on the positive real axis and for which

$$\int_0^\infty f(x)\, d\mu_t^1(x) = \int_{-\infty}^\infty f(x^2)\, d\mu_t(x),$$
$$\int_0^\infty f(x)\, d\mu_t^2(x) = \int_{-\infty}^\infty x^2 f(x^2)\, d\mu_t(x).$$

Using the recurrence relation twice, we have

$$\begin{aligned} x^2 p_n(x;t) &= a_{n+1}(t)xp_{n+1}(x;t) + a_n(t)xp_{n-1}(x;t), \\ &= a_{n+1}(t)a_{n+2}(t)p_{n+2}(x;t) + \left[a_{n+1}^2(t) + a_n^2(t)\right]p_n(x;t) \\ &\qquad + a_n(t)a_{n-1}(t)p_{n-2}(x;t), \end{aligned}$$

so that the recurrence relation for $q_n(x;t)$ becomes

$$xq_n(x;t) = A_{n+1}(t)q_{n+1}(x;t) + B_n(t)q_n(x;t) + A_n(t)q_{n-1}(x;t),$$

with $A_n(t) = a_{2n}(t)a_{2n-1}(t)$ and $B_n(t) = a_{2n+1}^2(t) + a_{2n}^2(t)$. For the $r_n(x;t)$ we have in a similar way

$$xr_n(x;t) = C_{n+1}(t)r_{n+1}(x;t) + D_n(t)r_n(x;t) + C_n(t)r_{n-1}(x;t),$$

with $C_n(t) = a_{2n+1}(t)a_{2n}(t)$ and $D_n(t) = a_{2n+2}^2(t) + a_{2n+1}^2(t)$. Observe that (1.7) implies that

$$\begin{cases} \dot{A}_n = \frac{1}{2}A_n(B_{n-1} - B_n), \\ \dot{B}_n = A_n^2 - A_{n+1}^2, \end{cases} \qquad n \geq 1,$$

so that the new recurrence coefficients (A_{n+1}, B_n) satisfy the chain of differential equations (1.6) for the Toda lattice. The same is true for (C_{n+1}, D_n). The transformations $(a_{n+1} : n \geq 0) \mapsto (A_{n+1}, B_n : n \geq 0)$ or $(C_{n+1}, D_n : n \geq 0)$ are known as Miura transformation and correspond, in the theory oforthogonal polynomials, to the well known transformation from a symmetric system of orthogonal polynomials on $(-\infty, \infty)$ to a system of orthogonal polynomials on $(0, \infty)$, the classical example being the relationship between Hermite polynomials and Laguerre polynomials:

$$H_{2n}(x) = (-1)^n 2^{2n} n! L_n^{(-1/2)}(x^2),$$
$$H_{2n+1}(x) = (-1)^n 2^{2n+1} n! x L_n^{(1/2)}(x^2).$$

2. Multiple Orthogonal Polynomials

For multiple orthogonality we will spread the orthogonality conditions over several measures. Let $r \geq 1$ be an integer and let $\mu_1, \mu_2, \ldots, \mu_r$ be r positive measures for which all the moments exist and each with infinitely many points in their supports $E_1, E_2 \ldots, E_r$. We will use multi-indices $\vec{n} = (n_1, n_2, \ldots, n_r) \in \mathbb{N}^r$. The multiple orthogonal polynomial $p_{\vec{n}}$ is the monic polynomial of degree $|\vec{n}| = n_1 + n_2 + \ldots n_r$ which satisfies the orthogonality conditions

$$\int_{E_1} p_{\vec{n}}(x) x^k \, d\mu_1(x) = 0, \qquad k = 0, 1, \ldots, n_1 - 1, \tag{2.1}$$

$$\int_{E_2} p_{\vec{n}}(x) x^k \, d\mu_2(x) = 0, \qquad k = 0, 1, \ldots, n_2 - 1, \tag{2.2}$$

$$\vdots$$

$$\int_{E_r} p_{\vec{n}}(x) x^k \, d\mu_r(x) = 0, \qquad k = 0, 1, \ldots, n_r - 1. \tag{2.3}$$

This polynomial need not be unique; we need to put some extra conditions on the r measures $\mu_1, \ldots, \mu_r$ in order that $p_{\vec{n}}$ is uniquely defined by (2.1)–(2.3). Two useful systems of measures are:

- Angelesco systems [**1**] for which each E_i is a subset of an interval Δ_i with all Δ_i mutually disjoint: $\Delta_i \cap \Delta_j = \emptyset$ for $1 \leq i \neq j \leq r$. For such a system one can easily show that $p_{\vec{n}}$ has exactly n_j zeros in the interval Δ_j, for $j = 1, 2, \ldots, r$.
- An AT system is such that $d\mu_j(x) = u_j(x)\, d\mu_0(x)$, where μ_0 has support $E_0 \subset \Delta_0$ and $u_j \geq 0$ on Δ_0. Furthermore we require that each linear combination
$$\sum_{j=1}^{r} u_j(x) P_{n_j - 1}(x),$$
with P_{n_j-1} a polynomial of degree at most $n_j - 1$ has at most $|\vec{n}| - 1$ zeros on Δ_0. Hence for an AT system all the measures are supported on one interval Δ_0 but we require that $\{(1, x, x^2, \ldots, x^{n_1-1})u_1(x), \ldots, (1, x, x^2,$

$\ldots, x^{n_r-1})u_r(x)\}$ is a Chebyshev system. In this case $p_{\vec{n}}$ is unique and all its zeros are in Δ_0.

2.1. Recurrence relation. We will take a closer look at the multiple orthogonal polynomials for which all the indices in the multi-index $\vec{n}$ are nearly the same. In particular for $n = kr + j$ we will consider the multi-index $\vec{s}(n) = (\underbrace{k+1, k+1, \ldots, k+1}_{j \text{ times}}, k, k, \ldots, k)$ and investigate the polynomials $P_n = p_{\vec{s}(n)}$. A very useful relation for these polynomials is that they satisfy a linear recurrence relation of order $r+1$, generalizing the second order linear recurrence relation (1.3) for ordinary monic orthogonal polynomials (for which $r = 1$). Indeed, we have

$$xP_n(x) = P_{n+1}(x) + \sum_{i=0}^{r} a_i(n)P_{n-i}(x), \quad n \geq 0, \tag{2.4}$$

with initial conditions $P_0 = 1$ and $P_{-1} = P_{-2} = \cdots = P_{-r} = 0$. To see why this is true, we will expand $xP_n(x)$ using the basis functions P_k, $(k = 0, 1, \ldots, n+1)$:

$$xP_n(x) = \sum_{i=0}^{n+1} a_{i,n}P_i(x), \tag{2.5}$$

with $a_{n+1,n} = 1$. Suppose $n = kr + j$, then we will show that $a_{mr+\ell,n} = 0$ for $m = 0, 1, \ldots, k-2$ and $\ell = 0, \ldots, r-1$, and when $m = k-1$ also for $\ell = 0, 1, \ldots, j-1$ so that the sum reduces to $\sum_{i=n-r}^{n+1} a_{i,n}P_i(x)$, which is of the form (2.4). We will use induction on m. First suppose $m = 0$, then we want to prove that $a_{\ell,n} = 0$ for $\ell = 0, 1, \ldots, r-1$. Integrate both sides of the equality (2.5) with respect to μ_1, then (2.1) implies that the right hand side reduces to $a_{0,n}\mu_1(E_1)$, and the left hand side is 0 whenever $n > r$, hence we see that $a_{0,n} = 0$. To see that $a_{1,n} = 0$ we integrate both sides of (2.5) with respect to μ_2, and in general integration with respect to $\mu_{\ell+1}$ will show that $a_{\ell,n} = 0$ for $\ell = 0, 1, \ldots, r-1$. Now suppose that $a_{Mr+\ell} = 0$ for $M \leq m-1$ and $\ell = 0, 1, \ldots, r-1$, with $m \leq k-2$, then we will show that $a_{mr+\ell} = 0$ for $\ell = 0, 1, \ldots, r-1$ as well. Multiply both sides of (2.5) by x^m and integrate with respect to μ_1, then the right hand side only contains $a_{mr,n} \int_{E_1} x^m P_{mr}(x)\, d\mu_1(x)$. The integral can't be zero, since otherwise we would have an orthogonality condition too much which would imply that $P_{mr} \equiv 0$. The left hand side is also zero, which implies that $a_{mr,n} = 0$. In a similar way, we can integrate with respect to $\mu_{\ell+1}$ to find that $a_{mr+\ell,n} = 0$ for $\ell = 0, 1, \ldots, r-1$. Now finally, assume that $m = k-1$, then we multiply (2.5) by x^{k-1} and integrating with respect to $\mu_{\ell+1}$ still gives 0 on the left hand side, but only for $\ell = 0, 1, \ldots, j-1$, whereas the right hand side is proportional to $a_{(k-1)r+\ell,n}$, which implies that $a_{(k-1)r+\ell,n} = 0$ for $\ell = 0, 1, \ldots, j-1$. This is what we wanted to prove.

2.2. Banded Hessenberg matrix. We can now introduce the banded Hessenberg matrix

$$H_n = \begin{pmatrix} a_0(0) & 1 & & & & & & \\ a_1(1) & a_0(1) & 1 & & & & & \\ a_2(2) & a_1(2) & a_0(2) & 1 & & & & \\ \vdots & & \ddots & \ddots & \ddots & & & \\ a_r(r) & a_{r-1}(r) & & a_1(r) & a_0(r) & 1 & & \\ & \ddots & & & \ddots & \ddots & \ddots & \\ & & a_r(n-2) & & & & & 1 \\ & & & a_r(n-1) & \cdots & & a_1(n-1) & a_0(n-1) \end{pmatrix}, \tag{2.6}$$

and the recurrence relation (2.4) now becomes

$$H_n \begin{pmatrix} P_0(x) \\ P_1(x) \\ \vdots \\ P_{n-1}(x) \end{pmatrix} = x \begin{pmatrix} P_0(x) \\ P_1(x) \\ \vdots \\ P_{n-1}(x) \end{pmatrix} - P_n(x) \begin{pmatrix} 0 \\ \vdots \\ 0 \\ 1 \end{pmatrix},$$

so that each zero of P_n is an eigenvalue of H_n. But the matrix H_n is not symmetric and there is no simple way to make it symmetric (unless $r = 1$), so there is no reason in general that the eigenvalues (zeros) should be real. Nevertheless in many cases (e.g., for an Angelesco system or for an AT system) the zeros turn out to be real, which already gives some indication that there is some relationship between the sequences $a_0, a_1, \ldots, a_r$. In general we do not believe that every banded Hessenberg matrix will lead to a system of multiple orthogonal polynomials with r measures on the real line, but that one needs extra conditions on these recurrence coefficients for a bona-fide Favard theorem.

An important case is a banded Hessenberg matrix with entries only on the extreme diagonals, i.e.,

$$H_n = \begin{pmatrix} 0 & 1 & & & & & & \\ 0 & 0 & 1 & & & & & \\ 0 & 0 & 0 & 1 & & & & \\ \vdots & & \ddots & \ddots & \ddots & & & \\ a_r & 0 & & 0 & 0 & 1 & & \\ & a_{r+1} & 0 & & \ddots & \ddots & \ddots & \\ & & \ddots & & & & & 1 \\ & & & a_{n-1} & \cdots & & 0 & 0 \end{pmatrix}, \tag{2.7}$$

with $a_k > 0$ for $k \geq r$. In this case, Aptekarev *et al.* [**4**] proved that the corresponding polynomials P_n will be the multiple orthogonal polynomials $p_{\vec{s}(n)}$ on a $(r+1)$-star in the complex plane, i.e., the measures $\mu_1, \ldots, \mu_r$ are supported on

$$\Gamma_r = \bigcup_{j=0}^{r} \{x e^{2ij\pi/(r+1)} : x \geq 0\}, \tag{2.8}$$

and they are all radially symmetric in the sense that they are invariant under a rotation over an angle $2\pi/(r+1)$. It is therefore sufficient to restrict the measures $\mu_1, \mu_2, \ldots, \mu_r$ to $[0, \infty)$ and to consider the measures $d\nu_j(x) = d\mu_j(x^{1/(r+1)})$.

2.3. Some examples. As a typical example we will take a generalization of the Jacobi polynomials, first considered by Pineiro [**22**], see also Nikishin and Sorokin [**20**, p. 162]. The r orthogonalizing measures are given by

$$d\mu_j(x) = x^{\alpha_j}(1-x)^{\alpha_0}\,dx, \quad x \in (0,1),$$

where $\alpha_i > -1$ for $i = 0, 1, \ldots, r$ and $\alpha_i - \alpha_j \notin \mathbb{Z}$ for $1 \leq i \neq j \leq r$. The multiple orthogonal polynomials can be found explicitly using a Rodrigues type formula

$$(-1)^{|\vec{n}|} \prod_{j=1}^{r} (|\vec{n}| + \alpha_0 + \alpha_j + 1)_{n_j} p_{\vec{n}}(x) = (1-x)^{-\alpha_0} \prod_{j=1}^{r} [x^{-\alpha_j} D^{n_j} x^{n_j+\alpha_j}] (1-x)^{|\vec{n}|+\alpha_0},$$

where the product of the differential operators can be taken in any order. For $r = 1$ we have the Jacobi polynomials. For $r = 2$ we have the recurrence relation

$$xP_n(x) = P_{n+1}(x) + b_n P_n(x) + c_n P_{n-1}(x) + d_n P_{n-2}(x),$$

with rather involved recurrence coefficients [**28**]. Asymptotically they behave as

$$\lim_{n\to\infty} b_n = 3\left(\frac{4}{27}\right), \quad \lim_{n\to\infty} c_n = 3\left(\frac{4}{27}\right)^2, \quad \lim_{n\to\infty} d_n = \left(\frac{4}{27}\right)^3.$$

There is a similar generalization of the Laguerre polynomials by taking

$$d\mu_j(x) = x^{\alpha_j} e^{-x}\, dx, \quad x \in (0, \infty),$$

where $\alpha_j > -1$ for $j = 1, 2, \ldots, r$ and $\alpha_i - \alpha_j \notin \mathbb{Z}$ for $1 \leq i \neq j \leq r$. The recurrence relation for these multiple Laguerre polynomials is

$$xP_n(x) = P_{n+1}(x) + b_n P_n(x) + c_n P_{n-1}(x) + d_n P_{n-2}(x),$$

with [**28**]

$$\begin{aligned} b_{2n} &= 3n + \alpha_1 + 1, & b_{2n+1} &= 3n + \alpha_2 + 2, \\ c_{2n} &= n(3n + \alpha_1 + \alpha_2), & c_{2n+1} &= (n+1)(3n + \alpha_1 + \alpha_2) - \alpha_1 + 1, \\ d_{2n} &= n(n + \alpha_1)(n + \alpha_1 - \alpha_2), & d_{2n+1} &= n(n + \alpha_2)(n + \alpha_2 - \alpha_1). \end{aligned}$$

2.4. Spectral theory and perturbations. The operator $H : \ell^2 \to \ell^2$ associated with the infinite matrix

$$\begin{pmatrix} a_0(0) & 1 & & & & & \\ a_1(1) & a_0(1) & 1 & & & & \\ a_2(2) & a_1(2) & a_0(2) & 1 & & & \\ \vdots & & \ddots & \ddots & \ddots & & \\ a_r(r) & a_{r-1}(r) & \cdots & a_1(r) & a_0(r) & 1 & \\ 0 & \ddots & & & & & \ddots \end{pmatrix},$$

is an interesting object for further study. One would hope that the spectral theory of this operator will give useful information about the orthogonalizing measures $(\mu_1, \ldots, \mu_r)$, in particular about the support of these measures. Partial results in this direction have been obtained by Kalyagin [**11**], [**12**], but there are still various open problems. The main open problem is to find conditions on the sequences $a_0, a_1, \ldots, a_r$ on the diagonals that lead to multiple orthogonal polynomials on the real line. The next problem is to find out from the operator H when one is dealing with an Angelesco system or an AT-system (in particular a Nikishin system), or any of the mixtures which where recently analyzed by Gonchar *et al.* [**7**].

The main difference with the spectral analysis of the Jacobi matrix for orthogonal polynomials is that there is no way to symmetrize H and that H therefore is essentially nonsymmetric. We can therefore not use the spectral theorem for selfadjoint operators. Furthermore, there is not a cyclic vector, but one needs to work with a cyclic set, which can be taken as $\{e_0, e_1, \dots, e_{r-1}\}$. If all the recurrence coefficients are bounded then H is a bounded operator and

$$\text{span}\{H^{n_1}e_0, H^{n_2}e_1, \dots, H^{n^r}e_{r-1} : \vec{n} \in \mathbb{N}^r\} = \ell^2.$$

A possible way to do the spectral analysis is to work with the resolvent $(zI - H)^{-1}$ and the resolvent functions

$$f_j(z) = \langle (zI - H)^{-1}e_{j-1}, e_0 \rangle, \quad j = 1, 2, \dots, r.$$

If one looks for the Hermite–Padé approximation (type II) for the vector $(f_1, f_2, \dots, f_r)$, i.e., the rational approximation with common denominator $p_{\vec{n},0}$ of degree $|\vec{n}|$ and numerators $(p_{\vec{n},1}, p_{\vec{n},2}, \dots, p_{\vec{n},r})$ for which

$$p_{\vec{n},0}(z)f_j(z) - p_{\vec{n},j}(z) = \mathcal{O}\left(\frac{1}{z^{n_j+1}}\right), \quad z \to \infty,$$

then this common denominator is precisely the multiple orthogonal polynomial $p_{\vec{n}}$ that we are interested in.

The connection between the limit behaviour of the zeros of multiple orthogonal polynomials and the spectrum of the operator H is not as strong as in the case of Jacobi operators. For Jacobi operators in the class $\mathrm{M}(a,b)$ the limit set of the zeros coincides with the spectrum of the operator. For $r > 1$ this is no longer true, as can be seen by the following extension of the class $\mathrm{M}(a,b)$. Suppose we take $r = 2$ and that we examine the multiple orthogonal polynomials for which the recurrence coefficients have the following asymptotic behaviour

$$\lim_{n\to\infty} a_0(n) = 3a, \quad \lim_{n\to\infty} a_1(n) = 3a^2, \quad \lim_{n\to\infty} a_2(n) = a^3. \tag{2.9}$$

This class is not empty since the Jacobi–Pineiro polynomials belong to it for $a = 4/27$. If we consider the constant Hessenberg operator

$$H_0 = \begin{pmatrix} 3 & 1 & & & & \\ 3 & 3 & 1 & & & \\ 1 & 3 & 3 & 1 & & \\ & 1 & 3 & 3 & 1 & \\ & & \ddots & \ddots & \ddots & \ddots \end{pmatrix}, \tag{2.10}$$

then $H - H_0$ will be a compact operator so that the essential spectrum of H and H_0 is the same. The spectrum of H_0 has been studied before, since H_0 is a Toeplitz matrix associated with a Laurent polynomial (see, e.g., Schmidt and Spitzer [**23**] and Hirschman [**8**]). Without loss of generality we take $a = 1$, then the corresponding Laurent polynomial is $g(w) = w + 3 + 3/w + 1/w^2 = w(1 + 1/w)^3$. Consider the sets $S_t = \{g(w) : |w| = t\}$ and the sets S_t^+ consisting of the interior and the boundary of S_t, then for t large the curves S_t look like circles and are analytic. As t decreases these curves will schrink until $t = 2$ in which case the curve has a cusp at the real point $27/4$. If t decreases, then the curve S_t has a double point on the interval $[0, 27/4]$ and this double point takes any value on the interval $(0, 27/4)$ starting from $27/4$ for $t = 2-$ and ending at 0 for $t = 1+$. The winding number of the curve S_t is one for each point in the interior of S_t^+. For $t = 1$ the curve S_1 has an inward cusp at 0 and for $0 < t < 1$ the curves S_t have a double point to the left of 0. The winding number of the curve S_t is one or two for each point in the interior of S_t^+. The spectrum $\sigma(H_0)$ is precisely S_1^+ (the shaded area

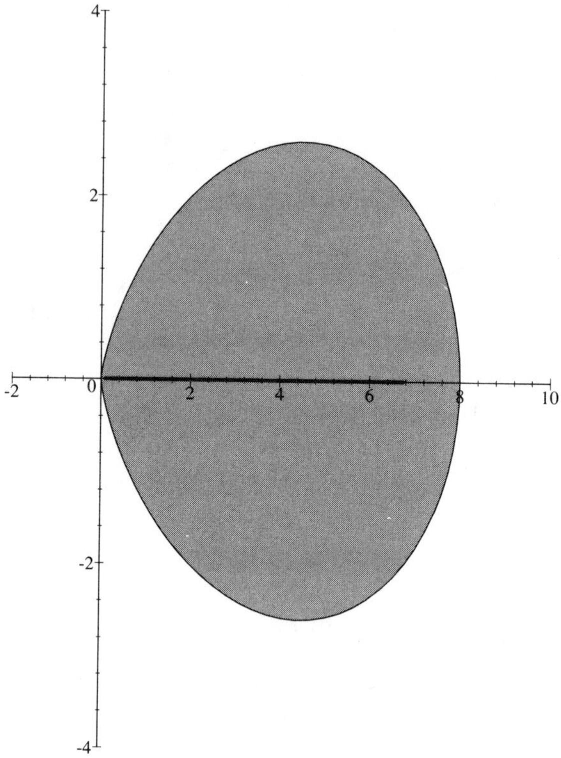

FIGURE 1. Spectrum for the Toeplitz matrix H_0 and the limit set of the zeros.

in Fig. 1). On the other hand, the set of limit points of eigenvalues $\lambda_{j,n}$ $(1 \leq j \leq n)$ of the truncated $n \times n$ matrices H_n, i.e., the set

$$B = \{\lambda_{j,n} : 1 \leq j \leq n, \quad n \in \mathbb{N}\}',$$

is a strict subset of $\sigma(H_0)$ and is given by $\bigcap_{t>0} S_t^+$, which turns out to be the interval $[0, 27/4]$. Another description of the set B is to look at the zeros of $z^2\big(g(z) - \lambda\big) = (1+z)^3 - z^2\lambda$. There are three zeros $z_1(\lambda), z_2(\lambda), z_3(\lambda)$ and we order them so that $|z_1(\lambda)| \leq |z_2(\lambda)| \leq |z_3(\lambda)|$. These zeros can be found explicitly using Cardano's formula. Then the set B is also equal to the set $\big\{\lambda : |z_2(\lambda)| = |z_3(\lambda)|\big\}$. In order to determine the supports of the orthogonalizing measures (μ_1, μ_2), the set B seems to be more relevant than the spectrum of the operator.

The asymptotic distribution of the zeros of multiple orthogonal polynomials P_n for which the recurrence coefficients satisfy (2.9) with $a = 4/27$ is given by

$$\lim_{n\to\infty} \frac{1}{n} \sum_{k=1}^{n} f(x_{k,n})$$
$$= \frac{\sqrt{3}}{4\pi} \int_0^1 f(t) t^{-2/3}(1-t)^{-1/2}\big[\big(1+\sqrt{1-t}\big)^{1/3} + \big(1-\sqrt{1-t}\big)^{1/3}\big]\, dt.$$

This limiting measure appeared earlier in [**14**]. Observe that the density has a stronger singularity at 0 than at 1 so that the zeros are shifted towards 0 as compared to the zero distribution in the class M(1/4, 1/2) which also lives on [0, 1]. If we take the balayage of this measure to the boundary S_t for $t > 0$, then this balayage measure is the equilibrium measure for the set S_t^+. So even though the limit distribution is supported on $B = [0,1]$, it is still characterized as an equilibrium measure for the spectrum S_1^+ of H_0.

2.5. Application: Bogoyavlenskii lattice. Consider the higher order chain of differential equations for the Bogoyavlenskii lattice [**5, 6**]

$$\dot{a}_n = a_n\Bigl(\sum_{j=1}^{r} a_{n-j} - \sum_{j=1}^{r} a_{n+j}\Bigr), \quad n \geq 1, \tag{2.11}$$

with $a_0 \equiv 0$ and initial data $a_n(0)$, $n \geq 1$. If the initial data is such that $a_n(0) > 0$ then one can find the solution of this dynamical system by using banded Hessenberg operators for multiple orthogonal polynomials, much as for the Toda lattice and the Langmuir lattice [**3, 4, 21, 24**]. Indeed, start with the Hessenberg operator for the initial data by putting them on the rth subdiagonal under the main diagonal, with r zero diagonals up to the subdiagonal just above the main diagonal which contains ones:

$$H(0) = \begin{pmatrix} 0 & 1 & & & & & \\ 0 & 0 & 1 & & & & \\ 0 & 0 & 0 & 1 & & & \\ \vdots & & \ddots & \ddots & \ddots & & \\ a_1(0) & 0 & & \ddots & \ddots & 1 & \\ & a_2(0) & 0 & & \ddots & \ddots & 1 \\ & & \ddots & \ddots & & \ddots & \ddots & \ddots \end{pmatrix}.$$

This operator corresponds to a system of multiple orthogonal polynomials on the $(r+1)$-star Γ_r, given by (2.8), for r measures $(\mu_{1,0}, \mu_{2,0}, \ldots, \mu_{r,0})$ which are invariant under a rotation over an angle $2\pi/(r+1)$, i.e., for $j = 1, 2, \ldots, r$ we have the orthogonality conditions

$$\int_{\Gamma_r} p_{\vec{n}}(x;0) x^k \, d\mu_{j,0}(x) = 0, \quad k = 0, 1, 2 \ldots, n_j - 1.$$

We will now apply the dynamics and consider the measures

$$d\mu_{j,t}(x) = e^{-x^{r+1}t} \, d\mu_{j,0}(x), \quad j = 1, 2, \ldots, r,$$

then the system $(\mu_{1,t}, \mu_{2,t}, \ldots, \mu_{r,t})$ lives again on the $(r+1)$-star Γ_r and these measures are again invariant under the rotation over an angle $2\pi/(r+1)$, thanks to the presence of x^{r+1} in the exponential function. Let $p_{\vec{n}}(x;t)$ be the multiple orthogonal polynomials for these new measures, then the polynomials $P_n(x;t)$ satisfy a recurrence relation (2.4) for which all the recurrence coefficients, except for the last one, are identically zero. Hence the Hessenberg operator is of the form

$$H(t) = \begin{pmatrix} 0 & 1 & & & & & \\ 0 & 0 & 1 & & & & \\ 0 & 0 & 0 & 1 & & & \\ \vdots & & \ddots & \ddots & \ddots & & \\ a_1(t) & 0 & & \ddots & \ddots & 1 & \\ & a_2(t) & 0 & & \ddots & \ddots & 1 \\ & & \ddots & \ddots & & \ddots & \ddots & \ddots \end{pmatrix},$$

and the $a_n(t)$ on the lower diagonal are the solution of the differential equations (2.11). Observe that the case $r = 1$, corresponds to the Langmuir lattice if we replace a_n by a_n^2.

References

1. M. A. Angelesco, *Sur deux extensions des fractions continues algébriques*, C. R. Acad. Sci. Paris **18** (1919), 262–263.
2. A. I. Aptekarev, *Multiple orthogonal polynomials*, J. Comput. Appl. Math. **99** (1998), 423–447.
3. A. I. Aptekarev and V. A. Kaliaguine, *Complex rational approximation and difference equations*, Proc. of the Third International Conference on Functional Analysis and Approximation Theory, vol. I (Acquafredda di Maratea, 1996), Rend. Circ. Mat. Palermo (2) Suppl. no. 52, vol. I, 1998, pp. 3–21.
4. A. Aptekarev, V. Kaliaguine, and J. Van Iseghem, *Genetic sum representation for the moments of a system of Stieltjes functions and its application*, Constr. Approx. (to appear)
5. O. I. Bogoyavlenskii, *Some constructions of integrable dynamical systems*, Izv. Akad. Nauk SSSR, Ser. Mat. **51** (1987), no. 4, 737–766 (Russian); Transl. Math. USSR Izv. **31** (1988), no. 1, 47–75.
6. ———, *Integrable dynamical systems associated with the KdV equation*, Izv. Akad. Nauk, Ser. Mat. **51** (1987), no. 6, 1123–1141 (Russian); Transl. Math. USSR Izv. **31** (1988), no. 3, 435–454.
7. A. A. Gonchar, E.'A. Rakhmanov, and V. N. Sorokin, *Hermite–Padé approximants for systems of Markov-type functions*, Mat. Sb. **188** (1997), 33–58 (Russian); Sb. Math. **188** (1997), 671–696.
8. I. I. Hirschman Jr., *The spectra of certain Toeplitz matrices*, Illinois J. Math. **11** (1967), 145–159.
9. M. Kac, and P. Van Moerbeke, *On an explicitly soluble system of nonlinear differential equations related to certain Toda lattices*, Adv. Math. **16** (1975), 160–169.
10. V. A. Kalyagin, *On a class of polynomials defined by two orthogonality relations*, Mat. Sb. **110** (1979), 609–627 (Russian); Math. USSR Sb. **38** (1981), 563–580.
11. ———, *Hermite–Padé approximants and spectral analysis of nonsymmetric operators*, Mat. Sb. **185** (1994), 79–100 (Russian); Sb. Math. **82** (1995), 199–216.
12. V. A. Kaliaguine, *On operators associated with Angelesco systems*, East J. Approx. **1** (1995), no. 2, 157–170.
13. V. A. Kaliaguine and A. Ronveaux, *On a system of classical polynomials of simultaneous orthogonality*, J. Comput. Appl. Math. **67** (1996), 207 217.
14. A. B. J. Kuijlaars, *Chebyshev quadrature for measures with a strong singularity*, J. Comput. Appl. Math. **65** (1995), 207–214.
15. A. Máté, P. Nevai, and W. Van Assche, *The supports of measures associated with orthogonal polynomials and the spectra of the related self adjoint operators*, Rocky Mountain J. Math. **21** (1991), 501–527.
16. J. Moser, *Three integrable Hamiltonian systems connected with isospectral deformations*, Adv. Math. **16** (1975), 197–220.
17. P. G. Nevai, *Orthogonal polynomials*, Memoirs Amer. Math. Soc. **213**, Providence, RI, 1979.

18. E. M. Nikishin, *A system of Markov functions*, Vestnik Moskov. Univ. Ser. I Math. Mekh (1979), no. 4, 60–63 (Russian); Moscow Univ. Math. Bull. **34** (1979), 63–66.
19. ———, *On simultaneous Padé approximants*, Math. Sb. **113** (1980), 499–519 (Russian); Math. USSR Sb. **41** (1982), 409–425.
20. E. M. Nikishin and V. N. Sorokin, *Rational approximations and orthogonality*, Transl. Math. Monogr. **92**, Amer. Math. Soc., Providence, RI, 1991.
21. A. S. Osipov, *A discrete analog of the Korteweg-de Vries equation: Integration by the inverse problem method*, Mat. Zametki **56** (1994), no. 6, 141–144 (Russian); English transl., Math. Notes **56** (1994) no. 6, 1312–1314.
22. L. R. Pineiro, *On simultaneous approximations for a collection of Markov functions*, Vestnik Moskov. Univ., Ser. I Math. Mekh. (1987), no. 2, 67–70 (Russian); Moscow Univ. Math. Bull. **42** (1987), no. 2, 52–55
23. P. Schmidt and F. Spitzer, *The Toeplitz matrices of an arbitrary Laurent polynomial*, Math. Scand. **8** (1960), 15–38.
24. V. N. Sorokin, *Integrable nonlinear dynamical systems of Langmuir lattice type*, Mat. Zametki **62** (1997), no. 4, 588–601; English transl. Math. Notes **62** (1997), no. 4, 488–500.
25. G. Szegő, *Orthogonal polynomials*, 4th ed., American Mathematical Society, Colloquium Publications, vol. XXIII, Amer. Math. Soc., Providence, RI., 1975.
26. W. Van Assche, *Asymptotics for orthogonal polynomials*, Lecture Notes in Mathematics **1265**, Springer-Verlag, Berlin-New York, 1987.
27. ———, *Orthogonal polynomials in the complex plane and on the real line*, Special Functions, q-Series and Related Topics (Toronto, 1995), Fields Inst. Commun., vol. 14, Amer. Math. Soc., Providence, RI, 1997, pp. 211–245.
28. W. Van Assche and E. Coussement, *Some classical multiple orthogonal polynomials*, J. Comput. Appl. Math. (to appear)

Department of Mathematics, Katholieke Universiteit Leuven, Celestijnenlaan 200 building B, B-3001 Heverlee, Belgium.

E-mail address: walter@wis.kuleuven.ac.be

Centre de Recherches Mathématiques
CRM Proceedings and Lecture Notes
Volume **25**, 2000

Asymptotics for Solution to the Cauchy Problem for Volterra Lattice with Step-like Initial Values

V. L. Vereschagin

The Volterra lattice

(1) $$D_t c_n = c_n(c_{n+1} - c_{n-1}), \quad c_n = c_n(t), \ n \in \mathbf{Z},$$

is known as an interesting example of integrable difference-differential model with various applications to plasma physics, crystallography, zoology. The Inverse Scattering Data Method (ISDM) allows to investigate in details Cauchy problem for the Volterra lattice in case of rapidly decreasing initial data (see [**4**]) and in the periodic case [**5, 8**]. Further interest is stimulated by problems with more complex behavior of solutions to (1) as $n \to \pm\infty$.

Let us examine the Cauchy problem for Eq. (1) with initial data (i.d.) $c_n(0)$ which rapidly tend to constant values as $n \to \pm\infty$:

(2) $$\begin{cases} u_n(0) = c_{2n}(0) \to u^{\pm}, \\ v_n(0) = c_{2n+1}(0) \to v^{\pm}, \end{cases} \qquad n \to \pm\infty,$$

The goal of this paper is to construct asymptotics as $t \to \infty$ for the solution of the problem (1)–2.

Let us examine the linear problem

(3) $$c_{n-1}^{1/2}\psi_{n-1} + c_n^{1/2}\psi_{n+1} = \lambda\psi_n, \quad c_n \to c_{\pm}, \ n \to \pm\infty, \quad c_+ > c_- > 0,$$

c_n are solutions to Volterra lattice, λ is a spectral parameter. Define Riemann surface of genus 0 $\Gamma_{\pm}(z_{\pm}, \lambda)$: $2\lambda c_{\pm}^{-1/2} = z_{\pm} + z_{\pm}^{-1}$. Involution $\sigma(z_{\pm}) = z_{\pm}^{-1}$ near circle $|z| = 1$ induces change of sheet of the surface $\Gamma_{\pm}$ near the cut $[0, c_{\pm}] = E_{\pm}$.

Fix solutions of problem (3) by asymptotics

(4) $$\psi_n^{\pm}(\lambda) \sim e_n^{\pm}(z_{\pm}) = z_{\pm}^{\pm n} \exp\left(\frac{1}{8}tc_{\pm}(z_{\pm}^2 - z_{\pm}^{-2})\right), \quad n \to \pm\infty,$$

$\psi_n^+(P)$ is analytic on the lower sheet $\Gamma_{\pm}^- \setminus \infty^-$ while $\psi_n^-(P)$ is analytic on the upper sheet $\Gamma_{\pm}^+ \setminus \infty^+$. In the spectral domain E_+ there is a scattering correlation

(5) $$\psi_n^+(P) = a(P)\psi_n^-(P) + b(P)\psi_n^-(\sigma P), \quad P \in E_+.$$

2000 *Mathematics Subject Classification.* 35P25.

This is the final form of the paper.

Quantities $a(P)$, $b(P)$ are connected by condition

$$(6)\qquad a(P)a(\sigma P)-b(P)b(\sigma P)=\frac{c_-^{1/2}W(\psi_n^+(P),\psi_n^+(\sigma P))}{c_+^{1/2}W(\psi_n^-(P),\psi_n^-(\sigma P))}=-\frac{z_+-z_+^{-1}}{z_--z_-^{-1}},$$

$W(\psi_n,\phi_n)=c_n^{1/2}(\psi_n\phi_{n+1}-\psi_{n+1}\phi_n)$ is a Wronskian.

There is a general scheme for investigation of integrable systems via solving the inverse scattering problem in the form of matrix Riemann problem (MRP).

(1) Let $\Psi_n(P)$ be 2×2-matrix function piecewise analytic in $\mathbf{C}$ with asymptotics under $P\to\infty_\pm$

$$(7)\ \Psi_n(P)=A_\pm(I+\lambda^{-1}V_n^\pm+O(\lambda^{-2}))\begin{pmatrix}(\lambda c_\pm^{-1/2})^{\mp n} & 0\\ 0 & (\lambda c_\pm^{-1/2})^{\pm n}\end{pmatrix},\qquad \lambda\to\infty_\pm,$$

where

$$A_\pm=\begin{pmatrix}0 & 1\\ \frac{1}{\sqrt{c_\pm/c_\mp-1}} & 0\end{pmatrix}.$$

(2) Let the function $\Psi_n(P)$ be analytic outside of the spectrum E_+ and on the cut E_+ have a jump

$$(8)\qquad \Psi_n(P-i0)=\Psi_n(P+i0)G(P),\quad P\in E_+,$$

where

$$(9)\qquad G(P)=-\begin{pmatrix}r^-(P) & \Delta^-\\ \frac{r(P)r(\sigma P)-1}{\Delta^-} & r^-(\sigma P)\end{pmatrix},\quad \Delta=c_-^{1/2}(z_--z_-^{-1}).$$

The upper index $(-)$ denotes value on lower side of the cut E_+.

(3) The nondegenerate matrix $\Psi_n(P)$ is a regular function in $\mathbf{C}$.

STATEMENT 1. Solution c_n of Cauchy problem (1)–(2) for Volterra lattice is determined by solution $\Psi_n(P)$ of formulated above MRP (7)–8 via the following formula:

$$(10)\qquad c_n=\left[\frac{(V_{n-1}^+)^{21}}{V_n^+)^{21}}\right]^2,$$

where V^+ is matrix coefficient of expansion (7).

The exact formulas for the real solutions for Volterra lattice were written out in paper [**5**] and read:

$$(11)\qquad c_{2n}=u_n=u(\tau)=\zeta(2\omega\tau)-\zeta(2\omega\tau-a_-)-\zeta(a_+)-\zeta(a_--a_+),$$

$$(12)\qquad c_{2n+1}=v_n=v(\tau)=\zeta(2\omega\tau-a_+)-\zeta(2\omega\tau)+\zeta(a_-)-\zeta(a_--a_+),$$

where $\tau=1/2(a_--a_+)n+t+\omega'$; $\zeta(x)$ is Weierstrass zeta-function:

$$\zeta'(x)=-\wp(x),$$

$$x=\int_\infty^{\wp(x)}\frac{d\nu}{\sqrt{4\nu^3-g_2\nu-g_3}};\quad \wp(x+2\omega)=\wp(x);\quad \wp(x+2\omega')=\wp(x);$$

$a_\pm$, ω, $i\omega'$ are four real parameters completely determining formulas (11). Let us define the following four quantities:

$$(13)\qquad r_j=2\zeta\left(\frac{1}{2}(a_--a_+)+\varpi_j\right)-2\eta_j+\zeta(a_+)-\zeta(a_-),\quad j=1,2,3,4,$$

where $\omega_1=0$, $\omega_2=\omega$, $\omega_3=\omega+\omega'$, $\omega_4=\omega'$, $\eta_j=\zeta(\omega_j)$.

The four parameters $\overline{r} = (r_1, r_2, r_3, r_4)$ specify completely the solutions (11) $u_n = u(\tau, \overline{r})$, $v_n = v(\tau, \overline{r})$.

Let us suppose that the quantities $\overline{r}$ depend on the variables $x \in \mathbf{R}$, t: $\overline{r} = \overline{r}(x,t)$.

DEFINITION. On the Riemann surface $\widetilde{\Gamma}(w, \lambda)$: $w^2 = R_4(\lambda) = (\lambda - r_1)(\lambda - r_2)(\lambda - r_3)(\lambda - r_4)$ we define a pair of Abel differentials Ω_0, Ω_1:

$$\Omega_0 = \frac{1}{2}(\lambda + d_0)w^{-1}d\lambda, \Omega_1 = \frac{1}{2}\left(\lambda^2 - \frac{1}{2}\Lambda\lambda + d_1\right)w^{-1}d\lambda, \tag{14}$$

where $\Lambda = r_1 + r_2 + r_3 + r_4$; d_0, d_1 are constants fixed by the conditions

$$\int_{r_2}^{r_3} \Omega_0 = \int_{r_2}^{r_3} \Omega_1 = 0. \tag{15}$$

The equation

$$D_t\Omega_0 = D_x\Omega_1, \tag{16}$$

is called Whitham equation or modulation equation.

Formula (16) can be rewritten in the form of a system of quasilinear differential equations for which the quantities r_j, $j = 1, 2, 3, 4$ are Riemann invariants:

$$D_t r_j = W_j(\overline{r}) D_x r_j, \quad j = 1, 2, 3, 4, \tag{17}$$

where $W_j(\overline{r}) = \Omega_1/\Omega_0(r_j)$. In the paper [**6**] the characteristic velocities W_j were written out explicitly in terms of complete elliptic integrals. In the same paper self-similar solutions to system (17) were computed. Here we are interested in the solutions that depend on the variable $\xi = x/t$: $\overline{r} = \overline{r}(\xi)$.

Let us define the Riemann surface $\Gamma(w, \lambda; \xi)$:

$$\Gamma(w,\lambda;\xi) = \begin{cases} w^2 = (\lambda - r_1)(\lambda - r_2^-), & \xi \le \xi^-, \\ w^2 = (\lambda - r_1)(\lambda - r_2^-)(\lambda - r_3^-(\xi))(\lambda - r_4), & \xi^- < \xi \le \xi_0^-, \\ w^2 = (\lambda - r_1)(\lambda - r_2^-(\xi))^2(\lambda - r_4), & \xi_0^- < \xi \le \xi_0^+, \\ w^2 = (\lambda - r_1)(\lambda - r_2^+(\xi))(\lambda - r_3^+)(\lambda - r_4), & \xi_0^+ < \xi \le \xi^+, \\ w^2 = (\lambda - r_3^+)(\lambda - r_4), & \xi > \xi^+, \end{cases} \tag{18}$$

where $r_1 \le r_2^+(\xi) \le r_3^+ \le r_2^0(\xi) \le r_2^- \le r_3^-(\xi) \le r_4$; r_1, r_3^+, r_2^-, r_4 are constants, and

$$\xi^- = -\frac{[(r_4 - r_1)(r_4 - r_2^-)]^{1/2}}{\log((\sqrt{r_4 - r_1} + \sqrt{r_4 - r_2^-})/(\sqrt{r_4 - r_1} - \sqrt{r_4 - r_2^-}))},$$

$$\xi^+ = \frac{1}{2}(r_3^+ + r_4 - 2r_1) - 2\frac{(r_3^+ - r_1)(r_4 - r_1)}{r_4 + r_3^+ - 2r_1} + \frac{r_3^+ - r_1}{\log 1/2(r_4 - r_3^+)/(r_4 - r_1)},$$

$$\xi_0^- = \frac{1}{2}(r_1 + r_4) - r_2^-, \quad \xi_0^+ = \frac{1}{2}(r_1 + r_4) - r_3^+.$$

The dependence of the quantities $r_3^-(\xi)$, $r_2^0(\xi)$, $r_2^+(\xi)$ on the variable ξ is determined by the self-similar solutions to the system (16):

$$\xi + W_3(r_1, r_2^-, r_3^-(\xi), r_4) = 0, \quad \xi^- < \xi \le \xi_0^-, \tag{19}$$

$$\xi + W_3(r_1, r_2^0(\xi), r_2^0(\xi), r_4) = 0, \quad \xi_0^- < \xi \le \xi_0^+, \tag{20}$$

$$\xi + W_2(r_1, r_2^+(\xi), r_3^+, r_4) = 0, \quad \xi_0^+ < \xi \le \xi^+. \tag{21}$$

The holomorphic differential $\Omega = Dw^{-1}\,d\lambda$, D is a constant, fixed by the condition $\oint_a \Omega = 1$. Let us examine the following ansatz for the solutions of the Volterra lattice:

$$\begin{pmatrix} u(\tau) \\ v(\tau) \end{pmatrix} = \bar{c}(\tau \mid \Gamma(w,\lambda;\xi))_{x=n} + O(t^{-\delta}), \quad \delta > 0, \tag{22}$$

$$\bar{c} = \begin{cases} \begin{pmatrix} u \\ v \end{pmatrix}^{\pm}, & \xi > (<)\xi^{\pm}, \\ \begin{pmatrix} u(\tau, \bar{r}(\xi)) \\ v(\tau, \bar{r}(\xi)) \end{pmatrix}, & \xi^- \le \xi \le \xi^+, \end{cases}$$

where $\bar{r}(\xi)$ are branch points of the curve $\Gamma(w,\lambda;\xi)$. The quantities $u^\pm$, $v^\pm$, $\bar{r}$ are connected in the following way:

$$4r_1 = u^- + v^- - 2\sqrt{u^- v^-}, \quad 4r_2^- = u^- + v^- + 2\sqrt{u^- v^-}, \tag{23}$$

$$4r_3^+ = u^+ + v^+ - 2\sqrt{u^+ v^+}, \quad 4r_4 = u^+ + v^+ + 2\sqrt{u^+ v^+}, \tag{24}$$

whereas one suppose that $r_1 \le r_3^+$.

Let us deal now with the case $u^\pm = v^\pm = c^\pm$, $r_1 = r_3^+ = 0$, $a_+ = -a_-$, which does not change the scheme but considerably simplifies the calculations.

In the paper [**7**] explicit formulas for the case of the one-gap spectrum of the linear problem (3) were obtained. These formulas represent the Baker-Akhiezer function in terms of elliptic functions:

$$\tilde{e}_n^{\pm}(t,z) = \gamma_n^{\pm 1} \exp\left[\pm\pi i\left(t\mathcal{V}(z) + n\mathcal{P}(z)\right)\right] \frac{\theta(\omega'\tau \pm z + P)}{\theta(\pm z + P)\theta(\omega'\tau - a + P)}, \tag{25}$$

where $a = a_+ = -a_-$, $P = \omega + a$, $\theta(\tau) = \sigma(\tau + \omega)\exp(-\tau\eta - (\tau^2\eta')/(2\omega'))$, $\tau = an + Vt$, $(\sigma'(\tau))/(\sigma(\tau)) = \zeta(\tau)$; the variable z is connected with the parameter λ via the following uniformization

$$\begin{gathered} \lambda(z) = \zeta(z-a) - \zeta(z+a) + 2\zeta(a), \\ \mathcal{V}(z) = \int_{r_1}^{\lambda(z)} \Omega_1, \quad \mathcal{P}(z) = \int_{r_1}^{\lambda(z)} \Omega_0, \\ \gamma_n = \beta^n \left[\frac{\sigma(\omega'\tau + \omega)\sigma(2\omega' a + \omega)}{\sigma(\omega'(\tau + 2a) + \omega)\sigma(\omega)}\right]^{1/2}, \quad \beta = \sigma(2\omega' a + \omega)/V. \end{gathered} \tag{26}$$

Let us examine the function

$$\Phi_n(P) = \begin{cases} a^{-1}(P)\psi_n^+(P), & P \in \Gamma^-, \\ \psi_n^-(P), & P \in \Gamma^+. \end{cases} \tag{27}$$

It is analytic on each sheet of the curve and has a jump on contour $\partial\Gamma^+ = E_+$ equal to $r(P)\psi_n^-(\sigma P)$. So the function $\Phi_n(P)$ can be restored by the use of the formula

$$\Phi_n(P) = e_n^-(P) + \frac{1}{2\pi i}\int_{\partial\Gamma^+} M(P,Q) r(Q)\psi_n^-(\sigma Q), \tag{28}$$

where $M(P,Q)$ is the Cauchy kernel:

$$M(P,Q) = -\frac{W(e_n^-(P), e_n^-(\sigma Q))}{W(e_n^-(Q), e_n^-(\sigma Q))}\frac{d\lambda(Q)}{\lambda(P) - \lambda(Q)}. \tag{29}$$

Let us suppose now that $\xi \in [\xi^-, \xi_0^-]$ and the dynamics of the branch point $r_3^-(\xi)$ is unknown. Then the function $\Phi_n(P)$ is explicitely specified by a Cauchy integral on the curve $\widetilde{\Gamma}$ (18):

$$\Phi_n(P) = e_n^-(P) + \frac{1}{2\pi i}\int_{\partial\widetilde{\Gamma}^+\cup L} \widetilde{M}(P,Q)\widetilde{f}(Q), \tag{30}$$

where $\partial\widetilde{\Gamma}^+ = [r_1, r_2^-] \cup [r_3^-(\xi), r_4]$; $L = [r_2^-, r_3^-(\xi)] \cup \sigma([r_3^-(\xi), r_2^-])$, σ is an involution that interchanges sheet on $\widetilde{\Gamma}$,

$$\widetilde{M}(P,Q) = -\frac{W(\widetilde{e}_n^-(P), \widetilde{e}_n^-(\sigma Q))}{W(\widetilde{e}_n^-(Q), \widetilde{e}_n^-(\sigma Q))}\frac{d\lambda(Q)}{\lambda(P)-\lambda(Q)}. \tag{31}$$

The choice of the function $\widetilde{f}$ is inspired by paper [**9**] in analogy with Zakharov-Manakov ansatz:

$$\psi_n^-(P) = A(P)\widetilde{e}_n^-(P) + B(P)\widetilde{e}_n^-(\sigma P). \tag{32}$$

Let us substitute this ansatz into the linear equation and get:

$$\begin{aligned} A(P)\widetilde{e}_n^-(P) &+ B(P)\widetilde{e}_n^-(\sigma P) \\ &= e_n^-(P) + \frac{1}{2\pi i}\int_{\partial\widetilde{\Gamma}^+\cup L} \widetilde{M}(P,Q)r(Q)\big[A(\sigma Q)\widetilde{e}_n^-(\sigma Q) + B(\sigma Q)\widetilde{e}_n^-(Q)\big]. \end{aligned} \tag{33}$$

Let us examine equation (33) as $n \to \infty$ and use the known formula ($t \to \infty$)

$$\begin{aligned} \frac{1}{2\pi i}\int_{-\infty}^{\infty} &\frac{f(\eta)}{\eta - \mu + i0}\exp\big(it\gamma(\eta)\big)\,d\eta \\ &= -f(\mu)\exp\big(it\gamma(\mu)\big)H\big(-\gamma'(\mu)\big) + O\Big(\frac{1}{\sqrt{t}(\mu - \mu_{CT})}\Big), \end{aligned}$$

where $\gamma'(\mu_{CT}) = 0$ and $H(x) = (1 + \operatorname{sign} x)/2$ is the Heaviside function. The role of $\gamma(\mu)$ is played by the differential $\Omega_0(P)\xi - \Omega_1(P)$. Therefore one obtains:

$$B(P) = -r(P)A(\sigma P)H\big(\lambda(P) - \lambda(P_{CT})\big), \quad P \in \partial\widetilde{\Gamma}^+, \tag{34}$$

where point P_{CT} is determined by the condition

$$\Omega_0(P_{CT})\xi = \Omega_1(P_{CT}). \tag{35}$$

Let us examine now (33) as $n \to -\infty$ and get the integral equations for $A(P)$:

$$\begin{aligned} A(P)&\widetilde{e}_n^-(P) \\ &= e_n^-(P) + \frac{1}{2\pi i}\int_{\partial\widetilde{\Gamma}^+\cup L} \widetilde{M}(P,Q)r(Q)r(\sigma Q)A(Q)\widetilde{e}_n^-(Q)H\big(\lambda(P) - \lambda(P_{CT})\big), \\ &\qquad P \in \partial\widetilde{\Gamma}^+, \end{aligned} \tag{36}$$

$$A(P)\widetilde{e}_n^-(P) = e_n^-(P) + \frac{1}{2\pi i}\int_{\partial\widetilde{\Gamma}^+\cup L} \widetilde{M}(P,\sigma Q)r(\sigma Q)A(Q)\widetilde{e}_n^-(Q), \quad P \in L. \tag{37}$$

Thus the function $\widetilde{f}$ in (30) must have the following form:

$$\widetilde{f}(P) = \begin{cases} -r(P)\big[A(\sigma P)\widetilde{e}_n^-(\sigma P) + B(\sigma P)\widetilde{e}_n^-(P)\big], & P \in \partial\widetilde{\Gamma}^+, \\ \big(1 + r^+(\sigma P)\big)A(P)\widetilde{e}_n^-(P), & P \in [r_2^-, r_3^-(\xi)], \\ \big(1 - r^+(P)\big)A(P)\widetilde{e}_n^-(P), & P \in \sigma([r_2^-, r_3^-(\xi)]). \end{cases} \tag{38}$$

The quantity $A(P)$ should be obtained from eqs (36)–(37). They indicate that the function $A(P)$ is analytic on the sheets $\Gamma_{\pm}$ and has a jump on the contour $\partial\widetilde{\Gamma}^+ \cup L$:

$$A_n^-(P) = A_n^+(P)g(P), \quad P \in \partial\widetilde{\Gamma}^+ \cup L, \tag{39}$$

where conjugation matrix $g(P)$ has the form:

$$g(P) = \begin{cases} 1 - r(P)r(\sigma P)H(\lambda(P) - \lambda(P_{CT})), & P \in \partial\widetilde{\Gamma}^+, \\ -r^+(\sigma P), & P \in [r_2^-, r_3^-(\xi)], \\ r^+(P), & P \in \sigma([r_2^-, r_3^-(\xi)]). \end{cases} \tag{40}$$

Solution to such a problem is given by the following formula:

$$A(P) = R(P)\exp\Big(-\frac{1}{2\pi i}\int_{\partial\widetilde{\Gamma}^+\cup L} \widehat{M}(P,Q)\log g(Q)\Big), \tag{41}$$

where $\widehat{M}(P,Q)$ is an analogue of Cauchy kernel:

$$\widehat{M}(P,Q) = \int_{\infty_+}^{P} m(P',Q), \tag{42}$$

$m(P,Q)$ is a meromorphic bidifferential fixed by conditions

$$\oint_a m(P,Q) = 0, \quad m(P,Q) \sim \frac{dq(P)dq(Q)}{(q(P)-q(Q))^2}, \quad P \sim Q, \tag{43}$$

q is a local parameter. It ensues from norm conditions (43) that $\oint_b m(P,Q) = 2\pi i\Omega(P)$.

The multiplier $R(P)$ is specified by the condition of uniqueness of $A(P)$ and the asymptotics $A(P) \to c_-^{1/2}$, $P \to \infty_{\pm}$:

$$R(P) = \text{const.}\frac{\theta\Big(\int_{\infty_+}^P \Omega + D\Big)}{\theta\Big(\int_{\infty_+}^P \Omega + D_0\Big)}, \tag{44}$$

where $D_0 = -(\int_{\infty_+}^{r_1} + \int_{\infty_+}^{r_4})\Omega - K$, K is a vector of Riemann constants (see [**8**]).

The conjugation conditions (40) provide a formula for the phase shift:

$$D = D_0 - \frac{1}{2\pi i}\int_{\partial\widetilde{\Gamma}^+\cup L} \Omega(Q)\log g(Q). \tag{45}$$

So we have constructed the function $\Psi_n(P)$ which, in the case $r_3^-(\xi) = \lambda(P_{CT})$, asymptotically satisfies the conjugation condition (8) and the estimate (7).

STATEMENT 2. Under $r_3^-(\xi) = \lambda(P_{CT})$ the following condition holds:

$$\Psi_n(P+i0)G(P)\Psi_n(P-i0) = I + \delta(P,\xi,t),$$

where small parameter δ has the form

$$\delta(P,\xi,t) = \begin{cases} O(t^{-\varepsilon}(P - P_{CT})), & |P - P_{CT}| \geq t^{-\varepsilon}, \\ O(1), & |P - P_{CT}| < t^{-\varepsilon}, \quad \varepsilon > 0. \end{cases}$$

In an analogous way one studies the Riemann problem (8) for all other values of the parameter ξ.

Using formula (10) we obtain the following result:

STATEMENT 3. The main term of the asymptotics as $t \to \infty$ of the solution to the Cauchy problem (1)–(2) for the Volterra lattice under the condition that no solitons are present is described by Whitham-modulated one-gap solution $c(n,t \mid \Gamma(\xi), D(\xi))$ (11), (13), where the Riemann curve $\Gamma(\xi)$ is determined by formulas (18) and the phase shift $D(\xi)$—by formula (45). The remaining terms of the asymptotic series decrease as powers in t.

References

1. R. F. Bikbaev, *On Korteweg-de Vries equation with finite-gap boundary values*, preprint BNT Ural Branch Academy of Sciences, Ufa., 1988.
2. ———, *Whitham transformation and temporal asymptotics for solution to nonlinear Schrodinger equation with finite-gap behavior as $x \to \pm\infty$*, Algebra i Analiz **2** (1990), no. 3, 131–143 (Russian); English transl., Leningrad Math. J. **2** (1991), no. 3, 577–588.
3. ———, *Shock waves in one-dimensional models with cubic nonlinearity*, Teoret. Mat. Fiz. **97** (1993), no. 2, 191–212 (Russian); English transl., Theoret. and Math. Phys. **97** (1993), no. 2, 1236–1249.
4. S. V. Manakov, *On complete integrability and stochastization in discrete dynamical systems*, Soviet Physics JETP **67** (1974), no. 8, 543–555.
5. V. L. Vereschagin, *Hamiltonian structure of averaged difference systems*, Math. Notes **44** (1988), no. 5, 584–595.
6. ———, *Whitham equations for one-phase solutions of Volterra lattice*, Nonlinear Anal. **19** (1992), no. 2, 177–185.
7. ———, *Spectral theory for one-gap solutions of Volterra lattice*, Math. Notes **48** (1990), no. 2, 145–148.
8. A. P. Veselov, *Integration of stationary problem for classical spin lattice*, Theor. and Math. Phys. **71** (1987), no. 1, 154–159.
9. V. L. Zakharov and S. V. Manakov, *Asymptotic behavior of nonlinear wave systems integrable via inverse problem method*, Soviet Physics JETP **71** (1976), no. 1, 203–215.

IRKTUSK COMPUTING CENTER, SIBERIAN BRANCH OF THE RUSSIAN ACADEMY OF SCIENCES, P.O. BOX 1233, IRKUTSK 664033, RUSSIA.

E-mail address: vadim@icc.ru

Centre de Recherches Mathématiques
CRM Proceedings and Lecture Notes
Volume **25**, 2000

Complexity and Integrability

Claude Viallet

ABSTRACT. We describe the relations between singularity, complexity, and integrability of discrete systems with a rational evolution.

1. Introduction

Characterizing the integrability of evolution equations is still an open problem. For continuous systems, apart from the analysis of the hamiltonian structure and the direct proof of integrability in the sense of Liouville [**12, 14**], there exist other approaches like the singularity analysis [**11, 13**]. For discrete systems, which are the ones we are interested with, an analog of the Painlevé analysis was proposed in [**8, 15**].

We describe here a method one can use for discrete systems having a rational evolution, i.e: the state of the system belongs to some finite dimensional space (R^n or preferably projective space $\mathbb{P}_n$, real or complex), and the state at time $t+1$ is expressible in terms of the state at time t through rational functions, or polynomial functions if one uses the homogeneous coordinates of projective space.

We associate to any such system an entropy (algebraic entropy) which measures the complexity of the evolution map, and propose to use a nonvanishing entropy as an index of nonintegrability. In such a general setting the proper notion of integrability will be that the motion reduces to a shift on some abelian variety. Basic considerations linking complexity with singularity and integrability of rational maps go back to [**7**]. Further elaborations followed, and we describe some of them here.

2. Definitions

Let φ be a rational evolution map. The action of φ represents one time-step. We use homogeneous coordinates $[x_0, x_1, \ldots, x_n]$ over $\mathbb{P}_n$, so that φ is given in terms of $n+1$ homogeneous polynomials of degree d. If any common factor appears among these polynomials, we ought to factor it out. The degree d is thus well defined, but of course not invariant by changes of coordinates (see later).

2000 *Mathematics Subject Classification.* Primary: 37-XX; Secondary: 39-XX.

This work is supported by Centre National de la Recherche Scientifique, Unité Mixte de Recherche 7589, CNRS et Universités Paris VI et Paris VII.

This is the final form of the paper.

DEFINITION 1. Consider the sequence $\{d_k\}$ of degrees of the iterates $\varphi^{(k)}$ of φ. We shall use as an index of complexity, and call algebraic entropy of φ the limit:

$$\varepsilon(\varphi) = \lim_{k\to\infty} \frac{1}{k} \operatorname{Log}(d_k). \tag{2.1}$$

CLAIM. This limit always exists. It is a straightforward consequence of the inequality on the degree of composed rational maps:

$$d(f\cdot g) \leq d(f)\cdot d(g). \quad \square \tag{2.2}$$

DEFINITION 2. We shall make use of the generating function of the sequence of degrees

$$\eta_\varphi(z) = \sum_{k=0}^{\infty} d_k z^k. \tag{2.3}$$

REMARK 1. The degree d_k of the $k-th$ iterates of φ is bounded above by d^k, and the entropy is thus bounded by $\operatorname{Log}(d)$. The possible simplifications of the fractions (or equivalently the presence of common factors in the homogeneous description) are responsible for the drop of the degree. This drop may be so important that the growth falls from its generic exponential rate to a polynomial rate.

REMARK 2. The entropy of φ is the logarithm of the smallest modulus of the poles of η_φ. If the entropy is non vanishing, the growth of the degree is exponential. If the entropy vanishes, the growth is polynomial, and the degree of this polynomial may be taken as a secondary index.

REMARK 3. There is a similarity between η and the dynamical ζ function introduced by Artin-Mazur and Smale [**3, 16**]. Some precaution however has to be taken with the invariance of all these objects under changes of coordinates. If one studies homeomorphisms (or diffeomorphisms) like in [**3, 16**] or [**2**], the (natural) changes of coordinates should themselves be diffeomorphisms. The multiplicity of fixed points for example is unchanged by such coordinate transformations. If, and this is what we do here, one is interested in rational transformations, the natural group of coordinate transformations is the group of birational transformations, and neither the order of fixed points, nor the degree are invariant under such changes. Neither the function ζ nor the function η are invariant. However, and this is one of its important features, our entropy is left unchanged by these coordinate transformations. There are also relations between the asymptotic behaviours of degrees, and number of fixed points, if the latter is finite [**1**]. The drawback of the ζ function is that it is not defined if the number of fixed points of some order is infinite. This happens quite often when the map we consider is integrable, in particular if there exists any rational invariant [**4**].

REMARK 4. The definition may straightforwardly be extended to sequences of maps, allowing to consider nonautonomous maps, provided for example the degree of the successive maps is bounded.

3. Examples

We will study a number of elementary quadratic maps in P_2. Here is a list:

(a) The Hénon map: $[x, y, z] \to [x^2, x^2 + xz - \alpha y^2, \beta xy]$.
(b) ϕ: $[x, y, z] \to [-xy + \gamma z^2 + x^2, x^2, xz]$.

(c) Alternating sequence with odd terms $\varphi_o\colon [x,y,z] \to [-xy+\gamma z^2+\beta xz+x^2, x^2, xz]$ and even terms $\varphi_e\colon [x,y,z] \to [-xy+\gamma z^2+\beta xz, x^2, xz]$, with $\beta \neq 0$.
(d) The same as (c) with an exchange of odd and even
(e) $\varphi\colon [x,y,z] \to [x^3+\alpha z^3 - yx^2, x^3, x^2 z]$.

4. A Phenomenological Approach

Calculating the full expressions of the iterates of φ is beyond reach, even with a good machine. We rely on the simple fact that their degree is also the degree of the curve obtained by successive transformations of a generic straight line in $\mathbb{P}_n$ [**7**]. The rest of the task will be to analyze the sequences of integers (see [**6**] for an extensive study).

We get the following sequence of numbers:

(a) $1, 2, 4, 8, 16, 32, 64, 128, 256, 512, \dots$
(b) $1, 2, 4, 8, 13, 20, 28, 38, 49, 62, 76, 92, 109, \dots$
(c) $1, 2, 4, 7, 14, 22, 44, 70, 140, 224, 448, 716, \dots$
(d) $1, 2, 4, 8, 14, 27, 42, 84, 133, 266, 426, 852, 1362, 2724, 4352, 8704, 13908, \dots$
(e) $1, 3, 9, 27, 73, 195, 513, 1347, \dots$

The dots mean of course: guess the next terms, like in IQ tests. Sequences of integers are known to have arbitrary continuations if no hint is given about their construction rule. What is remarkable is that we can actually infer what the rule is, because all these sequences have common stringent features.

To evaluate the growth of the degrees, it is sometimes sufficient to calculate the discrete derivatives of the sequence (d_k).

To do better it is useful to compute a generating functional, e.g using Pade-type methods, i.e. approximate the (truncated) series η with a rational fraction.

What is remarkable is that in all the cases which have been studied, what comes out is *a rational generating function with integer coefficients*, i.e. the Pade approximation stabilizes, and the untruncated series is given by a rational function. This has two consequences (yet to be proven in general, but no counterexample is available at this moment):

(1) The algebraic entropy is the logarithm of an algebraic number.
(2) The sequences of degrees verify a *finite recurrence relation with integer coefficients.*

For the above list of examples, we gave a sufficient information in the listings of the various sequences, by cutting them after they stabilize.

(a) Elementary: the sequence was $d_k = 2^k$.
(b) Call (b) the sequence of degrees. If we compute the the discrete derivatives b1, and b2 of b, we get:
    b1: $1, 2, 4, 5, 7, 8, 10, 11, 13, 14, 16, 17, \dots$
    b2: $1, 2, 1, 2, 1, 2, 1, 2, 1, 2, 1, \dots$
    The degree d_k grows like a polynomial of degree 2 in k, and the entropy vanishes. The answer to the test was:
$$\dots, 109, 128, 148, 170, 193, 218, 244, 272, 301, \dots$$
    The generating function has poles on the unit circle.
$$\eta_b = -(1+2s^3)/(s+1)/(s-1)^3$$

(c) Calculating the discrete derivatives yields:
c1: $1, 2, 3, 7, 8, 22, 26, 70, 84, 224, 268, \dots$
c2: $1, 1, 4, 1, 14, 4, 44, 14, 140, 44, \dots$
c3: $0, 3, -3, 13, -10, 40, -30, 126, -96, \dots$
etc, and we do not see much of the structure, but, if we use the Pade method, we immediately get the generating function:

$$\eta_c = -\frac{(s^2 - s + 1)(s + 1)^3}{2s^6 + 3s^2 - 1}$$

This can be checked against the next calculable terms:

$$\dots, 716, 1432, 2288, 4576, 7312, 14624, 23368, 46736, 74680.$$

Entropy is non vanishing, and is given by the root of the denominator of η_c with smallest modulus.

(d) The discrete derivatives are :
d1: $1, 2, 4, 6, 13, 15, 42, 49, 133, 160, 426, 510, 1362, 1628, 4352, 5204, \dots$
d2: $1, 2, 2, 7, 2, 27, 7, 84, 27, 266, 84, 852, 266, 2724, 852, \dots$
etc, and is again not illuminating. Pade gives $\eta_d = P/Q$, with

$$P = -1 - 2s - s^2 - 2s^3 - 2s^4 - 3s^5 + 2s^6 + s^7 + s^8 + 2s^9 + s^{10},$$

$$Q = 2s^6 + 3s^2 - 1.$$

This agrees with the next computed terms:

$$\dots, 13908, 27816, 44448, 88896.$$

REMARK 5. The two cases (c) and (d) have the same entropy, because they are related to the same nonautonomous iterations.

The only integrable map in that list is case (b). Our experience from numerous examples (it is indeed an experimental fact, but we turned it into a conjecture) is that if the entropy is non vanishing, the map is not integrable. We have found no counterexample yet.

In other words, the Pade approximation should always stabilize to a rational fraction with integer coefficients. Phrased differently: there always exists a finite recurrence relation with integer coefficients governing the sequence of degrees of the iterates (see Section 6). If the characteristic polynomial of this recurrence relation has a root of modulus smaller than 1, then the map is not integrable.

5. Degree Growth Versus "Singularity Confinement"

It is simple to understand the relation between the lowering of the degree and the singularity structure. It suffices to look at the geometry behind the factorization process, and more precisely the blow-downs and blow-ups familiar to algebraic geometers. Fig. 1 explains the process (this is a *virtual* but plausible example).

Suppose a variety Σ of codimension 1 is blown down by φ. Further transformations by φ may blow the image down again (if $n \geq 3$), reaching a singular point of φ (i.e. a point which blows up). This means that all points of Σ are sent to $[0, 0, \dots, 0]$ by φ^4. In other words the equation of Σ factorizes in the expression of φ^4, implying a drop of the degree. In the above picture, there is more: there is a "resolution of the singularity" (alias confinement) by φ^5 since points of Σ have an image on some other hypersurface Σ', after all common factors, coming from the successively encountered singularities, have been removed. The removal of factors plays the role

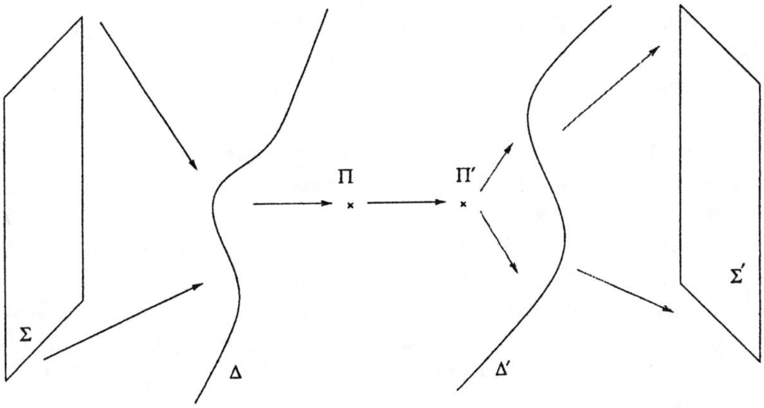

FIGURE 1. A possible blow-down blow-up scheme in $\mathbb{P}^3$.

of the regularization and local exploration performed in the singularity confinement approach [**8, 10, 15**].

One should point out that there is no equivalence between resolution of singularities and absence of entropy. There is room between the two [**5, 9, 10**]. The main result is that there exist nonintegrable (chaotic) maps which verify singularity confinement. There also exist integrable maps which do not verify confinement. It seems that the singularity confinement is not strong enough if it happens "too late" or if the number of dimensions of the phase space of the system is too large.

6. The Meaning of η

The fact that η is rational says that the degrees verify, at least after a certain rank, a recurrence relation. Indeed suppose this function is

$$\eta = \frac{P}{Q} = \frac{\sum_{i=0}^{p} \alpha_i z^i}{\sum_{j=0}^{q} \beta_j z^j}.$$

Then since by definition

$$\eta_\varphi(z) = \sum_{k=0}^{\infty} d_k z^k,$$

the degrees verify at least for large enough r,

$$\sum_{j=0}^{q} \beta_j d_{r-j} = 0,$$

and this recurrence controls their asymptotic behaviour.

7. Comments

In some instances, we have been able to prove the recurrence relation between the degrees [**5**].

The algebraic entropy is exactly calculable by constructing the generating function of the sequence.

The measure of entropy is more sensitive to nonintegrability than singularity confinement [**9, 10**].

The algebraic entropy exists even in cases where the Kolmogorov–Sinai entropy cannot be defined (as it does not need the knowledge a priori of an ergodic measure).

From Section 5 we see that the resolution of singularities is a reversibility requirement for the system: one requests that the indeterminations created by the possible noninvertibilities of the evolution map be lifted after a finite number of steps. What appears in [**5, 9, 10**] is that the process helps not having entropy, but is not the end of the story.

A final comment: it was pointed out by M. Kruskal that the Painlevé criterion is by essence an infinitesimal one (not a local one nor a global one). The infinitesimal scale seems indeed out of reach of discrete systems. However, if one wants to recover this scale, one has to think of smaller and smaller time steps, of which the number increases indefinitely. One thus has to examine large (discrete) time behaviour, and this is precisely what we do.

References

1. N. Abarenkova, J.-C. Anglès d'Auriac, S. Boukraa, S. Hassani, and J.-M. Maillard, *Topological entropy and complexity for discrete dynamical systems*, preprint LPTHE, 1998.
2. V. I. Arnold, *Dynamics of complexity of intersections*, Bol. Soc. Brasil Mat. (N.S.) **21** (1990), 1–10.
3. M. Artin and B. Mazur, *On periodic points*, Ann. of Math. (2) **21** (1965), 82–99.
4. M. P. Bellon, J.-M. Maillard, and C.-M. Viallet, *Integrable Coxeter groups*, Phys. Lett. A **159** (1991), 221–232.
5. M. P. Bellon and C.-M. Viallet, *Algebraic entropy*, Comm. Math. **204** (1999), 425–437.
6. S. Boukraa, J.-M. Maillard, and G. Rollet, *Integrable mappings and polynomial growth*, Physica A **209** (1994), 162–222.
7. G. Falqui and C.-M. Viallet, *Singularity, complexity, and quasi-integrability of rational mappings*, Comm. Math. Phys. **154** (1993), 111–125.
8. B. Grammaticos, A. Ramani, and V. Papageorgiou, *Do integrable mappings have the Painlevé property?*, Phys. Rev. Lett. **67** (1991), 1825–1827.
9. J. Hietarinta and C.-M. Viallet, *Singularity confinement and chaos in discrete systems*, Phys. Rev. Lett. **81**(2) (1998), pp. 325–328.
10. ———, *Discrete Painlevé and singularity confinement in projective space*, Integrability and Chaos in Discrete Systems (Bruxelles, 1997) (to appear).
11. S. V. Kovalevska, *Sur le problème de la rotation d'un corps solide autour d'un point fixe*, Acta Math. **12** (1889), 177–232.
12. J. Liouville, *Sur l'intégration des équations différentielles de la dynamique*, J. Math. Pures Appl. (9) **XX** (1855), 137–138.
13. P. Painlevé, *Sur les équations différentielles d'ordre quelconque à points critiques fixes*, C. R. Acad. Sc. Paris **130** (1900), 1112–1115.
14. H. Poincaré, *Les méthodes nouvelles de la mécanique céleste*, Gauthier-Villars, Paris, 1892.
15. A. Ramani, B. Grammaticos, and J. Hietarinta, *Discrete versions of the Painlevé equations*, Phys. Rev. Lett. **67** (1991), 1829–1832.
16. S. Smale, *Differentiable dynamical systems*, Bull. Amer. Math. Soc. **73** (1967), 747–817.

Laboratoire de Physique Théorique et des Hautes Énergies, UMR 7589, CNRS et Universités Paris VI et Paris VII, France.

Current address: LPTHE, Boîte postale 126, 4 Place Jussieu F–75252 Paris, Cedex 05, France.

E-mail address: `viallet@lpthe.jussieu.fr`